appalachian geomorphology

appalachian geomorphology

Edited by

T.W. Gardner

Department of Geosciences, Pennsylvania State University, University Park, PA 16802, U.S.A.

and

W.D. Sevon

Pennsylvania Geological Survey, P.O. Box 2357, Harrisburg, PA 17105, U.S.A.

Reprinted from *Geomorphology*, Volume 2, No. 1–3

ELSEVIER, Amsterdam — Oxford — New York — Tokyo 1989

ELSEVIER SCIENCE PUBLISHERS B.V.
Sara Burgerhartstraat 25
P.O. Box 211, 1000 AE Amsterdam, The Netherlands

Distributors for the United States and Canada:

ELSEVIER SCIENCE PUBLISHING COMPANY INC.
655, Avenue of the Americas
New York, NY 10010, U.S.A.

ISBN 0-444-88326-6

Printed in The Netherlands

CONTENTS

PREFACE

The long, even ridgecrests of the Appalachians stretching from horizon to horizon and broken only occasionally by spectacular water and wind gaps have long intrigued the wondering mind of many geologists. In 1889 William Morris Davis offered in his paper "The Rivers and Valleys of Pennsylvania" an elegant set of arguments for the origin of the ridges and gaps and went on to propose a model for the drainage development and geomorphic evolution of the Appalachians. His work served to focus the geomorphic community into an intense, extended debate on the nature and fundamental controls of the long-term erosion of the Appalachians. If some of his ideas have not withstood the acid test of time, it may be because he did not have at his disposal the wealth of data available to us today from diverse subdisciplines within the geosciences.

The paradigms with which we construct our models, theories, and geologic histories have changed significantly in the past 100 years. The cycle of erosion has been vigorously re-evaluated in terms of plate tectonics and our understanding of surfaces processes and rates of crustal deformation. The wealth of geologic information from disparate fields that can now be brought to bear on the "Appalachian problem" offers the opportunity to make significant advances in our understanding of the geomorphic history of this fascinating mountain range. In addition, we now recognize that the geomorphic evolution of the Appalachians involves more than the Ridge and Valley Province which was the focus of Davis' work. It involves the complex interaction of all parts of the passive continental margin from the Appalachian Plateau to the offshore basins.

The 100th anniversary of William Morris Davis' provocative paper on the rivers and valleys of Pennsylvania seemed to us an appropriate time to convene a symposium to address some of the fundamental questions about the Appalachians. To attain that goal we have gathered a diverse group of papers bound together by a common theme – the geomorphic evolution of the Appalachians.

In the introductory paper, Morisawa reviews the paradigms of Davis' day, contrasts them to those of today, and briefly discusses other ideas about Pennsylvania's geomorphic evolution. To initiate the current point of view, Slingerland and Furlong model part of the landscape generated by the Alleghanian orogeny and hypothesize mountains of Andean proportions on the eastern side of the former Appalachian basin. This model is supported by the fission-track thermochronology work of Roden and Miller. They show that considerable variation in time of unloading occurs regionally and that much of the unloading occurred prior to the development of eastward drainage. The passive margins of both sides of the Atlantic are compared by Battiau-Queney who emphasizes the present landscape inheritance from deep-seated and ancient structures. Gardner concludes the tectonic aspect with an extensive review of late Cenozoic crustal activity which indicates that the passive margin may be undergoing accelerated deformation at the present time.

Barron discusses the considerable climatic variation from the Paleozoic to the present noting its probable effect on erosion rates. Large variations in long-term Appalachian erosion rates are demonstrated by Poag and Sevon through the sediment-accumulation rate in the Baltimore Trough–Salisbury Embayment–Hatteras Basin area. A long-term decline in sediment accumulation started in the Aalenian and reversed in the Middle Miocene is attributed mainly to tectonics. Cleaves attempts to recreate the appearance of the Piedmont landscape at five points in time between the Permian and the present through use of block diagrams. He also introduces the problem of the age of Piedmont saprolite. The saprolite problem is further addressed by Pavich who concludes that Piedmont saprolite is not old, but is continually being formed at the bottom of the weathering zone and stripped off at the top.

Craig demonstrates the applicability of computer modelling to landscape evolution, but finds the end results very dependent upon initial conditions. Clark, in contrast, approaches the evolution of Appalachian drainage from the field-oriented viewpoint and notes several specific reasons, such as ensnarement, why streams are located where they are. The controversial topic of the amount of glacial erosion is tackled by Braun who concludes that the high offshore sediment-accumulation rate during the Quaternary necessitates large amounts of ice erosion in addition to the large amount of periglacial erosion in areas beyond the ice margin. The vagrancies of rainfall are related to geology and landscape by Jacobson, Miller and Smith who demonstrate that the generation of landslides and catastrophic flooding is variable with respect to rainfall duration, intensity, and local geology. The character of soil genesis on glacial and nonglacial materials is treated by Ciolkosz, Waltman, and Simpson. The volume concludes with a review by Sevon of various modern erosion rates calculated for the Appalachians of which many are comparable to a new rate which he calculated from rivergauge data.

We hope that these papers will not only provide an appropriate summary of the current state of knowledge, but will also stimulate and focus future research. To that end we have encouraged some speculation in the papers. We regretfully had to omit some topics and geographical areas not because they are unimportant, but because the volume was limited in size. The real credit for this volume belongs with the authors. They were invited to contribute from their respective areas that research which addressed the problem at hand. To them and the other members of the 1989 Geomorphology Symposium Steering Committee, Duane Braun, Noel Potter Jr., and Glenn Thompson, who offered insightful contributions to the selection of topical areas, we are most grateful.

This volume has been rigorously reviewed and revised. An external referee (below) and at least one of the editors reviewed each manuscript. The manuscripts benefitted enormously from the input of several anonymous and informal reviewers as well as the following. To them we are most grateful.

Christopher Beaumont, Department of Oceanography, Dalhousie University, Halifax NS, B3H4J1 Canada

Thomas J. Crowley, ARC Technologies, 305 Argvello Dr., College Station, TX 77840

Helen L. Delano, Pennsylvania Geological Survey, P.O. Box 2357, Harrisburg, PA 17120

Leon R. Follmer, Illinois Geological Survey, 615 E. Peabody Drive, Champaign, IL 61820

Teri Kaktins, Barron Hill Consultants, R.D. #2 Box 158, Johnstown, PA 15904

Dorothy J. Merritts, Department of Geology, Franklin and Marshall College, Lancaster, PA 17604

D.W. O'Leary, U.S. Geological Survey, Woods Hole, MA 02543

John B. Ritter, Department of Geosciences, Pennsylvania State University
University Park, PA 16802

R. Scott Snow, Department of Geology, Ball State University, Muncie, IN 47306

John E. Costa, U.S. Geological Survey, Water Resources Division, Cascades Volcano Observatory, 5400 MacArthur Blvd., Vancouver, WA 98661

Edward B. Evenson, Department of Geological Sciences, Lehigh University, Bethlehem, PA 18015

Donald M. Hoskins, Pennsylvania Geological Survey, P.O. Box 2357, Harrisburg, PA 17120

Timothy M. Lutz, Department of Geology, University of Pennsylvania, Philadelphia, PA 19104

Hugh H. Mills, Department of Earth Sciences, Tennessee Technological University, Cookeville, TN 38505

Noel Potter Jr., Department of Geology, Dickinson College, Carlisle, PA 17013

James P. Reger, Maryland Geological Survey, 2300 St. Paul St., Baltimore, MD 21218

June, 1989

THOMAS W. GARDNER (University Park, Pa.)
WILLIAM D. SEVON (Harrisburg, Pa.)

Geomorphology, 2 (1989) 1-22
Elsevier Science Publishers B.V., Amsterdam — Printed in The Netherlands

Rivers and Valleys of Pennsylvania, Revisited

MARIE MORISAWA

Department of Geological Sciences, SUNY at Binghamton, Binghamton, NY 13901 (U.S.A.)

(Received March 16, 1989; accepted after revision June 8, 1989)

Abstract

Morisawa, M., 1989. Rivers and valleys of Pennsylvania, revisited. In: T.W. Gardner and W.D. Sevon (Editors), Appalachian Geomorphology. Geomorphology, 2: 1-22.

The 1889 paper by William Morris Davis on the "Rivers and Valleys of Pennsylvania" is a landmark in the history of geomorphology. It was in this manuscript that he set forth what came to be known as the Davisian system of landscape. It is important to understand that Davis' interpretation of landforms was restricted by the geologic paradigms of his day. Uniformitarianism was strongly entrenched and Darwin's theory of evolution had become popularly accepted. The concept of the landmass Appalachia and the then current theories on mountain building affected the approach that Davis took in hypothesizing the origin and development of the Folded Appalachian drainage. All of these geologic precepts influenced the formulation and explanation of his theories.

In his exposition he adapted, synthesized and embellished on ideas he derived from fellow geologists such as Gilbert, Dutton, Powell, and McGee. A number of the concepts he proposed in the 1889 paper quickly became the bases for geomorphic studies by others: the cycles of river erosion and landscape evolution and the peneplain (here called base level erosion). The cycle of erosion became the model for subsequent geomorphic analyses, and peneplain hunting became a popular sport for geomorphologists.

Davis' hypothesis of the origin and development of Pennsylvanian drainage stimulated subsequent discussion and further hypotheses by others. In fact, many of the later theories were refinements and/or elaborations of ideas mentioned in this paper of Davis. He proposed the origin of the drainage as consequent streams, then antecedence, superposition, headward extension of divides by piracy, erosion along lines of weaknesses (faults, easily erodible beds) through resistant ridges and normal fluvial erosion. Thus, the hypotheses of regional superposition (Johnson), extended consequents (Ruedemann), consequents and local superposition (Meyerhoff and Olmstead), the utilization of structural weaknesses in development of transverse drainage (Thompson; Meyerhoff; Oberlander, among others), and migration of divides (Thompson), all had been suggested by Davis in 1889.

Although the concepts of erosion cycles and peneplaination have waned in popularity in recent geomorphic research, the principles of formation of water and wind gaps, headward migration of divides, stream piracy and adjustment to streams to structure, so clearly and minutely explained in his 1889 publication, are still viable today.

Introduction

1989 marks the hundredth anniversary of the publication of William Morris Davis' classic paper "The rivers and valleys of Pennsylvania" in the National Geographic Magazine. For more than fifty years after that the Folded Appalachians were fertile ground for geomorphologists to study. In the paper Davis set forth what came to be known as the Davisian system of landscape analysis. It is appropriate at this symposium to review both Davis' hypothesis of the origin of the drainage systems of the Ridge and Valley Province of Pennsylvania, as well as the basic principles of landform evolution introduced in the 1889 paper.

The work of Lesley (1856) and others of the early Pennsylvania Geological Survey stimulated interest in the physiography of the Folded Appalachians. The noticeable accordance of summits, the transverse courses of major streams, the aligned wind and water gaps, and the striking adjustment to structure aroused questions as to the geomorphic development of the region. How can the present drainage system be explained in terms of the geologic history and structure of the area? If the original drainage was to the northwest (actually a problem in itself), how did the streams reverse their flow to the southeast? Why are the master streams transverse to the structure, cutting across resistant rock ridges? How were the aligned wind and water gaps formed? These problems had to be solved in forming a satisfactory hypothesis for the origin of Appalachian drainage.

In order to understand the basis for Davis' interpretations and the restrictions by which he was bound it is necessary to examine the geological paradigms of his day. Then his theory of the drainage evolution and chronology of the Folded Appalachians of Pennsylvania, as well as the other contributions he made to geomorphic concepts in his 1889 paper will be discussed. Finally hypotheses of Appalachian drainage development as proposed by Johnson (1931), Ruedemann (1932), Meyerhoff and Olmstead (1934, 1938), Thompson (1939), and Oberlander (1985) will be examined. This is not intended to be a comprehensive review of the development of the drainage of the whole Appalachian Mountains. Rather, it deals specifically with the Folded Appalachians of Pennsylvania, as did Davis.

Geologic paradigms of 1889

Davies (1988) suggested that geologists interpret geologic phenomena in terms of the paradigms accepted for their day. What were the geologic "truths" that were the basic principles of geologic interpretation at the time when Davis wrote his paper?

By 1889 the uniformitarianism of Hutton and Playfair (Hutton, 1795) had become accepted dogma:

> "No one now regards a river and its valley as ready-made features of the earth's surface. All are convinced that rivers have come to be what they are by slow processes of natural development."
>
> Davis, 1889a, in Essays p. 413

Another concept that had been established by the time of Davis' paper was Darwin's theory of evolution. Davis applied it to his scheme of landscape development and he did so consciously (Judson, 1960). This emphasis on evolution in the erosion cycle, as well as the terminology "youth", "maturity" and "old age" was a reflection of the scientific community immersed in Darwinism (Chorley, 1965).

> "Being fully persuaded of the gradual and systematic evolution of topographical forms, it is now desired, in studying the rivers and valleys of Pennsylvania, to seek the causes of the location of the streams in their present courses; to... trace the development of the several river systems.... from their ancient beginning to the present time."
>
> ibid., p. 413

In terms of stratigraphy and sedimentation, Hall (1882) had examined the thick Paleozoic sedimentary sequence in the northern Appalachian and it was generally assumed that the deposits were derived from a landmass to the east (later named Appalachia). Because of the widespread acceptance of this idea, Davis found it difficult to get away from the thesis that the original drainage was from the southeast to the northwest.

> "The region in which the Susquehanna and the neighboring rivers are now located is built...of marine sediments derived in Paleozoic time from a large land area to the southeast..."
>
> ibid, p. 418

> "We may picture the drainage of the southeastern land area wandering slowly across...to the still submerged land to the west..."
>
> ibid, p. 420

Davis synthesized the then current theories

on mountain building: eustasy, subsidence, and vertical uplift, as well as the effects of compressive forces as advanced by Suess (1875).

"...the great Permian deformation...was characterized by a resistless force of compression..."

ibid, p. 421

He pictured episodic periods of mountain building and folding as well as intermittent subsidence and elevation.

"...at some later date, as Permian, a second period of mountain growth arrived, and further folding was effected...The depression of the Newark belt was after a time reversed into elevation...."

ibid, p. 422, 423

"The Jurassic deformation was probably accompanied by a moderate elevation..."

ibid, p. 425

Davisian geomorphology

Davis integrated these basic geologic tenets into his system of landscape evolution. In the 1889 paper he set forth a number of concepts which quickly became accepted as bases for geomorphic studies: the cycles of river erosion and landscape evolution, and the peneplain. By his own admission he was indebted to American geologists on the Geological Survey with whose work he was familiar and whom he knew personally.

"It is proper that I should add that the suggestion of base-levelling...is due largely to personal conferences with Messrs. Gilbert and McGee of the Geological Survey..."

ibid, p. 427

He adapted, synthesized, embellished and popularized many of their ideas into his system of river and landscape development (Higgins, 1975). He adopted Powell's concept of base-level and his genetic classification of stream valleys.

"The Permian Mountains are seen to have been once greatly reduced... (to) a Cretaceous base-level lowland...we may well conclude with Powell that "mountains cannot remain long as mountains."

ibid, p. 430

"These will then be called antecedent streams in accordance with the nomenclature introduced by Powell."

ibid, p. 444

Although he had earlier (1885) proposed the general idea of the cycle of erosion, Davis elaborated and gave a full exposition of it in the 1889a paper. Apparently, in detailing the origin and history of the rivers of Pennsylvania, Davis was compelled to sort out, systematize, and formulate concisely his perception of river evolution as well as the progress of overall landscape development. It was here that he first used the term "cycle in the life of a river", "cycle of development" and "cycle of erosion" (Chorley et al., 1973).

"The complete *cycle of river life*:
youth, adolescence, maturity and old age...."

ibid, p. 430

"The lapse of time will be called a *cycle in the life of a river*. A complete cycle is a long measure of time..."

ibid, p. 431

"At the close or at any time...the drainage area of a river system may be bodily elevated. The river is then turned back to a new youth and enters a *new cycle of development*."

ibid, p. 441

"Lowl's first essay...took no sufficient account of the several *cycles of erosion* through which the region has certainly passed..."

ibid, p. 417

As he outlined the cycle of erosion of a landscape (1889a, p. 422–427, 430–434) Davis applied tectonic paradigims of his time (mentioned previously) in tracing the geomorphic history.

Elevation of the land was needed to provide for a renewed period of erosion. A long period of tectonic quiescence took place between each uplift during which erosion reduced the area to a low peneplain. Thus Davis developed a de-

Fig. 1. View of the flat accordant ridge tops of the Folded Appalachians in Pennsylvania. Photograph by M. Morisawa.

nudation chronology of alternative periods of uplift and erosion to a peneplain, as indicated by the accordance of summits over the region. First was the Cretaceous base-level lowland, "a low undulating plain of wide extent", recognized regionally as the level of the highest hilltops (Fig. 1).

> "In consequence of this elevation a new cycle of erosion was entered upon, which I shall call the Jura–Cretaceous cycle. It...ended in the production of a general lowland of denudation, a wide area of faint relief, whose elevated remnants are now to be seen in the even ridge crests...the extraordinarily persistent accordance among the crest-line altitudes...the generally corresponding elevation of the western plateau surface...the generally uniform and consistent altitude of the uplands in the crystalline highlands...and the extension of the same general surface...over the even crest lines of the Newark trap ridges."
>
> ibid, p. 425

Uplift in the early Tertiary led to renewed erosion to form a late Tertiary peneplain.

> "The elevation seems to have taken place early in Tertiary time...opportunity was then given for the revival of the...forces of denudation, and as a consequence we now see the formerly even surface of the plain greatly roughened by the incision of deep valleys and the opening of broad lowlands on the softer rocks...these areas have been...almost base-leveled again in the Tertiary cycle. They will be referred to as the Tertiary base-level lowlands".
>
> ibid, p. 427

A post-Tertiary uplift led to the present trenching.

> "Again at the close of Tertiary time, there was an elevation...and to this may be referred the trenches that are distinctly cut across the Tertiary base-level lowland by the larger river..."
>
> ibid, p. 427

The progression of regional denudation outlined in 1889a was detailed and expanded in subsequent publications (Davis, 1899, 1905). Each of the cycles mentioned in the denudational history of the Folded Appalachians ended in a "base-level lowland" later named a peneplain (Davis, 1889b, 1895). The concept was not original with Davis but rather came from the writings of Powell and Dutton. Powell (1875) introduced the term base-level of erosion and

Dutton (1882) broadened the concept to include a regional base-level. Davis, then, used base-level lowland for the end form of a denudational cycle and, in a discussion of the degradation of the Triassic (1889b) gave the form a name.

The peneplain has been the most extolled and most criticized of the Davisian concepts (Flemal, 1971). Articles for and against peneplains abound in the literature (Cole, 1934, 1941; Rich, 1938, for example, and see Thornbury, 1969). So many peneplains have been described that there was a reaction against them by some geomorphologists (Tarr, 1898; Flemal, 1971).

Other end-forms of denudation have since been proposed. Penck (1924) used both Primarrumpf and Endrumpf for surfaces of low relief. The Primarrumpf is a land surface which maintains an original low relief because of slow uplift, whereas an Endrumpf is the peneplain equivalent: a final base-level surface of erosion, reached primarily by the backwearing of slopes. King (1953) used the term pediplain for a surface of low relief resulting from the headward retreat of scarp slopes. Neither of these terms replaced the peneplain concept despite its controversial history (Thornbury, 1969).

Before examining in detail the evolution of the drainage, Davis outlined his conception of the life history of a river, taking it through his now famous cycle of erosion from youth to maturity to old age.

"Once established, an original river advances through its long life, manifesting certain peculiarities of youth, maturity and old age, by which its successive stages of growth may be recognized without difficulty...In its infancy the river drains its basin imperfectly...The channels followed by the streams that compose the river as a whole are narrow and shallow, and their number is small...The divides by which the side streams are separated are poorly marked...As the river becomes adolescent, its channels are deepened and all the larger ones descend close to base-level...with the deepening of the channels, there comes an increase in the number of gulleys on the slopes of the channel; the gulleys grow into ravines and these into side valleys...With their continued development the maturity of the system is reached...marked by an almost complete acquisition of every part of the original...surface by erosion...the waterfalls of adolescence have been worn back...it is at maturity that the river receives and carries the greatest load; indeed, the increase may be carried so far that the lower trunk stream...is unable to carry the load brought to it by the upper branches, and therefore resorts to the temporary expedient of laying it aside in a flood plain...In the later and quieter old age of a river system...the headwater streams deliver less detritus to the main channel...The flood plains of maturity are carried down to the sea, and at last the river settles down to an old age of well-earned rest with gentle flow and light load, little work remaining to be done."

ibid, p.431–434

At the same time he laid the ground for his later exposition by describing changes from consequent drainage via superimposition, adjustment to structure as subsequents, and headward erosion.

Both proposed systems of cyclic development were immediately successful and dominated geomorphic research for decades. The concept of cycles was readily accepted as it had been ingrained into geologic thought after its proposal by Hutton (Higgins, 1983). Indeed, it had a pre-Hutton tradition since the 16th century (Davies, 1969). Flemal (1971) stated that the Davisian scheme of cyclic erosion formed the framework for landform studies and teaching in North America and much of Europe. According to Judson (1960) "the Geographic Cycle has been the single most influential concept in the field of geomorphology" and that almost every report on regional geology contained a description of the erosion cycle and peneplaination. Davis' cycle was the basis for denudation chronology which became the popular type of geomorphic analysis worldwide (Beckinsale and Chorley, 1968).

Despite early critics such as Tarr (1898) and the introduction of another scheme of landscape development by Penck (1924), Davis' system held sway in geomorphology until the 1940's (Higgins, 1975). It still is offered as a popular concept in general textbooks and courses in geology (Flemal, 1971) despite two other systems: pediplaination as proposed by King (1962) and Hack's (1960) dynamic equilibrium. However, since World War II geomorphologists in general have turned away from descriptive, historical analyses of landscapes to more quantitative, process-oriented analyses (Judson, 1960). Recently, a neo-Davisian plea for the erosion cycle in terms of "sequential development" of landform has been advanced by Bloom (1978).

Davis' cycle of river erosion and landscape denudation, as espoused in the 1889a paper, has thus left an indelible imprint on geomorphology. The Davisian model of landscape evolution dominated geomorphic research for the first half of the twentieth century (Gregory, 1978). Despite its demise today, no one can deny that the cycle of erosion was the first successful attempt to develop a powerful theoretical model in geomorphology (Chorley, 1978). And, according to Ritter (1978, p. 5) "there can be no argument that Davis influenced geomorphic thinking more than any other person".

Davis' hypothesis of Pennsylvania drainage

Davis first reviewed the hypotheses of others before him. He dismissed the proposal of Peschel (1880) that streams formed water and wind gaps by eroding along fractures at those points. Peschel's theory smacked of catastrophism, was therefore unacceptable, and was arbitrarily dismissed, although it is now recognized as the most likely explanation (Hoskins, 1987). Davis agreed with Tietze (1878) that antecedent rivers must have created the transverse drainage. Although he had previously (Davis, 1883) discussed and dismissed Lowl's description of drainage development by headward erosion, in this 1889a paper he adapted the suggestion in his own hypothesis.

"...he (Lowl) says that in mountain systems of very great age the original arrangement of the longitudinal valleys often becomes entirely confused by means of their conquest by transverse erosion gaps...Its thesis is the essential element of my present study."

ibid, p. 417

..."all the while the headwater streams of the region were gnawing at the divides, seeking to develop the most perfect arrangement of waterways."

ibid, p. 461

Davis reconstructed the original drainage by restoring the topography of the folded region of Pennsylvania during the Permian, using the outline of formations exposed at the present time. Thus, the area where the youngest sediments are now at the surface would represent a downsag in Permian times; places where the oldest formations are present would indicate anticlinal structures. This reconstruction showed a large synclinal basin in eastern Pennsylvania and a large anticlinal area in the west with a small syncline represented by the Broad Top coal field.

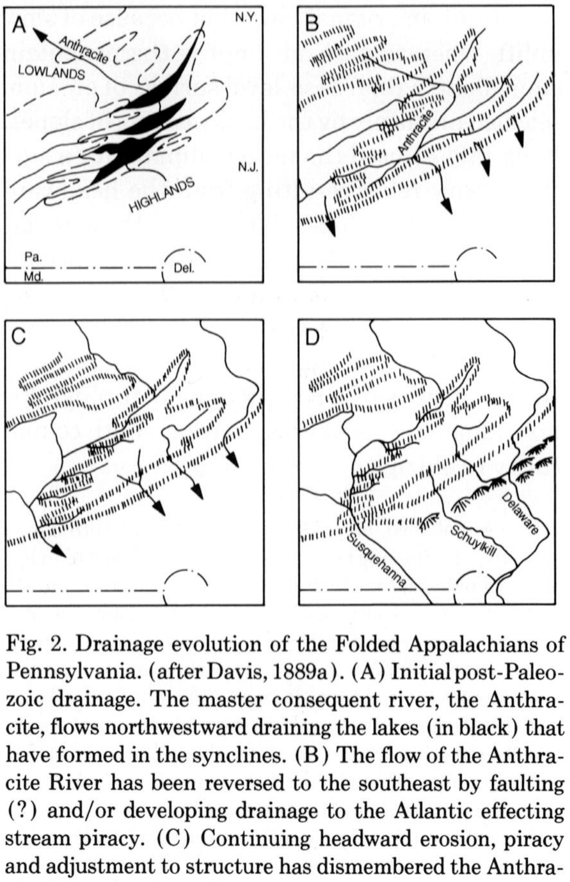

Fig. 2. Drainage evolution of the Folded Appalachians of Pennsylvania. (after Davis, 1889a). (A) Initial post-Paleozoic drainage. The master consequent river, the Anthracite, flows northwestward draining the lakes (in black) that have formed in the synclines. (B) The flow of the Anthracite River has been reversed to the southeast by faulting (?) and/or developing drainage to the Atlantic effecting stream piracy. (C) Continuing headward erosion, piracy and adjustment to structure has dismembered the Anthracite River, leaving only vestiges of its former course. (D) Present drainage.

Davis postulated that the original master stream of the Folded Appalachians, the Anthracite River, drained to the northwest as consequent flowing from the highlands to the east (Fig. 2a). Initial stream patterns would follow the synclinal axes, flowing down the pitch toward the inner basin where a lake would form (Fig. 2a).

"What would be the drainage of such a country?...it consisted of numerous streams...following synclinal axes until some master stream led them across the intervening anticlinal ridges at the lowest points of their crests and away into the...northwest. All the enclosed basins would hold lakes, overflowing at the lowest part of the rim...the master stream, the great Anthracite River."

ibid, p. 451, 453

As the mountains rose across its path the Anthracite maintained its course across the ridges and became antecedent.

"...the first step in our problems consists in deciding...whether the streams held their courses after the antecedent fashion, or whether they were thrown into new courses by the growing folds...Possibly both conditions prevailed, the larger streams holding their courses little disturbed, and the smaller ones disappearing."

ibid, p. 448

The major northwest flowing consequent, the Anthracite, reversed its flow and was gradually dismembered by adjustment and capture, (Fig. 2b, 2c) to the present configuration (Fig. 2d).

"The drainage...retained the original courses of the streams...but the great Anthracite river is drawn as if it had been controlled by the Newark depression and reversed in the direction of its flow."

ibid, p. 462

"one early stream alone found a favorable opportunity for conquest, and thus grew to be the master river - the Susquehanna of today. The head of the Anthracite was carried away by this captor, and its beheaded lower portion remains in our Schuylkill."

ibid, p. 466

By example and in minute detail he showed how adjustment to structure on the Juniata and Susquehanna Rivers occurred by headward erosion and capture.

"...there is a process of adjustment...by which the...courses that were chosen in early life...unadvisedly and with the heedlessness and little foresight of youth, are given up for others better fitted for the work of the mature river system...a part of the main channel may...be deserted by the withdrawal of its upper waters to a lower course by way of a side stream."

ibid, p. 434, 435

"A stream flowing along a syncline of hard beds...develops side streams which breach the adjacent anticlines and open lowlands in the underlying softer beds...a contest ensues as to possession of the drainage territory. The divides are pushed away...this conquest goes until the upper course of the main stream is diverted to a new and easier path.

ibid, p. 467, 468

He considered two alternatives for the reversal of drainage to the southeast. The first was that due to the subsidence and tilting of the Newark belt during the Triassic faulting and the submergence of Appalachia, the middle course of the northwestward flowing stream was turned backward toward the southeast.

"The depression and submergence of the broad Newark belt may at this time have broken the continuity of the streams that once flowed across it...The middle courses were perhaps turned from the central part of the state back to the Newark belt."

ibid, p. 459

There is, however, objection to the assumption that tilting, even extreme tilting, could bring about the reversal of flow of a well established river. Indeed, the very principle of antecedence is that rivers are able to maintain an original channel and direction of flow against formidable opposition. Thus Davis, himself, realized the inadequacy of this explanation.

"At the same time I am constrained to admit that this is, to my own view, the least satisfactory of the suggestions here presented."

ibid, p. 460

The alternative suggestion was capture of the headwaters on the southeast by the Susquehanna and other southeast flowing Triassic rivers which had the advantage of flowing on more erodible beds and had a shorter distance to the sea.

> "The only other method of reversing the original northwestward flow of the streams...is by capture of their headwaters by Atlantic rivers."
>
> ibid, p. 460

Davis sums up the history of the rivers as they developed from a "system of ancient consequent drainage".

> "...most of our rivers took their courses between the primitive folds of our mountains...the changes they have suffered are due to...their own mutual adjustment...The middle Susquehanna and its branches and the upper portion of the Schuylkill and Lehigh are descendents of original Permian rivers consequent on the constructional topography of that time; Newark depression reversed the flow of some of the transverse streams...The shifting of courses in the progress of mature adjustment has had more to do with determining the actual location of our rivers and streams than any other process."
>
> ibid, p. 484

The major problem of drainage reversal which Davis felt he had to solve has become less problematic as more has been learned about plate tectonics of the Paleozoic and Mesozoic. The landmass, Appalachia, has been replaced by island arcs in the Lower Paleozoic (Kay, 1951; Hatcher and Odom, 1980), and by closure of the Iapetus Ocean and continental collision to form Pangaea in the upper Paleozoic (Van der Voo, 1983; Williams and Hatcher, 1983). During plate convergence in the Carboniferous, extreme westward thrusting took place in the Appalachians (Crowley, 1976). Triassic rifting and Mesozoic-Cenozoic opening of the Atlantic (LePichon and Fox, 1971) completes the plate tectonic picture.

Regional superimposition

A second problem which Davis faced in developing a system of drainage evolution was that of transverse drainage. Why should a river cut across a resistant ridge? And even more problematic – why do the transverse courses cross successive ridges in aligned water gaps (Von Engeln, 1942)? Davis (1889a) proposed superimposition of the Susquehanna River from a widespread flood plain sediment cover to account for its transverse course through two Pocono sandstone ridges at Susquehanna Gap. He had also recognized several peneplains in the Ridge and Valley of Pennsylvania. He thought the upland surface, denoted by the accordance of ridge tops was Cretaceous in origin. However, he later (Davis, 1899) acknowledged the existence of an earlier peneplain, underlying the Coastal Plain sediments. The older peneplain came to be known as the Fall Zone surface of Cretaceous age; the ridge summit level was the Schooley peneplain, developed in the early Tertiary.

In 1931 Douglas Johnson proposed the regional superposition of Appalachian rivers from a Coastal Plain cover which extended some 320 km inland and thus had destroyed the pre-existing drainage. This is probably the hypothesis that was accepted as most plausible by mid-twentieth century geomorphologists (Von Engeln, 1942). Johnson based this idea on two fundamental assumptions: (1) that there are two peneplains, represented by the Fall Zone and Schooley surfaces and (2) that Cretaceous deposits extended far inland. He believed that normal dissection could not explain the aligned wind and water gaps. Ruling out antecedence as unsatisfactory, he arrived at the conclusion that the transverse streams must indicate superposition.

Chronologically, Johnson postulated an extensive peneplain developed during the Jurassic on the uplifted, folded and faulted, Paleozoic sediments. The Fall Zone surface was subsequently submerged and covered by Cretaceous sediments far inland across the full width of the Folded Appalachians (Fig. 3). A doming followed, enabling the southeastward flowing consequent streams to strip off the sedimentary

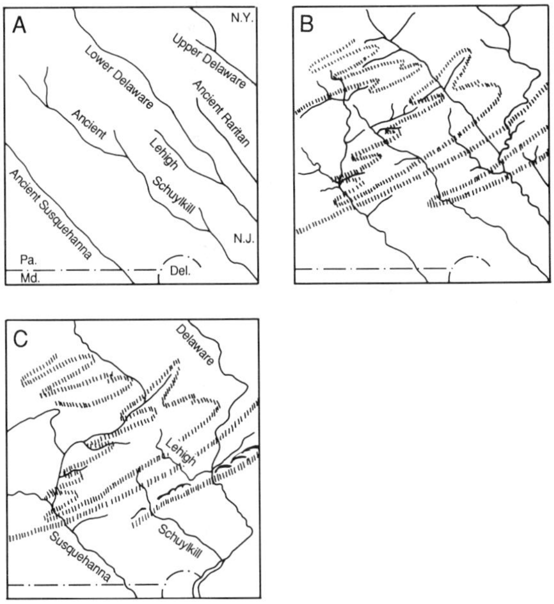

Fig. 3. Appalachian drainage evolution after Johnson, 1931. (A) Initial consequent drainage flowing southeast on the Cretaceous coastal plain sediments. (B) Stripping of the coastal plain cover has exposed the underlying folds. Courses of the rivers have been superimposed on the resistant ridges. (C) Present drainage.

cover, become superimposed, and erode down to the underlying folded rocks; thus forming the Schooley surface in the early Tertiary. Renewed uplift in the Tertiary led to erosion and entrenchment of the superposed streams onto the underlying structures. Continuing degradation removed all the Coastal Plain cover. During these successive erosion cycles, brought on by renewed uplift, adjustment of the streams to the underlying structure was achieved.

Johnson believed that the parallelism of the transverse rivers and the alignment of the water and wind gaps could best be explained by superposition of extended consequents. Moreover, he thought that the Upland Schooley peneplain was too well preserved to have been formed as far back as the Cretaceous. He also believed that there was not enough time since the Cretaceous to permit the present adjustment of drainage. Peneplaination of the Fall Zone in the Jurassic period and subsequent erosion since then would allow sufficient time for adjustment to take place.

The validity of Johnson's fundamental assumptions is the main objection to accepting the hypothesis of regional superposition. First is the question of peneplaination. Although identifying erosion surfaces (peneplains?) in the Appalachians became popular among geomorphologists, the reality and number (if any) is still in doubt. In addition, the problem of an original or lowered peneplain occupied Appalachian geomorphologists for some time. The second of Johnson's assumptions has even less factual foundation. There is no proof that the Coastal Plain sediments extended so far inland as Johnson postulated. In fact, Sevon (1985) argued that no peneplaination or substantive Cretaceous transgression had taken place in the Pennsylvanian Appalachians.

So although the hypothesis of regional superposition accounts reasonably for many of the drainage features, such as alignment of gaps, parallelism of major rivers, and transverse drainage, the fundamental assumption of an extended coastal plain cover is in doubt.

Extended consequents

The concept of original consequent streams on the uplifted Permian surface proposed by Davis (see previous quote of Davis, 1889a, p. 451) was applied by Ruedemann to explain the origin of the Susquehanna River. According to Ruedemann (1932) the restored Permian surface would give a New York highland in the Catskill area with a lowland to the south and southwest. He postulated a master consequent flowing southwest from this raised Catskill dome. This primitive Susquehanna drained an area far to the northeast and flowed down the gently southwestward dipping beds. This river had a very long course to the sea and a very low gradient, therefore it was susceptible to capture by younger, more vigorous streams.

The drainage development took place in a number of stages, Fig. 4. Stage A is the initial

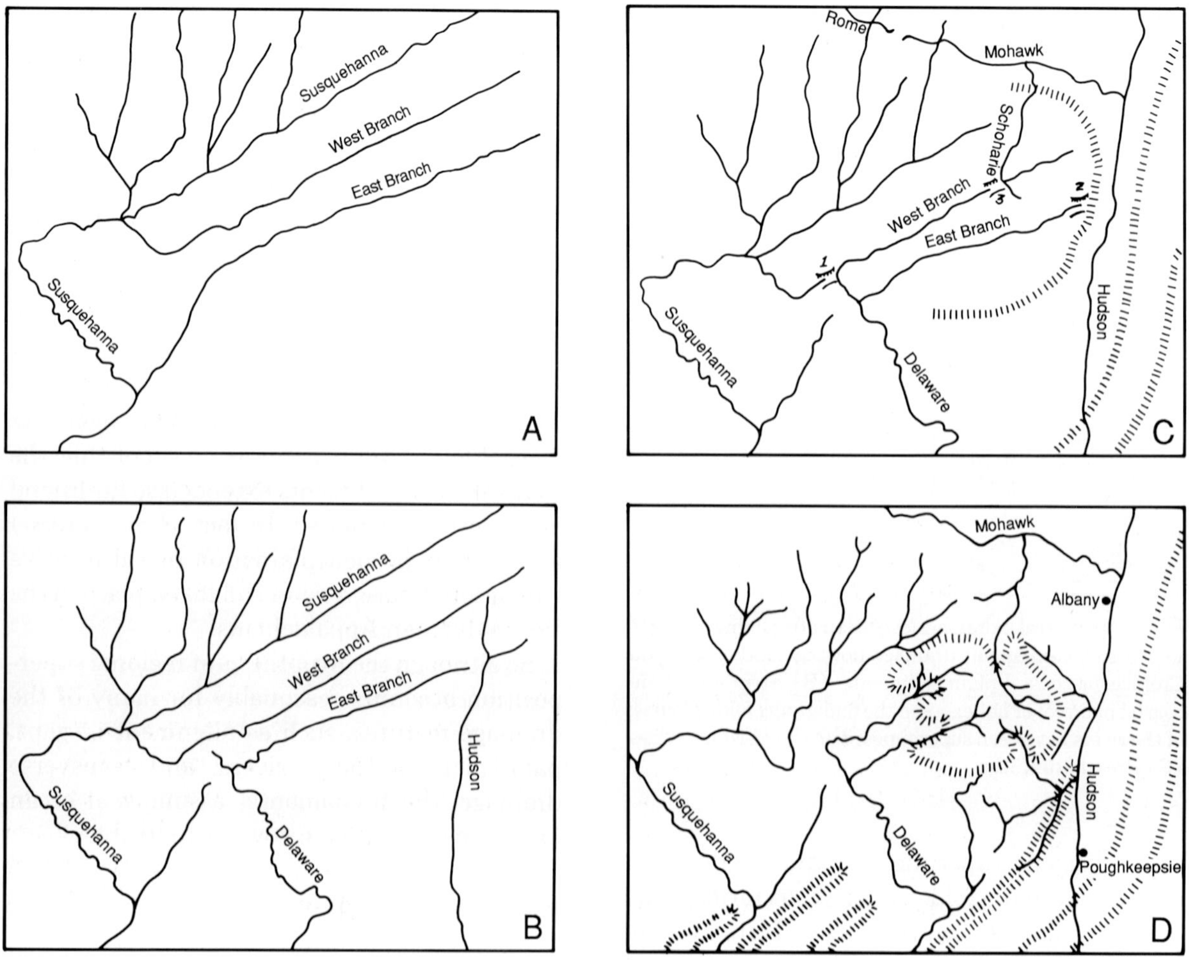

Fig. 4. Evolution of the Susquehanna River, after Ruedemann, 1932. (A) Early Mesozoic drainage, showing southwestward-flowing streams consequent on the Catskill dome (west side). Rocks dip southwest. (B) Capture of the East Branch by headward erosion of the Delaware and Hudson Rivers. (C) Capture of the West Branch by the headward erosion of the Delaware and Schoharie Rivers. The Mohawk and Rome Rivers are capturing tributaries on the north. Gorges at points of capture are: *1* between Lanesboro and Deposit; *2* at East Windham; and *3* Grand Gorge. (D) Present drainage.

drainage with one master consequent and its tributaries. By Stage B the Hudson River has captured the East Branch, as has the Delaware a little farther to the west. By Stage C the Delaware has eroded its headwaters back to capture the West Branch and the Hudson–Mohawk–Schoharie system has eaten back on the Catskill plateau to capture the West Branch at another locality. Thus the present drainage has resulted from the dismemberment of the primitive Susquehanna.

Ruedemann evolved the primitive Susquehanna by projecting the longitudinal sections of the present streams to give his master consequent. Mackin (1938) pointed out that both the southwest flowing reaches of the West and East Branch Delaware and the upper segment of the West Branch Susquehanna River were recognized by Meyerhoff and Olmstead (1936) as subsequent on weaker sediments. Thus, although they flow in the direction of the regional dip, they may not necessarily be consequents. There are a number of valleys which Ruedemann considers abandoned channels. One is the valley between Lanesboro and Deposit which once connected the present West Branch Delaware with the present Susquehanna. Another is the pass at East Windham on the north side of the Catskills which is in line with the East Branch Delaware. A third is the broad notch known as Grand Gorge which indicates the former channel where (present) West Branch Delaware was connected to the Schoharie. These channels (Fig. 4C) have since been identified as cut by glacial meltwaters (Rich, 1934), casting doubt on Ruedemann's hypothesis.

Consequent and local superposition

Another version of extended consequents was offered by Meyerhoff and Olmstead (1938). They proposed the Permo–Triassic origin of the main streams that drain the central and northern part of the Folded Appalachians as the Appalachian orogeny would have destroyed previous drainage. The geographic distribution of Paleozoic clasts in Triassic sediments and the nature of the fold and fault structures indicated to them that the post-Alleghenian drainage divide was west or northwest of the Folded Appalachian belt. Thus master streams established original southeastward courses and despite many Mesozoic and Cenozoic adjustments, still follow their early Triassic routes to the southeast.

The eastward direction of the drainage was assumed from the fold structures many of which (but not all) are overturned to the northwest. The folds with a long southeast slope and shorter western slope would result in long southeast flowing consequents down the dip slope. These cut through the ridges at points of structural weakness or sags.

As denudation proceeded, capture across ridges took place on lines of structural weakness. This was indicated by the intimate relation of the streams to the structure. Where there was no overturning there were no gaps and where there was overturning gaps were formed. A consistent utilization of structural weakness by the streams was shown by faults they found at the Delaware Water Gap, Lehigh Gap, Schuylkill Gap, Pen Argyl Gap, Harrisburg Gaps, and at Wind Gap.

They had no proof of the extreme thrusting which would place the drainage divide far to the northwest of even its present position. More recently, basing his hypothesis on new structural interpretations that a large thrust sheet once covered the Anthracite basins of southeastern Pennsylvania, Sevon (1986) proposed that the Susquehanna River originally flowed northeast along the margin of the thrust sheet. Reversal of the flow direction occurred over time as the streams ate headward. Capture and adjustment to structure took place by normal stream processes as outlined by Davis (1889a).

Further objections to Meyerhoff and Olmstead's theory (and indeed, to all theories of preserved consequents) are: major streams do not follow synclinal axes; in fact, part of the Susquehanna flows against the pitch of the syncline above Sunbury. There is not instance of capture by transverse streams. On the contrary, those captures which are indicated are piracy of a transverse stream by a subsequent (Davis, 1889a; Ver Steeg, 1930; Meyerhoff and Olmstead, 1934).

Meyerhoff (1972) elaborated the case of both local superposition and original southeast flowing consequents. Taking advantage of structural and stratigraphic studies that had been done in the Appalachians since the early 1900's, he emphasized the idea that Appalachia had not always been present in the Paleozoic and certainly not in the Mesozoic. Since there was no highland to impede southeast flowing drainage there was no need to postulate northwestward flowing rivers. He also refined the process of superposition offering two types of instances: stratigraphic and structural. Stratigraphic superimposition occurred when a stream incised its channel from broad outcrops of a thick sedimentary unit onto the subjacent layers. This could be recognized where a stream maintained its course along the strike of the folds but shifted

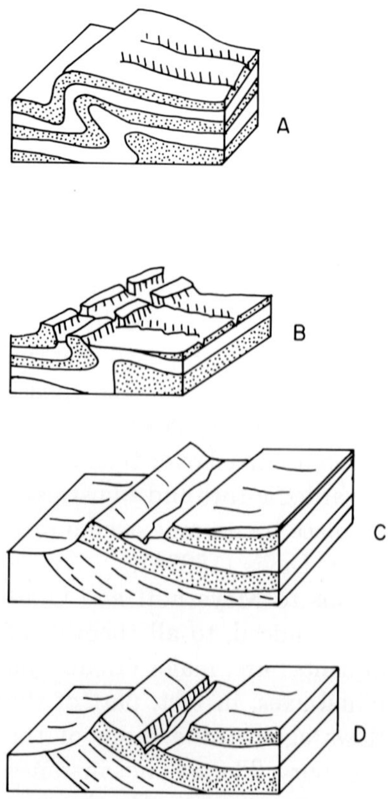

Fig. 5. (A) and (B) Superposition from an overturned anticline (after Thompson, 1939); (C) and (D) Superposition on a synclinal structure.

from one formation to another. Structural superimposition occurred from downcutting into lower stratigraphic units from superjacent layers in a fold (Fig. 5). He reiterated structural weaknesses found at wind and water gaps.

Progressive piracy and local superposition

The Atlantic–Gulf drainage divide is an irregular, sinuous line that does not conform to any one geomorphic province or structural unit (Fig. 6). Southwest of the Roanoke River the divide is in the crystalline Blue Ridge; the Roanoke–New River divide is in the Great Valley; the James–Kanawha River divide is farther west in the Great Valley; the Potomac drainage begins at the margin of the Appalachian Plateau; the Hudson not only invades the Plateau but also carries the drainage from the Ontario lowland via the Mohawk. So as we proceed northward the divide moves farther from the crystalline bedrock. From the Chattahoochee River northeast to the Hudson River the Atlantic slope streams seem to have been successful in the conquest of the Appalachian highlands. The divide has been migrating westward as pointed out by Davis and, according to Thompson (1939), is, in places, as much as 100 mi west of its original position. In Georgia the Altamaha–Chattahoochee divide has been moved 10–20 mi, the James–New divide 40–75 mi, the Potomac–Monongahela 100 mi, and the Hudson approximately 120–140 mi.

Thompson (1939) used headwater piracy along an asymmetrical watershed to account for this westward migration of the main drainage divide. The process of drainage reversal has been slow and long-continued. It has taken place by stream adjustments, during which initial courses were given up for other easier channels (Davis, 1889a). Such an adjustment may take place where a stream flowing in a strike valley on easily eroded sediments may capture a stream cutting across a resistant ridge (Fig. 7). The adjustment may also happen when the course of the main stream is much longer than the length of headward erosion of an adjacent stream (Davis, 1889a).

In locating the original drainage divide Thompson assumed that it was somewhere in Appalachia. The fundamental cause of the migration was shorter distance to the Atlantic than to the Gulf. A long, low-gradient stream, flowing northwest even if on weak rock, was unable to compete successfully with a stream having the advantage of a much shorter and steeper course to the southeast.

The process has long been recognized to explain the evolution of Appalachian drainage but Thompson applied it in detail. In discussing piracy and migration of divides Tarr (1918) wrote that:

"divides may slowly march onward, crossing more inexorably onward giving increased opportunity for piracy as they

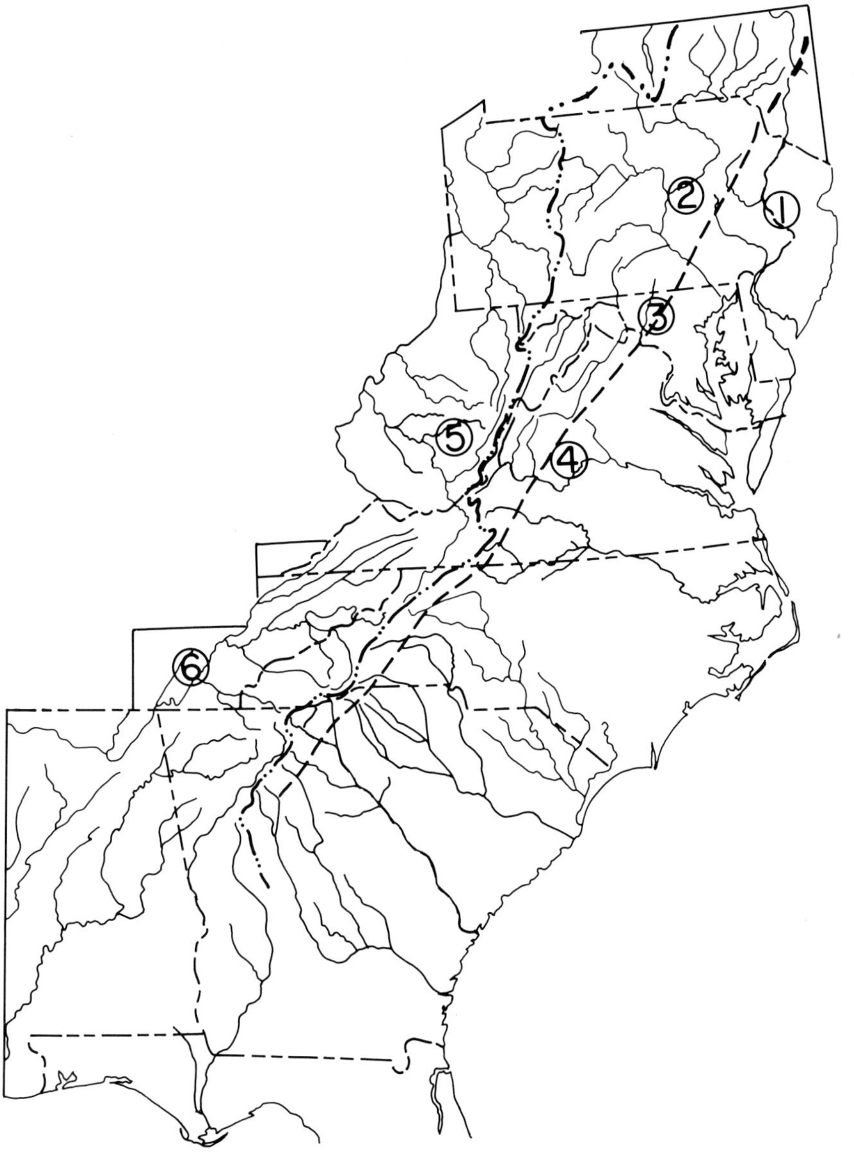

Fig. 6. Appalachian drainage showing the original divide according to Thompson, 1939 and the present divide. Major rivers are: *1* Delaware; *2* Susquehanna; *3* Potomac; *4* James; *5* New; *6* Tennessee. Original divide = dashed line; present divide = dash-dot line.

do so and that this is what seems to have occurred in the Appalachians".

There has been more than enough time for such a slow migration. Thompson calculated that the Potomac would have to cut back 6.4 km Ma^{-1} to move its headwaters to the present position.

The retreat of the drainage divide was shown by three factors. First, most of southeast flowing streams have steeper gradients than the westward flowing rivers. Secondly, the highest summits are southeast of the present divide. Thirdly, Thompson pointed out cases of recent piracies. Earlier Wright (1934) had described the diversion of some 320 km of drainage from

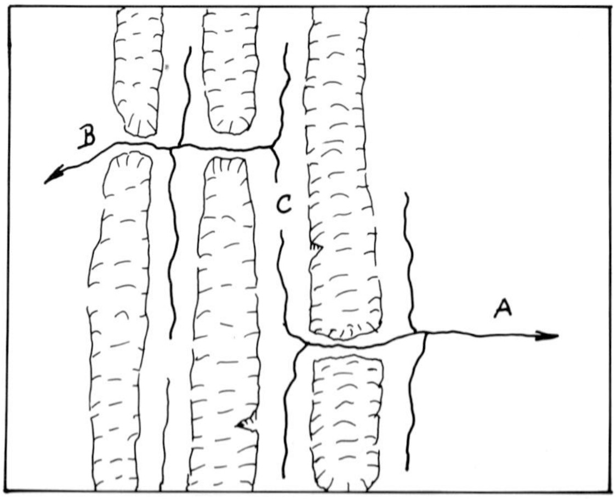

Fig. 7. Piracy and migration of a divide. Stream A (flowing eastward) will capture stream *B* (flowing westward) at *C*. The divide will then be shifted to the west.

the west-flowing New River to the east-flowing Roanoke.

In the migration by headward erosion, resistant rock barriers slowed but did not prevent the process. Moreover, streams took advantage of favorable rock structure, they crossed ridges where resistant beds had a weakness of some kind. For example, Thompson (1939) examined the Potomac as it crossed the Blue Ridge at Harpers Ferry (Fig. 8). Here the lower Cambrian formations and the Triassic diabase are the resistant rocks. In crossing these barriers the river has chosen points of comparative weakness. At Harpers Ferry and Weverton the outcrops are narrow, crumpled and folded (Fig. 8); at Point of Rocks the sandstone is discontinuous; east of Point of Rocks the river is on weak shales. Throughout its entire course in the Ridge and Valley west of Harpers Ferry (not shown on the map), the Potomac zigzags and avoids the two ridge-making formations, the Tuscorora and Pocono. In a similar way Thompson used the geologic map to show that the James and Roanoke Rivers have avoided resistant formations wherever possible, and when it could not avoid them entirely, the streams chose relatively weak or narrow outcrops. Ashley (1935) and Thompson (1949) demonstrated that the gaps on Kittatiny Ridge in Pennsylvania are all at points of weakness.

Thompson (1949) felt that a study of the gaps would help in the interpretation. If gaps were made by superposed streams, those in the flanking ridges should correspond in number and position with those on the main ridge i.e. there should be an alignment of gaps. He found that this was, in general, not true. Coves and notches in smaller, flanking ridges did not lead to gaps in the main ridge, indicating that they were made by headward growth of small streams. He classified the gaps into four groups: (1) gaps located where resistant rocks have relatively narrow outcrops; (2) gaps located where rocks are thinly layered, vesicular and weak; (3) gaps located at faults; and (4) gaps formed by solution action of the underlying limestone. He maintained that these gaps were made not by superposition but by normal processes of erosion at points of weakness.

Thompson (1949) examined fold valleys of Pennsylvania and found that, in general, the synclinal valleys drained through the eastern limb and the anticlinal valleys drained through the western rim (for example, see Fig. 9). He found it difficult to imagine a method of regional superposition whereby streams would be superposed on the west limb of anticlines and the east limb of synclines. However, because of the greater thrust from the east, the western limb of anticlines and the eastern limb of synclines are steeper and have narrower outcrops. Consequently, gaps were cut on the steeper limb, as suggested by Davis.

> "The location of the successful lateral stream...may be determined by the dip of the beds, gaps being cut quicker on steep than on gentle dips."
>
> Davis, 1889a, p. 436.

Strahler (1945) critically examined all these hypotheses advanced for the origin of Appalachian drainage, limiting himself to the problem in the Folded Appalachians of Pennsylvania. Contrary to the statements of others, Strahler

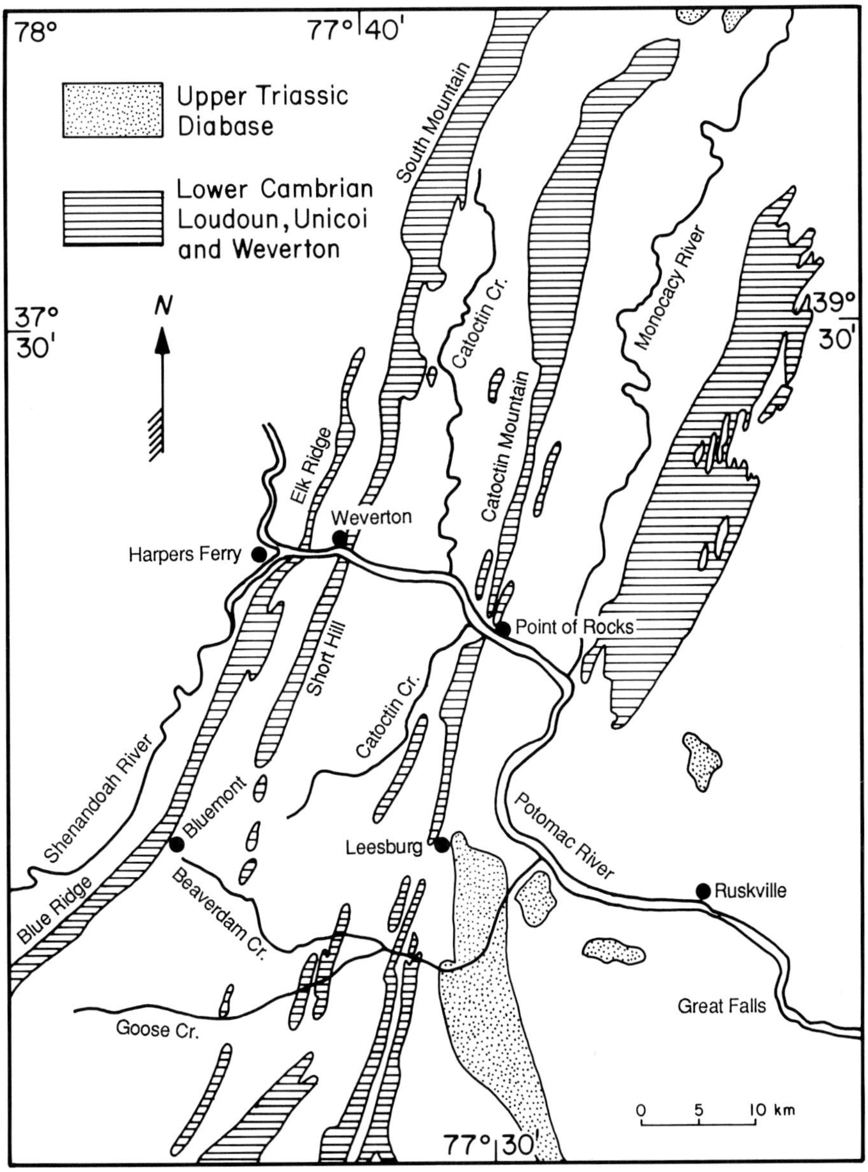

Fig. 8. Map of the Potomac River, showing its path through resistant rock ridges. After Thompson, 1939.

could find no evidence of faulting at either wind or water gaps, with only a couple of exceptions. However, more recent, detailed field work by Epstein (1966) indicates there are structural weaknesses at many of the major water gaps in the Stroudsburg area. One or more of the following occurred at the six gaps studied: the dying out of the fold a short distance away from the gap, an abrupt change in strike of the beds, a steep dip and narrowing of the outcropping bed, closely spaced joints and extensive fracturing, and overturned beds or more intense folding. Detailed structural and stratigraphic studies by Dyson (1963), Hoskins (1967, 1976) and Theisen (1983) show the presence of weaknesses which control the position of the Susquehanna River gaps north of Harrisburg (Hoskins, 1987) (Fig. 10). Hoskins (1976) found a strike-parallel fault as well as overturned and faulted beds at the Mahantango gap.

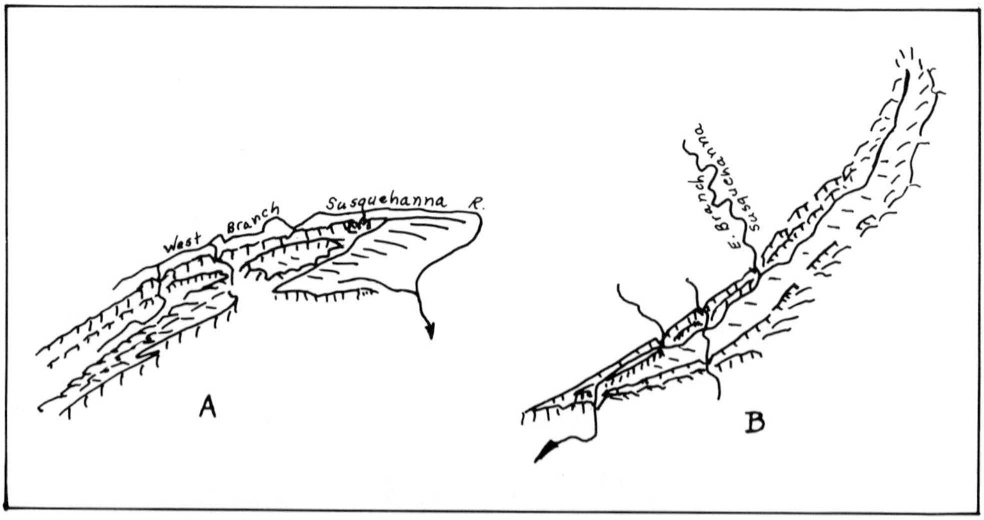

Fig. 9. (A) An example of an anticline draining to the west (Nittany Valley); (B) an example of a syncline draining to the east (Lackawanna syncline).

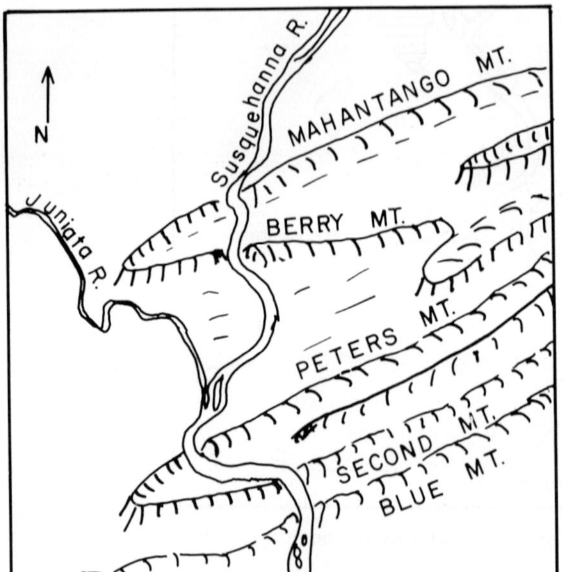

Fig. 10. The Susquehanna River gaps north of Harrisburg. Each of the gaps that have been studied in detail exhibit structural weaknesses that may have provided an easy path for stream erosion.

A major fracture filled with Triassic diabase is present at both the Peters Mt. (Dyson, 1967) and Berry Mt. gaps (Hoskins, 1976). At Second Mt. the strike of the Pocono beds changes from the east to the west side of the river and Hoskins (1987) interprets local fracturing from sharp bends in the Pocono Formation. Finally, at Blue Mt. the Susquehanna cuts through the ridge which is weakened by wedge faults, kink bands, strike parallel thrusts, and extensional faults observable at the gap (Thiessen, 1983). These data supports Thompson's proposal for the presence of gaps.

On the other hand, Strahler found five locations where there were evident faults along the ridges and streams had not utilized the weakness to cut through the resistant rocks in water gaps. Why did they not do so? He stated also that processes of superposition described by Meyerhoff and Olmstead (1938) were dubious since strata of differing competence would not bend conformably as shown in Fig. 5. The process of westward shifting of divides by headward erosion and piracy he believed to be untenable, as they were contrary to principles of fluvial action, since he could not conceive how a small stream could cut through a resistant ridge and capture a larger subsequent flowing on softer rock. Even if the processes were valid, it would not, in his opinion, explain linear transverse courses, abundant wind gaps, southeast alignment of gaps, or close dual gaps on the same ridge; all present in the drainage system. He concluded that all these arguements as well as the absence of faults in water gaps and the presence of faults not utilized by rivers to cut gaps strengthened the hypothesis of regional superposition from a widespread Coastal Plain cover of sediments. Thus, Johnson's hypothesis could best explain all the drainage features, according to Strahler's analysis. It was not a positive proof (i.e. proving a coastal plain cover) but a result reached by eliminating other hypotheses.

The Zagros streams

In 1965 Oberlander presented his interpretation of the origin of transverse drainage in the Zagros Mountains of Iran, an active orogenic zone similar to the Appalachians. As in the Folded Appalachians, there is some degree of summit accordance. Many drainage features similar to those in the Appalachians are present. He found basins diagonal to structural trends, elongated strike valleys, transverse cuts across resistant ridges, wind and water gaps some of which were aligned, others not. Some streams ignored structures; some were deflected by a ridge and then would erode through the same ridge elsewhere without apparent reason. There were anticlines and synclines in all stages of dissection. It seemed an ideal place to study the evolution of drainage in a folded terrain.

Many consequent streams took long and circuitous courses, flowing in sags between anticlines (Fig. 11A), or down a transverse axis. Other consequent streams flowed down the axis of synclines, finding an exit from the structural basin at sags located at the broadest part of the structure. Many transverse reaches were consequent on the flanks of anticlines. Thus consequence could result in both aligned and unaligned gaps. Stream could later become antecedent or superimposed (Fig. 11B).

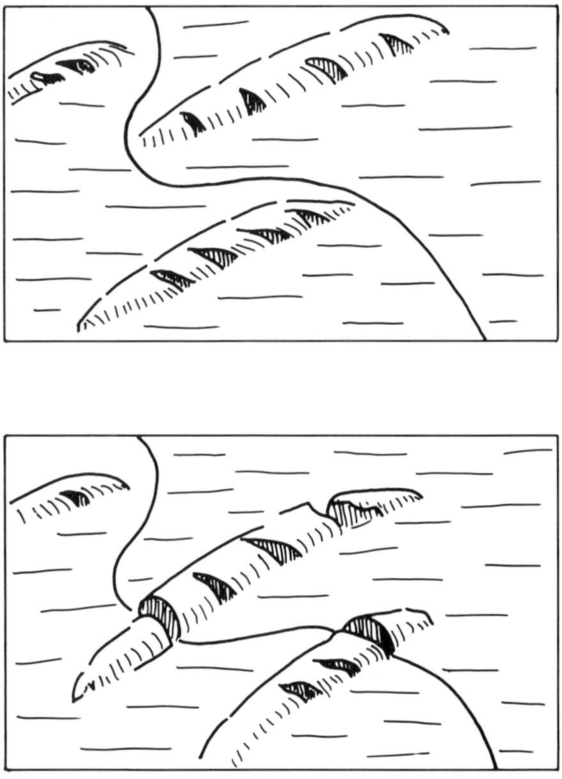

Fig. 11. (A) A consequent river winding its way around anticlinal noses. (B) The same river after either antecedence or superimposition. After Oberlander, 1965.

Antecedent streams would maintain the consequent courses after uplift. Such rivers would be indicated by a lack of deflection where the river crosses a resistant ridge. Other indications of antecedence, he proposed, would include: little deviation in overall course, watergaps would vary in location as well as orientation along the ridge and gaps would generally be aligned. He found few antecedent reaches, although this tectonically young region seemed ideal for antecedence to occur.

Oberlander believed that a theory of superimposition often evades the basic question of origin, since every superposed reach was previously an antecedent, consequent, or subsequent stream. Evidence of superposition is similar to that of antecedence, so much so that it is difficult to distinguish between them. According to Oberlander indications of superimposition include disregard of geologic structure and rock type, alignment of gaps in series, lack of consistency in orientation of approach to transverse stretches, and differences in orientation of the reach through the barrier. A meandering gorge through a ridge is not indicative of superposition, only of rejuvenation, unless there are remnants of either upraised terrace deposits or another covermass. Stream deflection at geologic contacts denies superposition.

He found anticlinal gaps may be formed by structural imposition of consequent streams (Fig. 11). Although Oberlander observed that none of the major through flowing rivers were superimposed, a large number of transverse gaps had been created by local superposition from discordant superjacent beds (Meyerhoff's, 1972, stratigraphic superposition).

He saw many evidences of structural control of stream reaches.

> "Denudation of...anticlines is guided by bedding planes and joint systems. Intersecting shear joints fix the figures of the flatirons produced by downstripping of the anticlinal arches, and occasionally extension jointing is spectacularly displayed, facilitating the dissection of the fold by closely spaced lateral gullies."
>
> Oberlander, 1965, p. 41

> "The anticlines being stripped by joint control give way bed by bed."
>
> ibid, p. 42

Thus, he found transverse stretches of subsequent origin everywhere because of the extensive jointing. Long strike valleys resulted from erosion of the less resistant beds. Headward erosion occurred along with development of subsequents to result in trellis drainage patterns, non-aligned gaps, offset of gaps at barriers, and other deflections of course. Further evidences of headward extension of drainage, primarily by subsequent erosion, included sharp turns out of a longitudinal valley, gorges parallel to shear lines, barbed tributaries, and the location of reaches near anticlinal crests. In fact, he came to the conclusion that "the ordinary processes of denudation will produce drainage lines transverse to a fold belt" (ibid, p. 150).

The results of this detailed study was that different transverse reaches were developed by different processes. *No one process* produces the drainage as we now see it.

A new look at the folded Appalachians of Pennsylvania

In the light of his Zagros study, Oberlander (1985) took a cursory look at the Susquehanna River (Fig. 12). He pointed out that the main stream had a consistently close relation to structure or is located on erodible formations. For example, above and below Sunbury the Susquehanna River is flowing in, or parallel to, the sag between the structural axes of the anticlinorium and synclinorium. He suggested that the transverse gorges north of Harrisburg might have evolved from a Mauch Chunk covermass which produced a broad band of erodible sediments between resistant outcrops on the fold limbs.

The results of recent studies of plate tectonics in the history of the Appalachians and to the Susquehanna River drainage have been applied by Judson (1975) who revised Davis' conception of the history and drainage of the Appalachians. In the southern Appalachians, folding and thrusting in the Devonian raised the Appalachian region leaving a small ocean to the east between North and South America and Africa (Hatcher and Odom, 1980). The area remained high as Africa became welded to the American continents and as the Valley and Ridge folds and thrusts occurred in the Carboniferous. Farther to the north the same schedule of events was taking place between northeastern North America and Europe (Williams and Hatcher, 1983). Then in the Triassic, rifting of the super continents began with major faulting in the present Triassic basins. This resulted in uplift of the region flanking the rift valleys, due to thermally induced arching (Judson, 1975) so that rivers flowed both east and west from a divide along the thermal ridge. Subsidence of the

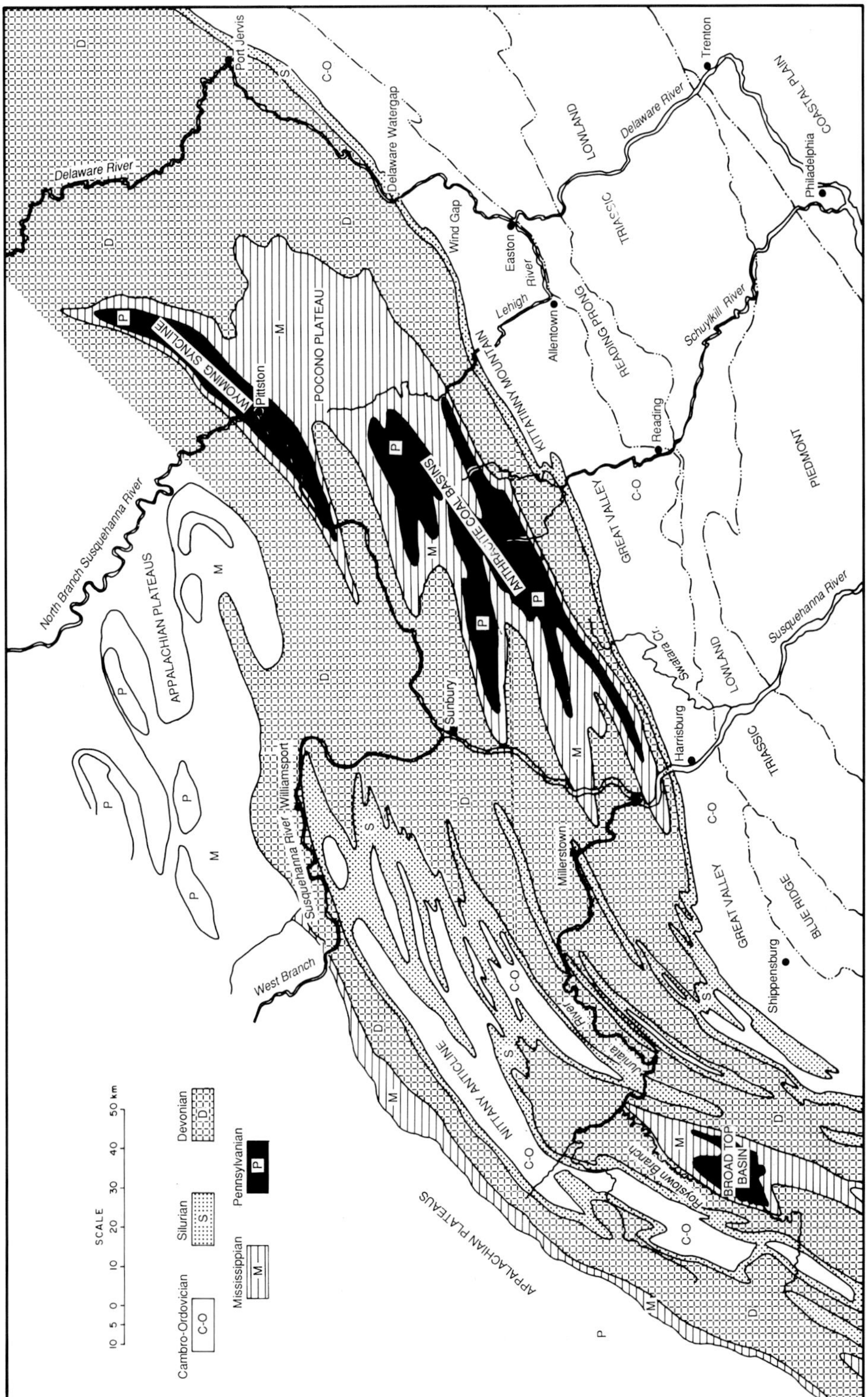

Fig. 12. A generalized map of the structure and stratigraphy of the Folded Appalachians of Pennsylvania.

continental margin as the Atlantic widened during the Jurassic reversed the tilt of the surface toward the southeast. The uplift and tilting allowed the vigorous southeastward flowing streams to erode headward quickly, shifting the divide westward by piracy (Judson, 1975) according to principles elucidated by Davis (1889a). Judson's explanation changed the timing of drainage reversal from the Triassic (Davis, 1889a) to the Jurassic.

Summary

Before and since Davis' 1889a paper the explanation of the drainage of the Appalachians has intrigued geomorphologists. Those hypotheses advanced since Davis have relied primarily on various solutions he suggested. Despite the constriction on him by geologic paradigms of his day, Davis' agile mind managed to conceive basic principles which affected future studies of geomorphology. Although the concepts of erosion cycles and peneplains have waned in popularity recently, the principles of formation of water and wind gaps, headward migration of divides, stream piracy, and adjustment of streams to structure, so clearly explained in his 1889a paper are still viable today. Hypotheses advanced by Johnson, Ruedemann, and Thompson are embellishments of statements in the 1889a publication. His observation that the rivers of Pennsylvania were complex and composite in form and origin is echoed by Oberlander. The present drainage is a result of many processes, still ongoing.

Perhaps the best closing for this paper is to end as did Judson (1975) with a modest quote from Davis' paper. Perhaps it contains the best model he could offer to geomorphologists in presenting papers.

"If the postulates...seem unsound and the arguments overdrawn, error may at least be avoided by not holding too fast to the conclusions that are presented, for they are presented only tentatively. I do not feel by any means absolutely persuaded of the correctness of the results, but at the same time deem them worth giving out for discussion."

Davis, 1889a in Geographical Essays, p. 447

References

Ashley, G.H., 1935. Studies in Appalachian mountain sculpture. Geol. Soc. Am., Bull., 46: 1398-1436.

Beckinsale, R.P. and Chorley, R.J., 1968. History of Geomorphology. In: R. Fairbridge (Editor), Encyclopedia of Geomorphology. Reinhold, New York, N.Y., pp. 410-416.

Bloom, A.L., 1978. Geomorphology. PrenticeHall, Englewood Cliffs, N.J., 510 pp.

Chorley, R.J., 1965. A re-evaluation of the geomorphic system of William Morris Davis. In: R.J. Chorley and P. Haggett (Editors), Frontiers and Geographical Teaching. Methuen, London, pp. 21-38.

Chorley, R.J., 1978. Bases for theory in geomorphology. In: C. Embleton, D. Brunsden, and D.K.C. Jones (Editors), Geomorphology, Present Problems and Future Prospects. Oxford Univ. Press, Oxford, pp. 1-13.

Chorley, R.J., Beckinsale, R.P. and Dunn, A.J., 1973. The Story of the Study of Landforms, Vol. 2, The Life and Work of William Morris Davis. Methuen, London, 874 pp.

Cole, W.S., 1934. Identification of erosion surfaces in eastern and southern Ohio. J. Geol., 42: 285-294.

Cole, W.S., 1941. Nomenclature and correlation of Appalachian erosion surfaces. J. Geol., 49: 129-148.

Crowley, W.P., 1976. The geology of the crystalline rocks near Baltimore and its bearing on the evolution of the eastern Maryland Piedmont. Maryland Geol. Surv., Rep. Invest., 27, 40 pp.

Davies, G.H.L., 1988. Whiggishness in the history of geomorphology. Paper presented at the 19th Binghamton Geomorphology Symposium "History of Geomorphology", St. Catherine's, Can.

Davies, G.H.L., 1969. The earth in decay. Elsevier, New York, N.Y., 390 pp.

Davis, W.M., 1883. Origin of cross valleys. Science, 1: 325-327 and 356-357.

Davis, W.M., 1885. Geographic classification, illustrated by a study of plains, plateaus and their derivatives. Am. Assoc. Adv. Sci., Proc., 33: 428-432.

Davis, W.M., 1889a. The rivers and valleys of Pennsylvania. Natl. Geogr. Mag., 1: 183-253; reprinted in Geogr. Essays, Ginn and Co., 1909, pp. 413-484.

Davis, W.M., 1889b. Topographic development of the Trassic formations of the Connecticut Valley. Am. Jour. Sci., 3rd Ser., 37: 423-434.

Davis, W.M., 1895. The physical geography of southern New England. Natl. Geogr. Soc., Monogr., 1 (9): 269-304.

Davis, W.M., 1899. The geographical cycle. Geogr. J., 14: 481-504.

Davis, W.M., 1905. Complications of the geographical cycle.

Rep. Eighth Int. Geogr. Congr., Washington, D.C., pp. 150-163.

Dutton, C.E., 1882. Tertiary history of the Grand Cañon district. U.S. Geol. Survey., Mem., 2, 76 pp.

Dyson, J.L., 1963. Geology and mineral resources of the northern half of the New Bloomfield Quadrangle. Pa. Geol. Surv., 4th Ser., Atlas 137ab, 63 pp.

Epstein, J.B., 1966. Structural control of wind gaps and water gaps and of stream capture in the Stroudsburg area, Pennsylvania and New Jersey. U.S. Geol. Surv., Prof. Pap., 55B: 1380-1386.

Flemal, R.C., 1971. The attack on the Davisian system of geomorphology: a synopsis. J. Geol. Educ., 19: 3-13.

Gregory, K.J., 1978. Fluvial processes in British basins. In: C. Embleton, D. Brunden and D.K.C. Jones (Editors), Geomorphology: Present Problems and Future Aspects. Oxford Univ. Press, Oxford, pp. 40-72.

Hack, J.T., 1960. Interpretation of erosional topography in humid temperate regions. Am. J. Sci., 258A: 80-97.

Hall, J., Jr., 1882. Contributions to the geological history of the North American continent. Am. Assoc., Adv. Sci., Proc., 31: 29-71.

Hatcher, R.D. and Odom, A.L., 1980. Timing of thrusting in the southern Appalachians, USA: model for orogeny? Geol. Soc. London, J. 137: 321-327.

Higgins, C.G., 1975. Theories of landscape development: a perspective. In: W.N. Melhorn and R.C. Flemal (Editors), Theories of Landform Development. Publ. Geomorphology, Binghamton, New York, N.Y., pp. 1-28.

Higgins, C.D., 1983. Biological models in geomorphology. In: S. Boardman (Editor), Revolution in the Earth Sciences. Kendall/Hunt, Dubuque, Iowa, pp. 181-187.

Hoskins, D.M., 1976. Geology and mineral resources of the Millersburg 15 minute Quadrangle, Dauphin, Juniata, Northumberland, Perry, and Snyder counties, Pennsylvania. Pa. Geol. Surv., 4th Ser., Atlas 146, 38 pp.

Hoskins, D.M., 1987. The Susquehanna River water gaps near Harrisburg, Pa. In: D.C. Roy (Editor), Centennial Field Guide-Northeastern Section. Geol. Soc. Am., 5: 47-50.

Hutton, J., 1795. Theory of the Earth with Proofs and Illustrations. William Creech, Edinburgh, 620 pp.

Johnson, D.W., 1931. Stream Sculpture on the Atlantic Slope. Columbia Univ. Press, New York, N.Y., 142 pp.

Judson, S., 1960. William Morris Davis - an appraisal. Z. Geomorphol., 4: 193-201.

Judson, S., 1975. Evolution of Appalachian topography. In: W.N. Melhorn and R.C. Flemal (Editors), Theories of Landform Development. Publ. Geomorphol., Binghamton, New York, N.Y., pp. 29-44.

Kay, M., 1951. North American geosyclines. Geol. Soc. Am., Mem., 48: 143 pp.

King, L.C., 1953. Canons of landscape evolution. Geol. Soc. Am., Bull., 64: 721-752.

King, L.C., 1962. The Morphology of the Earth: a study and synthesis of world scenery. Hafner Pub. Co., New York, 699 pp.

Le Pichon, X. and Fox, P.J., 1971. Marginal offsets, fracture zone and the early opening of the North Atlantic. J. Geophys. Res., 76: 6294-6308.

Lesley, J.P., 1856. Manual of coal and its topography. Lippincott, Philadelphia, Pa.

Mackin, J.H., 1938. The origin of Appalachian drainage - a reply. Am. J. Sci., 5th Ser., 36: 27-53.

Meyerhoff, H.A., 1972. Postorogenic development of the Appalachians. Geol. Soc. Am., Bull., 83: 1709-1727.

Meyerhoff, H.A. and Olmstead, E.W., 1934. Wind gaps and water gaps in Pennsylvania. Am. J. Sci., 5th Ser., 27: 410-416.

Meyerhoff, H.A. and Olmstead, E.W., 1936. The origins of Appalachian drainage. Am. J. Sci., 232: 21-42.

Meyerhoff, H.A. and Olmstead, E.W., 1938. Evolution of the northern Appalachian drainage. Geol. Soc. Am., Bull., 49: 1938 (abstr.).

Oberlander, T.M., 1965. The Zagros streams. Syracuse Univ., Geogr. Ser., 1: 168 pp.

Oberlander, T.M., 1985. Origin of drainage transverse to structures in orogens. In: M. Morisawa and J. Hack (Editors), Tectonic Geomorphology. Unwin Hyman, London, pp. 155-182.

Penck, W., 1924. Die Morphologische Analyse. Ingelhorns, Stuttgart, 283 pp. Translated 1953 as Morphological Analysis of Landforms, by H. Czech and K.C. Boswell, Macmillan, London, 429 pp.

Peschel, O.F., 1880. Physische Erdkunde. 11, 442. Referenced in Davis (1889a).

Powell, J.W., 1875. Exploration of the Colorado River of the West and its tributaries. Washington, D.C., 291 pp.

Rich, J.L., 1934. Glacial geology of the Catskills. New York State Museum, Bull., 299.

Rich, J.L., 1938. Multiple erosion surfaces. Geol. Soc. Am., Bull., 49: 1695-1722.

Ritter, D.F., 1978. Process Geomorphology. Brown, Dubuque, Iowa, 603 pp.

Ruedemann, R., 1932. Development of drainage of the Catskills. Am. J. Sci., 5th Ser., 23: 337-349.

Sevon, W.D., 1985. Pennsylvania polygenetic landscape. Guideb. 4th Annu. Field Trip, Harrisburg Area Geol. Soc., 55 pp.

Sevon, W.D., 1986. Susquehanna River water gaps: many years of speculation. Pa. Geol., 17(3): 4-7.

Strahler, A.N., 1945. Hypotheses of stream development in the Folded Appalachians of Pennsylvania. Geol. Soc. Am., Bull., 56: 45-88.

Suess, E., 1875. Die Entstehung der Alpen. Braumuller, Vienna, Referenced in Greene, M.T., 1982. Geology in the 19th Century. Cornell Univ. Press, Ithaca, N.Y., 324 pp.

Tarr, R.S., 1898. The peneplain. Amer. Geol., 21: 351-370.

Tarr, R.S., 1918. College Physiography. Macmillan, New York, N.Y.

Theisen, J.P., 1983. Is there a fault in our gap? Pa. Geol. 14(3): 5-11.

Thompson, H.D., 1939. Drainage evolution in the southern

Appalachians. Geol. Soc. Am., Bull., 50: 1323–1355.

Thompson, H.D., 1949. Drainage evolution in the Appalachians of Pennsylvania. N.Y. Acad. Sci., Ann., 52(2): 31–62.

Thornbury, W.D., 1969. Principles of Geomorphology (2nd edition). John Wiley and Sons, New York, 594 pp.

Tietze, E., 1878. Jahrb. Geol. Reichsanstalt, XXVII, 600. Referenced by Davis, 1889a.

Van der Voo, R., 1983. A plate tectonics model for the Paleozoic assembly of Pangea based on paleomagnetic data. In: R.D. Hatcher, Jr., H. Williams and I. Zietz (Editors), Contributions to the Tectonics and Geophysics of Mountain Chains. Geol. Soc. Am., Mem. 158: 19–24.

Ver Steeg, K., 1930. Wind gaps and water gaps of the northern Appalachians, their characteristics and significance. N.Y. Acad. Sci., Ann., 32: 87–220.

Von Engeln, O.D., 1942. Geomorphology: Systematic and Regional. Macmillan, New York, 655 pp.

Williams, H. and Hatcher, R.D., Jr., 1983. Appalachian suspect terranes. In: R.D. Hatcher, Jr., H. Williams and I. Zietz (Editors), Contributions to the Tectonics and Geophysics of Mountain Chains. Geol. Soc. Am., Mem., 158: 33–54.

Wright, F.J., 1934. The new Appalachians of the south, Part I, Between the Potomac and New Rivers. Dennison Univ., Bull., 34(13).

Geomorphology, 2 (1989) 23-37
Elsevier Science Publishers B.V., Amsterdam — Printed in The Netherlands

Geodynamic and Geomorphic Evolution of the Permo–Triassic Appalachian Mountains

RUDY SLINGERLAND and KEVIN P. FURLONG

Department of Geosciences, The Pennsylvania State University, University Park, PA 16801 (U.S.A.)

(Received February 12, 1989; accepted after revision May 21, 1989)

Abstract

Slingerland, R. and Furlong, K.P., 1989. Geodynamic and geomorphic evolution of the Permo-Triassic Appalachian Mountains. In: T.W. Gardner and W.D. Sevon (Editors), Appalachian Geomorphology, 2: 23-37.

The basic topography of a steady state, accretionary mountain range is a function of the critical taper of the accretionary wedge, which in turn depends upon the convergence flux at the toe, internal rock strength, basal slope, and erosion rate off the top. It is possible to obtain a first-order estimate of this topography for ancient mountain ranges by constraining these parameters using sediment thicknesses in adjacent, flexurally-produced sedimentary basins. Here we apply the technique to the Early Permian central Appalachians, by first obtaining the steady-state mass of the wedge from a flexural loading model of Beaumont and others. Then, the equations of a critically tapered wedge are solved using a Monte Carlo method of parameter selection. Those solutions that match the observed mass of the wedge suggest that the Early Permian Appalachians possessed a central Andean topography with average relief of from 3.5-4.5 km and width of 250-300 km.

Introduction

Elevation and width are two primary geomorphical attributes of a mountain range, and it is of some interest to speculate on their controlling factors. Recent advances in our understanding of crustal convergence now make it possible to offer a first order explanation as to why modern mountain ranges, at the same stage in their evolution, exhibit a wide variation in these basic attributes. This understanding raises the possibility that ancient ranges, now only vestiges of their former selves, can be reconstructed to their once lofty heights by an application of the same principles. Of all the mountain ranges that have existed none would seem more suited to bring back to life than the ancestral Appalachians. They are one of the best studied ranges in the world and the literature is rich in speculation on their geomorphic evolution (e.g., Judson, 1975; Sevon, 1985; Morisawa, this volume). Maybe most importantly, the Appalachian sedimentary basin preserves important, albeit circumstantial, information.

Our intention here is to use geodynamic principles and the results of recent basin modelling to better rationalize the evolution of Appalachian topography during the Permo–Triassic. We have attempted to answer two questions: (1) what variables and feedback loops control the topographies of orogenic mountain belts in general, and the Appalachians in particular? and (2) how high and wide were the Appala-chian Mountains in the central Atlantic at the end of the Alleghanian Oroge

Controls of the topography of mountain ranges

To first order, the topography of an orogenic mountain range – that is, its mean elevation, relief, and width – is controlled by the processes of crustal thickening and denudation (e.g., Molnar, 1988; Molnar and Lyon-Caen, 1988). Crustal thickening can be produced by a variety of mechanisms including thrusting-related shortening and magmatic addition. In convergent orogenic belts such as the Appalachian or Alpine chains, crustal thickening via folding and thrusting predominates. Such crustal thickening can occur on length scales ranging from short (0–5 km) imbricate thrusts and buckle folds seen in accretionary wedges, to the mega-thrusts of continental convergence (sometimes referred to as continental subduction) where 'thrust' sheets of greater than 10 km in thickness may be displaced tens or hundreds of km. The resultant mountain belt geometry is additionally "filtered" by the response of the lithosphere to this added mass of crust which acts to deflect the underlying lithosphere. This deflection will normally be flexural in nature, that is, it will be a broad regional downwarp rather than a mirror image of the surface topography that would be expected if local isostatic compensation occurs. The magnitude and wavelength of this lithospheric deflection is controlled by the strength or flexural rigidity of the lithosphere and thus the net topographic shape of the mountain range and adjacent sedimentary basins will depend on both the size and shape of the load (i.e., the region of crustal thickening), and on the strength of the underlying plate.

Several approaches use these basic ideas to evaluate the evolving and steady-state geometry of a mountain belt. For thickening via folding and thrusting, the width and height of the region of thickened crust is limited by the magnitude of thickening (Molnar and Lyon-Caen, 1988), because the horizontal driving forces must work against gravity and the gravitational potential energy of a range grows as the square of its height. When this work is greater than the work necessary to break through the crust in an adjacent lowland area, thickening commences there and the range grows wider. Davis et al. (1983) and Dahlen and Suppe (1988) have combined this energy view with a failure criteria for the crust to calculate the wedge geometries of many convergent orogens. Alternatively, one may evaluate the lithospheric response to mountain loads. Often, the amplitude and shape of the peripheral sedimentary basin can be used via a flexural model to place constraints on the excess mass present in the thickened crust of the mountain range (e.g. Beaumont et al., 1987, 1988).

Our contribution here is to combine the approaches mentioned above in order to develop a model for the evolution of the Appalachian system that has its mass distribution at any time controlled by the response of the lithosphere to the load of the mountain range, and its geometry controlled by the combination of energy considerations, an assumed failure criterion for the region, and erosion. Below we describe the models used, starting first with the wedge model which provides us with the geometry (including effects of erosion), and then the flexural loading model which, through an analysis of the basin geometry, constrains the excess mass in the range. Secondly, we describe the application of this method to the geometry of the Appalachians. Given a lack of hard data to constrain many of the parameters, we use a Monte Carlo approach to evaluate the most probable width and heigth of the Permian Appalachians in the region of Pennsylvania.

Theory

Wedge model

Dahlen (1984), Dahlen et al. (1984), and Dahlen and Suppe (1988) consider the crustal thickening in active fold and thrust belts to be analogous to the wedge of snow or soil in front of a bulldozer blade (Fig. 1). As material is

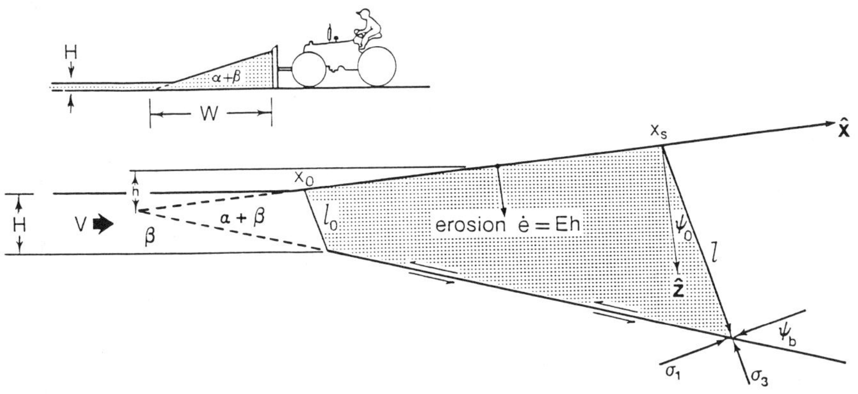

Fig. 1. Schematic of the Dahlen and Suppe (1988) critically tapered wedge. Material enters the toe (x_0) at the rate of HV km^2 Myr^{-1}, where H is the thickness of the incoming material and V is the convergence velocity. The material is stretched along the axis of least principal stress, σ_3, and narrowed along σ_1 as it passes back through the wedge. The basal slope of the wedge is β and the topographic slope is α. Ψ_0 and Ψ_b are the acute angles between the axis of principal compressive stress and the topographic and basal surfaces, respectively. Erosion proceeds perpendicular to x at a rate of $\dot{e}$ km Myr^{-1} which is a function of h, the height of the wedge surface above the vertex. A steady state occurs for some width W (equal to $x_s - x_0$) when $\dot{e}$ equals HV (modified from Dahlen and Suppe, 1988).

scraped up, the wedge grows in height until it reaches a critical surface slope or taper governed by the relative magnitudes of the sliding resistance along the base and the strength of the wedge material. After reaching critical taper, the wedge grows self-similarly as new material is added at the toe, becoming wider and thicker but maintaining the same surface slope. This equilibrium taper is called critical because in thinner wedges, compressive stress will exceed the rock strength, as given say, by the Coulomb failure law, and the wedge will deform by thickening until the critical taper is achieved. Finally, the wedge reaches a steady state when accretionary influx is balanced by erosive efflux. The Coulomb wedge model has been successfully applied to several active fold and thrust belts and accretionary wedges (Dahlen and Suppe, 1988) with good agreement between observed wedge geometries and theory for reasonable values of rock strengths.

The wedge width and height can be calculated from a mass balance of the material entering and leaving the wedge (Fig. 1) (Dahlen and Suppe, 1988). In Figure 1 a column of rock aligned with the principal stresses moves back through the wedge, elongating in the σ_3 direction from l_0 to l while conserving its cross-sectional area. The flux into the wedge due to convergence is:

$$\dot{A} = (1 - \Delta\rho/\rho) HV \tag{1}$$

and the flux out the back of the wedge at position x_s is:

$$l\dot{x}_s \cos\psi_0 = x_s \dot{x}_s \frac{\sin(\alpha+\beta)\cos\psi_0}{\cos\psi_b} \tag{2}$$

where the dot denotes differentiation with respect to time, ρ is the density of the material, $\Delta\rho$ is the change in density as material enters the wedge, and the remaining variables are defined in Fig. 1.

The flux off the surface of the wedge due to erosion is rationalized as follows. Regression equations of modern erosion rate as a function of relief (Ruxton and McDougall, 1967; Ahnert, 1970; William Hay, pers. commun., 1988) seem to show a linear dependence, and thus as a first approximation Dahlen and Suppe (1988) let:

$$\dot{e} = Eh \tag{3}$$

where $\dot{e}$ is the erosion rate perpendicular to the surface of the wedge, E is a constant with units of Myr^{-1}, and h is height (km) of the surface above the vertex of the wedge. Note that because of self-similarity, the surface slope of the wedge doesn't change, and therefore in the model the increase in erosion rate must be assumed to arise from complex changes in vegetation, elevation-dependent processes such as glaciation, and from increasing slopes of streams entrenched into the wedge. The total erosive flux off the surface of the wedge from x_o to x_s is:

$$\int_{x_0}^{x_s} \dot{e} dx = \frac{1}{2} E \sin \alpha (x_s^2 - x_0^2) \tag{4}$$

By conservation of mass eqn. (1) must equal 2 plus 4. Solving for $x_s(t)$, the location of a rock column in the wedge, yields:

$$x_s(t) = (W + x_0)[1 - \{1 - (x_0/(W + x_0))^2\} \exp(-\eta t)]^{1/2} \tag{5}$$

where $x_s(0) = x_0$ when $t = 0$, and:

$$w = \left[\frac{2\dot{A}}{E \sin\alpha} + x_0^2\right]^{1/2} - x_0 \tag{6}$$

$$\eta = \frac{E \sin\alpha \cos \psi_b}{\sin(\alpha + \beta) \cos \psi_0} \tag{7}$$

$$x_0 = \frac{H \cos \psi_b}{\sin(\alpha + \beta) \cos(\psi_0 + \alpha)} \tag{8}$$

As t approaches infinity, x_s approaches $x_0 + W$, and thus W is the steady-state width of an eroding, critically tapered wedge. By construction in Fig. 1:

$$h = x \sin \alpha \tag{9}$$

Dahlen and Suppe (1988) have applied eqn. (5) to the accretionary wedge of which Taiwan is a part, using the parameters in Table 1. The observed steady state width, W, is 87 km which is predicted from eqn. (6) only if $E = 1.27$ Myr^{-1}. This yields an erosion rate $\dot{e}$, averaged over the length of the wedge, of 5.7 km Myr^{-1}. Taiwan streamload data indicate a mean erosion rate of 5–6 km Myr^{-1} over the last 30 years (Li, 1976) and fission track data indicate a rate of 3–10 km Myr^{-1} over the last million years (Liu, 1982), values that are surprisingly close to the predicted. The steady-state height above the point x_0 at a distance $x_0 + W/2$ from the origin is:

$$h = \frac{W}{2} \sin \alpha \tag{10}$$

or 2.2 km, also in reasonable agreement with

TABLE 1

Values of independent parameters in modern wedges (from Westbrook et al., 1982; Davis et al., 1983; Dahlen, 1984; Dahlen and Suppe, 1988)

Wedge	α (°)	β (°)	$(\alpha+\beta)$ (°)	ψ_0 (°)	ψ_b (°)
Taiwan	2.9∓0.3	6∓1	–	2.6	11.5
Barabdos	0 to 3	2	3	–	–
Makran (Gulf of Oman)	–	–	4	–	–
Eastern Aleutians	–	–	7	–	–
off Oregon Coast	–	–	8	–	–

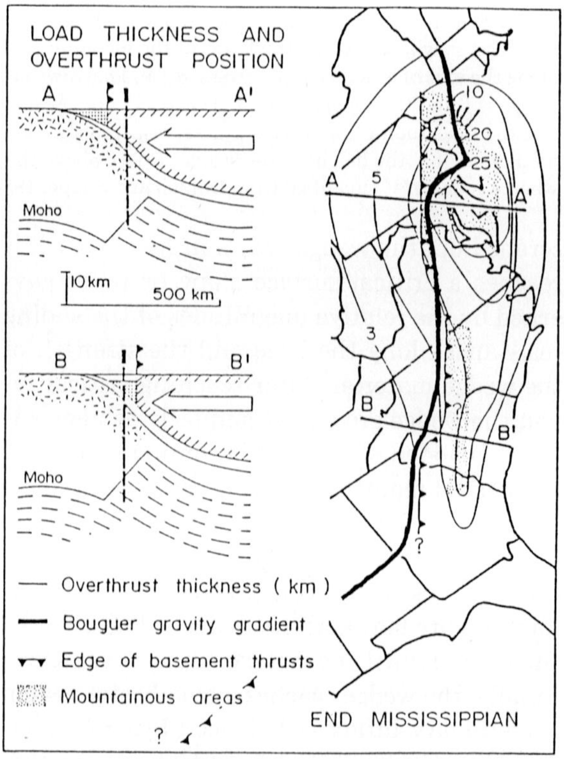

Fig. 2. Results of the Beaumont et al. model at the end of the Mississippian, showing overthrust thickness (km) in relation to the Bouguer gravity gradient. See text for interpretation (from Beaumont et al., 1987).

the geomorphology of Taiwan, where the central ranges average 2–3 km in elevation with a few peaks reaching 4 km.

Implicit in this theory is the assumption that fold and thrust belt mountains grow in height and width until their erosion, given by $W \langle \dot{e} \rangle$, balances $\dot{A}$, at which time they are in steady-state. Thus high erosion rates (on the order of 10^1 km Myr^{-1}) should produce narrower (50–200 km) and lower (1–5 km) mountain ranges than low erosion rates (on the order of 10^0 km Myr^{-1}) which should produce wider (>500 km) and higher (ca. 9 km) ranges (Dahlen and Suppe, 1988). As Dahlen and Suppe (1988) point out, this is substantiated by the Andes, whose width is greater in the arid zone between 10° and 30°S latitude than to the north or south.

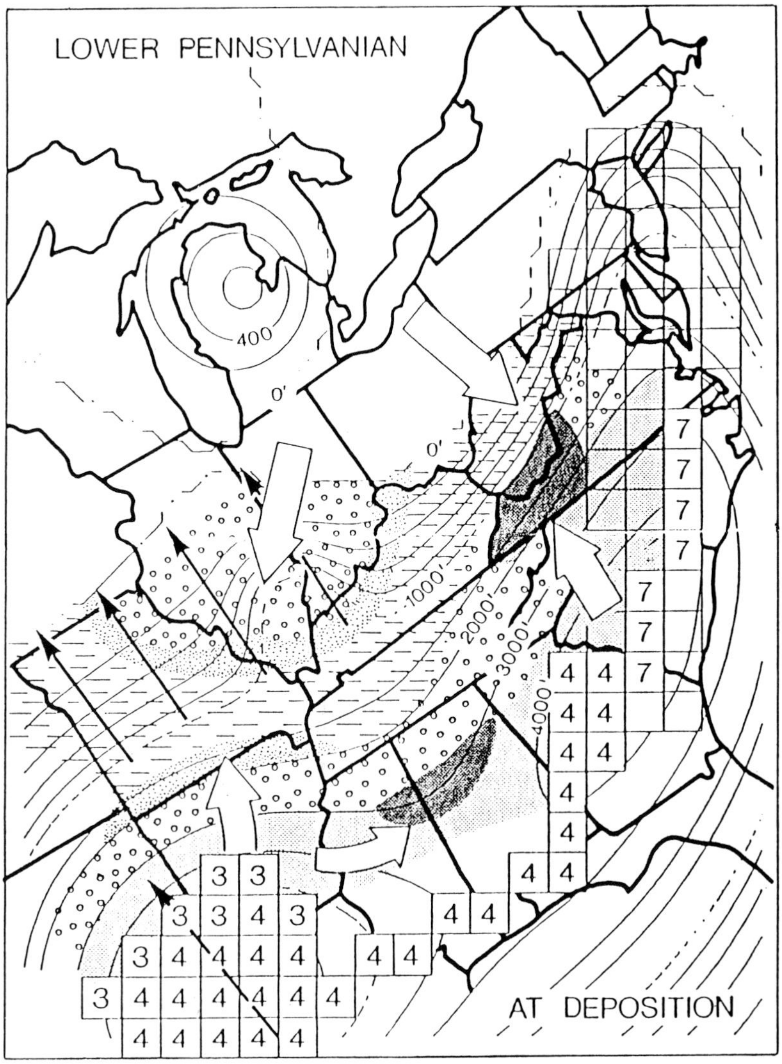

Fig. 3. Predicted sedimentary isopachs (in ft for comparison with original data) from the Beaumont et al. model for the Early Pennsylvanian. Numbers in squares are the load thicknesses in km that generate the necessary subsidence in the foreland basin. Dash and dot patterns denote observed shales and sandstones, respectively; large arrows show the major dispersal directions (from Beaumont et al., 1987).

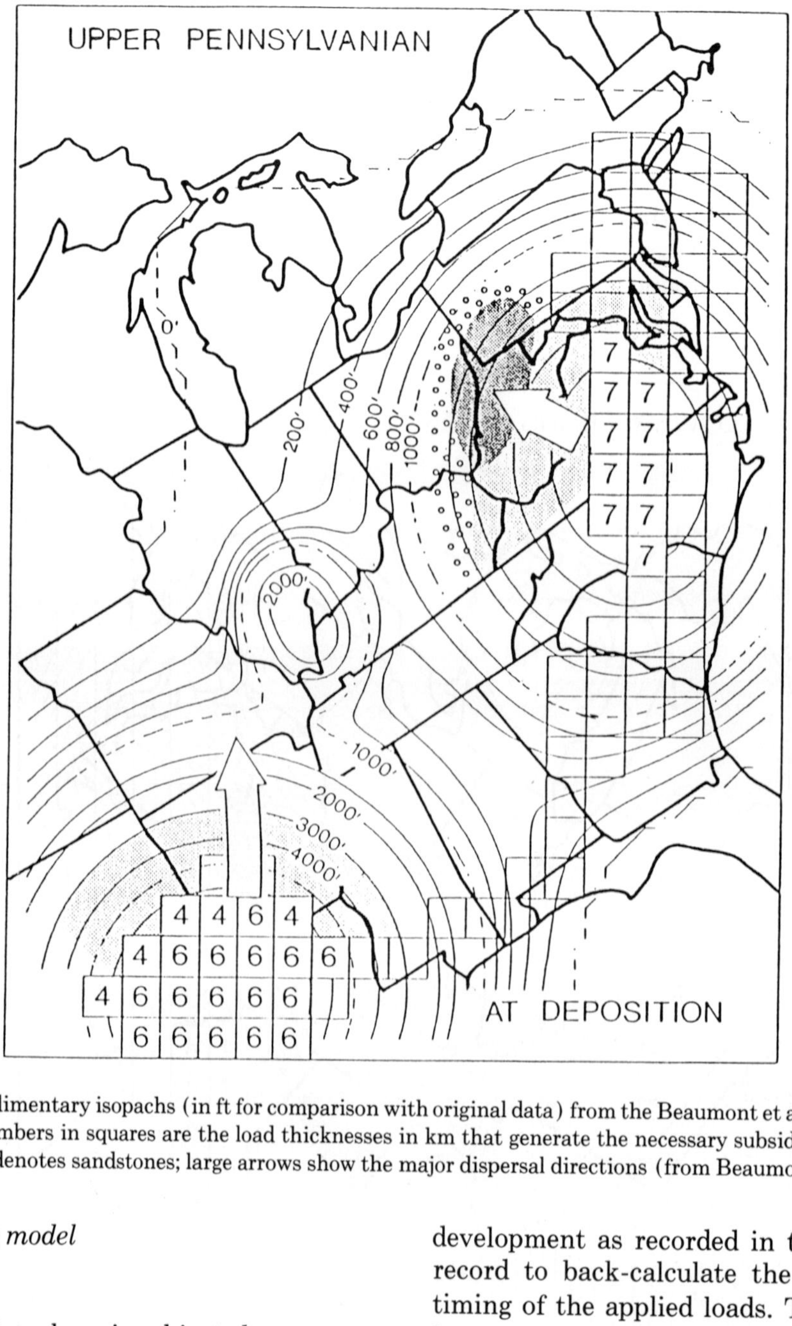

Fig. 4. Predicted sedimentary isopachs (in ft for comparison with original data) from the Beaumont et al. model for the Late Pennsylvanian. Numbers in squares are the load thicknesses in km that generate the necessary subsidence in the foreland basin. Dot pattern denotes sandstones; large arrows show the major dispersal directions (from Beaumont et al., 1987).

Flexural loading model

When the lithosphere is subjected to supracrustal loading it flexes downward, producing a sedimentary basin adjacent to the load. The width and depth of the basin are functions of the magnitude and rate of application of the load, and the rigidity of the crust. This simple relationship between basin geometry and load suggests that one could use the history of basin development as recorded in the stratigraphic record to back-calculate the magnitude and timing of the applied loads. This method has been elegantly pursued by Quinlan and Beaumont (1984), Stockmal et al. (1986), Beaumont et al. (1987), Beaumont et al. (1988), and Jamieson and Beaumont (1988), and summarized for the Appalachians in Slingerland and Beaumont (1989). Insofar as the loads represent a thickening of the crust, their magnitudes and rate of application should be some measure

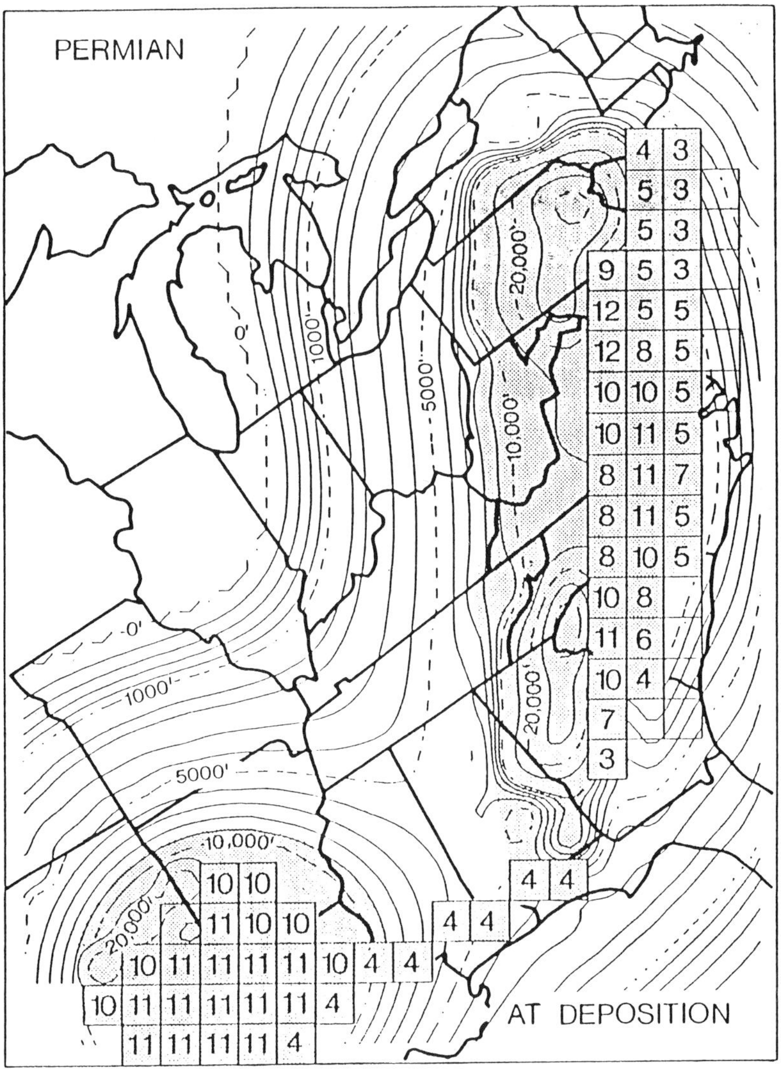

Fig. 5. Predicted sedimentary isopachs (in feet for comparison with original data) from the Beaumont et al. model for the Permian. Numbers in squares are the load thicknesses in km that generate the necessary subsidence in the foreland basin. Sediment thicknesses are inferred, for the most part from coal moisture, vitrinite reflectance, fluid inclusion data, and thermal maturation of shales (from Beaumont et al., 1987).

of the width and height of any mountain range they create.

To back-calculate the magnitude and timing of the crustal thickening, Beaumont and coworkers model the lithosphere as a linearly viscoelastic body in which the viscosity is temperature dependent. The lithosphere is supported by an inviscid or low viscosity fluid half-space (for details, see Quinlan and Beaumont, 1984). The area of the Appalachians is subdivided into a 20×31 grid, each a square 88.9 km on a side. There are twenty timesteps of non-uniform length during which loads are applied along the eastern margin of the Paleozoic Appalachians and the crust allowed to flex. Overthrust loads and sedimentary fill have a density of 2400 kg m^{-3} except for one timestep when the effect of higher density sediment is included. Load thickness scales linearly with density; therefore loads of 2800 kg m^{-3} would be 14% thinner. The

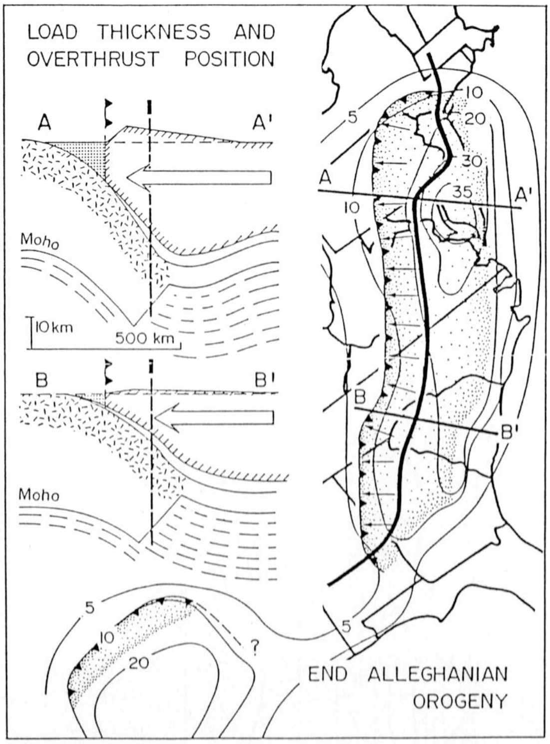

Fig. 6. Results of the Beaumont et al. model at the end of the Alleghanian orogeny, showing overthrust thickness (km) in relation to the Bouguer gravity gradient. See text for interpretation (from Beaumont et al., 1987).

thicknesses of overthrust loads are constrained in two ways – results from the mathematical model of lithospheric flexure must match both the Paleozoic and present sedimentary isopachs. In the case of the Permian strata, now largely eroded, the thickness of the eroded strata is estimated from the moisture content and rank of the coals (Beaumont et al., 1988).

The results (Figs. 2–6) show the loads that best fit the sedimentary isopachs and depths of burial. At the end of the Mississippian, the cumulative thicknesses of overthrusts and basin fill reached a maximum of over 25 km in the Delmarva area, outboard of the present Bouguer gravity gradient (Fig. 2). Insofar as this gradient represents the transition from continental crust to stretched and oceanic crust, these loads on thinned crust can be accommodated without significant topography (Beaumont et al., 1988). During the Early Pennsylvanian (Fig. 3), little if any thrusting occurred in the central Appalachian region. This is inconsistent with previous interpretations of the Pottsville gravels, but one must remember that although those gravels are coarse, they also are quite thin. Levine and Slingerland (1987) suggest they are more directly a result of increased precipitation than of increased elevation in the source terrain.

By the Late Pennsylvanian, thrusting had moved northward to the central Appalachian region (Fig. 4), although the major loads remained south of New Jersey. At the end of the Permian, thrusting had moved into southern New England and also cratonward across eastern Pennsylvania and Maryland (Fig. 5). These results are corroborated by vitrinite reflectance and fluid inclusion studies of the eastern anthracite fields of Pennsylvania (MacLachlan, 1985; Orkan and Voight, 1985; Levine, 1986), suggesting tectonic emplacement of up to 10 km of overburden there during the Permian. The cumulative thicknesses of model loads and foreland basin deposits during the Alleghanian orogeny (Fig. 6) suggest that overthrusts extended significantly westward onto the unthinned craton, and therefore may have had significant topography (Beaumont et al., 1988). Unfortunately these results are not unique – thicker loads further outboard (that is, wider mountains) – could match the observed sediment thicknesses equally well). Also, the effects of erosion are not taken into account.

Application to the ancestral Appalachians

By combining the critical wedge theory with the loading history proposed by Beaumont and coworkers, we hope to develop a self-consistent model of topography and load which uniquely satisfies the available data from the central Appalachian region. We use a Monte Carlo simulation because few of the parameters in eqns. (1)–(10) are well constrained. The results of this modelling are distributions of solutions, each solution simultaneously satisfying the physical constraints of the wedge theory and the observational constraints of the basin geometry and mountain load. The physical constraints of the wedge theory are that the steady state width and average height should be given by eqns. (6) and (10), respectively. These equations along with (1), (3), and (8), are solved using various combinations of the parameters described below. The observational constraint is that the steady-state load of the wedge per unit length of mountain range (stippled area in Fig. 1), given by:

$$L=\tfrac{1}{2}(W^2+2Wx_0)\frac{\tan(\alpha+\beta)}{1-\tan(\alpha+\beta)\tan\Psi_0} \tag{11}$$

must be equal to the Permian loads calculated by Beaumont and coworkers (Fig. 5). These amount to 5800 ± 580 km^3 per unit length of mountain range in the central Atlantic region. Thus, wedge model solutions are obtained for a very large number of random combinations of the independent parameters, V, H, and so on, but only those that produce a load of 5800 ± 580 km^2 are retained.

Parameter estimation

Table 2 lists the ranges of parameters used in the simulations. The convergence velocity, V, is obtained from the plate convergence velocities at compressive margins today, estimated by Dahlen and Suppe (1988) to be $10 < V < 100$ km Myr^{-1}. Paleogeographic reconstructions (Ziegler et al., 1979) yield a value of roughly 66 km Myr^{-1} during the Late Pennsylvanian, consistent with past and present velocities in the region of the Alps (Le Pichon et al., 1988), and within the range of modelled values. Sediment thicknesses entering present-day mountain belts are in the range $1 < H < 10$ km (Dahlen and Suppe, 1988). It is difficult to estimate H for the late Paleozoic Appalachian wedge, but the convergence flux $\dot{A}$, equal to VH, can be constrained.

The minimum possible value of $\dot{A}$ is obtained from the thicknesses of the loads predicted by Beaumont and others (Table 3) by dividing the km^3 loaded (per km along the thrust belt) by the interval over which the loading occurred. In table 3 it is assumed that the Permian loading occurred over the first 28 my of that period as suggested by the timing of Alleghanian remagnetization (Miller and Kent, 1988). The remagnetization age of Appalachian redbeds and limestones youngs northwards. The assumption that the remagnetization reflects the influence of fluids driven ahead of thrust sheets allows Miller and Kent (1988) to infer that major thrusting and large-scale fluid migration ended near the Permo–Carboniferous boundary in the southern Appalachians but continued on through a part of the Permian in the central Appalachians. Note that this corroborates the northward progression of loading inferred by

TABLE 2

Ranges of values used in Monte Carlo simulations of the wedge model

$\dot{A}$	10-1000	km^2 Myr^{-1}
V	10-100	km Myr^{-1}
H	1-10	km
E	0.1-10	Myr^{-1}
$(1-\Delta\rho/\rho)$	0.9	-
ψ_b	11.5	degrees
ψ_0	2.6	degrees
α	1-3	degrees
β	2-6 or 2-12	degrees

TABLE 3

Cross-sectional compressive fluxes, A, into the toe of the Alleghanian wedge (km^2 Myr^{-1}), estimated from the loads in Figs. 2-6

Interval	Northern Virginia	Southern Pennsylvania
Early Pennsylvanian	none	none
Late Pennsylvanian	57	none
Early Permian	80	54

Beaumont and co-workers (Figs. 2–6) and is consistent with the observation that this secondary natural remnant magnetization is independent of bedding attitude, indicating it is post-tectonic (Van der Voo and French, 1977; Scotese et al., 1982). This timing also is consistent with a Rb–Sr whole-rock age of 273 my for Brevard zone ultramylonite from near Rosman, North Carolina (Hatcher et al., 1988). The values obtained in this manner (Table 3) are comparable to the Barbados wedge away from the Orinoco delta which has a compressive influx of 20 $km^2\ Myr^{-1}$. Taiwan, by contrast has a compressive influx of 500 $km^2\ Myr^{-1}$. The values in Table 3 are minimum fluxes however, because they do not account for the material entering the wedge that is eroded during the interval. Thus, the range of $\dot{A}$ from 10 to 1000 $km^2\ Myr^{-1}$, obtained from the product of HV, appears to cover the expected values for the late Paleozoic Appalachians.

Determining the best estimates of E, the erosion constant in eqn. (3) requires considering its range of values today as a function of climate and estimating the paleoclimate of the central Appalachians in Late Pennsylvanian–Early Permian time. Data of Whitehouse (1988) show that regional erosion rates across the Central Southern Alps of New Zealand are proportional to mean elevation. Given the quality of the data, when $\langle \dot{e} \rangle$ is plotted versus h as in eqn. (3), the proportionality constant E, lies between 3 and 6 Myr^{-1} for this wet region where precipitation values range between 4 and 12 m yr^{-1}. Dahlen and Suppe (1988) derive $E = 1.3\ Myr^{-1}$ for the mountain belt of central Taiwan, another wet climate. Dry regions such as the central Andean Altiplano, experience average erosion rates of 0.5 km Myr^{-1} (Dahlen and Suppe, 1988), yielding an $E = \mp 0.15\ Myr^{-1}$, an order of magnitude lower (although see Flemings and Jordan, 1989, for higher values based on volumes of foreland basin fill). This contradicts the results of Ahnert's study (1970) where there was no obvious variation of E with precipitation. However, we feel this arises because Ahnert's database consists only of 20 mid-latitude small drainage basins. More recently, Pinet and Souriau (1988) determined E using a worldwide data set from continent-scale drainage basins. Denudation rates were determined from the total sediment load of rivers (dissolved and suspended) and h was taken as the mean elevation within the basin. As expected, they found that the dissolved load was mainly correlated with precipitation and not with relief. In the case of the suspended sediment load, two correlations with mean elevation arise, one for basins associated with recently active orogenies and one not. For the former, the linear regression equation is:

$$\dot{e} = 0.419h - 0.245 \qquad (12)$$

where $\dot{e}$ is in km Myr^{-1} and h is in km. The non-zero intercept can be interpreted to mean that a threshold elevation of about 0.58 km (at which value $\dot{e}$ equals 0) separates an erosive from a depositional terrain. Above the threshold elevation, E in eqn. (3) is approximately 0.42. They found no significant correlation with mean annual precipitation, which ranged from 0.76 to 2.03 m yr^{-1} among the basins. Based on these studies, we conclude that E varies from 0.1 to 1.0 for regions with less than about 2 m yr^{-1} annual precipitation, and from 1 to 10.0 for wetter regions.

The Permian climate of the central Appalachians can be estimated by two methods: (1) mapping paleoclimatic indicators such as evaporites, and (2) paleoclimatic modelling using atmospheric general circulation models. Most researchers (Ziegler et al., 1979; Crowley et al., 1987; Parrish et al., 1986; Barron, this volume) who have summarized the paleoclimatic indicators agree that the Pangean climate generally was one of widespread warmth and aridity. Numerical climate models corroborate this (Kutzbach and Gallimore, in press), calculating year-round aridity for all regions of Pangea but the eastern coastal regions, the tropical west coast, and the regions poleward of 40° latitude. This is true for various CO_2 concentrations, solar luminosities, and two different continental mean elevations (0 and 1 km). The model does not include the ancestral Appalachians, and the possibility arises that they created a local orographic climate wetter than surrounding areas (e.g., Hay et al., 1982). Computer simulations similar to Kutzbach and Gallimore's, but including crude estimates of mountains, did not indicate this however (E. Barron, pers. commun., 1989). Thus, we conclude that the appropriate erosion constant, E, is most probably in the range of 0.1–1 Myr^{-1}, but we include simulations with $1.0<E<10.0$ Myr^{-1} for comparison.

Ψ_0 and Ψ_b the acute angles between the axes of principal compressive stress and the top and bottom of wedge, must be consistent with Byerlee's law. Here we take these as constant at the Taiwan values of 2.6 and 11.5°.

The values of α and β vary today from 0–3° to 2–6°, respectively (Table 1). But, as Jamieson and Beaumont (1988) point out, the geometry of the wedge on the edge of a rifted continental margin is likely to be more complex than a simple triangle. The basement ramp is likely to have slopes of up to 12° in the region above the transition from attenuated to unattenuated crust, leveling off cratonward to around 2°. It should be expected that this steeply dipping buttress would produce a region of high topography not accounted for in the simple wedge model. While we are not prepared to construct a wedge model with multiple basal slopes, we can examine the influence of varying β through shallow and steep slopes. Thus, β will take on the ranges of either 2–6° or 2–12° in the various numerical simulations.

Results of the Monte Carlo simulation

The results of the Monte Carlo simulation are presented in Figs. 7–9 as frequency by number versus mean height and width of the steady-state mountain range. It is important to note that in all simulations, only a few percent of the many thousand trials came within 10% of the

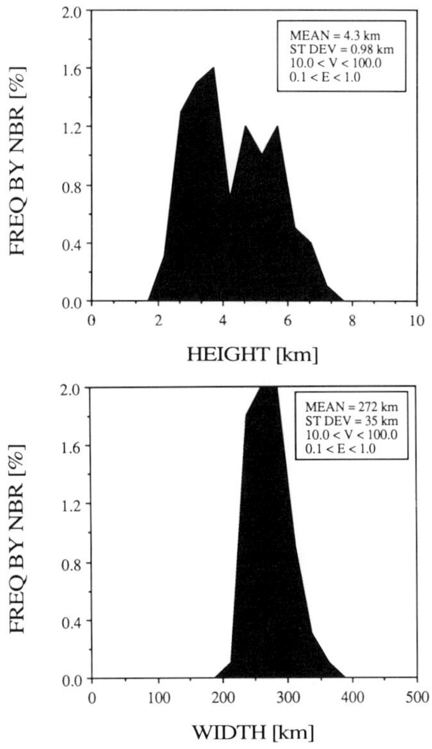

Fig. 7. Results of the Monte Carlo simulation for a dry climate and wedge basal slope of 2–6°. (a) Predicted mean height of the wedge (with respect to the ground surface at its toe) for the Permian central Appalachians; (b) predicted steady state width. The histogram gives the frequency of occurrence of height or width computed for various combinations of the parameters described in the text. V is convergence velocity (km Myr^{-1}) and E is the erosion constant (Myr^{-1}),.

loads required by eqn. (11). This indicates to us a certain robustness of the method in that there are relatively few combinations of parameter values that match the observed load. Also, the average convergence velocities of these acceptable simulations are all near 65–70 km Myr^{-1}, values remarkably close to the 66 km Myr^{-1} estimated from paleogeographic reconstructions as mentioned earlier.

Assuming a dry climate and shallow basal slope of the accretionary wedge (2–6°), the most probable width and cross-sectional average height of the central Appalachians are 272 km and 4.3 km, respectively (Fig. 7). Note that this height is measured not from sea level or the

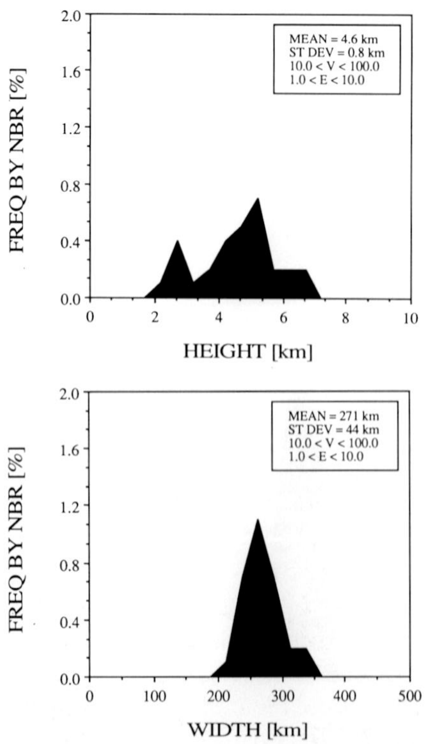

Fig. 8. Results of the Monte Carlo simulation for a wetter climate and wedge basal slope of 2–6°. (a) Predicted mean height of the wedge (with respect to the ground surface at its toe) for the Permian central Appalachians; (b) predicted steady-state width. The histogram gives the frequency of occurrence of height or width computed for various combinations of the parameters described in the text. V is convergence velocity (km Myr^{-1}) and E is the erosion constant (Myr^{-1}).

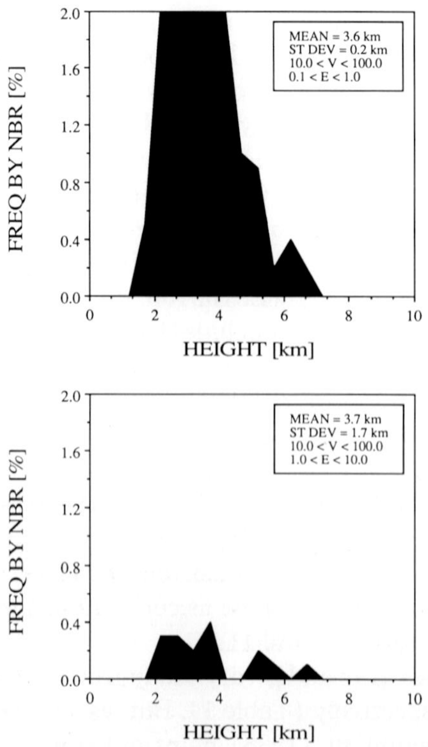

Fig. 9. Predicted mean height of the wedge (with respect to the ground surface at its toe) for the Permian central Appalachians for (a) a dry and (b) a wetter climate and wedge basal slope of 2–12°. The histogram gives the frequency of occurrence of height computed for various combinations of the parameters described in the text. V is convergence velocity (km Myr^{-1}) and E is the erosion constant (Myr^{-1}).

vertex of the wedge (Fig. 1) but from the elevation of point x_0 (Fig. 1). In the Permian the seas had withdrawn from the Appalachian basin, and therefore x_0 almost certainly was at or above sea level. Thus the heights are crude estimates of elevation or more precisely, estimates of relief. For comparison, the central Andes from 10° to 30°S comprise about a 300 km wide zone, all above 3 km and much above 4 km, with peaks rising to 5 km. Assuming a wetter climate (Fig. 8), does not vary the topography substantially because the wedge must still produce a load sufficient to flex the foreland basin. Allowing a steeper basal slope to the wedge (Fig. 9), only reduces the mean height from 4.3 to 3.6 km for a dry climate. Thus, with some confidence we can say that the Early Permian central Appalachians were similar to, or slightly less majestic than, the central Andes.

Although it confirms our well entrenched intuition (Rodgers, 1987; Slingerland and Beaumont, 1989; Cleaves, this volume), this conclusion is dependent upon so many premises that independent tests still seem warranted. The least demanding test is that eqn. (3) and the erosion constant used to calculate the steady state height should provide a subsequent rate of denudation consistent with observations. Fortunately, the Mesozoic basins of the central Atlantic states allow us to test this requirement. All available evidence indicates that the rocks immediately beneath the *unconformity* of the Newark basin in eastern Pennsylvania are identical to those adjacent to the basin margins (Manspeizer and Cousminer, 1988), an observation first made by Davis (1889). For example, the Furlong fault near Doylestown, Pa., fully 25 km into the basin, brings Cambro–Ordovician formations to the surface identical to those exposed on the north side of the basin. This indicates to us that the ancestral Appalachians must have been worn down to roughly their present stratigraphic level of erosion by earliest Carnian time (230 Myr ago using the DNAG time scale). Forty kilometers to the north in the Southern Anthracite Field, vitrinite reflectance and fluid inclusion data (Orkan and Voight, 1985; Levine, 1986) indicate that an additional 9–10 km of overburden lay atop Middle Pennsylvanian rocks presently exposed at the surface. Thus, if we assume that this was the overburden thickness at the end of thrusting, roughly 9–10 km of the Appalachians must have been eroded over the interval from 258 to 230 Myr.

Although it is not the point of this paper, this observation that the Appalachians experienced substantial pre-Triassic erosion, bears further discussion. Two problems are noteworthy: (1) where were the temporary and ultimate repositories of the erosional debris, and (2) if the central Appalachians reached their present degree of exhumation in Early Triassic time, what is the source of the siliciclastics on the present eastern United States passive margin? In answer to the first question, the paleo-lithological maps of Ronov et al. (1984) show shales of Late Permian age in the west-central United States. It is possible that a vast alluvial wedge of debris spread westward from the Appalachians, much as the present apron of the Colorado Rocky Mountains spreads eastward, only to be eroded at a later time by the early Mississippi River. In answer to the second question, the isopach maps of Poag and Sevon (this volume) indicate that 4 km of denudation of the Appalachians is necessary to account for the post-Early Jurassic sediment of the Atlantic passive margin. Roughly one-third to one-half of this comes from the Adirondack and New England terrains, leaving 2–3 km of denudation of the central Appalachians. Additionally, apatite fission-track thermochronology across Pennsylvania (Roden and Miller, this volume) indicates removal of only 5–6 km of overburden from Late Permian through Early Triassic. Thus, we can refine our estimate of pre-Carnian erosion to conclude that of the 9–10 km of overburden present in southeastern Pennsylvania in Early Permian time, roughly 6 km was removed prior to the Carnian, and the remaining 3 km removed subsequently. The pre-Carnian denudation yields an erosion rate of about 0.2 km Myr^{-1}. This can be compared to the erosion rate predicted from eqn. (3) for mountains about 4 km in average height (at the start of the interval) and an erosion constant of 01–1 Myr^{-1}. The erosion rates are similar and also in line with the higher rates for the post-Permian Appalachians presented in Table 1 of Sevon (this volume). Thus the model is at the least, internally consistent.

Another test of the model arises from the *P–T–t* histories of metamorphic and plutonic rocks in the orogen (Jamieson and Beaumont, 1988). Based on the closure temperatures of various mineralogie systems, they conclude that the southeastern Pennsylvania region experienced minimal uplift and exhumation during the Pennsylvanian. This changed during the Permian into significant net uplift relative to sea level with little exhumation, consistent with our postulate of mountain growth at that time.

Conclusions

Determining the ancient geometry of a mountain range requires knowledge of the mechanisms which formed the crustal thickening, the rate at which mass was eroded from the system, and the response of the lithosphere to the loading event itself. In the case of the ancestral Appalachians, the loading effect is reasonably constrained by the resulting size and shape of the Appalachian basin; the distribution of the load in space is less well constrained. By coupling a model of crustal response to thickening (given by the critical wedge theory) with a model of the basin evolution, we can determine not only the mass of the Appalachians but their shape as well. The technique indicates a central Andean topography with average relief on the order of 3.5–4.5 km and width of 250–300 km. Denudation rates generated by the high relief, and possibly Mesozoic extension, were sufficient to remove approximately 6 km of the orogen prior to Carnian time, suggesting that over half of the 10 km total of post orogenic unloading in southeastern Pennsylvania (Jamieson and Beaumont, 1988, Fig. 25) was accomplished early.

Acknowledgements

We thank Lee Kump and Don Fisher for patiently listening to, and commenting on, these ideas. This research was supported by the Department of Geosciences and the Earth System Science Center, The Pennsylvania State University.

References

Ahnert, F., 1970. Functional relationships between denudation, relief, and uplift in large mid-latitude drainage basins. Am. J. Sci., 268: 243-263.

Barron, E., 1989. Climate variations and the Appalachians from the Late Paleozoic to the Present - Results from model simulations. In: T.W. Gardner and W.D. Sevon (Editors), Appalachian Geomorphology. Geomorphology, 2: 99-118 (this volume).

Beaumont, C., Quinlan, G.M. and Hamilton, J., 1987. The Alleghanian orogeny and its relationship to the evolution of the eastern interior, North America. Mem. Can. Soc. Pet. Geol., 12: 425-446.

Beaumont, C., Quinlan, G.M. and Hamilton, J., 1988. Orogeny and stratigraphy: Numerical models of the Paleozoic in the eastern interior of North America. Tectonics, 7: 389-416.

Cleaves, E.T., 1989. Appalachian Piedmont landscapes from the Permian to the Holocene. In: T.W. Gardner and W.D. Sevon (Editors), Appalachian Geomorphology. Geomorphology, 2: 159-179 (this volume).

Crowley, T.J., Mengel, J.G. and Short, D.A., 1987. Gondwanaland's seasonal cycle. Nature, 329: 803-807.

Dahlen, F.A., 1984. Noncohesive critical Coulomb wedges: An exact solution. J. Geophys. Res., 89: 10125-10133.

Dahlen, F.A. and Suppe, John, 1988. Mechanics, growth, and erosion of mountain belts. In: S.P. Clark, Jr., B.C. Burchfiel and John Suppe (Editors), Processes in Continental Lithospheric Deformation. Geol. Soc. Am. Spec. Pap., 218: 161-178.

Dahlen, F.A., Suppe, John, and Davis, D., 1984. Mechanics of fold and thrust belts and accretionary wedges; Cohesive Coulomb theory. J. Geophys. Res., 89: 10087-10101.

Davis, D., Suppe, J. and Dahlen, F.A., 1983. Mechanics of fold-and-thrust belts and accretionary wedges. J. Geophys. Res., 88: 1153-1172.

Davis, W.M., 1889. The rivers and valleys of Pennsylvania. Natl .Geogr. Mag., I: 183-253.

Flemings, P.B. and Jordan, T., 1989. A synthetic stratigraphic model of foreland basin development. J. Geophys. Res., 94 (B4): 3851-3866.

Hatcher, R.D., Jr., Hooper, R.J., McConnell, K.I., Heyn, T. and Costello, J.O., 1988. Geometric and time relationships between thrusts in the crystalline southern Appalachians. In: G. Mitra and S. Wojtal, (Editors), Geometries and Mechanisms of Thrusting with Special Reference to the Appalachians. Geol. Soc. Am., Spec. Pap., 222: 185-196.

Hay, W.W., Behensky, J.F., Jr., Barron, E.J. and Sloan, J.L., 1982. Late Triassic-Liassic paleoclimatology of the proto-central north Atlantic rift system. Palaeogeogr. Palaeoclimat., Palaeoecol., 40: 13-30.

Jamieson, R.A. and Beaumont, C., 1988. Orogeny and metamorphism: A model for deformation and pressure-temperature-time paths with applications to the central and southern Appalachians. Tectonics, 7: 417-445.

Judson, S., 1975. Evolution of Appalachian topography. In: W.N. Melhorn and R.C. Flemal (Editors), Theories of Landscape Development. Proc. Vol. Sixth Annu. Geomorphol. Symp. Ser., Publ. Geomorphol., State University of New York, Binghamton, N.Y., pp. 29-42.

Kutzbach, J.E. and Gallimore, R.G., in press. Pangean climates: Megamonsoons of the megacontinent. J. Geophys. Res.

Le Pichon, X., Bergerat, F. and Roulet, M.-J., 1988. Plate kinematics and tectonics leading to the Alpine belt formation; a new analysis. In: S.P. Clark, Jr., B.C. Burchfiel, and J. Suppe (Editors), Processes in Continental Lithospheric Deformation. Geol. Soc. Am., Spec. Pap., 218: 111-131.

Levine, J.R., 1986. Deep burial of coal-bearing strata, anthracite region, Pennsylvania - sedimentation or tectonics? Geology, 14: 577-580.

Levine, J.R. and Slingerland, R., 1987. Upper Mississippian to Middle Pennsylvanian stratigraphic section, Pottsville, Pennsylvania. Geol. Soc. Am. Cent. Field Guide, Northeastern Section, pp. 59-63.

Li, Y.H., 1976. Denudation of Taiwan island since the Pliocene Epoch. Geology, 4: 105-107.

Liu, T.-K., 1982. Tectonic implications of fission-track ages from the Central Range, Taiwan. Geol. Soc. China. Proc., 25: 22–37.

MacLachlan, D.B., 1985. Pennsylvania anthracite as foreland effect of Alleghanian thrusting. Geol. Soc. Am., Abstr. Programs, 17; p. 53.

Manspeizer, W. and Cousminer, H.L., 1988. Late Triassic–Early Jurassic synrift basins of the U.S. Atlantic margin. In: R.E. Sheridan and J.A. Grow (Editors), The Geology of North America, V. I-2, The Atlantic Continental Margin, U.S. Geol. Soc. Am., Boulder, Colo., pp. 197–216.

Miller, J.D. and Kent, D.V., 1988. Regional trends in the timing of Alleghanian remagnetization in the Appalachians. Geology, 16: 588–591.

Molnar, P., 1988. Continental tectonics in the aftermath of plate tectonics. Nature, 335: 131–137.

Molnar, P. and Lyon-Caen, P., 1988. Some simple physical aspects of the support, structure, and evolution of mountain belts. In: S.P. Clark, Jr., B.C. Burchfiel, and J. Suppe (Editors), Processes in Continental Lithospheric Deformation. Geol. Soc. Am. Spec. Pap., 218: 179–207.

Morisawa, M., 1989. Origin of Appalachian drainages. In: T.W. Gardner and W.D. Sevon (Editors), Appalachian Geomorphology. Geomorphology, 2: 1–22 (this volume).

Orkan, N. and Voight, B., 1985. Regional joint evolution in the valley and ridge province of Pennsylvania in relation to the Alleghany orogeny. In: D.P. Gold, M.R. Canich, R.J. Cuffey, and others (Editors), Central Pennsylvania Geology Revisited. Guideb., 50th Annu. Field Conf. Pa. Geol. State College, Pa. Pa. Geol. Surv., pp. 144–163.

Parrish, J.M., Parrish, J.T. and Ziegler, A.M., 1986. Permian–Triassic paleogeography and paleoclimatology and implications for therapsid distribution. In: N. Holton, III, P. MacLean, RJ. Roth and E. Roth (Editors), The Ecology and Biology of Mammal-like Reptiles. Smithsonian Institution Press, Washington, D.C., 109–145.

Pinet, P. and Souriau, M., 1988. Continental erosion and large-scale relief. Tectonics, 7: 563–582.

Poag, W. and Sevon, W.D., 1989. A record of Appalachian denudation in postrift Mesozoic and Cenozoic sedimentary deposits of the U.S. middle Atlantic continental margin. In: T.W. Gardner and W.D. Sevon (Editors), Appalachian Geomorphology. Geomorphology, 2: 119–157 (this volume).

Quinlan, G.M. and Beaumont, C., 1984. Appalachian thrusting, lithospheric flexure, and the Paleozoic stratigraphy of the Eastern interior of North America. Can. J. Earth Sci., 21: 973–996.

Roden, M.K. and Miller, D.S., 1989. Apatite fission-track thermochronology of the Pennsylvania Appalachian basin. In: T.W. Gardner and W.D. Sevon (Editors), Appalachian Geomorphology. Geomorphology, 2: 39–51 (this volume).

Rodgers, J., 1987. The Appalachian–Ouachita orogenic belt. Episodes, 10: 259–266.

Ronov, A., Khain, V. and Seslavinsky, K., 1984. Atlas of Lithological Paleogeographical Maps of the World. Late Precambrian and Paleozoic of Continents. U.S.S.R. Acad. Sci., Leningrad, 70 pp.

Ruxton, B.P. and McDougall, I., 1967. Denudation rates in northeast Papua from potassium-argon dating of lavas. Am. J. Sci., 265: 545–561.

Sevon, W.D., 1985. Pennsylvania landscape development. Geol. Soc. Am. Abstr., Programs, 17: p. 713.

Sevon, W.D., 1989. Erosion in the Juniata River drainage basin. In: T.W. Gardner and W.D. Sevon (Editors), Appalachian Geomorphology. Geomorphology, 2: 303–318 (this volume).

Scotese , C.R., Van der Voo, R. and McCabe, C., 1982. Paleomagnetism of Upper Silurian and Lower Devonian carbonates of New York State: Evidence for secondary magnetizations residing in magnetite. Phys. Earth Planet. Interiors, 30: 385–395.

Slingerland, R. and Beaumont, C., 1989. Tectonics and sedimentation of the Upper Paleozoic foreland basin in the central Appalachians. In: R. Slingerland, K.P. Furlong, W. Manspeizer, and others (Editors), IGC Field Trip T152: Sedimentology and Thermal-mechanical History of Basins in the Central Appalachian Orogen. Am. Geophys. Union, 89 pp.

Stockmal, G.S., Beaumont, C. and Boutlier, R., 1986. Geodynamic models of convergent margin tectonics: transition from rifted margin to overthrust belt and consequences for foreland-basin development. Am. Assoc. Pet. Geol. Bull., 70: 181–190.

Van der Voo, R. and French, R.B., 1977. Paleomagnetism of the Late Ordovician Juniata Formation and the remagnetization hypothesis. J. Geophys. Res., 82: 5796–5802.

Westbrook, G.K., Smith, M.J., Peacock, J.H. and Poulter, M.J., 1982. Extensive underthrusting of undeformed sediment beneath the accretionary complex of the Lesser Antilles subduction zone. Nature, 300: 625–628.

Whitehouse, I.E., 1988. Geomorphology of the central Southern Alps, New Zealand: the interaction of plate collision and atmospheric circulation. Z. Geomorphol. Supp. Bd, 69: 105–116.

Ziegler, A.M., Scotese, C.R., McKerrow, W.S., Johnson, M.E. and Bambach, R.K., 1979. Paleozoic paleogeography. Annu. Rev. Earth Planet. Sci., 7: 473–502.

Geomorphology, 2 (1989) 39–51
Elsevier Science Publishers B.V., Amsterdam — Printed in The Netherlands

Apatite Fission-Track Thermochronology of the Pennsylvania Appalachian Basin

MARY K. RODEN and DONALD S. MILLER

Department of Geology, Rensselaer Polytechnic Institute, Troy, NY 12180-3590 (U.S.A.)

(Received December 12, 1988, accepted after revision May 1, 1989)

Abstract

Roden, M.K. and Miller, D.S., 1989. Apatite fission-track thermochronology of the Pennsylvania Appalachian Basin. In: T.W. Gardner and W.D. Sevon (Editors), Appalachian Geomorphology. Geomorphology, 2: 39–51.

Thirty-four apatite fission-track apparent ages and twenty-four track length distributions for ash bed samples from the Valley and Ridge Province and Upper Devonian to Upper Pennsylvanian sedimentary samples from the Allegheny Front and Allegheny Plateau of Pennsylvania suggest that these regions represent different thermal (uplift) regimes as well as different structural provinces. The Valley and Ridge Province Tioga and Kalkberg ash bed samples yield apatite fission-track apparent ages and track length distributions that indicate early post-Alleghanian (285–270 Ma) cooling and unroofing that began at ~250 Ma. Assuming a geothermal gradient of 25°C km^{-1}, a burial depth of at least 3.4 km can be estimated for all the Pennsylvania samples.

At the Allegheny structural front and on the western Allegheny Plateau, the apatite fission-track apparent ages (<150 Ma) and track length measurements indicate a Late Jurassic–Early Cretaceous thermal event for these samples possibly resulting from a higher geothermal gradient coinciding with kimberlite intrusion at this time along the Greene–Potter Fault Zone.

In northeast Pennsylvania on the Allegheny Plateau, the Upper Paleozoic sedimentary samples yield apatite fission-track apparent ages ≤180 Ma. Narrow track length distributions with long mean lengths (13–14 μm) and small standard deviations (1.3 μm) suggest rapid cooling from temperatures >110°C during the Middle Jurassic–Early Cretaceous for this part of Pennsylvania. This is consistent with the suggested uplift history of the Catskill Mountain region in adjacent New York State.

Introduction

Estimation of paleoburial depths in the Appalachian Basin have been based on conodont alteration indices (Harris, 1979), vitrinite reflectance anisotropy (Levine, 1986), fluid inclusions (Orkan and Voight, 1983), apatite fission-track analysis (Lakatos and Miller, 1983; Miller and Duddy, 1989), sedimentary compaction (Paxton, 1983), and geodynamic lithospheric flexure modelling (Beaumont et al., 1987). The results have indicated a general eastward-thickening (i.e., burial temperature increasing eastward) depositional basin.

Johnsson (1986), in an apatite fission-track and clay mineral diagenesis study of the Middle Devonian Tioga Ash Bed in New York State, found evidence of a thermal anomaly west of Syracuse that could not be explained solely by burial depth. The existence of this thermal anomaly suggested re-evaluation of the concept of a simple eastward-thickening depositional basin.

This paper presents a thermal history for the central Appalachian Basin of Pennsylvania based on apatite fission-track thermochronology of samples of primarily the Middle Devonian Tioga Ash Bed, and Upper Devonian Catskill Formation, as well as a few samples of other Upper Paleozoic sedimentary formations. Samples were taken from the Allegheny Plateau, Allegheny Front, and the Valley and Ridge Province of central and eastern Pennsylvania (Fig. 1). This sampling procedure was designed to test the concept of increasing temperatures in the eastern Appalachian Basin due to increasing burial depth and to detect any thermal anomalies that exist across the basin.

The fission-track method is based on the formation of damage-zones in uranium-bearing minerals resulting from the spontaneous fission of ^{238}U over geologic time. Apatite contains appropriate levels of uranium for fission-track analysis. A measured fission-track age is a result of both the time interval of track retention and the amount of annealing that has occurred during that time (Green, 1988). The fission-track age will be completely reset, if sufficient heat is applied to totally erase all existing fission-tracks in the apatite.

Below a certain critical temperature, known as the closure temperature (Dodson, 1973, 1979), fission-tracks will begin to accumulate. In isotope geochronology, closure temperature represents the time at which the radiogenic daughter product no longer diffuses out of the system but begins to be retained. In the case of fission-tracks in minerals, this is equivalent to the retention of the damaged regions referred to as tracks. For apatite, the closure temperature is estimated to be $100 \pm 20°C$ (Wagner, 1968; Naeser and Faul, 1969; Gleadow and Duddy, 1981; Naeser, 1981). Apatite fission-track apparent ages cited in this study (Table 1; Fig. 1) reflect the time those apatites cooled through the closure temperature of 100°C not their depositional age.

Fission-track length measurements on confined, spontaneous tracks in apatite combined with the apatite fission-track apparent ages have been used in thermal history interpretations (Gleadow et al., 1983, 1986a, b). Track length distributions from rapidly cooled volcanic apatites which have not been exposed to temperatures $>50°C$ subsequent to initial cooling have mean track length values between

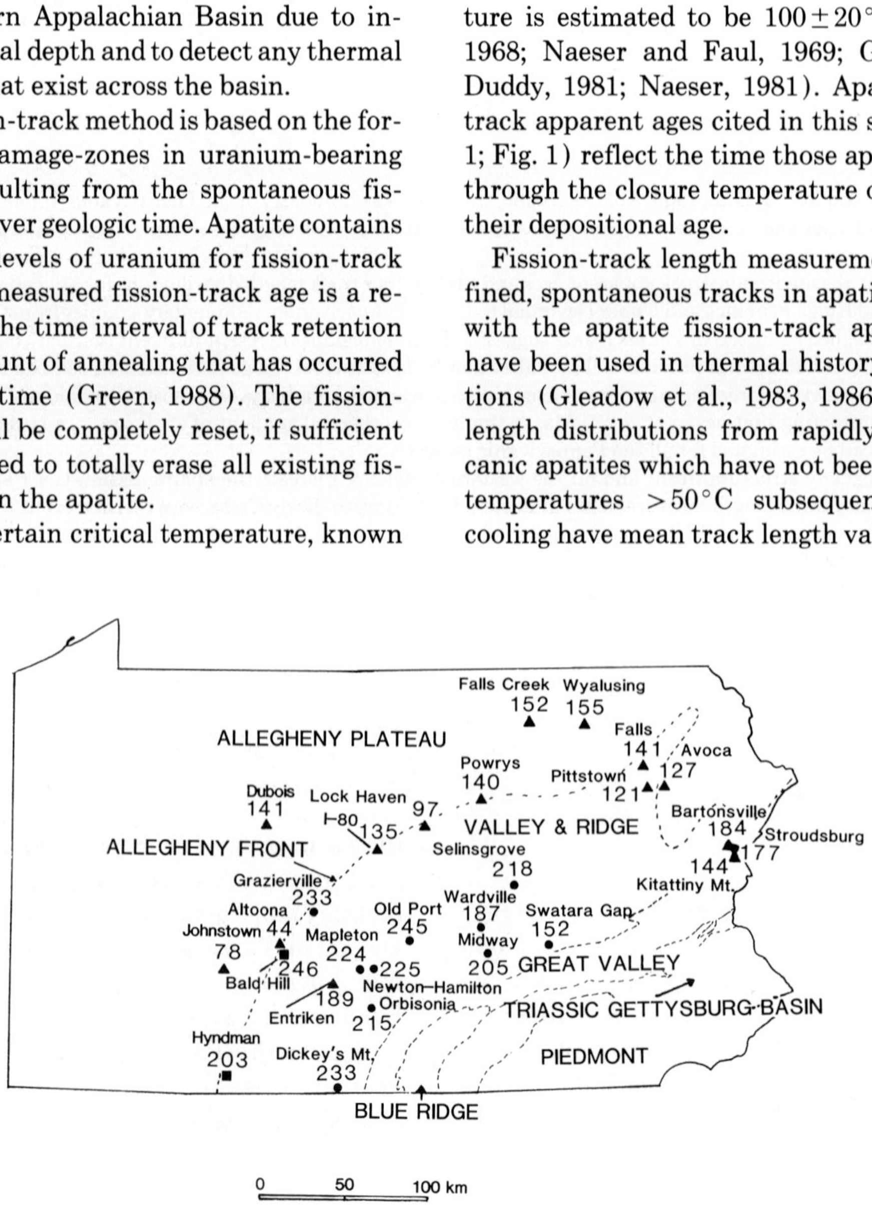

Fig. 1. Sample location names for Tioga (●), Kalkberg (■), and Upper Paleozoic sedimentary samples (▲), apatite fission-track apparent ages (Ma), and major tectonic provinces in Pennsylvania.

TABLE 1

Apatite fission-track apparent ages, mean track lengths, Cl/Cl+F ratios, and U concentrations for the Pennsylvania Appalachian samples

Sample number (elev.)	Formation	Sample location	Apparent apatite fission-track age in Ma (no. grains)	Mean track length in μm (no. tracks)	Cl/F+Cl	U (ppm)
Valley and Ridge Province samples						
84-92 (171 m)	Tioga	Swatara Gap	152 ± 8 (20)	13.23 ± 2.0 (98)	0.19	34
8500-1 (135 m)	Tioga	Old Port	245 ± 13 (15)	14.05 ± 1.7 (101)	0.19	33
8500-2 (208 m)	Tioga	Newton-Hamilton	225 ± 13 (20)	13.17 ± 1.9 (98)		28
8500-3 (104 m)	Tioga	Midway	205 ± 12 (19)	13.48 ± 1.6 (101)		24
8500-4 (190 m)	Tioga	Dickey's Mt.	233 ± 15 (19)	13.47 ± 1.4 (103)		28
8500-5 (239 m)	Tioga	Mapleton	224 ± 16 (16)	13.52 ± 1.6 (100)		29
8502-2 (294 m)	Tioga	Grazierville	233 ± 13 (19)	13.54 ± 1.8 (100)	0.16	33
8502-3 (141 m)	Tioga	Selingsgrove	218 ± 14 (20)	13.74 ± 1.5 (100)		39
8602-3 (184 m)	Tioga	Wardville	187 ± 12 (15)	13.78 ± 1.5 (101)		44
8602-4 (214 m)	Tioga	Orbisonia	215 ± 13 (12)	13.62 ± 1.6 (84)		37
8602-5 (141 m)	Tioga	Stroudsburg	177 ± 10 (18)	13.22 ± 1.6 (60)	0.19	47
8602-9 (294 m)	Kalkberg	Hyndman	203 ± 22 (15)	13.21 ± 1.6 (130)	0.21	10
8602-10 (306 m)	Kalkberg	Bald Hill	246 ± 26* (18)	13.34 ± 1.8 (114)	0.20	18
8602-52 (245 m)	Catskill	Entriken	189 ± 22* (19)	12.46 ± 1.7 (26)	0.06	27
Northeastern Allegheny Plateau samples						
84-79 (404 m)	Shawangunk	Kitattiny Mt.	144 ± 8 (21)	14.04 ± 1.2 (99)		
84-80 (263 m)	Catskill	Bartonsville	184 ± 17 (14)			
84-82 (367 m)	Catskill	Avoca	127 ± 10 (12)	13.25 ± 1.4 (100)		
84-83 (251 m)	Pottsville	Pittstown	121 ± 7 (20)	13.63 ± 1.1 (94)		
84-84 (184 m)	Catskill	Falls	141 ± 11 (10)	13.03 ± 1.5 (75)		
84-87 (208 m)	Pocono	Wyalusing	152 ± 17	13.60 ± 1.2 (103)		
84-89 (588 m)	Pottsville	Falls Creek	155 ± 12	13.60 ± 1.2 (100)		

TABLE 1 (continued)

Sample number (elev.)	Formation	Sample location	Apparent apatite fission-track age in Ma (no. grains)	Mean track length in μm (no. tracks)	Cl/F+Cl	U (ppm)
Western Allegheny Front and Allegheny Plateau samples						
8702-2 (203 m)	Catskill	Lockhaven	97±10 (10)	12.25±1.7 (103)	0.05	21
8702-5 (456 m)	Catskill	I-80	135±20* (9)	12.32±1.7 (23)	0.01	26
8702-7 (612 m)	Glenshaw	Dubois	141±12* (20)	12.38±1.7 (73)	0.02	18
8702-8 (685 m)	Burgoon	Altoona	44± 7 (10)		0.01	13
8702-11 (447 m)	Catskill	Johnstown	78± 8 (15)	12.81±1.5 (57)	0.02	31
8702-16 (220 m)	Catskill	Powrys	140±11* (20)	12.84±1.7 (100)	0.003	25

*The grain ages for these samples exceeded the limits defined by a chi-square test, thus mean ages are given rather than pooled ages.

14 and 15 μm and standard deviations that range from 0.8–1.3 μm (Gleadow et al., 1983, 1986a). Confined track length distributions for apatites from plutonic rocks have mean track lengths in the range of 12–14 μm and standard deviations between 1.0 and 2.0 μm (Gleadow et al., 1986a).

In studies of apatites from borehole samples of the Otway Formation in the Otway Basin (Gleadow and Duddy, 1981; Gleadow et al., 1983), the mean confined track length shows a progressive and continuous reduction with increasing down-hole temperature from 15 μm for surface and shallow samples to zero at temperatures of 120°C and greater. Gleadow et al. (1986b) show that small degrees of natural annealing produce narrow track length distributions which become progressively shorter and broader as annealing proceeds.

These empirically-determined and naturally-observed characteristics of confined track length distributions in apatite described in the previous two paragraphs will be compared to those track length distributions measured in this study of Pennsylvania rocks.

Composition (F and Cl concentration) has been shown to affect the individual fission-track grain ages of apatites (Green et al., 1986). In their study of apatites from the Otway Group sandstone, Green et al. (1986) found the Cl-rich grains which had been subjected to a thermal history identical to that of the F-rich grains, had fission-track ages closer to the depositional age of the formation. F and Cl concentrations have been determined for representative samples from this study to check the range of Cl variation.

Methods

Apatites and zircons for fission-track analysis were isolated after disaggregation from 1 to 2 kg samples by standard heavy liquid and magnetic separation techniques. Sample preparation procedures for age and trace length measurements were those of Gleadow (1984).

Suitable apatite were recovered from 34 out of 44 samples. Twenty-four samples had statistically significant confined track length measurements. Ten–twenty mg aliquots of apatite were mounted in epoxy on glass slides and polished to expose internal grain surfaces. Spontaneous fission-tracks in the apatites were revealed by etching for 20 s in $5M$ HNO_3 at 21°C.

The external detector method used in this study requires the mount to be covered by a thin sheet of low uranium muscovite secured by tape. The muscovite serves as a detector for the neutron-induced ^{235}U fission fragments.

Samples were irradiated with thermal neutrons at either the U.S. Geological Survey TRIGA reactor in Denver or the Oregon State University TRIGA reactor in Corvallis with two pieces of either NBS standard glass SRM 612 or the Corning Glass standard CN1 included at the ends of the irradiation package. A thermal neutron fluence on the order of 8×10^{15} n cm^{-2} was obtained for apatite samples. Mica detectors were etched in 40% HF for 15 min.

Ages were calculated using a weighted mean zeta calibration factor (Fleischer et al., 1975; Hurford and Green, 1983) for standard apatites from the Fish Canyon Tuff, Durango (Mexico), and the Mount Dromedary quartz monzonite (Miller et al., 1985). Track densities (tracks cm^{-2}) of both spontaneous and induced tracks were measured on a Leitz Ortholux microscope at 1600× (160× dry objective, 10× oculars). A chi-square statistic was calculated to discriminate between grains that belong to a single age population ($p>0.05$) and those that represent a mixed age population ($p\leq0.05$; indicated by an asterisk in Table 1).

Length measurements on confined horizontal tracks were made at 1563× (100× dry objective, 1.25× microscope tube, 12.5× oculars) using a Leitz Ortholux microscope with a drawing tube and a Houston Instruments 1011 digitizing pad interfaced with an IBM PC XT computer.

Results

Apatite fission-track apparent ages

The Tioga Ash Bed and Kalkberg Formation ash bed samples from west-central Pennsylvania display a range of apatite fission-track apparent ages of 215 ± 13–246 ± 26 Ma (Table 1; Fig. 1). Near the Southern Anthracite Field in east-central Pennsylvania, the Tioga Ash Bed samples yield a younger range of apatite fission-track apparent ages of 152 ± 8–218 ± 14 Ma. These ages are consistent with the 177 ± 10 Ma apatite fission-track apparent age from the easternmost sample at Stroudsburg. A concordant U–Pb date of 390.0 ± 0.5 Ma (Roden, 1989) for a monazite from bed B (Way et al., 1986) of a Pennsylvania Tioga sample tightly constrains the stratigraphic age for the formation. All apatite fission-track apparent ages determined in this study are significantly younger than this depositional age.

The sedimentary samples of Upper Devonian Catskill Formation and other Upper Paleozoic sedimentary formations from the Allegheny Front and Allegheny Plateau yield a range of apatite fission-track apparent ages, 44 ± 7–141 ± 12 Ma, significantly younger than the Tioga and Kalkberg samples located to the east (Table 1; Fig. 1). Sedimentary samples from the Allegheny Plateau of northeastern Pennsylvania show a range of apatite fission-track apparent ages of 121 ± 7–184 ± 17 Ma (Table 1; Fig. 1).

Track length distributions

The Tioga Ash Bed and Kalkberg Formation ash bed samples from the Valley and Ridge Province of Pennsylvania have measured track lengths that range from 13.2 to 14.0 μm with the mean standard deviation of the track length distributions ranging from 1.4 to 2.0 μm (Table 1). The mean track length is 13.5 ± 0.3 μm (probable error of the mean; Table 1). The frequency distributions are characterized by negative skewness and similar histogram shape (Fig. 2a).

The Upper Paleozoic sedimentary samples from the Allegheny Plateau of northeastern Pennsylvania (northeast plateau) yield a mean length of 13.5 ± 0.3 μm consistent with the mean track lengths for the ash bed samples but have a much smaller standard deviation of 1.3 μm (Table 1; Fig. 2b).

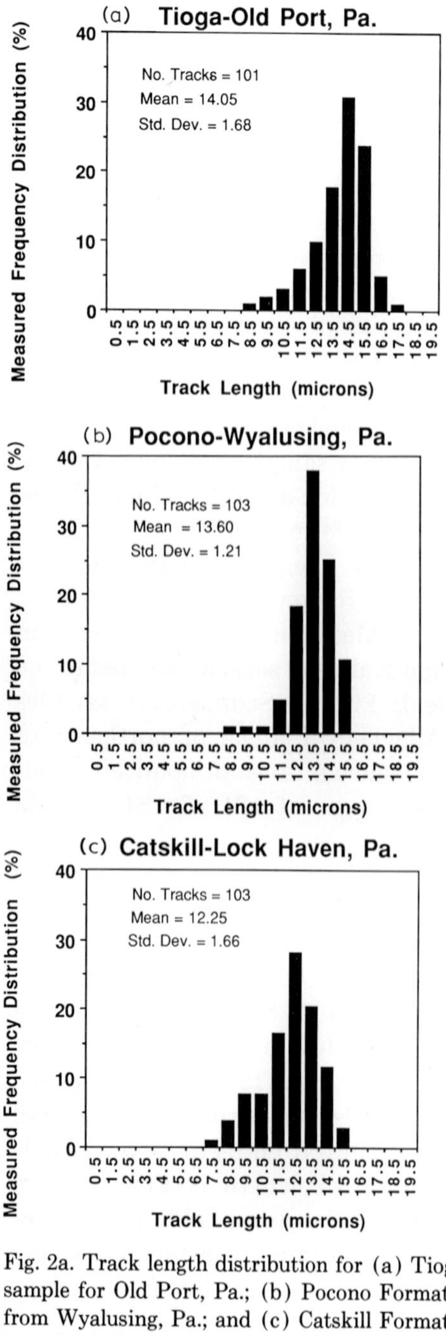

Fig. 2a. Track length distribution for (a) Tioga Ash Bed sample for Old Port, Pa.; (b) Pocono Formation sample from Wyalusing, Pa.; and (c) Catskill Formation sample from Lock Haven, Pa.

In contrast, the Upper Devonian and Upper Paleozoic samples from the Allegheny Front and western Allegheny Plateau yield a shorter mean track length of 12.5 ± 0.3 μm and a large standard deviation of 1.6 μm (Table 1; Fig. 2c). These differences, combined with the distinctly younger apatite fission-track apparent ages for the Allegheny Front and Allegheny Plateau samples (Table 1) suggest that they have experienced a different thermal history from that of the Valley and Ridge Province ash bed samples and the sedimentary samples from northeastern Pennsylvania.

Compositional variation

The Tioga Ash Bed and Kalkberg Formation apatites are monocompositional with a range of Cl/Cl+F composition values from 0.18–0.21 (Table 1). The Upper Devonian and Upper Paleozoic samples from the western Allegheny Plateau and Allegheny Front are fluorapatites with a limited range of Cl/Cl+F ratios (0.003–0.06; Table 1).

Because of the limited range of Cl/Cl+F content of the apatites studied, the fission-track apparent ages and track length reductions measured will be the result of thermal effects without the bias of compositional variation.

Discussion

Valley and Ridge Province

In Fig. 3, the correlation between coal rank, estimated burial depth (Beaumont et al., 1987) and apatite fission-track apparent age is shown. Apatite fission-track apparent ages for the Tioga samples decrease slightly from west to east as maximum estimated burial depth increases from 5 km at Grazierville (233 ± 13 Ma) to 9 km at Stroudsburg (177 ± 10 Ma). The Tioga samples with the youngest apatite fission-track apparent ages are located in the vicinity of the Southern Anthracite Field where both Beaumont et al. (1987) and Levine (1986) estimate burial depths to be as great as 9 km.

Using a geothermal gradient of 25°C km^{-1} (Beaumont et al., 1987) and a surface temperature of 15°C, a burial depth of at least 3.4 km can be estimated for all the Pennsylvania samples. This is a minimum estimate of burial depth using the apatite fission-track closure temper-

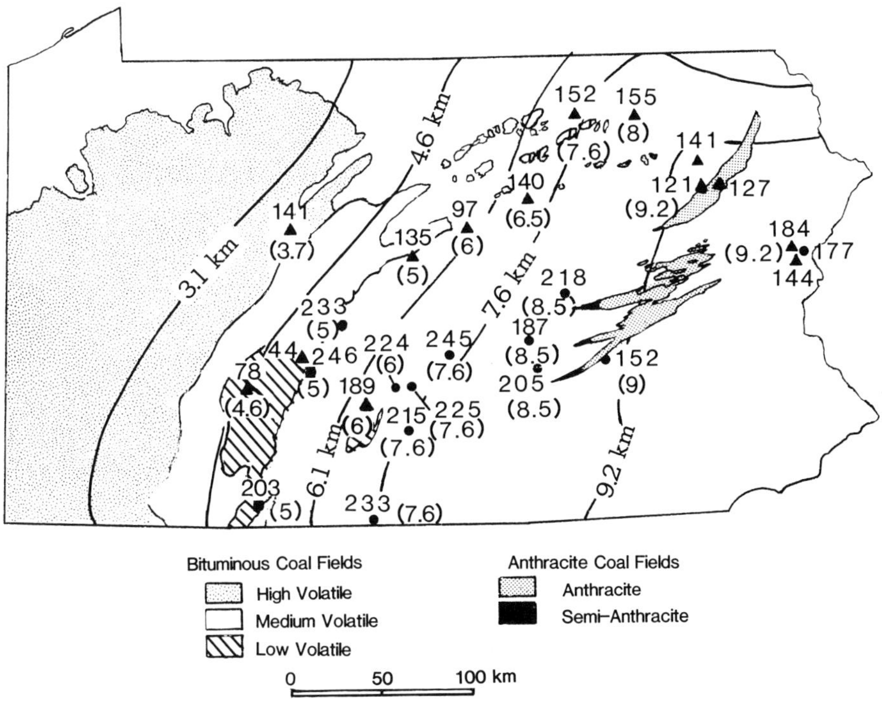

Fig. 3. Apatite fission-track apparent ages (Ma) for Tioga (●), Kalkberg (■), and Upper Paleozoic sedimentary samples (▲) from Pennsylvania, estimated maximum burial depth contours in parentheses (km) (Beaumont et al., 1987), and coal rank.

ature of 100° ±20°C. It is consistent with the models of Beaumont et al. (1987) and Levine (1986), however, their maximum burial depth estimates cannot be confirmed because the temperature at 5–9 km depth is ≫100°C (fission-track retention in apatite).

Both paleomagnetic ages of 285–270 Ma (Late Pennsylvanian–Early Permian, Van der Voo, 1979; Kent, 1979; Scotese et al., 1982) and vitrinite anisotropy data (Levine, 1986) limit the overburden emplacement and anthracite formation to a 10–15 Ma period. The apatite fission-track apparent ages for the ash bed samples from the Valley and Ridge Province (187±11–246±26 Ma), with the exception of the Swatara Gap and Stroudsburg samples (Table 1), suggest that uplift and erosion in this region began immediately after Alleghenian folding (285–270 Ma) which is consistent with the estimated short time interval (15 Ma) for maximum burial and deformation. The apatite fission-track apparent ages are also consistent with the calculated erosion period for the ancestral Appalachians (Slingerland and Furlong, this volume). The apatite fission-track apparent ages of ~220 Ma indicate an overburden of approximately 3–4 km (depending on estimated geothermal gradient) existed at that time. Slingerland and Furlong (this volume) suggest that the ancestral Appalachians would have been eroded to their present level by 230 Ma which may be consistent with the fission-track data within the errors of the model assumptions.

The Tioga samples from Swatara Gap and Stroudsburg with the youngest apatite fission-track apparent ages (152±8 and 177±10 Ma, respectively) are located on the southeastern edge of the Valley and Ridge Province near the Triassic Gettysburg Basin (Fig. 1). Their younger apatite fission-track apparent ages may reflect the effects of higher heat flow, higher paleogeothermal gradient, in this region during the Mesozoic as suggested by Kohn et al. (1988a, b).

Reset fission-track ages for sphene (mean age = 197 Ma) and zircon (mean age = 182 Ma) for samples of Grenville and Taconic crystalline rocks and the Mesozoic Newark Series from the Piedmont and southeastern Pennsylvania have been reported by Kohn et al. (1988a, b). The apatite fission-track apparent ages for their samples have a mean age of 146 Ma and mean track lengths of 13–14 μm. They attribute these fission-track ages to rapid cooling following a high thermal regime (50°C km^{-1}) during the Late Triassic–Early Jurassic extension along the Atlantic margin.

From the zircon and sphene fission-track apparent ages, Kohn et al. (1988a, b) estimate 6–10 km of uplift in the central Appalachian Piedmont since early Jurassic time. Their apatite fission-track apparent ages and track length distributions indicate a discrete phase of Late Jurassic-Early Cretaceous uplift. The apatite fission-track apparent ages for the Swatara Gap and Stroudsburg samples are consistent with a Middle to Late Jurassic uplift in southeastern Pennsylvania.

The measured track length distributions for the Valley and Ridge Province samples have long mean lengths (13.2–14.1 μm) and large standard deviations (1.3–2.0 μm; Table 1). This combined with a frequency distribution shape that is skewed toward short track lengths (Fig. 2a) suggests a slowly decreasing cooling path from temperatures of ~120°C to present surface temperature (15°C).

Western Allegheny Front and Allegheny Plateau

Figure 3 shows the apatite fission-track apparent ages for the sedimentary samples from the northwestern to northeastern Allegheny Plateau and the Allegheny Front of Pennsylvania and the predicted amounts of post-Alleghenian erosion (Beaumont et al., 1987). Maximum estimated burial depths vary from 3 km west of the Allegheny Front at Dubois to 9 km in the east near the Northern Anthracite Field.

The apatite fission-track apparent ages for all the Allegheny Front and Allegheny Plateau samples are significantly younger than those from the Valley and Ridge Province samples (Fig. 1; Table 1). This is inconsistent with a model of uniform uplift across an eastward deepening depositional basin because the samples in the western part of the basin (western Allegheny Plateau) should have apatite fission-track apparent ages that are the same as, or older than, those of the northeastern Plateau and Valley and Ridge Province samples ($\geq 246 \pm 26$ Ma). Instead, all the western Allegheny Plateau samples yield apatite fission-track apparent ages < 150 Ma.

The mean track lengths (12.5 ± 0.3 μm) for the western samples are also shorter than those of both the northeastern Plateau and the Valley and Ridge Province samples (Table 1). Mean track lengths in the 12 μm range with broad frequency distributions suggest slow-cooling with a relatively long time spent in the "track annealing zone" (70°–90°C).

A mechanism besides burial depth is needed to create a thermal perturbation in the western Allegheny Front and Allegheny Plateau region. There is recent evidence supporting the model of Oliver (1986) for orogenic dewatering due to thrust sheet emplacement which causes topographic highs and forces connate brines toward the craton. The fluid flow resulting from dewatering of the Appalachian Orogen has been suggested as a cause of a Kiaman remagnetization event in the Permo-Carboniferous (310–250 Ma) (Miller and Kent, 1988; Jackson et al., 1988), to form authigenic K-feldspars (322–278 Ma, $^{40}Ar/^{39}Ar$ ages) (Hearn and Sutter, 1985; Hearn et al., 1987), and to act as an illitization agent on mixed-layer illite–smectite in Middle Ordovician K-bentonites with K/Ar ages of illitization between 272–303 Ma (Elliott and Aronson, 1987; Altaner, 1985).

Fluids with temperatures in the range of 100°–200°C would have reset the apatite fission-track ages in sediments that they contacted. This means that the fission-track tT history from depositional age of the sediments (374–300 Ma; Palmer, 1984) to fluid infiltration age (~250–300 Ma) would be erased. The apatite fission-track apparent ages for the western Allegheny Front and Allegheny Plateau samples should be ≤245 Ma representing the time when the sediments had cooled to ≤100°C to allow fission-track accumulation.

However, the Allegheny Front and Allegheny Plateau samples have apatite fission-track apparent ages that range from 44±7 to 141±12 Ma (Fig. 1; Table 1). This indicates that there has been post-Permian thermal activity in this region and/or the sediments remained buried to depths where the temperature was 70°–90°C (track annealing zone) until the Late Jurassic–Early Cretaceous (≤150 Ma).

There is evidence for both these occurrences. The Allegheny Front and Allegheny Plateau samples from western Pennsylvania are located along an extension of the Rome Trough and near the cross-structural lineaments presented by Parrish and Lavin (1982) (Fig. 4). This corresponds to the keel-line trend of Dennison (1978, 1983).

Parrish and Lavin (1982) suggested that the coincidence of the Pennsylvania and New York kimberlite intrusions and the down-to-the-east basement faults and cross-structural lineaments of the Rome Trough indicate reactivation of these faults in response to the Jurassic extensional opening of the Atlantic Ocean. Basu et al. (1984) have presented K-Ar age determinations for the Syracuse and Ithaca, New York kimberlites which yield an age range of 113±11–176±17 Ma. This range of K-Ar ages coincides with the range of apatite fission-track apparent ages (44±7–141±12 Ma) determined for the Allegheny Front and Allegheny Plateau samples.

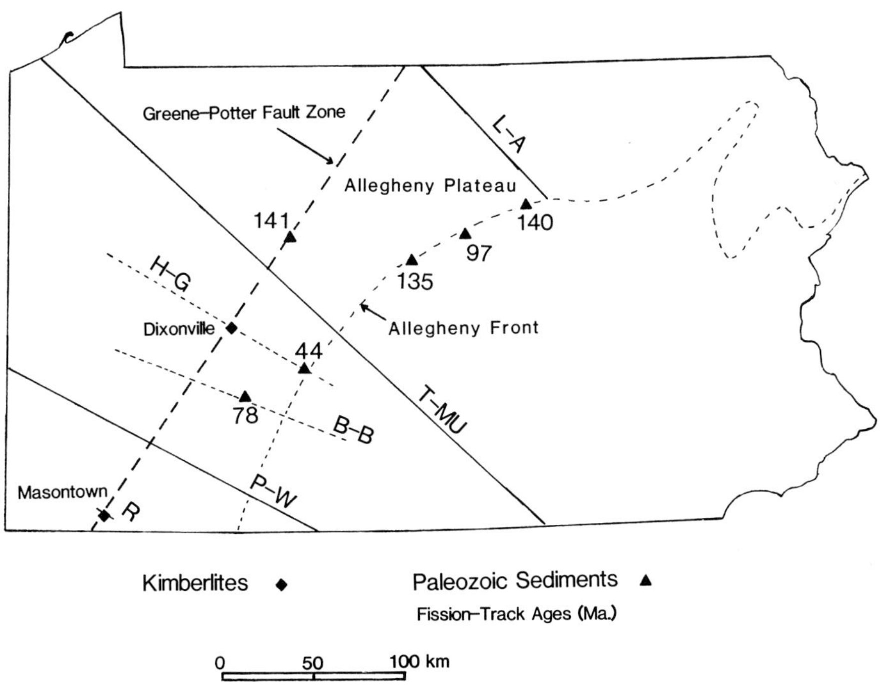

Fig. 4. Greene-Potter fault zone, cross-structural lineaments: *T-MU*=Tyrone-Mt. Union, *P-W*=Pittsburgh-Washington, *H-G*=Home-Gallitzen, *B-B*=Blairsville-Broad Top, *R*=Roen discontinuity, *L-A*=Lawrenceville-Attica (Parrish and Lavin, 1982), apatite fission-track apparent ages (Ma), and kimberlite locations.

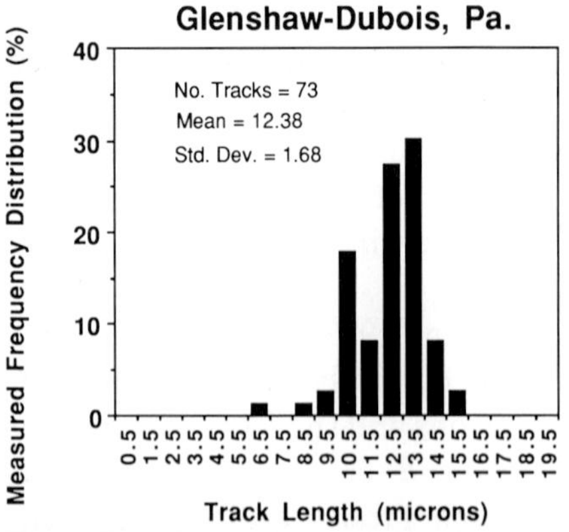

Fig. 5. Bimodal track length distribution for a Glenshaw Formation sample from Dubois, Pa.

The intrusion of the kimberlites themselves could not generate enough heat on a regional scale to overprint a post-Permian cooling age (245 Ma) in these sediments. However, the widespread occurrence of these kimberlites along this zone suggests the possibility of an increased thermal gradient associated with regional igneous intrusive activity in the Rome Trough during the Jurassic. The presence of an intersecting fault zone and lineaments provides tectonic pathways for fluid migration. If the fluids were in the temperature range of 70°–90°C, they could cause annealing of fission-tracks in the apatites they contacted.

Additional evidence for a post-Permian heating event on the western Allegheny Plateau of Pennsylvania can be seen in the track length distribution for a Glenshaw Formation from Dubois (Fig. 5). It has a bimodal track length distribution with a shorter peak at 10.5 μm. This type of length distribution is characteristic of apatites which have spent a prolonged period in the 80°C temperature range (Green et al., 1989). It was during the time at 80°C that the 10.5 μm peak in the bimodal distribution formed. The temperature was not hot enough to completely anneal the tracks formed since the initial post-Permian cooling, but it was sufficiently high to shorten the tracks to the 10 μm range. The higher peak at 13.5 μm formed after the temperature dropped below 80°C.

Northeastern Allegheny Plateau

More rapid cooling (uplift) is indicated for the Upper Devonian samples from the northeastern Allegheny Plateau in their narrow length distributions having a mean standard deviation of 1.3 μm and a long mean length (13.5 ± 0.3 μm). Apatite fission-track apparent ages (121 ± 17–184 ± 17 Ma) for these samples are consistent with a Middle Jurassic to Early Cretaceous uplift as suggested by Miller and Duddy (1989) based on apatite fission-track apparent ages for Upper Devonian samples from the Catskill Mountains to the northeast in New York State.

Cl/Cl+F ratios

It is doubtful that the apatite fission-track age difference between the Allegheny Front and Allegheny Plateau apatites and the Valley and Ridge Province apatites is due solely to compositional differences. All apatites analyzed in this study comprise a very narrow compositional range (Cl/Cl+F=0.003–0.21) compared to the Otway Group apatites analyzed by Green et al. (1986) (Cl/Cl+F=0.01–0.51). Because the Pennsylvania apatites lack a wide range of Cl concentrations, they do not show a significant correlation between increasing Cl/Cl+F ratio and increasing apatite fission-track apparent ages (Roden, 1989) as seen in the Otway Group apatites (Green et al., 1986).

Conclusions

Apatite fission-track apparent ages and track length distributions for the Pennsylvania sedimentary and ash bed samples from the Allegheny Front and Allegheny Plateau and Valley and Ridge Provinces suggest that these regions are not only different tectonic provinces but also represent different thermal (uplift) regimes. The Valley and Ridge Province ash bed samples yield apatite fission-track apparent ages and track length distributions that indicate early post-Alleghanian (285–270 Ma) cooling and unroofing beginning at ~250 Ma. This is in agreement with the calculated erosion times of Slingerland and Furlong (this volume).

Burial depths of at least 3.4 km can be calculated based on an estimated geothermal gradient of 25°C km^{-1} for all the Pennsylvania samples. This is consistent with vitrinite anisotropy measurements of Levine (1986) and geodynamic models of Beaumont et al. (1987) and Slingerland and Furlong (this volume) which suggest deep burial (6–10 km) in the anthracite region followed by rapid unroofing over a 15 Ma period.

Interpretation of the thermal history from apatite fission-track thermochronology west of the Allegheny structural front is complicated. The samples from western Pennsylvania along the Allegheny Front and Allegheny Plateau have been reset to give apatite fission-track apparent ages consistent with thermal histories beginning at <150 Ma. It is possible that these samples have experienced two heating events. One was caused by hot fluid infiltration (100°–200°C) during the time interval (~300–250 Ma) as a result of orogenic dewatering. These hot fluids would have reset the apatite fission-track ages for the sediments from depositional age (374–300 Ma) to ≤250 Ma. This event can not be detected by fission-track analyses of apatite due to the following thermal event.

A second thermal event at <150 Ma is indicated by the apatite fission-track apparent ages for these samples. Track length distributions with short mean lengths (12.5±0.3 mm), large standard deviations (1.6 μm), and evidence of bimodality indicate that the thermal event caused temperatures to remain in the 80°C temperature range for an extended period during the Late Jurassic–Early Cretaceous. It is suggested that the elevated temperatures may have been the result of a higher geothermal gradient during this time possibly related to kimberlite intrusion along the Greene–Potter Fault Zone and cross-structural lineaments.

For the samples from northeastern Pennsylvania on the Allegheny Plateau, the narrow track length distributions with small standard deviations (1.3 μm) indicate rapid cooling at <185 Ma (maximum apatite fission-track apparent age). This is consistent with a Middle Jurassic–Early Cretaceous uplift suggested by Miller and Duddy (1989) for the Catskill Mountain region of adjacent New York State. Rapid uplift in northeastern Pennsylvania as indicated here may correlate with that seen by Kohn et al. (1988a, b) in southeastern Pennsylvania. It may reflect, as Kohn et al. (1988a, b) suggest, post-rifting uplift caused by landward flexural bulging in response to sedimentary loading of the continental margin.

Acknowledgements

This research was supported by a Graduate Professional Opportunities Fellowship, 1985–1987, a Shell Foundation Fellowship, 1987–1988, a Texaco Research Grant, 1988–1989, a Geological Society of American Research Grant and a Sigma Xi Grant-in-Aid, 1986. We thank John Way of Lock Haven University and Robert C. Smith, II of the Pennsylvania Topographic and Geologic Survey for their help in the sample collection of the Tioga and Kalkberg ash bed samples and I.R. Duddy for measuring apatite fission-track ages and track lengths for the northeastern Pennsylvania samples. We thank W. Sevon, T. Gardner and an anonymous reviewer for their helpful comments and critical reviews.

References

Altaner, S.P., 1985. Potassium metasomatism and diffusion in Cretaceous K-bentonite from the disturbed belt, northwestern Montana, and in the Middle Devonian Tioga K-bentonite, eastern U.S.A. Ph.D. Thesis, University of Illinois, Urbana, Ill., 193 pp.

Basu, A.R., Rubury, E., Mehnert, H. and Tatsumoto, M., 1984. Sm–Nd, K–Ar and petrologic study of some kimberlites from eastern United States and their implication for mantle evolution. Contrib. Mineral. Petrol., 86: 35–44.

Beaumont, C., Quinlan, G.M. and Hamilton, J., 1987. The

Alleghanian orogeny and its relationship to the evolution of the Eastern Interior, North America. In: C. Beaumont and A.J. Tankard (Editors), Sedimentary Basins and Basin-forming Mechanisms. Can. Soc. Petrol. Geol. Mem., 12: 425–445.

Dennison, J.M., 1978. Origin of keel-line peridotites in the Appalachian basin. Geol. Soc. Am., Abstr. Program., 10: 387–388.

Dennison, J.M., 1983. Reply on "Tectonic model for kimberlite emplacement in the Appalachian Plateau of Pennsylvania". Geology, 11: 252–253.

Dodson, M.E., 1973. Closure temperature in cooling geochronological and petrological systems. Contrib. Mineral. Petrol., 40: 259–274,

Dodson, M.E., 1979. Theory of cooling ages. In: E. Jaeger and J.C. Hunziker (Editors), Lectures in Isotope Geology. Springer, Berlin, 329 pp.

Elliott, W.C. and Aronson, J.L., 1987. Alleghenian episode of K-bentonite illitization in the southern Appalachian basin. Geology, 15: 735–739.

Fleischer, R.L., Price, P.B. and Walker, R.M., 1975. Nuclear Tracks in Solids. University of California Press, Berkeley, Calif., 605 pp.

Gleadow, A.J.W., Duddy, I.R., Green, P.F. and Lovering, J.F., 1986a. Confined fission-track lengths in apatite-a diagnostic tool for thermal history analysis. Contrib. Mineral. Petrol., 94: 405–415.

Gleadow, A.J.W., Duddy, I.R., Green, P.F. and Hegarty, K.A., 1986b. Fission-track lengths in the apatite annealing zone and the interpretation of mixed ages. Earth Planet. Sci. Lett., 78: 245–254.

Gleadow, A.J.W., 1984. Fission-track dating methods-II: a manual of principles and techniques. Workshop of Fission Track Analysis: Principles and Applications. James Cook University, Townsville, Qld., 35 pp.

Gleadow, A.J.W., Duddy, I.R. and Lovering, J.F., 1983. Fission-track analysis: a new tool for the evolution of thermal histories and hydrocarbon potential. Aust. Petrol. Explor. Assoc. J., 23: 93–102.

Gleadow, A.J.W. and Duddy, I.R., 1981. A natural long-term annealing experiment for apatite. Nucl. Tracks, 5: 169–174.

Green, P.F., 1988. The relationship between track shortening and fission-track age reduction in apatite: combined influences of inherent instability, annealing anisotropy, and length bias and system calibration. Earth Planet. Sci. Lett., 89: 335–352.

Green, P.F., Duddy, I.R., Gleadow, A.J.W., Tingate, P.R. and Laslett, G.M., 1986. Thermal annealing of fission-tracks in apatite: 1. A qualitative description. Chem. Geol. (Isot. Geosci. Sect.), 59: 237–253.

Green, P.F., Duddy, I.R., Laslett, G.M., Hegarty, K.A., Gleadow, A.J.W. and Lovering, J.F., 1989. Thermal annealing of fission-tracks in apatite: 4. Quantitative modelling techniques and extension to geological time scales. Chem. Geol. (Isot. Geosci. Sect.), 79: 155–182.

Harris, A.G., 1979. Conodont color alteration, an organo-mineral metamorphic index and its application to Appalachian Basin geology. In: P.A. Scholle and P.R. Schluger (Editors), Aspects of Diagenesis. Soc. Econ. Paleontol. Mineral. Spec. Publ., 26: 3–16.

Hearn, P.P. and Sutter, J.F., 1985. Authigenic potassium feldspar in Cambrian carbonates: Evidence of Alleghenian brine migration. Science, 228: 1529–1531.

Hearn, P.P., Sutter, J.F. and Belkini, H.E., 1987. Evidence for Late-Paleozoic brine migration in Cambrian carbonate rocks of the central and southern Appalachians: Implications for Mississippi Valley type sulfide mineralization. Geochim. Cosmochim. Acta, 51: 1323–1334.

Hurford, A.J. and Green, P.F., 1983. The zeta age calibration of fission-track dating. Chem. Geol. (Isotope Geoscience Section), 1: 285–317.

Jackson, M., McCabe, C., Ballard, M.M. and Van der Voo, R., 1988. Magnetite authigenesis and diagenetic paleotemperatures across the northern Appalachian basin. Geology, 16: 592–595.

Johnsson, M.J., 1986. Distribution of maximum burial temperatures across northern Appalachian basin and its implication for Carboniferous sedimentation patterns. Geology, 14: 384–387.

Kent, D.V., 1979. Paleomagnetism of the Devonian Onondaga Limestone revisited. J. Geophys. Res., 84: 3576–3588.

Kohn, B.P., Wagner, M.E., Lutz, T.M. and Organist, G., 1988a. Regional thermal overprinting during Mesozoic rifting recorded by sphene and zircon fission-track dates, central Appalachian piedmont, USA. Abstr. 6th Int. Conf. Fission-Track Dating, C2–11.

Kohn, B.P., Wagner, M.E., Lutz, T.M., Organist, G. and Omar, G., 1988b. Mesozoic rift-related and post-rift uplift of the central Appalachian piedmont: constraints from fission-track analysis. Geol. Soc. Am., Abstr. Program., 20: A387.

Lakatos, S. and Miller, D.S., 1983. Fission-track analysis of apatite and zircon defines a burial depth of 4 to 7 km for lowermost Upper Devonian, Catskill Mountains, New York. Geology, 11: 103–104.

Levine, J.R., 1986. Deep burial of coal-bearing strata, Anthracite region, Pennsylvania: sedimentation or tectonics: Geology, 14: 577–580.

Miller, D.S. and Duddy, I.R., 1989. Early Cretaceous uplift and erosion of the Northern Appalachian Basin, New York, determined from apatite fission-track analysis. Earth Planet. Sci. Lett., 93: 35–49.

Miller, D.S., Duddy, I.R., Green, P.F., Hurford, A.J. and Naeser, C.W., 1985. Results of interlaboratory comparison of fission-track age standards: Fission-track Workshop, 1984. Nucl. Tracks, 10: 383–391.

Miller, J.D. and Kent, D.V., 1988. Regional trends in the timing of Alleghenian remagnetization in the Appalachians. Geology, 16: 588–591.

Naeser, C.W., 1981. The fading of fission-tracks in the geologic environment-data from deep drill holes (unrefereed abstract). Nucl. Tracks, 5: 248-260.

Naeser, C.W. and Faul, H., 1969. Fission-track annealing in apatite and sphene. J. Geophys. Res., 74: 705-710.

Oliver, J., 1986. Fluids expelled tectonically from orogenic belts: Their role in hydrocarbon migration and other geologic phenomena. Geology, 14: 99-102.

Orkan, N.I. and Voight, B., 1983. Fracture geometry and vein mineral paleothermometry-barometry from fluid inclusions, Valley and Ridge Province, Pennsylvania. Geol. Soc. Am., Abstr. Program., 15: 656.

Palmer, A.R., 1984. The Decade of North American Geology 1983 Time Scale. Geol. Soc. Am., Map Chart Ser., MC-50.

Parrish, J.B. and Lavin, P.M., 1982. Tectonic model for kimberlite emplacement in the Appalachian Plateau of Pennsylvania. Geology, 10: 344-347.

Paxton, S.T., 1983. Relationships between Pennsylvanian-age lithic sandstone and mudrock diagenesis and coal rank in the central Appalachians. Ph.D. Thesis, Pennsylvania State University, University Park, Pa., 503 pp.

Roden, M.K., 1989. Apatite fission-track thermochronology of the central and southern Appalachian Basin. Ph.D. Thesis, Rensselaer Polytech. Inst., Troy, N.Y., 214 pp.

Scotese, C.R., Van der Voo, R. and McCabe, C., 1982. Paleomagnetism of the Upper Silurian and Lower Devonian carbonates of New York State: Evidence for secondary magnetizations residing in magnetite. Phys. Earth Planet. Interiors, 30: 385-395.

Slingerland, R. and Furlong, K.P., 1989. Geodynamic and geomorphic evolution of the Permo-Triassic Appalachian Mountains. In: T.W. Gardner and W.D. Sevon (Editors), Appalachian Geomorphology. Geomorphology, 2: 23-37 (this volume).

Van der Voo, R., 1979. Age of Alleghenian folding in the central Appalachians. Geology, 7: 297-298.

Wagner, G.A., 1968. Fission-track dating of apatites. Earth Planet. Sci. Lett., 4: 411-415.

Way, J.H., Smith, R.C., II and Roden, M., 1986. Detailed correlations across 175 miles of Valley and Ridge of Pennsylvania using 7 ash beds in the Tioga zone. In: W.D. Sevon (Editor), Guidebook, 51st Annual Field Conference of Pennsylvania Geologists. Pa. Topogr. Geol. Surv., Huntington, pp. 55-72.

Geomorphology, 2 (1989) 53-70
Elsevier Science Publishers B.V., Amsterdam — Printed in The Netherlands

Constraints from Deep Crustal Structure on Long-Term Landform Development of the British Isles and Eastern United States

YVONNE BATTIAU-QUENEY
University of Lille, 59655 Villeneuve d'Ascq Cedex (France)

(Received February 29, 1989; accepted after revision June 6, 1989)

Abstract

Battiau-Queney, Y., 1989. Constraints from deep crustal structure on long-term landform development of the British Isles and Eastern United States. In: T.W. Gardner and W.D. Sevon (Editors), Appalachian Geomorphology. Geomorphology, 2: 53-70.

It is assumed that some major landforms not easily explained by surface geology could be controlled by deep crustal properties. The concept was first applied in the British Isles and then to the U.S. Atlantic provinces. Both regions were affected by similar orogenies in late Precambrian/Paleozoic times and presently belong to the passive margins of the North Atlantic Ocean. In both areas, the pattern of rifting was greatly influenced by preexisting tectonic fabric of the continental crust. In the British Isles, pre-Mesozoic structures are cut obliquely by the Atlantic margin, while in the U.S. border, the present margin is nearly parallel to the Paleozoic sutures.

The British Isles were affected by differential block-tectonics during the Tertiary. Examples are given of regional-scale landforms linked with crustal hinges. Recent data confirm the great heterogeneity of the crust beneath the British Isles. Each crustal unit has its own isostatic response according to the state of stress and its own dynamic equilibrium between rock resistance, erosion processes and uplift rate.

In the Atlantic provinces of the United States, the regional physiographic framework is simpler, but allochthonous terranes thrust westward over the Grenville basement and shelf-sediments have buried some major discontinuities in the deep crust. Reverse faults associated with the NW-SE compressional stressfield are numerous but cannot explain regional-scale landforms like the Blue Ridge escarpment above the Piedmont. Rock control is not the direct cause of the escarpment. The Blue Ridge front reaches its highest relief where discontinuities in the deep crust interfere with surficial geological features. Ocean to continent creep processes are suggested to explain the sustained uplift of the Appalachian highlands and the relatively stable position of the mountainous front since middle Mesozoic times.

Introduction

Because geomorphology is the study of surface landforms and processes, it seems normal to take into account first the surficial geological features and not the deep crustal structure when landform development is considered. In fact some major regional-scale landforms are not easily linked with surficial features, such as fault lines of sharp lithological contrasts, but are controlled by crustal heterogeneity. The implication for long-term development is important because geomorphic theories which are founded on earth surface processes and base-level changes are inadequate to explain these landforms.

The concept developed below were first applied to the British landscape. It seems interesting to test the same approach in the Appalachian highlands and Piedmont and the

Atlantic Coastal Plain because both areas belong to passive margins of the North Atlantic and were affected by the Iapetus spreading and closing and late Paleozoic continental collisions.

General outlines of the structural framework in the British Isles and eastern United States

The regional pattern of sedimentary basins and basement highs is much more complex in the British Isles (Fig. 1) than in the northeast American provinces (Fig. 2). The present continental margin of the Iceland basin and Porcupine Plain lies far from the British coast (from 200 to 500 km). The Rockall Plateau (Fig. 3) is a microcontinent isolated from Britain by a deep water embayment. The Rockall Trough is probably underlain by a thin intermediate crust rather than a true oceanic crust (Naylor and Shannon, 1982). The structural framework of the margin is due to several phases of rifting, the first one in Devonian and Carboniferous times, the second one in Permo–Jurassic times. The pattern of rifting was greatly influenced by preexisting tectonic fabric of the continental crust (Roberts, 1974; Dewey, 1982). Ocean spreading started in the Bay of Biscay in middle Cretaceous and then jumped along the Reykjanes Ridge in Paleocene (Bott, 1978). Thermal subsidence, regional warping, and uplifting or subsidence, in response to spreading at the mid-ocean ridge axis, were active in late Mesozoic and Cenozoic times. The main features of the structural framework of the British Isles are due to this relatively late epeirogenic history that was strongly controlled by fabric inherited from a long and complex structural development.

The British Isles were formed by juxtaposition of various continental units first separated from each other and successively accreted during several events of oceanic closing, subduction, continental collision, rifting, and oceanic spreading, from Precambrian to late Paleozoic times (Dewey, 1982). It must be emphasized that the northeast trending Caledonian structures and the east-southeast trending Hercynian front are cut obliquely by the Atlantic margin (Fig. 3). The Tertiary was a period of major faulting with reactivation of earlier Caledonian or Hercynian lines of weakness and formation of new faults along northwest or north trending lines (Battiau-Queney, 1989). The end result is that the continental crust beneath the British Isles can be compared to a mosaic of blocks which are variously shaped and generally small-sized (a few tens to less than 150 km).

In the Atlantic provinces of the United States, the regional structural pattern is somewhat simpler than in the British Isles. It consists of a series of northeast trending units roughly parallel to the edge of the continental shelf, including the Appalachian Plateau, Valley and Ridge, Blue Ridge, Piedmont, and Coastal Plain (Fig. 2). This relatively simple framework of the United States Atlantic border is dependent on its geologic history from Precambrian to middle Jurassic, before the opening of the Atlantic ocean (Battiau-Queney, 1988). During the late Paleozoic, crystalline rocks of the Blue Ridge and inner Piedmont were thrust over shelf sediments belonging to the late Precambrian/early Paleozoic continental edge of the North American craton (Cook and Oliver, 1981). Early Mesozoic rifting was closely controlled by the preexisting Paleozoic and Precambrian structural fabric. New sedimentary basins formed in an extensional regime in a 400 km wide belt roughly parallel to the Fall Zone and to the late Paleozoic suture between American and African plates (Nelson et al., 1985; Hutchinson et al., 1986). The Atlantic margin development started here 180 million years ago.

The fact that the present continental margin is nearly parallel to the Alleghenian suture and paleoedge of the proto-American craton in the United States and cuts obliquely the main structures of the British crust is fundamental to understanding the state of stress and mor-

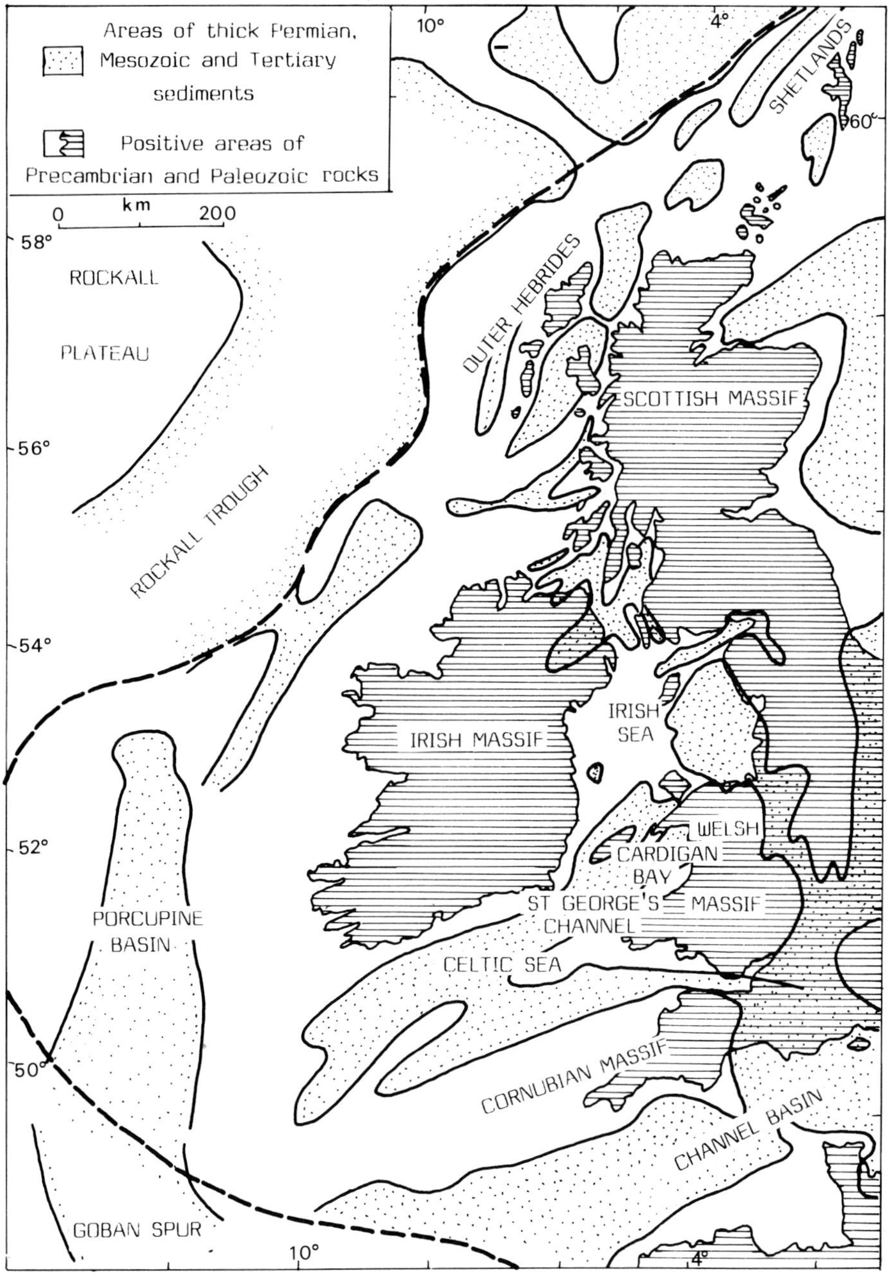

Fig. 1. Sedimentary basins west of Britain and Ireland. The thick, dashed line shows the approximate edge of the continental shelf. The crust beneath Rockall Trough is thought to be oceanic or, more probably, stretched continental crust. (After Naylor and Shannon, 1982, with modifications.)

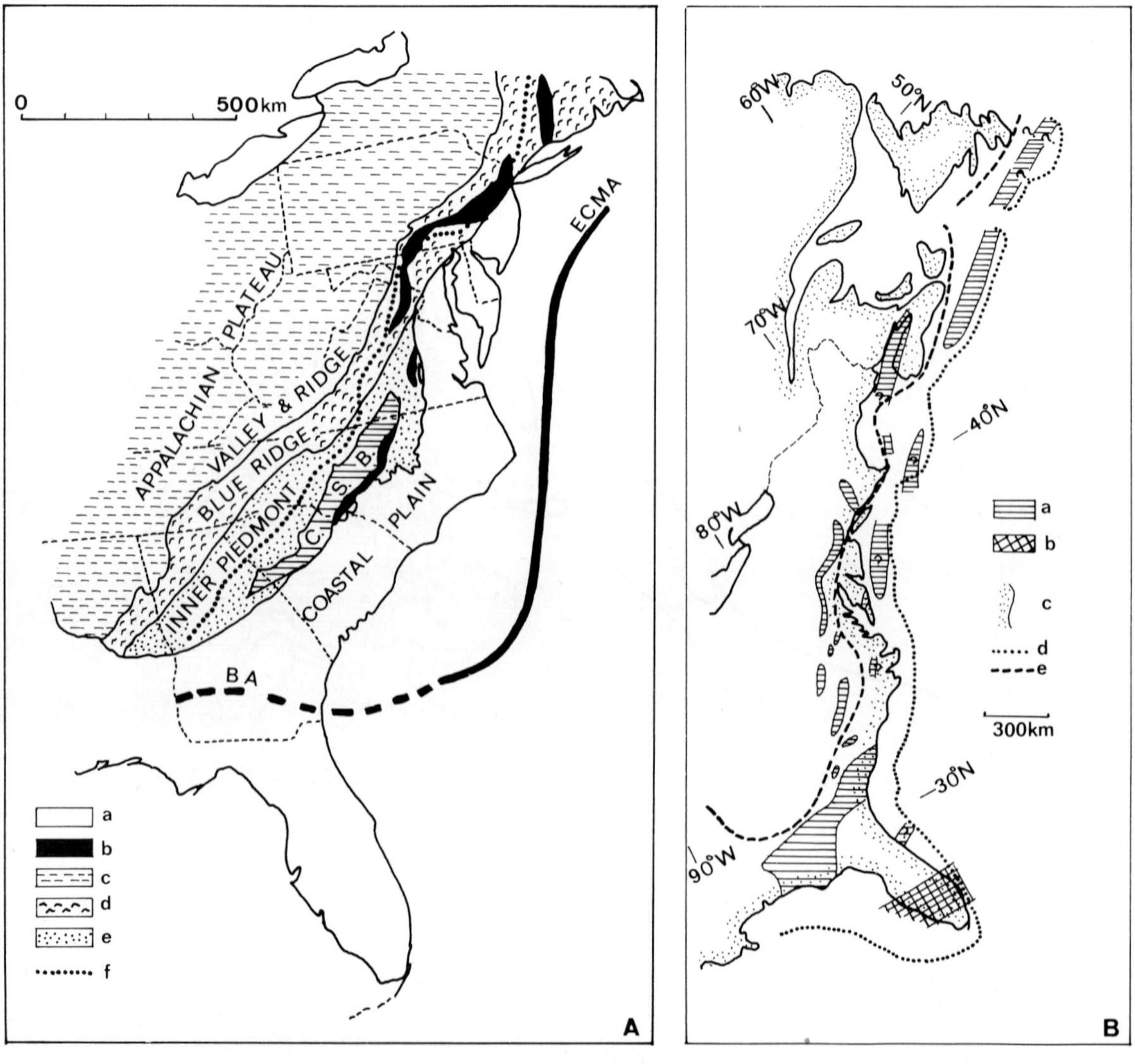

Fig. 2. Simplified geology of eastern United States. (A) Generalized geological map of Appalachian orogen (redrawn from Hutchinson et al., 1983) with the trend of the Piedmont gravity gradient. *a* = Unmetamorphosed Cenozoic and Mesozoic sedimentary rocks; *b* = Triassic–Jurassic rift related sedimentary and volcanic rocks *c* = unmetamorphosed Paleozoic sedimentary rocks; *d* = metamorphosed Paleozoic rocks; *e* = metamorphosed Paleozoic and Precambrian rocks; *f* = trend of Piedmont gravity gradient; *ECMA* = East Coast Magnetic Anomaly; *BA* = Brunswick Anomaly. *ECMA* and *BA* (from Nelson et al., 1985) could mark trend of late Paleozoic suture between North America and Africa as well as early Jurassic rifting that subsequently opened to form Atlantic Ocean (see Fig. 2B). (B) Approximate outline of the major known or inferred Triassic and Jurassic basins along the northwestern continental margin of Atlantic Ocean (redrawn from Owens, 1983). *a* = Triassic and Early or Middle Jurassic basins; *b* = Triassic and Jurassic volcanic rocks; *c* = coast; *d* = outer edge of Coastal Plain; *e* = inner edge of Coastal Plain.

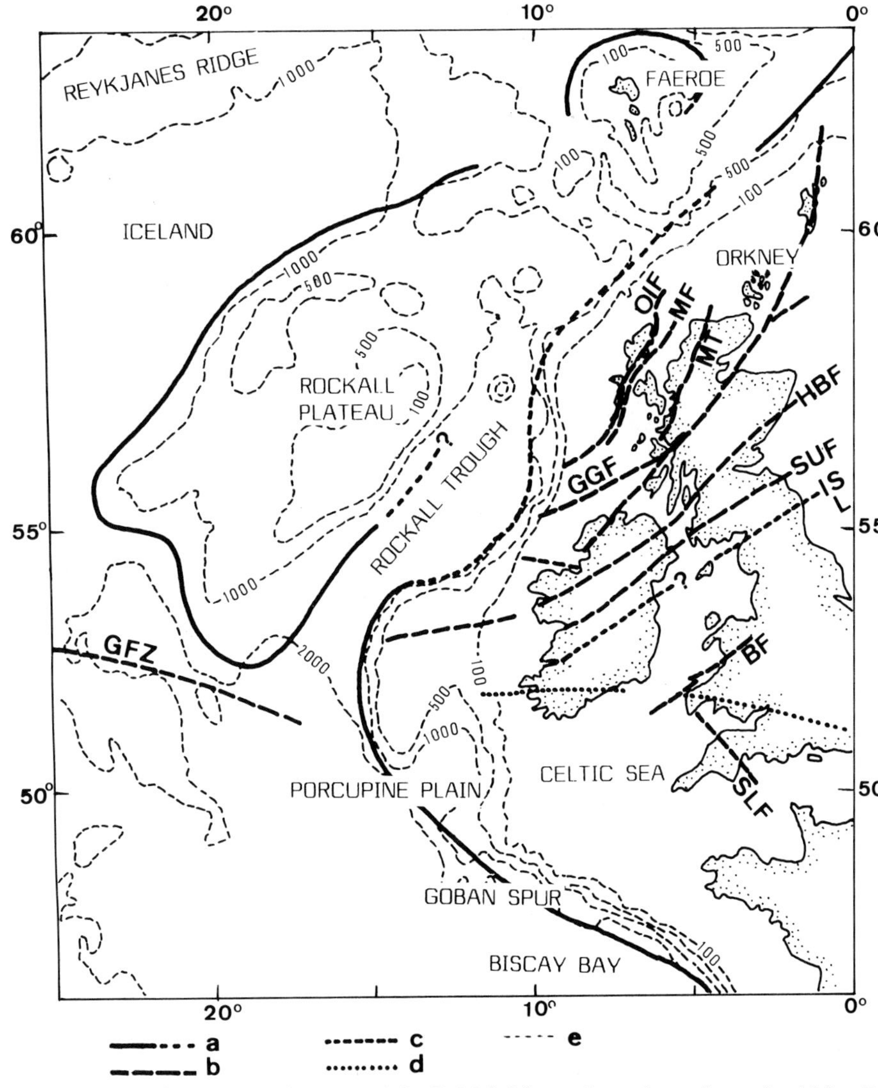

Fig. 3. Major bathymetric and structural elements of the British Isles and continental margin. (After Roberts et al., 1979; Bailey, 1979; Dunning, 1980.) a = Estimated position of boundary between oceanic and continental crust; b = main faults (*OIF*) = Outer Isles Thrust, *MF* = Minch Fault, *MT* = Moine Thrust, *GGF* = Great Glen Fault, *SLF* = Sticklepath-Lustleigh Fault, *GFZ* = Charlie Gibbs Fracture Zone, *SUF* = Southern Upland Fault, *HBF* = Highland Boundary Fault); c = *ISL*, Iapetus Suture Line; d = Hercynian Front; e = contours in m.

phogenic consequences on both sides of the Atlantic.

Block tectonics and large-scale landform development in the British Isles

The Davisian theory of the cycle of erosion (Davis, 1895) was widely adopted in Great Britain. For several decades, denudation chronology and recognition of ancient surfaces were the main topic of long-term landforms studies when most workers thought that the British Isles did not suffer major warping and faulting in Cenozoic times.

In the late 1960s and early 1970s, the earth sciences were undergoing revolutions in concepts with the plate tectonics paradigm, and much progress occurred in the geology of the

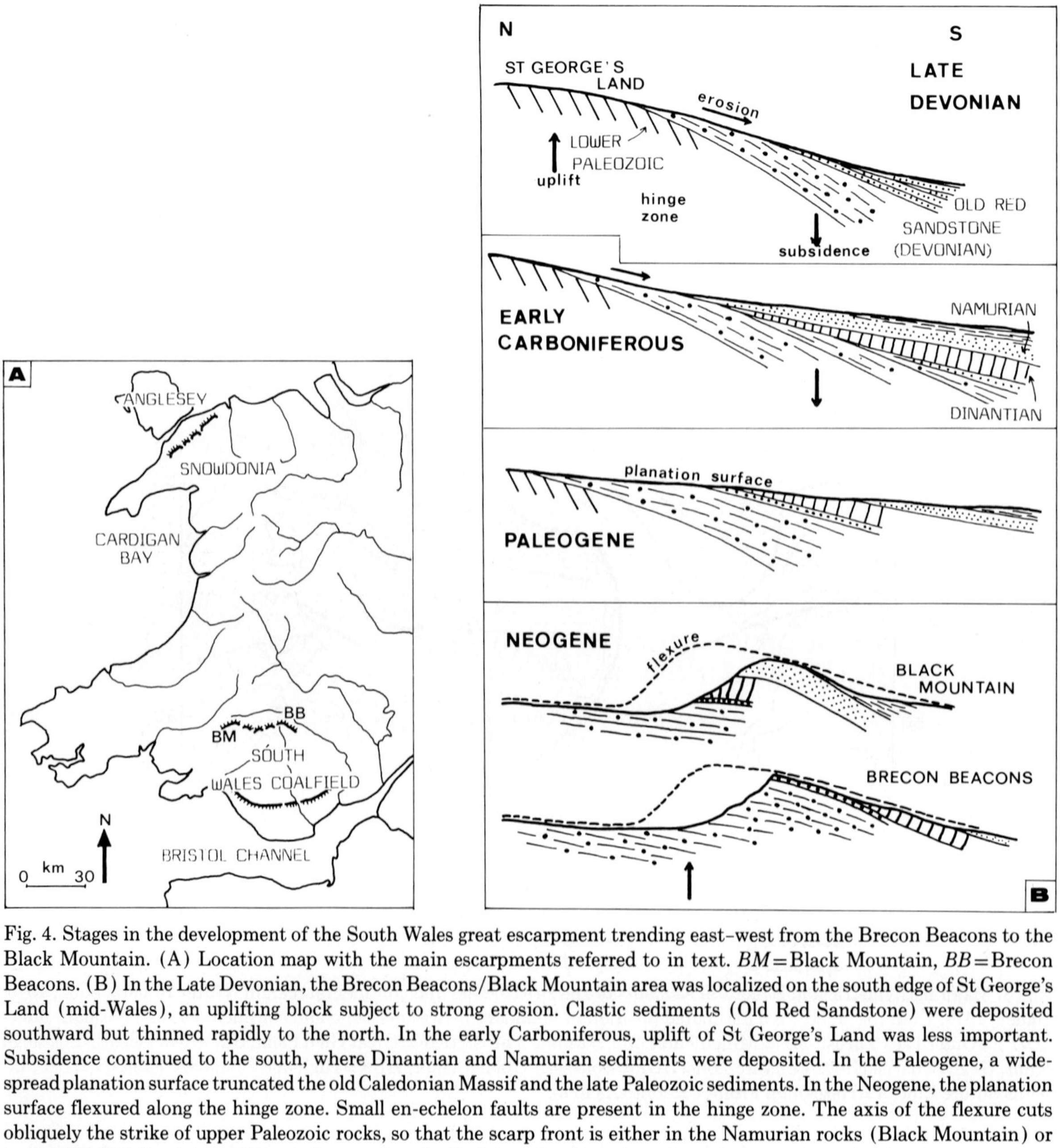

Fig. 4. Stages in the development of the South Wales great escarpment trending east–west from the Brecon Beacons to the Black Mountain. (A) Location map with the main escarpments referred to in text. *BM*=Black Mountain, *BB*=Brecon Beacons. (B) In the Late Devonian, the Brecon Beacons/Black Mountain area was localized on the south edge of St George's Land (mid-Wales), an uplifting block subject to strong erosion. Clastic sediments (Old Red Sandstone) were deposited southward but thinned rapidly to the north. In the early Carboniferous, uplift of St George's Land was less important. Subsidence continued to the south, where Dinantian and Namurian sediments were deposited. In the Paleogene, a widespread planation surface truncated the old Caledonian Massif and the late Paleozoic sediments. In the Neogene, the planation surface flexured along the hinge zone. Small en-echelon faults are present in the hinge zone. The axis of the flexure cuts obliquely the strike of upper Paleozoic rocks, so that the scarp front is either in the Namurian rocks (Black Mountain) or in the Old Red Sandstone (Brecon Beacons) or in the Dinantian limestone (not shown on Fig. 4). The steepness of the scarp depends upon the type of rocks.

sedimentary basins of the British shelf. The Llanbedr (Mochras Farm) borehole in Cardigan bay (Fig. 4 and Fig. 5), completed in 1969 to a depth of 1938 m was a fundamental milestone because it showed that Oligo–Miocene sediments directly overlie lower Jurassic deposits without intermediate Cretaceous chalk (Woodland, 1971), a fact that threw much doubt on the supposed role of a chalk cover as the initial datum plane of landscape evolution and development of drainage in Britain. It was also the first obvious proof of considerable tec-

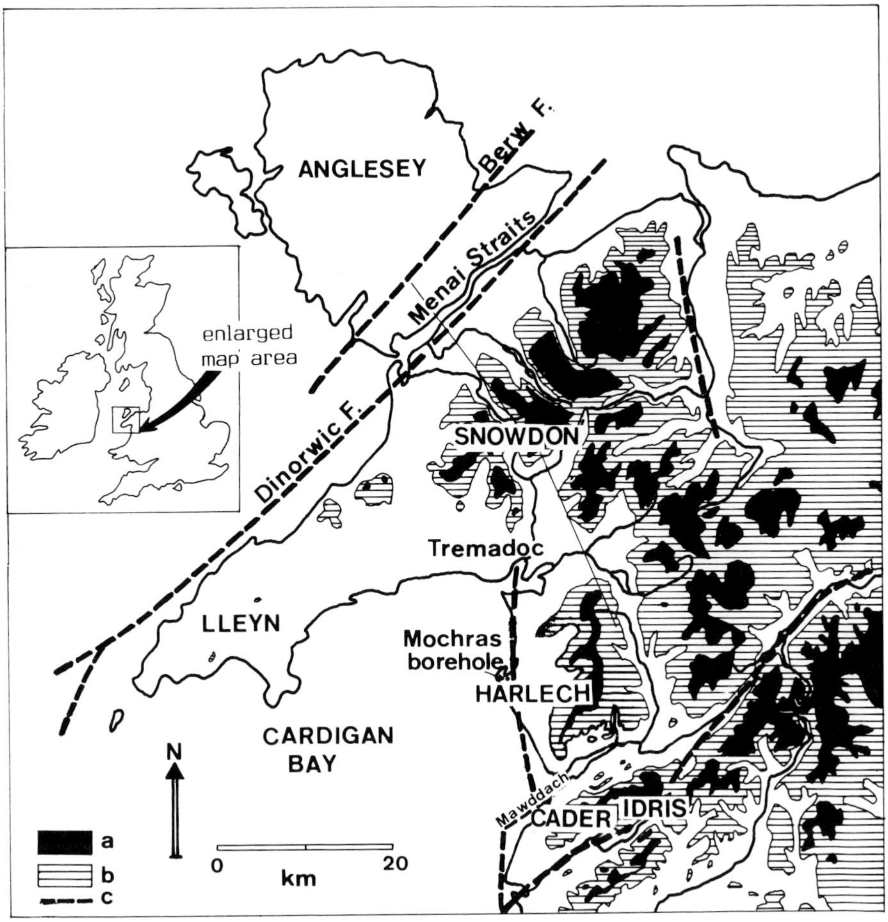

Fig. 5. Some topographic and structural features in North Wales. a = elevation above 427 m (1400 ft); b = elevation between 244 and 427 m (800 and 1400 ft); c = fault.

tonic activity in the Tertiary, because 524 m of predominantly middle Tertiary clastic deposits occur on the downthrown side of a fault which has a total apparent throw of at least 4500 m.

From various studies in northern England (King, 1976), Wales (Battiau-Queney, 1978, 1979), southern Pennines (Walsh et al.,1980), southeast Ireland (Mitchell, 1985) it became evident that the British Isles have been affected by active differential tectonics during the Tertiary. Some areas have been strongly uplifted (e.g., Snowdonia, South Wales Coalfield, southern Pennines) or have been subsiding (Cardigan Bay), while other areas were quite stable (Anglesey in North Wales, the Buchan in Scotland) (Hall, 1986).

Adjacent blocks are separated by a gentle slope or a sharp escarpment, but in most cases, it is not easy to explain the contact by marked fault lines of rock control. In South Wales. the great escarpment trending east–west from the Brecon Beacons to the Black Mountain cuts obliquely across the strike of Devonian strata so that the same rocks crop out either on the dip-slope, at the top of the scarp or below in front of the scarp (Fig. 4) (Battiau-Queney, 1979). This major landform results from the warping of an erosion surface with uplifting of the Black Mountain/Brecon Beacons block in the Neogene. The axis of the flexure, which is associated with short transverse en-echelon faults, follows approximately the south edge of

St George's Land, a tract of upland that occupied much of the central part of Wales in the Devonian while the present mountainous area was subsiding. The landscape which is seen today results from a complete reversal of regional slope along an axis which was already active in the Paleozoic. A similar conclusion has been proposed for the south scarp of the South Wales coalfield (Battiau-Queney, 1979).

Another significant feature is the northwest front of Snowdonia in North Wales. It is a straight northeast trending escarpment up to 800 m high above the low platforms of Anglesey and Arvon (Fig. 5). A northeast trending fault zone exists along the Menai Straits, but the apparent throw of the faults is too small to explain the huge escarpment. The relative positions of Snowdonia and Anglesey were inverted in Paleozoic and Cenozoic times: Snowdonia was a subsiding block until at least the early Carboniferous. The present position of Ordovician rocks of Snowdonia several hundred metres above the level of Carboniferous rocks of Anglesey is due to tectonic inversion (Fig. 6). Snowdonia began to uplift in the early or middle Cretaceous, but until late Cretaceous and Eocene, uplift was approximately balanced by denudation. Abundant clasts from Snowdonia rocks found in the Oligo–Miocene sediments of Cardigan Bay mark the onset of Snowdonia mountain building (Battiau-Queney, 1978, 1979).

During this same period, Anglesey was a stable area. Triassic detritus of Cardigan Bay came from the same Precambrian basement which crops out today, indicating that Anglesey was already eroded down to the present structural level in the Triassic. The low elevation of Anglesey compared to Snowdonia is not due to less resistant rocks. On the contrary, some of the most resistant rocks of Wales crop out in Anglesey. Furthermore, any interpretation of the front scarp of Snowdonia as the headward limit of a cyclic peneplain must be rejected. The existence of a major crustal hinge between a stable block (Anglesey) and a mobile block (Snowdonia) was suggested from geological and geomorphological data (Battiau-Queney, 1979). Several years later, geophysical techniques showed that the "small" faults of the Menai Straits are in fact discrete surficial tracks of a major crustal discontinuity revealed on a deep reflection profile down to 15–20 km (Brewer et al., 1983).

On the whole, it is evident from deep seismic reflection profiling and other geophysical data (Smythe et al., 1982; Brewer and Smythe, 1984; BIRPS and ECORS, 1986; Freeman et al., 1988) that the block tectonics assumed by long-term landscape analysis is closely controlled by the heterogeneity of the British crust. Sharp lateral variations are frequent in the middle and deep crust (Fig. 7). Furthermore, the crust appears on most profiles with two layers: a brittle upper

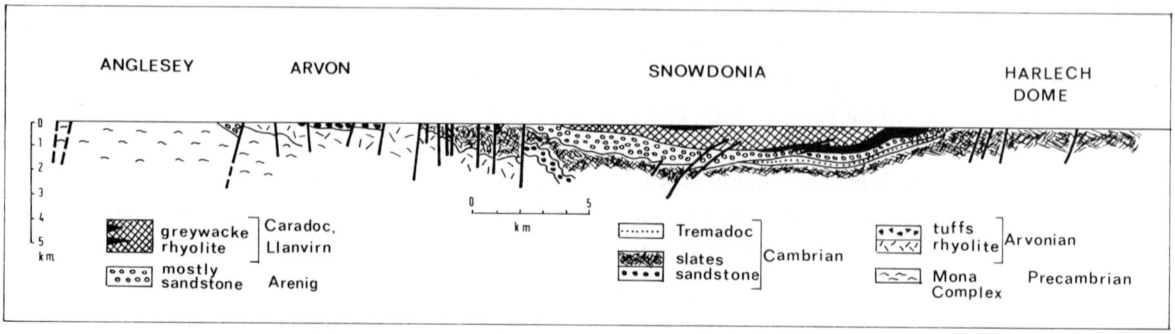

Fig. 6. Structural section of northwestern Wales (from Shackleton, 1953, simplified). The section line is drawn on Fig. 5. Ordovician rocks (Arenig, Llanvirn, Caradoc) were deposited in a subsiding basin, but presently crop out at 800 m and more above Anglesey/Arvon.

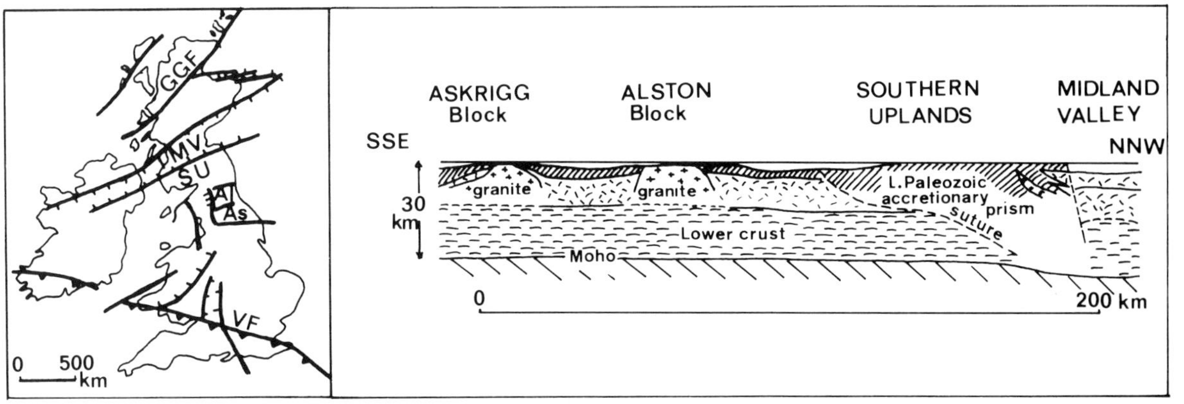

Fig. 7. Speculative cross section from the Midland Valley to the Northern Pennines (from Bott et al., 1985). The upper crust contains sharp horizontal discontinuities which disappear in the lower crust except along the supposed Iapetus suture. On the location map: *GGF* = Great Glen Fault; *MV* = Midland Valley; *SU* = Southern Uplands; *AL* = Alston Block; *AS* = Askrigg Block; *VF* = Variscan Front.

layer and a ductile lower one. The thickness of the brittle crust is generally 15–20 km (Fig. 8). The Moho appears remarkably flat, nearly horizontal even beneath sedimentary basins where 9 or 10 km of Mesozoic and Cenozoic sediments are preserved (e.g., beneath the north Celtic Sea basin and the Plymouth Bay basin). Lateral change in density of the brittle crust might be compensated by some dynamic mechanisms in the ductile crust. While the brittle crust records every major geological event of the past, the lower crust is in constant adjustment by lateral flows or chemical phase changes. Consequently a decoupling of the two layers is assumed.

The brittle crust is formed of different units which have their own density, heat flow, and rheological properties. In North Wales, Anglesey has a cooler and more rigid crust than Snowdonia (Battiau-Queney, 1989). Generally, isostatic adjustment differs from one block to another. Under the state of stress induced by plate tectonics, strain steadily dissipates by gradual release of energy along preexisting lines of weakness. Earthquakes are infrequent and weak. In each crustal unit, a dynamic equilibrium tends to occur between rock resistance, erosional processes, and uplift rate.

In North Wales, for example, Anglesey has been a very stable block since upper Paleozoic times. Denudation is estimated to be no more than several tens of metres during the last 200 million years (Battiau-Queney, 1978). For that reason thick saprolites have been preserved in some places. Very old paleoforms can survive. Mynydd Bodafon is a true inselberg made of Precambrian metaquartzite which was already exposed to erosion in Devonian times. Pebbles of quartzite are found in a Devonian conglomerate which crops out at the foot of Mynydd Bodafon. In contrast, in the uplifted block of Snowdonia, landform development was much faster with vigorous dissection developed in response to the rapid Neogene uplift.

Some general conclusions can be drawn from the Welsh example: (1) The state of stress is closely controlled by all the discontinuities inherited from the past and the crustal response differs from place to place according to the rheological properties of each crustal unit. (2) Old paleoforms can survive a very long time – 200 million years or more – close to regions which have known a rapid change in the last 30 million years. The rate of denudation can differ greatly from one area to another. (3) Some major escarpments, such as the northwest front of Snowdonia, have not retreated notably during the last 30 million years because their emplace-

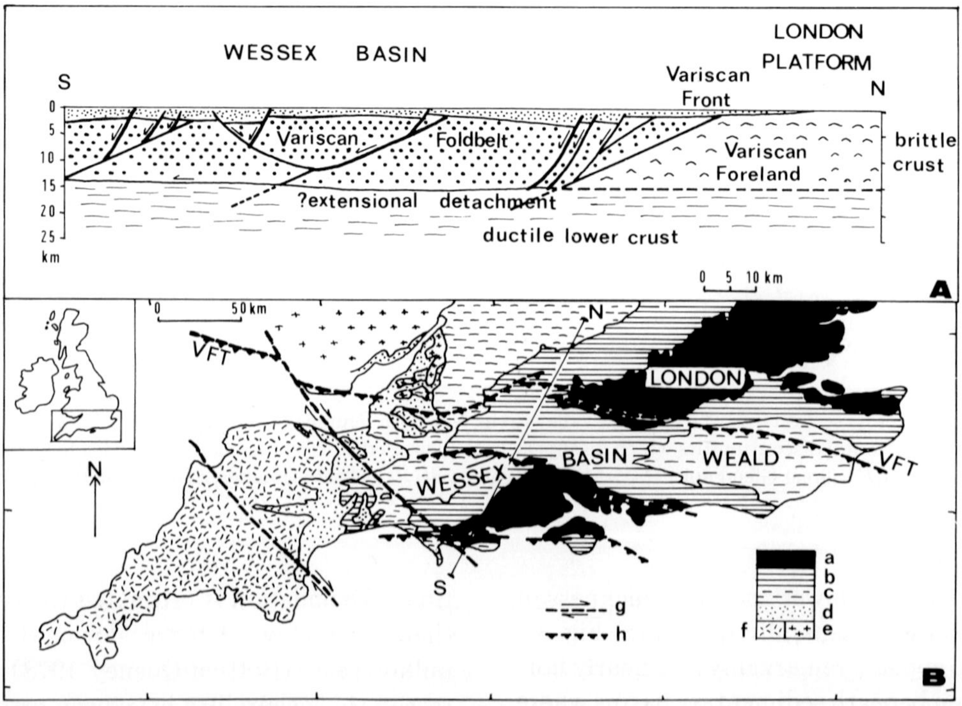

Fig. 8. The crust beneath the Wessex Basin (southern England). (A) Crustal section with an upper brittle layer and a lower ductile crust. Three major Variscan thrusts dipping southward have been reactivated with normal faults formed above them in Permian to Cretaceous times. Thrusts disappear at a depth of 15 km which marks the top of the ductile lower crust. (B) Sketch-map of southern England with simplified surface geology and some major structural features of the Variscan basement. Note that the south boundary of the Tertiary London Basin is close to the track of the Variscan Front. *a* = Tertiary sediments; *b* = Middle and Upper Cretaceous sediments; *c* = Jurassic and Lower Cretaceous sediments, *d* = Permo-Triassic sediments; *e* = Variscan foreland; *f* = Variscan foldbelt; *VFT* = Variscan Front Thrust; *g* = wrench fault; *h* = thrust (in the Variscan basement). (Adapted from Chadwick, 1986.)

ment is linked with crustal discontinuities and a hinge zone.

The cycle of erosion is a model of evolution that is not compatible with a heterogeneous crust. When two adjacent crustal units behave differently, one uplifted and the other stable, there is no reason that the long-term evolution would lead to the planation of both units to the same level.

Structural and physiographic framework of the Atlantic provinces of the United States

The history of Appalachian geomorphology (Sevon, 1985; Mills et al., 1987) presents some similarities but also differences with the British one. The concepts of the geographical cycle and peneplains were presented by Davis in "The Rivers and Valleys of Pennsylvania" (1889). The Davisian geomorphology was pervasive and dominating in the Appalachians until the 1940's (Sevon et al., 1983), when attention of geomorphologists turned to process and small-scale landforms. The problem of the origin of major rivers which drain the Appalachians to the Atlantic has long focused the attention of workers (e.g., Johnson, 1931). It is complicated by the presumed drainage reversal which occurred sometime after the Allegenian orogeny. The concept of dynamic equilibrium which was nearly unknown in Britain was strongly advocated in the Appalachians (Hack, 1960). Relation of landscape to lithology and structure was analysed by Hack (1957, 1979, 1982) who considered rock control generally more important than tectonics on a regional scale.

With new data provided by geophysics it is tempting to link the shaping of the Appalachian highlands and Piedmont with plate tectonics, Atlantic margin development, and deep crustal structure. Only the regional framework of the eastern United States arranged in belts roughly parallel to the Atlantic coast will be considered here. On that scale, the main problems concern the origin and significance of the Fall Zone at the edge of the Coastal Plain and the origin of the Blue Ridge escarpment which terminates the Appalachian Piedmont landwards.

The Fall Zone marks the landward edge of the sedimentary wedge which formed after the opening of the Atlantic Ocean. On COCORP profiles in Georgia the faults associated with the Fall Zone, for instance the Augusta fault, display east-dipping reflections which extend at a low angle some 80 km or more southeast of their surface position (Cook et al., 1981). Generally, large-scale thin-skinned thrusting occurred in the upper and middle crust of the U.S. Atlantic provinces during the late Paleozoic (Cook et al., 1979; Cook et al., 1981; Cook and Oliver, 1981; Oliver et al., 1983). The structure of the crust in these U.S. regions is therefore completely different from that of the British Isles.

The Valley and Ridge, Blue Ridge, and inner Piedmont are allochthonous terranes above a major detachment. The décollement was thrust westward over the Grenville basement and metasedimentary cover sequence (Fig. 9). The allochthonous upper crust thins from 15 km or more in the Piedmont to a few kilometres in the Valley and Ridge. The precise structure of the thrust sheets and allochthonous terrane and the exact geometry of the main Appalachian detachment are diversily interpreted (Cook et al., 1981; Iverson and Smithson, 1983; Hutchinson et al., 1983; Thomas, 1983; Farrar, 1985; Hooper and Hatcher, 1988), but the superposition of an upper allochthonous crust over an autochthonous basement is now well established in the Blue Ridge and inner Piedmont. Controversial interpretations concern the southeastward extension of the Appalachian décollement beneath the eastern Piedmont and Coastal Plain (Cook et al., 1981; Secor et al., 1986).

The Moho is also quite different here from the British Isles. Along the COCORP profile in Georgia, the Moho occurs at a 32 km depth beneath the Coastal Plain and eastern Piedmont (Cook et al., 1981), but it disappears west of the Elberton granite (Fig. 9). Additional geophysical data suggest that the crust thickens to 40–50 km beneath the inner Piedmont and Blue Ridge (Hutchinson et al., 1983; Iverson and Smithson, 1983). The strong gravity gradient observed in the inner Piedmont has been explained by a crustal thickening due to the buried passive margin of the proto-America continent (Cook and Oliver, 1981), combined with a southeast-dipping suture zone formed in the early Paleozoic and composed of high-density mafic rocks (Hutchinson et al., 1983). Whatever model is preferred, it is clear that the deep crust has different properties beneath the eastern and inner Piedmont.

Because the Atlantic Ocean opened along a rifting zone nearly parallel to the old sutures and the proto-American paleomargin, the stress field is relatively simple in these regions. The Appalachians highlands, Piedmont, and Coastal Plain belong to the same stress province (Zoback and Zoback, 1980) marked by compressional tectonism with dominantly reverse faults (Mixon and Newell, 1977, 1982; Prowell, 1988).

The state of stress is due to dynamic processes (as ridge-push forces) associated with oceanic spreading and lithostatic pressure differences at the transition between continental and oceanic crust (Vetter and Meissner, 1979) and to flexural effects linked with erosion and isostatic rebound of the Appalachians (Zoback and Zoback, 1980). Stresses are guided by preexisting weakness zones in the continental crust: reverse faults have been currently ob-

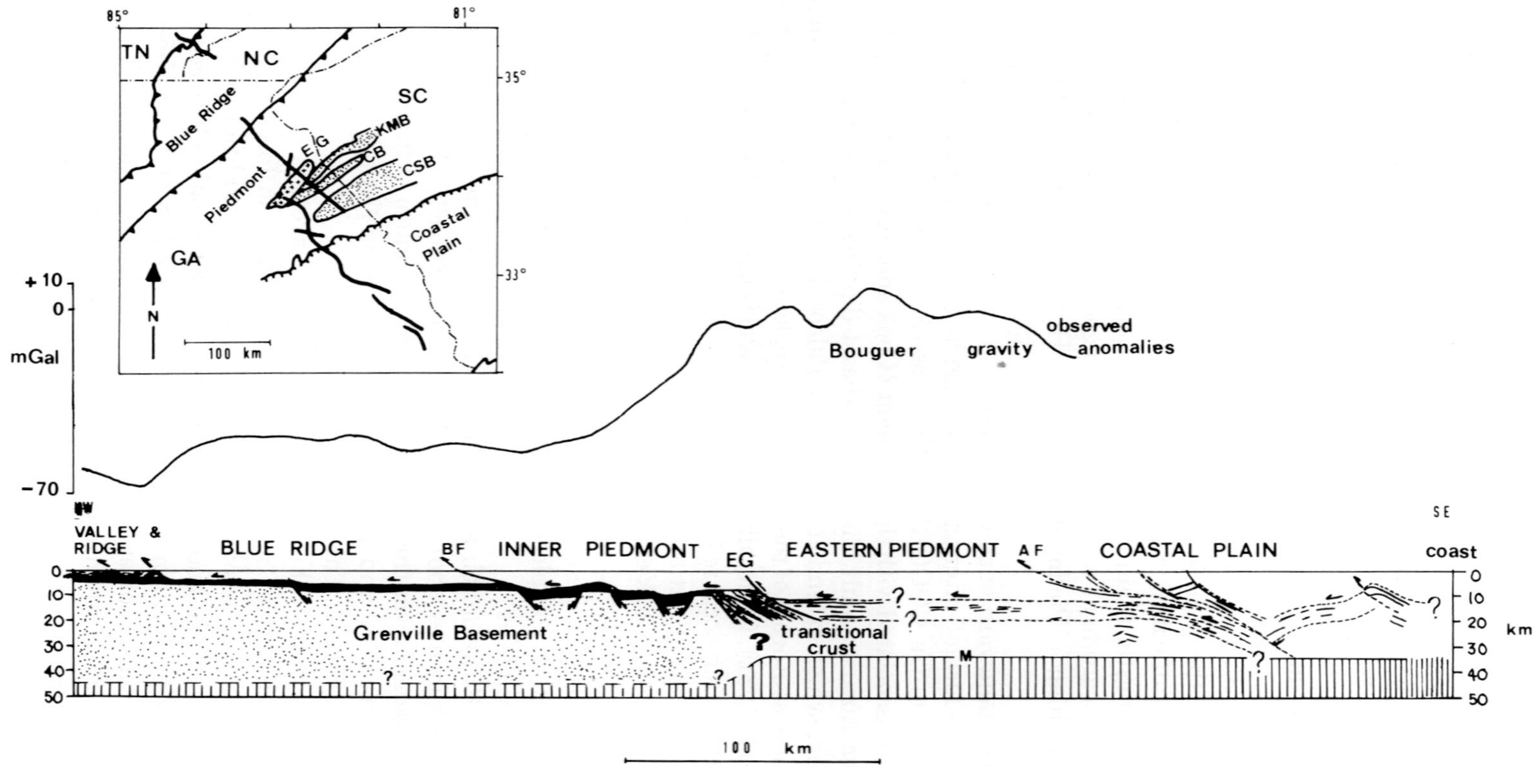

Fig. 9. COCORP southern Appalachian traverse: simplified tectonic cross-section (from Cook et al., 1979), with gravity profile superposed (from Thomas, 1983). *BF*=Brevard Fault Zone; *EG*=Elberton granite; *AF*=Augusta Fault; *M*=Moho. In this interpretation, a subhorizontal detachment is supposed to extend beneath the Coastal Plain. Another possible interpretation supposes the thrust surfaces of the inner Piedmont-Blue Ridge allochthon to "root" in the mantle. The map gives the location of COCORP lines and some important geologic features. *EG*=Elberton Granite; *KMB*=Kings Mountain Belt; *CB*=Charlotte Belt; *CSB*=Caroline Slate Belt. The gravity profile shows the strong gradient observed in the inner Piedmont.

served at the surface in the Coastal Plain, the Fall Zone and the Piedmont (Mixon and Newell, 1977; Reinhardt et al., 1984; Prowell, 1988).

Reverse faults and the NW–SE compressional pattern formed as early as the late Jurassic, soon after the opening of the Atlantic Ocean (Prowell, 1988). Field analysis indicates recurrent movement through much of Mesozoic and Cenozoic times. In some cases latest Tertiary or Quaternary movement is probable (Darton, 1950; Mixon and Newell, 1982).

At the surface, faults are relatively minor features and their morphological consequences are generally modest on a regional scale, even if they are linked with major deep faults inherited from Triassic rifting or more ancient tectonic events (Behrendt et al., 1981, 1983; Newell, 1985). So they cannot explain major landforms such as the Blue Ridge escarpment.

In Georgia, North Carolina, and Virginia, the Blue Ridge escarpment is several hundred metres high and was accurately described by White (1950) who concluded that it could be interpreted as a scarp caused by a series of en-echelon faults. In fact, for 50–60 km the Brevard Fault zone marks the edge of the Piedmont and one could be tempted to explain the Blue Ridge escarpment by this fault zone. Recently, the same escarpment was interpreted (Ollier, 1985) as a "Great Escarpment" similar to the "Randschwellen" of Jessen (1943) which are asymmetric bulges with a gentle landward tilt and a steep slope facing a low coastal plain. Typical "Randschwellen" separate "two regions of vastly different geomorphology: the tablelands with many paleoforms, low relief and slow process rates, and the coastal zone with few paleoforms, moderate relief and rapid process rates" (Ollier, 1985). This type of landform characterizes many passive continental margins, especially in South Africa and India. In the classic scheme, headward erosion and scarp retreat from the nearby continental margin creates the coastal plain. If the Blue Ridge front is interpreted as a "Great Escarpment", it implies that the Piedmont surface was formed by landward extension of an erosional surface and simultaneously westward retreat of the Blue Ridge scarp. Geological and geomorphological facts are not consistent with a long-distance westwards retreat of the Blue Ridge escarpment: sediments of the Coastal Plain prove that the distribution of relative exposure of the various Piedmont rock types has not changed appreciably since the Jurassic and early Cretaceous times, a fact difficult to reconcile with extensive backwearing of the escarpment and simultaneous erosion of 2–3 km of rocks in the Piedmont (Glaser, 1969). From the study of stream piracy and drainage evolution, Hack argued for a slow and spatially restricted migration of the escarpment from some position southeast of the present one (Hack, 1982).

The origin of the Blue Ridge escarpment has been thoroughly discussed by Hack (1982). He showed that the effect of bedrock on landforms is important in the Piedmont and Blue Ridge: for instance, the Catoctin formation which consists mostly of massive green metamorphosed basalt is a resistant rock making a ridge along much of its outcrop. The Brushy Mountains and South Mountains, two areas that have many peaks at 750–900 m are underlain by the more resistant rocks of the region (Hack, 1982, p. 24). On the other hand, Hack observed that "all these resistant units also occur in low relief areas" and suggested that "both tectonic forces and rock control have operated to produce the present forms, with tectonic processes playing a major role" (1982, p. 25). The Brevard zone, "a broad zone of faults and fault slices that extends from Virginia to Alabama", is itself a major topographic feature generally forming a narrow zone of low relief (Hack, 1982, p. 30). However, faulting is not the direct cause of the escarpment. At its northern end, near Winston–Salem, the Brevard fault zone is 35 km south of the escarpment. In the Morganton basin, it is at the foot of the scarp, but southeastward it rises to cross the Blue Ridge high-

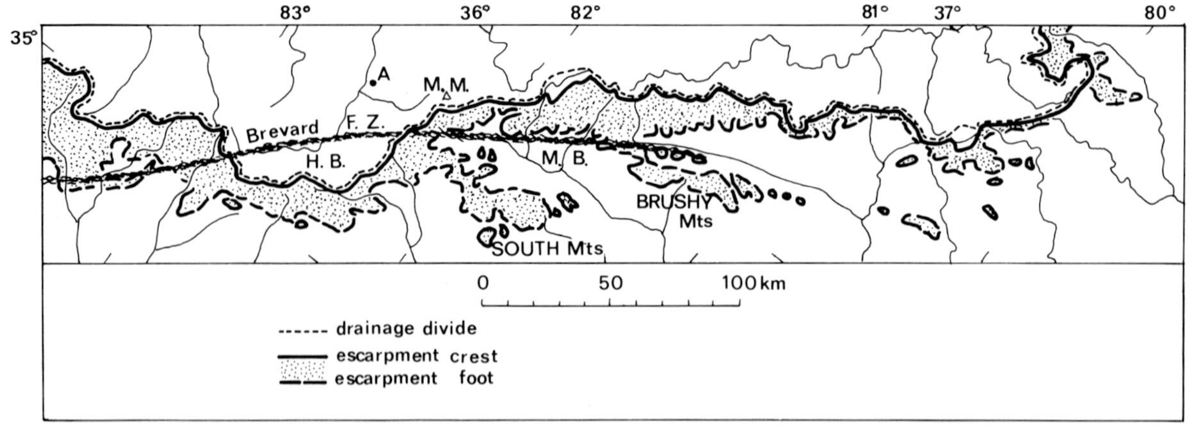

Fig. 10. Map of the Blue Ridge escarpment and foothills in the adjacent Piedmont in North Carolina and South Virginia (adapted from Hack, 1982). *A* = Asheville; *M.M.* = Mount Mitchell; *M.B.* = Morganton Basin; *Brevard F.Z.* = Brevard Fault Zone; *H.B.* = Hendersonville Bulge.

lands for a distance of about 120 km (Hack, 1982, p. 30–31) (Fig. 10).

More detailed geologic mapping (Hadley and Nelson, 1971; Rankin et al., 1972; Espenshade et al., 1975; Goldsmith et al., 1988) shows that it is very difficult to link a major landform such as the Blue Ridge escarpment with surficial geology. For Hack, the mountainous front is neither a simple fault scarp nor a differential erosion scarp. It is important to emphasize the relation of the Blue Ridge front with the steep gravity gradient trending northeast within the Piedmont from Georgia to south Virginia. The gradient closely parallels the main drainage divide between the Gulf of Mexico and the Atlantic Ocean. In Georgia and North Carolina, it borders the Blue Ridge, but northwards it shifts to the Valley and Ridge province where the Blue Ridge front is less sharp and high. This gravity gradient cannot be correlated with any single surficial geological feature and is more accurately linked with deep buried structures. At a time when COCORP and other geophysical data were not yet available, Hack wrote (1979: p. B15): "The steep gravity gradient that parallels the Blue Ridge front is suggestive of different crustal blocks that may separate areas having two rates of uplift". Two other observations are important: (1) the axis of maximum uplift in the Appalachian highlands, as indicated by repeated geodetic traverses (Brown and Oliver, 1976) and stream profiles (Hack, 1982), corresponds to the main drainage divide and the gravity low of the paired gravity anomalies referred to above, and (2) stress directions acting on the continental crust of the U.S. Atlantic provinces are normal to the trend of the same gravity gradient (Zoback and Zoback, 1980).

It is possible to present a speculative explanation for the origin of the blue Ridge escarpment taking into account the main structural discontinuities of the middle and deep crust, the compressional stress field in the continental crust bordering the passive margin of the Atlantic Ocean, and the track of the main drainage divide. On a regional scale, two structural systems interfere at two different levels: a deep system indicated by the regional Bouguer anomalies which are characterized by a crustal root beneath the inner Piedmont and Blue Ridge, and a surficial system related to the properties of the upper crust above the main Appalachian detachment. In the North American type of passive margin, major deep crustal discontinuities which were buried beneath allochthonous terranes still control large-scale landform development. The Blue Ridge escarpment reaches its highest relief where discontin-

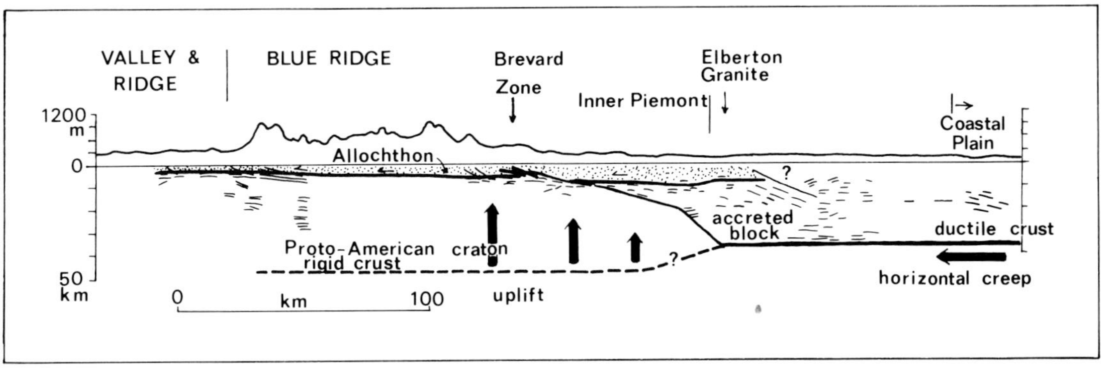

Fig. 11. Speculative mechanism of uplifting of the inner Piedmont and Blue Ridge along COCORP profile in southern Appalachians. Elevation profile from Thomas, 1983.

uities in the deep crust interfere with surficial geological features: for example, sharp lithological contrasts or the presence of lines of weakness such as the Brevard fault zone which is a major thrust diverted from a subhorizontal décollement. It is probably the youngest major fault in the region, but, as Hack pointed out, there is no direct evidence of motion after middle Mesozoic. On the other hand, if the uplift was not recent, the mean relief of the Blue Ridge should have been reduced to the same level as the Piedmont long ago.

As a working hypothesis, it is presumed that horizontal compressional stresses could leap to creep processes directed from the ocean to the continent in the lower ductile crust and upper mantle of the Coastal Plain and eastern Piedmont, a mechanism referred to as silent subduction (Vetter and Meissner, 1979). In contrast to the British Isles where a flat Moho is observed beneath Caledonian and Hercynian mountains, a thick crustal root exists beneath the Appalachian highlands. Creep movement active in the low-viscosity, deep crust of the Coastal Plain and eastern Piedmont might be blocked against the thicker and cooler Grenville crust. Uplift could result from this situation (Fig. 11). Because creep processes in the ductile crust and upper mantle are directly related to the development of the Atlantic passive margin, they could be the long-duration mechanism able to explain the sustained uplift of the Appalachian highlands and the relatively stable position of the mountainous front above the Piedmont, since middle Mesozoic times.

Summary

The continental lithosphere becomes more heterogeneous with time because it records every major tectonic event of the past. Stresses are closely guided by the anisotropy of the crust, especially in the upper brittle crust. In the British Isles, the Atlantic Ocean margin cuts obliquely the old structural pattern. The result is a complex mosaic of blocks which control long-term landform evolution. On the American side of the Atlantic, the new margin was close to the old paleosuture producing a simpler apparent structural framework. However, the existence of a major detachment in the middle crust is a source of complexity because, on a regional scale, major landforms depend upon deep crustal discontinuities buried beneath allochthonous terranes.

A basic assumption of Davis' model, which is implied also in the concept of "Great Escarpments", is that erosion of one cycle would remove parts of the landscape developed in any previous cycle. In that case, the Piedmont is interpreted as a Cenozoic surface produced by backwearing of the Appalachian highlands frontscarp at the expense of a previous landmass formed by asymmetric doming of the eastern United States. According to the same model, the low surface of Anglesey is interpreted as a Pliocene or early Pleistocene surface produced by backwearing of the Snowdonia front. The Davisian model of landscape evolution emphasizes time-dependent landforms. A different concept has been proposed here: regional-scale landforms are not fundamentally time-related but controlled mainly by the crustal properties. An important consequence would be that long-term landform evolution does not lead necessarily to a widespread planation surface. Some major escarpments have not retreated noticeably during the course of several tens of millions of years because they are closely related to deep crustal discontinuities.

Acknowledgements

Financial support from the Fulbright Program (in 1986), the CNRS and the University of Sciences and Techniques of Lille (in 1988) is gratefully acknowledged. Thanks are also due to many other colleagues and friends in the U.S.A. who provided me the documentation and data used in my work.

References

Bailey, R.J., 1979. The continental margin from 50N to 57N: its geology and development. In: Banner, F.T., Collins, M.B. and Massie, K.S. (Editors), The North-West European Shelf Sea 1: Geology and Sedimentology. Elsevier, Amsterdam, pp. 11-24.

Battiau-Queney, Y., 1978. Contribution à l'étude géomorphologique du Massif Gallois. Thèse d'Etat, Université de Brest, Honoré Champion, Paris, 792 pp. (publ. 1980).

Battiau-Queney, Y., 1979. Origine tectonique de quelques grands escarpements du Pays-de-Galles (G.B.). Rev. Géol. Dyn. Géogr. Phys., 21: 109-126.

Battiau-Queney, Y., 1988. L'évolution géomorphologique du Piedmont Appalachien et de la Plaine Côtière, du New Jersey à la Georgie (USA). Cah. Geogr. Phys., 6, Université des Sciences et Techniques de Lille, Lille, 82 pp.

Battiau-Queney, Y., 1989. Liens entre l'évolution géomorphologique des Iles Britanniques et les propriétés de la croûte sous-jacente. Rev. Geomorphol., Dyn., 37: in press.

Behrendt, J.C., Hamilton, R.M., Ackerman, H.D. and Henry, V.J., 1981. Cenozoic faulting in the vicinity of the Charleston, S.C., earthquake zone. Geology, 9: 117-122.

Behrendt, J.V., Hamilton, R.M., Ackerman, H.D., Henry, V.J. and Bayer, K.C., 1983. Marine multichannel seismic-reflection evidence for Cenozoic faulting and deep crustal structure near Charleston, S.C. In: Gohn, G.S. (Editor), Studies Related to the Charleston, S.C., Earthquake of 1886. Tectonics and Seismicity. U.S. Geol. Surv., Prof. Pap., 1313: J1-J29.

BIRPS and ECORS, 1986. Deep Seismic reflection profiling between England, France and Ireland. J. Geol. Soc. London, 143: 45-52.

Bott, M.H.P., 1978. The Interior of the Earth. Elsevier, London, 400 pp.

Bott, M.H.P., Long, R.E., Green, A.S.P., Lewis, A.H.J., Sinha, M.C. and Stevenson, D.L., 1985. Crustal structure south of the Iapetus suture beneath northern England. Nature, 314: 724-726.

Brewer, J.A. and Smythe, D.K., 1984. MOIST and the continuity of crustal reflector geometry along the Caledonian-Appalachian orogen. J. Geol. Soc. London, 141: 105—120.

Brewer, J.A., Matthews, D.H., Warner, M.R., Hall, J., Smythe, D.K. and Whittington, R.J., 1983. BIRPS deep seismic reflection studies of the British Caledonides. Nature, 305: 206-210.

Brown, L.D. and Oliver, J.E., 1976. Vertical crustal movements from leveling data and their relation to geologic structure in the Eastern United States. Rev. Geophys. Space Phys., 14: 13-35.

Chadwick, R.A., 1986. Extension tectonics in the Wessex Basin, southern England. J. Geol. Soc. London, 143: 465-488.

Cook, F.A. and Oliver, J.E., 1981. The Late Precambrian-Early Paleozoic continental edge in the Appalachian orogen. Am. J. Sci., 281: 993-1008.

Cook, F.A., Albaugh, D.S., Brown, L.D., Kaufman, S., Oliver, J.E. and Hatcher, R.D. Jr., 1979. Thin-skinned tectonics in the crystalline southern Appalachians; COCORP seismic reflection profiling of the Blue Ridge and Piedmont. Geology, 7: 563-567.

Cook, F.A., Brown, L.D., Kaufman, S., Oliver, J.E. and Petersen, T.A., 1981. COCORP seismic profiling of the Appalachian orogen beneath the Coastal Plain of Georgia. Geol. Soc. Am., Bull., 92: 738-748.

Darton, N.H., 1950. Configuration of the bedrock surface of the District of Columbia and vicinity. U.S. Geol. Surv., Prof. Pap., 217, 42 pp.

Davis, W.M., 1889. The rivers and valleys of Pennsylvania. Natl. Georg. Mag., 1: 183-253.

Davis, W.M., 1895. The development of certain English rivers. Geogr. J., 5: 127-146.

Dewey, J.F., 1982. Plate tectonics and the evolution of the British Isles. J. Geol. Soc. London, 139: 371-412.

Dunning, F.W., 1980. The geotectonic position of the British Isles in Northwest Europe. In: Geology of the European countries, 1. 26th Int. Geol. Congr., Bordas, Paris, pp. 331-334.

Espenshade, G.H., Rankin, D.W., Wier Shaw, K. and Neuman, R.B., 1975. Geologic map of the east half of the Winston-Salem Quad., N.C., VA. U.S. Geological Survey. Misc. Invest. Ser., Map I-7098.

Farrar, S.S., 1985. Tectonic evolution of the easternmost Piedmont, N.C. Geol. Soc. of Am., Bull., 96: 362-380.

Freeman, B., Klemperer, S.L. and Hobbs, R.W., 1988. The deep structure of northern England and the Iapetus suture zone from BIRPS deep seismic-reflection profiles. J. Geol. Soc. London, 145: 727-740.

Glaser, J.D., 1969. Petrology and origin of Potomac and Magothy (Cretaceous) sediments. Middle Atlantic Coastal Plain. Md. Geol. Surv., Rep. Invest., 11, 101 pp.

Goldsmith, R., Milton, D.J. and Horton, Jr., J.W., 1988. Geologic map of the Charlotte 1°2° Quad. N.C. and S.C. U.S. Geol. Surv., Misc. Invest. Ser., Map I-1251-E.

Hack, J.T., 1957. Studies of longitudinal stream profiles in Virginia and Maryland. U.S. Geol. Surv., Prof. Pap., 294-B: 45-97.

Hack, J.T., 1960. Interpretation of erosional topography in humid temperate regions. Am. J. Sci., 258 A: 80-97.

Hack, J.T., 1979. Rock control and tectonism. Their importance in shaping the Appalachian Highlands. U.S. Geol. Surv., Prof. Pap., 1126-B: 1-17.

Hack, J.T., 1982. Physiographic divisions and differential uplift in the Piedmont and Blue Ridge. U.S. Geol. Surv., Prof. Pap., 1265, 49 pp.

Hadley, J.B. and Nelson, A.E., 1971. Geologic map of the Knoxville Quad., N.C., Tenn. and S.C. U.S. Geol. Surv., Misc. Geol. Invest., Map I-654.

Hall, A., 1986. Deep weathering patterns in north-east Scotland and their geomorphological significance. Z. Geomorphol., 30: 407-422.

Hooper, R.J. and Hatcher, Jr., R.D., 1988. Pine Mountain terrane, a complex window in the Georgia and Alabama Piedmont; evidence from the eastern termination. Geology, 16: 307-310.

Hutchinson, D.R., Grow, J.A. and Klitgord, K.D., 1983. Crustal structure beneath the southern Appalachians: non uniqueness of gravity modeling. Geology, 11: 611-615.

Hutchinson, D.R., Klitgord, K.D. and Detrick, R.S., 1986. Rift basins of the Long Island Platform. Geol. Soc. Am., Bull., 97: 688-702.

Iverson, W.P. and Smithson, S.B., 1983. Reprocessing and reinterpretation of COCORP southern Appalachian profiles. Earth Planet. Sci. Lett., 62: 75-90.

Jessen, O., 1943. Die Randschwellen der Kontinente. Petermans Geographische Mitteilungen, Ergänzung. 241 pp.

Johnson, D., 1931. A theory of Appalachian geomorphic evolution. J. Geol., 39: 497-508.

King, C.A., 1976. Northern England. Methuen, London, 213 pp.

Mills, H.H., Brakenridge, G.R.,Jacobson, R.B., Newell, W.L., Pavich, M.J. and Pomeroy, J.S., 1987. Appalachian mountains and plateaus, In: W.L. Graf (Editor), The Geomorphic Systems of North America. Geol. Soc. of Am., Cent. Spec. Vol. 2: 5-50.

Mitchell, G.F., 1985. The preglacial landscape. In: The Quaternary history of Ireland. Academic Press, London, pp. 17-37.

Mixon, R.B. and Newell, W.L., 1977. Stafford fault system: structures documenting Cretaceous and Tertiary deformation along the Fall Line in northeastern Virginia. Geology, 5: 437-440.

Mixon, R.B. and Newell, W.L., 1982. Mesozoic and Cenozoic compressional faulting along the Atlantic Coastal Plain margin, Virginia: central Appalachian geology, NE-SE. Geol. Soc. Am., Guideb., Field Trip, 2: 29-54.

Naylor, D. and Shannon, P., 1982. Geology of offshore Ireland and West Britain. Graham and Trotman, London, 161 pp.

Nelson, K.D., McBride, J.H., Arnow, J.A., Oliver, J.E., Brown, L.D. and Kaufman, S., 1985. New COCORP profiling in the southeastern United States, part II: Brunswick and east coast magnetic anomalies, opening of the North-central Atlantic Ocean. Geology, 13: 718-721.

Newell, W.L., 1985. Architecture of the Rappahannock estuary. Neotectonics in Virginia. In: M. Morisawa and J.T. Hack (Editors), Tectonic Geomorphology. Allen and Unwin, London, pp. 321-342.

Ollier, C.D., 1985. Morphotectonics of passive continental margins: introduction. Z. Geomorphol., N.F., Suppl. Bd., 54: 1-9.

Oliver, J., Cook, F. and Brown, L., 1983. COCORP and the continental crust. J. Geophys. Res., 88 (B4) 3329-3347.

Owens, J.P., 1983. The northwestern Atlantic Ocean margin. In: The Phanerozoic Geology of the World, II, The Mesozoic B. Elsevier, Amsterdam, pp. 33-60.

Prowell, D.C., 1988. Cretaceous and Ceonozoic tectonism on the Atlantic coastal margin. In: Sheridan, R.E. and Grow, J.A. (Editors), The Geology of North America, I,2: The Atlantic margin. Geol. Soc. Am., pp. 557-564.

Rankin, D.W., Espenshade, G.H. and Neuman, R.B., 1972. Geologic map of the west half of the Winston-Salem Quad., N.C., VA., Tenn. U.S. Geol. Surv., Misc. Geol. Invest., Map I-709 A.

Reinhardt, J., Prowell, D.C. and Christopher, R.A., 1984. Evidence for Cenozoic tectonism in the southwest Georgia Piedmont. Geol. Soc. Am., Bull., 95: 1176-1187.

Roberts, D.G., 1974. Structural development of the British Isles, the continental margin and the Rockall Plateau. In: Burk, C.A. and Drake, C.L. (Editors), The Geology of Continental Margins. Springer, New York, pp. 343-359.

Roberts, D.G., Hunter, P.M. and Laughton, A.S., 1979. Bathymetry of the Northeast Atlantic continental margin around the British Isles. Deep Sea Res., 26 A: 417-428.

Secor, D.T., Snoke, A.W. and Dallmeyer, R.D., 1986. Character of the Alleghanian orogeny in the southern Appalachians: Part III, regional tectonic relations. Geol. Soc. Am., Bull., 97: 1345-1353.

Sevon, W.D., 1985. Pennsylvania's polygenetic landscape. Harrisburg Area Geol. Soc., 4th Annu. Field Trip, Guideb., 55 pp.

Sevon, W.D., Potter, N., Jr. and Crowl, G.H., 1983. Appalachian peneplains: a historical review. Earth Sc. Hist., 2: 156-164.

Shackleton, R.M., 1953. The structural evolution of North Wales. Liverpool Manchester Geol. J., 1: 261-297.

Smythe, D.K., Dobinson, A., Mcquillin, R., Brewer, J.A., Matthews, D.H., Blundell, D.J. and Kelk, B., 1982. Deep structure of the Scottish Caledonides revealed by the MOIST reflection profile. Nature, 299: 338-340.

Thomas, M.D., 1983. Tectonic significance of paired gravity anomalies in the southern and central Appalachians. Geol. Soc. Am., Mem., 158: 113-124.

Vetter, V.R. and Meissner, R.D., 1979. Rheologic properties of the lithosphere and applications to passive continental margins. Tectonophysics, 59: 367-380.

Walsh, P.R., Collins, P., Ijtaba, M., Newton, J.P., Scott, N.H. and Turner, P.R., 1980. Palaeocurrent directions and their bearing on the origin of the Brassington formation (Miocene-Pliocene) of the southern Pennines, Derbyshire, England. Mercian Geol., 8: 47-62.

White, W.A., 1950. Blue Ridge front: a fault scarp. Geol. Soc. Am., Bull., 61: 1309-1346.

Woodland, A.W. (Editor), 1971. The Llanbedr (Mochras Farm) borehole. Inst. Geol. Sci., H.M.S.O., London, Rep. 71/18, 115 pp.

Zoback, M.L. and Zoback, M.D., 1980. State of stress in the conterminous United States. J. Geophys. Res., 85 (B11): 6113-6156

Geomorphology, 2 (1989) 71-97
Elsevier Science Publishers B.V., Amsterdam — Printed in The Netherlands

Neotectonism along the Atlantic Passive Continental Margin: A Review

THOMAS W. GARDNER

Department of Geosciences, Pennsylvania State University, University Park, PA 16802 (U.S.A.)

(Received May 20, 1989; accepted after revision June 13, 1989)

Abstract

Gardner, T.W., 1989. Neotectonism along the Atlantic passive continental margin: A review. In: T.W. Gardner and W.D. Sevon (Editors), Appalachian Geomorphology. Geomorphology, 2: 71-97.

An extensive body of geologic data including the modern state of stress, historical seismicity, surface and subsurface stratigraphy, numerical models of crustal deformation, surficial geomorphic systems, and historical precise leveling and tidal gauge records constrain the style and rate of neotectonic deformation for the Appalachians and Atlantic passive continental margin. There are two major styles of neotectonism in the eastern United States. The northeastern United States is dominated by isostatic uplift and northward migration of peripheral bulge collapse in response to deglaciation. This locally rapid, but decreasing rate of deformation is superimposed upon slower, long-term deformation along the Atlantic margin. Most of the long-term, continental margin deformation is attributed to lithospheric flexuring in response to sediment loading in sedimentary basins (especially the Baltimore Canyon Trough and Carolina Trough), isostatic deformation in response to continental denudation and water loading of the shelf, and stress from far-field plate tectonic sources.

Significant deformational features include an uplift anomaly near Cape Fear, N.C.; northward and southward tilting of the Coastal Plain into the Salisbury and Southeast Georgia Embayments respectively; seaward tilting of the Coastal Plain/Piedmont, and a complex pattern of postglacial uplift and later subsidence in the northeast. Estimates of vertical crustal velocities for similar locations vary over several orders of magnitude. Measurement interval bias and systematic leveling errors may account for some of the discrepancies. Evidence for periodic deformation in the eastern United States is substantial and it is possible that historic data indicate a period of accelerated deformation along the Atlantic continental margin.

Introduction

"In the ideal case [of the geographic cycle], which provides the general scheme with respect to which all other cases are classified, the land mass once uplifted is supposed to stand still until it is worn down flat. This supposition is so artificial and does so great violence to much that is known as to the behavior of the earth's crust that some students are therefore disposed to discard the scheme of the cycle ..."

William Morris Davis, 1902.

With that statement Davis acknowledged the dynamic nature of the earth's crust. In a later article Davis (1932) allowed for slow, continuous deformation of the earth's crust, suggesting eastern Canada as a possible location. Lacking the concept of plate tectonics (and therefore passive continental margins) as well as our increasingly detailed measurements of vertical elevation changes and seismicity in eastern North America, Davis could only conjecture about the style of neotectonic deformation that characterizes the eastern United States. Nevertheless, he was well aware of the existence and significance of Cenozoic tectonism in the evolution of Appalachian landscapes (Davis, 1889).

It is the purpose of this paper to review our

understanding of neotectonic activity in eastern North America. Topical areas that will be explored include the modern state of stress; historic seismicity and its inferred cause(s); historic deformation as inferred from releveling surveys and tidal gauge records; and late Cenozoic crustal deformation inferred from geomorphic, structural, and stratigraphic data. The final section will examine the spatial relationship between areas of historical tectonism and seismicity and areas of inferred, longer-term Cenozoic deformation.

In addressing the topic of Appalachian neotectonics, where Appalachian in the broadest sense refers to that land area draining to the Atlantic Ocean (Fig. 1), several problems of varying significance must be resolved. The first problem involves a definition of neotectonics, and as such, determines the scope of the paper, especially its temporal boundaries. Most authors have given arbitrary age limits by defining neotectonics as the study of all late Cenozoic deformation (Vita-Finzi, 1986; Summerfield, 1987), or the post-Miocene structural history which is generally limited to epeirogenic movements (American Geological Institute, 1977), or even further restricted to distinguish post-Pleistocene isostatic adjustment from neotectonics (Doornkamp, 1986). A less restrictive definition establishes neotectonism as beginning with the onset of the contemporary stress field (Blenkinsop as reported in Hancock and Williams, 1986), or with establishment of the present-day plate boundaries and motions (Hancock and Williams, 1986).

The implicit cause and effect relationship in the latter definition is intuitively appealing, but may be difficult to employ if the history of the contemporary stress field(s) or plate motion(s) is poorly constrained. This is generally the case for eastern North America where initiation of the modern stress field and even its principle stress directions are currently debated. Although the general properties of the modern stress regime could have been established soon after the opening of the North Atlantic around 200 Ma (Poag and Valentine, 1988), changes in spreading directions at about 80 and 60 Ma (Pitman and Talwani, 1972; Condie, 1982) and thus the strike of oceanic transform faults (Sykes, 1978), suggest that the modern stress field was established some time after the Late Cretaceous. Cretaceous and Cenozoic Atlantic Coastal Plain faulting (Prowell, 1988) indicates that regional compression has existed since Late Cretaceous, although the exact magnitude and direction may have varied. Analysis of faulting in eastern North America led York and Oliver (1976) to conclude that the stress field has varied spatially and/or temporally since the Cretaceous. Metcalf (cited in De Boer et al., 1988) used plate reconstructions and analyses of post-Paleozoic fractures in eastern North America to suggest the last major variation in the regional stress field could have occurred post 7 Ma. It apparent that the existing data are insufficient to resolve fully either the modern state of stress or its change through time for all parts of the eastern United States. Therefore, I will limit discussion of Appalachian neotectonism to the late Cenozoic (the last suspected variation in the stress field) except for the reverse faulting along the Coastal Plain. I will also include data for Pleistocene and Holocene isostatic, glacial rebound because neotectism should be inclusive as to cause.

A second major problem for Appalachian neotectonic studies involves the determination of vertical movement relative to a nonstationary datum, such as sea-level or adjacent rock bodies as in the case of faults. This problem is especially acute when the magnitude of vertical elevation change of the datum is nonlinear with respect to time and is equal to or greater than the soughafter neotectonic movement. For example, do elevated transgressive–regressive marine sequences reflect local or regional tectonically/isostatically induced movement or eustatic sea-level fluctuations or both? Relative sense of motion of adjacent rock bodies or tilting of geomorphic surfaces may be somewhat easier to infer from structural, strati-

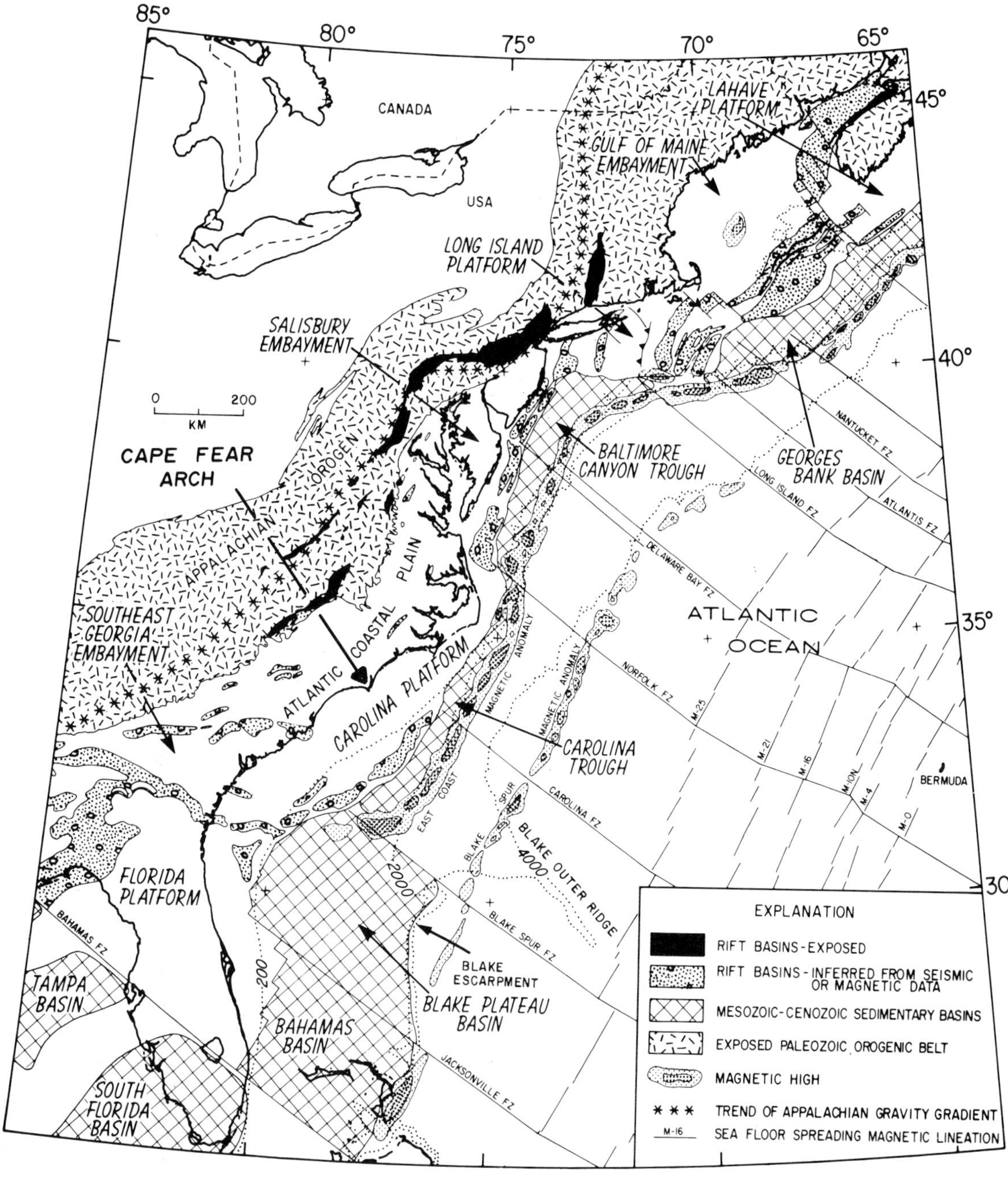

Fig. 1. Generalized geologic map of the eastern United States illustrating the primary structural elements of the Atlantic passive continental margin. Major Mesozoic and Cenozoic basins and arches along the Atlantic margin that play an important role in neotectonic deformation are identified (modified Grow and Sheridan, 1988).

graphic, or topographic data, but often the timing and rate of deformation are poorly constrained. In essence, it is difficult to determine the tectonic deformation signal from other "geologic noise" on passive continental margins. The problem becomes even more complex when attempts are made to apportion vertical deformation to various causes ranging from long-term thermal contraction and lithospheric flexuring to isostatic adjustments from (de)glacation and erosion and sediment loading along the continental margin.

Neotectonic movements have been inferred from several types of data that will be discussed in detail later. These include direct evidence such as fault displacements (although no surface displacement has been directly related to historic earthquakes in the eastern United States), topographic releveling surveys, tidal gauge records, deformed marine and fluvial terraces, and marine/estuarine deposits. More indirect evidence has also been used to infer neotectonic movement including, but not limited to, historic seismic activity and geomorphic properties such as landslides, stream profiles, hillslope curvature, drainage basin morphometry, and denudation rates. Finally, thermal and mechanical models have explored crustal deformation in response to thermal events, sediment and ice loading/unloading and applied stress from plate motions. These lines of evidence, however consistent or ambiguous, constrain the nature of neotectonism along the passive continental margin of the Appalachians.

Several fundamental questions about Appalachian neotectonism will be addressed systematically here. What is the modern state of stress in the crust and its cause(s) and is it consistent with the historic seismic record? What are the distribution and cause(s) of historic seismicity? What are the limitations of that data and how do those limitations affect questions concerning earthquake hazards and long-term stationarity of seismicity? Is the short-term vertical deformation record from releveling surveys and tidal gauge data spatially consistent with generally defined seismic zones? Finally and most importantly, is there consistency in the spatial pattern, magnitude and rate estimates of neotectonic deformation among the long-term record as inferred from stratigraphic and geomorphic data, the short-term elevation change data, and historical seismicity? Analysis of the modern state of stress offers an appropriate starting point.

Modern state of stress

Data for determining the principal stress orientation for the modern stress field in the eastern United States include four distinct types: (1) earthquake focal mechanism solutions for some instrumentally recorded earthquakes; (2) wellbore breakouts and elongation measurements; (3) in-situ stress measurements; and (4) geologic indicators such as fault slip orientations. There is a general consensus that the principle stress is compressive and nearly horizontal in the eastern United States. However, the strike of this maximum horizontal compression is more controversial. The maximum horizontal compression has been determined to strike generally northeast to east-northeast (Fig. 2) for the eastern United States (Sbar and Sykes, 1973; Sykes, 1978; Plumb and Cox, 1987; Zoback and Zoback, in press) and New England (Gephart and Forsyth, 1985). Alternatively, a zone of generally west-northwest maximum horizontal compression (Fig. 2) has been delineated along the Atlantic seaboard (Zoback and Zoback, 1980, before revision by their in press paper; Behrendt et al., 1981; Yang and Aggarwal, 1981; Wentworth and Mergner-Keefer, 1983) extending as far inland as the Appalachian Front (Seeber and Armbruster, 1988).

Determinations of the fundamental cause(s) of the modern stress field in the eastern United States has been elusive, although several reasonable causes have been proposed. Remnant stress, although a possible source of stress, is not considered to greatly influence modern stress measurements (Sbar and Sykes, 1973). Zoback and Zoback (1980, in press) have suggested that the overall uniformity of the midplate stress pattern indicates a far-field plate

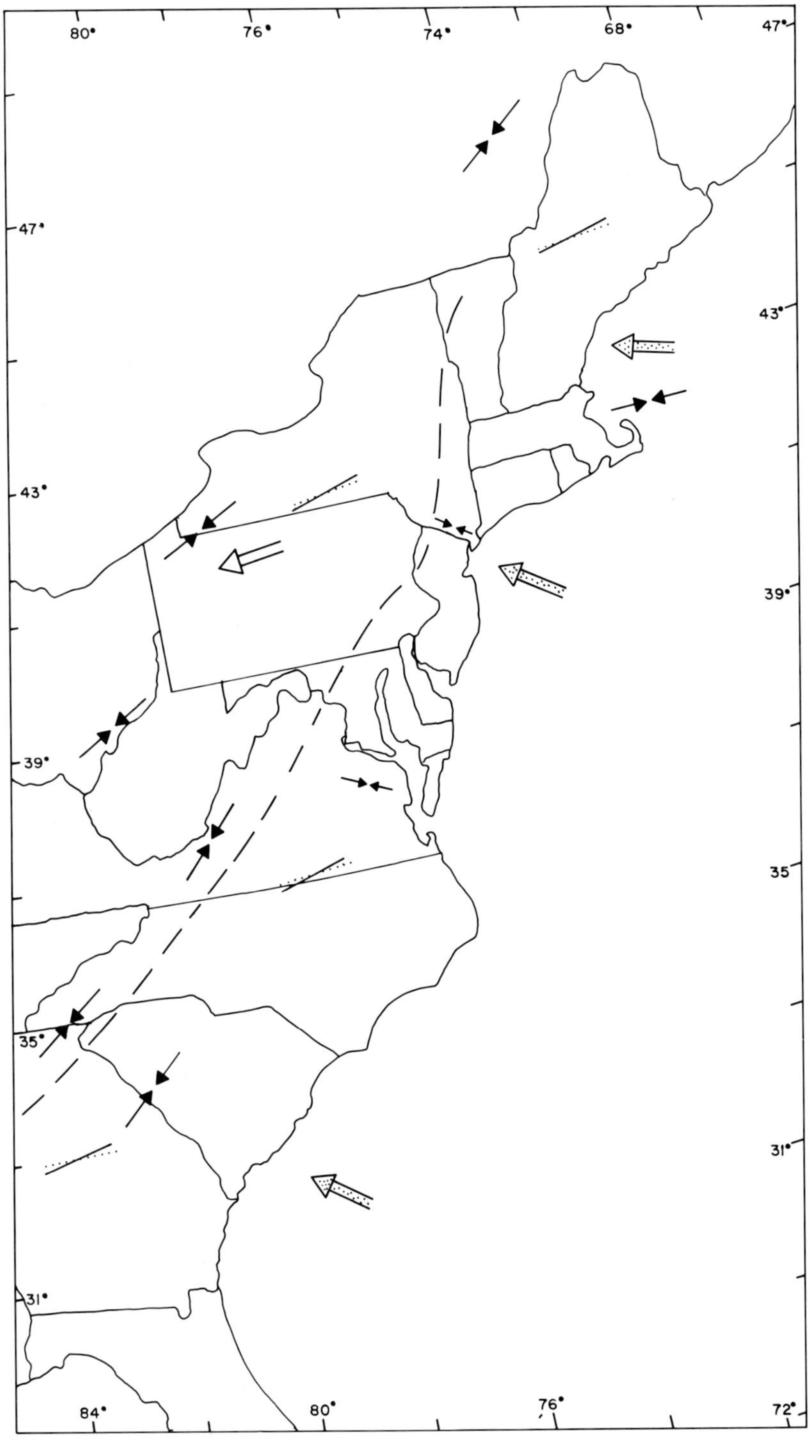

Fig. 2. Map of the eastern United States (Atlantic drainage) illustrating the orientation of the maximum horizontal compression (inward pointing arrows), direction of absolute plate motion of North America (large open arrow), and direction perpendicular to offshore magnetic anomalies (large stippled arrows). For comparative purposes maximum horizontal compressive stress predicted from ridge-push (dotted lines) and drag resistance to absolute plate motion (solid lines) are also shown. The controversial northwest–southeast oriented maximum horizontal stress province along the Atlantic seaboard (heavy-dashed line) is also indicated (modified from data in Zoback and Zoback, 1980, in press; Yang and Aggarwal, 1981; Seeber and Armbruster, 1988).

tectonic source, most likely ridge-push or asthenospheric drag. Asthenospheric drag is caused by resistance from the viscous mantle to motion of the thick, continental lithosphere and is driven by mantle flow patterns. Asthenospheric drag could produce the observed horizontal compressive stress that is in excess of stress from lithostatic loading (Voight, 1969) in the eastern United States (Sbar and Sykes, 1973). The ridge-push force is a gravitational force derived from lateral changes in lithospheric thickness and density that result from change in elevation, composition, and temperature of the lithosphere away from the ridge. Although both ridge-push and asthenospheric drag give similarly oriented stress fields that are compatible with the modern stress field in the mid-continent (Fig. 2), Zoback and Zoback (in press) suggest ridge-push as the primary source for the uniform, northeasternly directed stress field. However, ridge-push is also invoked by Yang and Aggarwal (1981) and Wentworth and Mergner-Keefer (1983) to support a separate northwesternly-oriented stress field along the Atlantic margin. This controversy arises because interpretation of earthquake focal mechanisms can be ambiguous (e.g., Muncey and Bollinger, 1985) and because of the importance, timing, and inferred sense of motion on northeast striking, Cretaceous and Cenozoic faults along the Atlantic seaboard. In noting that the maximum horizontal stress is better correlated with the orientation of the Paleozoic structural fabric, Seeber and Armbruster (1985) suggested that the Appalachians and Atlantic seaboard could have unique stress regimes. In essence, a compressive stress field oriented obliquely to the Appalachian fabric could produce a maximum compression that is rotated by the anisotropic structure to be perpendicular to the compliant direction, generally west-northwest (Zoback and Zoback, 1980). It is apparent that the available data is insufficient to determine the precise orientation of the modern, maximum horizontal compression for all parts of the eastern United States.

Historic seismicity – Distribution, trends, and causes

Compilations of historic seismicity are based on records of variable lengths; 1534–1959 for eastern Canada (Smith, 1962, 1966); 1534–1971 for the eastern United States (York and Oliver, 1976); 1754–1970 for the southeastern United States (Bollinger, 1973a); 1800–1972 for the eastern United States (Hadley and Devine, 1974); 1925–1980 for the central and eastern United States (Dewey and Gordon, 1984); and 1970–1985 for the eastern United States (Seeber and Armbruster, 1988). These maps include many types of data with varying reliability and accuracy ranging from felt reports from archival data in newspapers for pre-instrumentation events to relocated hypocenters from regional seismic networks. Attempts to establish long-term trends in seismicity have justified inclusion of these diverse data.

Although the uneven coverage of seismicity may be partly responsible for the observed distribution, it is apparent that earthquakes are not evenly scattered along the Appalachians. The distribution of earthquakes has been interpreted to indicate variable numbers and intensities of discrete seismic zones and orientations of seismic trends. Generally acknowledged seismic zones (Fig. 3) include: Charlevoix (Basham et al., 1979); western Quebec–Ottawa (Sykes, 1978; Basham et al., 1979); Attica, N.Y. (Basham et al., 1979); Passamaguoddy Bay, Maine (Barosh, 1981; Ebel, 1985). Boston–Cape Ann, Mass. (Sykes, 1978); Moodus, Conn. (Block et al., 1979; Ebel et al., 1982); Ramapo, N.Y.–N.J. (Ratcliffe, 1971; Aggarwal and Sykes, 1978); Lancaster County, Pa. (Stockar, 1986; Seeber and Armbruster, 1988); northern Virginia–Maryland (Bollinger, 1973a); central Virginia (Bollinger, 1973a); Giles County, Va. (Bollinger and Wheeler, 1983); southern Appalachian (Bollinger, 1973a); and Charleston, S.C. (Bollinger, 1973a; Talwani, 1982; Seeber and Armbruster, 1988).

In the eastern United States, all instrumen-

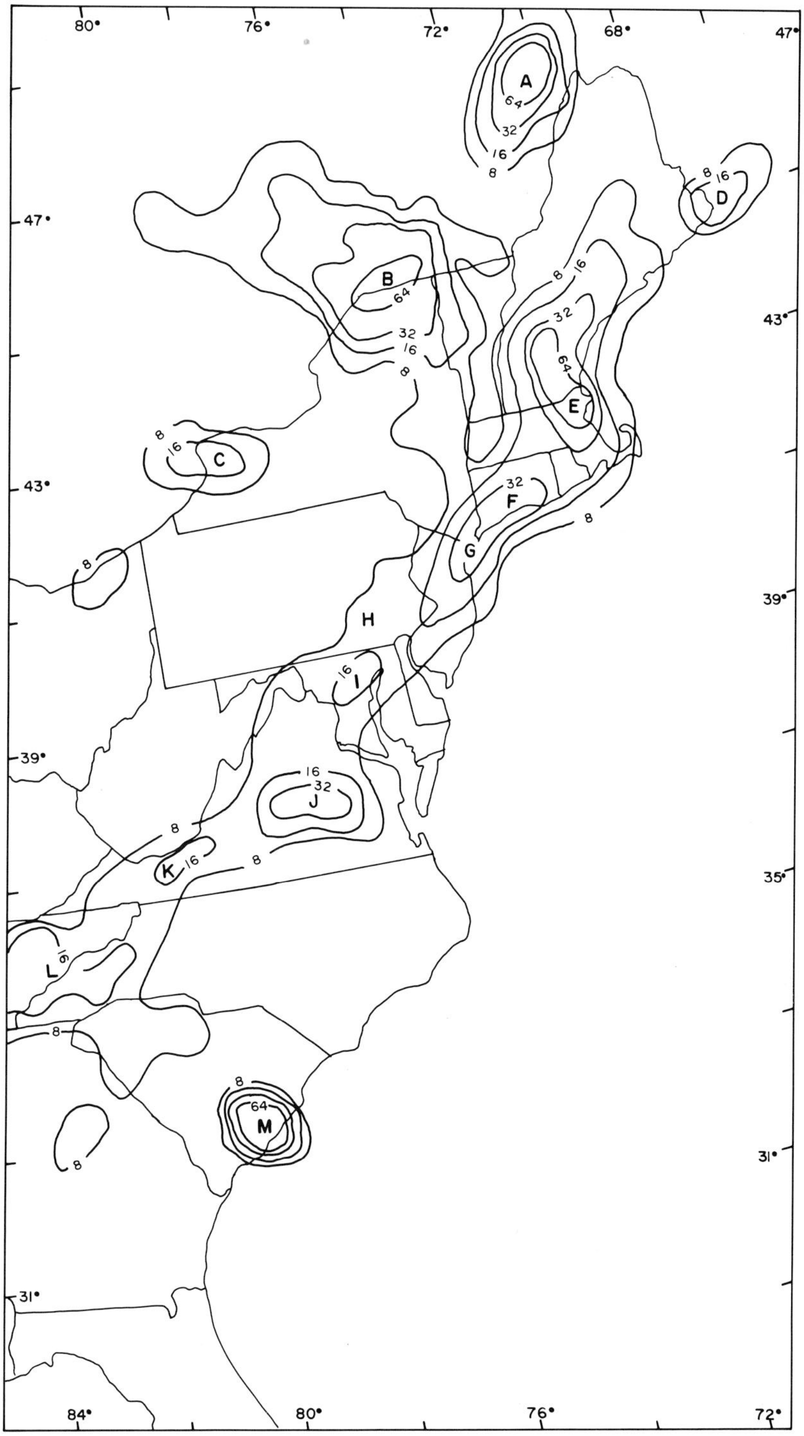

Fig. 3. Map showing seismic frequency for the eastern United States and adjacent Canada. Generalized seismic frequency contours (solid lines) represent the areal distribution of earthquake epicenters with epicentral intensity of Modified Mercalli III and greater, as indicated by the total number per 10^4 km^2 during the period 1800–1972 (from Hadley and Devine, 1974). Location of seismic zones indicated by letter, Charlevoix (*A*), western Quebec-Ottawa including the Adirondacks (*B*), Attica (*C*), Passamaquoddy Bay (*D*), Boston-Cape Ann (*E*), Moodus (*F*), Ramapo (*G*), Lancaster (*H*), northern Virginia-Maryland (*I*), central Virginia (*J*), Giles County (*K*), southern Appalachian (*L*), and Charleston (*M*).

tally recorded earthquakes have occurred in the brittle crust within the uppermost 25–30 km of the lithosphere (Hatcher et al., 1987; Seeber and Armbruster, 1988). For example, focal-depth distributions indicate that 90% of the earthquakes occurred above depths of 11, 13, and 17 km in the Charleston, central Virginia, and Giles County seismic zones, respectively (Costain et al., 1987).

Earthquake intensity and frequency are more variable across seismic zones. Large earthquakes such as the 1886 Charleston earthquake (Modified Mercalli intensity IX–X) and the 1929 Grand Banks earthquake (M=7.2) have not occurred in all seismic zones, but most seismic zones have had earthquakes as large as Modified Mercalli intensity VI or VII. Furthermore, analysis of earthquakes in the Charleston and Cape Ann–Boston area (Seeber and Armbruster, 1988) and movement along Mesozoic and Cenozoic faults in the southeastern United States (Wentworth and Mergner-Keefer, 1983) suggest that seismicity is not stationary at the time scale of the historic record or for the long-term. Conversely, a review of historical seismicity in the northeastern United States (Yang and Aggarwal, 1981) and analysis of seismic network data in the southeastern United States (Wheeler and Bollinger, 1984) suggest that on a time scale of decades, seismicity is spatially stationary. Thus, potential variation in large-scale patterns of seismicity, the infrequency of large events, and the relatively short record create uncertainty in earthquake hazard prediction as well as inferences about the relationships among historical seismicity, seismic source regions, and long-term, neotectonic deformation.

Seismic source for earthquakes in some seismic zones is much more tightly constrained than in others. For example, the Ramapo fault zone, a northeast trending boundary fault on the northwest side of the Newark Basin (Ratcliffe, 1971; Aggarwal and Sykes, 1978), has been convincingly demonstrated to be the seismic source (with a northwest–southeast maximum horizontal compression) for many of the earthquakes in that seismic area. Similar, but more poorly constrained, northeast trending, Mesozoic basin boundary faults in the Delmarva Peninsula defined from seismic reflection (Hansen, 1988) and aeromagnetic anomalies (Higgins et al., 1974) have been invoked as seismic sources for earthquakes in that area (Sbar et al., 1975). The western boundary of the northeast trending Richmond Basin is spatially consistent with one seismic area in the Central Virginia seismic zone. Seismic source for the Charleston seismic zone is more poorly constrained, ranging from buried northeast trending (possibly Mesozoic) faults (Behrendt et al., 1981; Bollinger, 1983) or a basement decollement (Seeber and Armbruster, 1981) to northwest trending steeply dipping reverse faults identified from aeromagnetic, gravity and geomorphic data (Talwani, 1982). Recent work in other areas suggests that some earthquakes are generated by structures trending across the strike of the Gettysburg and Newarks basins (Quittmeyer et al., 1985; Stockar, 1986). Reactivation of Iapetus (Precambrian or Cambrian) normal faults that trend 20° to the detached structure in the southern Appalachian may be the seismic source in the Giles County, Va. seismic zone (Bollinger and Wheeler, 1983).

Convincing evidence has also been demonstrated for a spatial association between seismicity and mafic/ultramafic igneous bodies as identified from outcrop, gravity, and magnetic surveys (Kane, 1977; McKeown, 1978). Stress amplification along boundaries between contrasting lithologic materials (especially serpentinized masses) is invoked as the causal mechanism. Induced, shallow seismicity from reservoir loading (Seeber and Armbruster, 1988, among others) and hydroseismicity up to 20 km deep caused by a transient increase in hydraulic head in groundwater recharge basins proximal to fractured crust (Costain et al., 1987) also contribute to some seismicity in the southeastern United States. Unfortunately, lack of surface rupture associated with any historical earthquake activity in the eastern United States has made correlation to specific structure difficult.

There is a definite correlation between seismicity and some Mesozoic structures which have been implicated in neotectonic movements along the Appalachians (Wentworth and Mergner-Keefer, 1983). However, a lack of seismicity along other Mesozoic structures (e.g., the Gettysburg Basin) may simply reflect an insufficiently long sample interval for seismicity or that not all unhealed Mesozoic normal faults are seismically active. Furthermore, an apparent lack of Mesozoic structural sources for some earthquakes indicate that those types of structures may not be a necessary condition for seismicity.

In addition to identifying seismicity with specific source regions or faults, numerous researchers have extended seismic zones into elongate trends in an attempt to provide fundamental structural and tectonic explanations for intraplate seismicity in the eastern United States. These trends can be generally grouped into a set parallel or subparallel to the northeast structural grain of the Appalachians (Woollard, 1958, 1965; Bollinger, 1973a; Wheeler and Bollinger, 1984; Hatcher et al.,

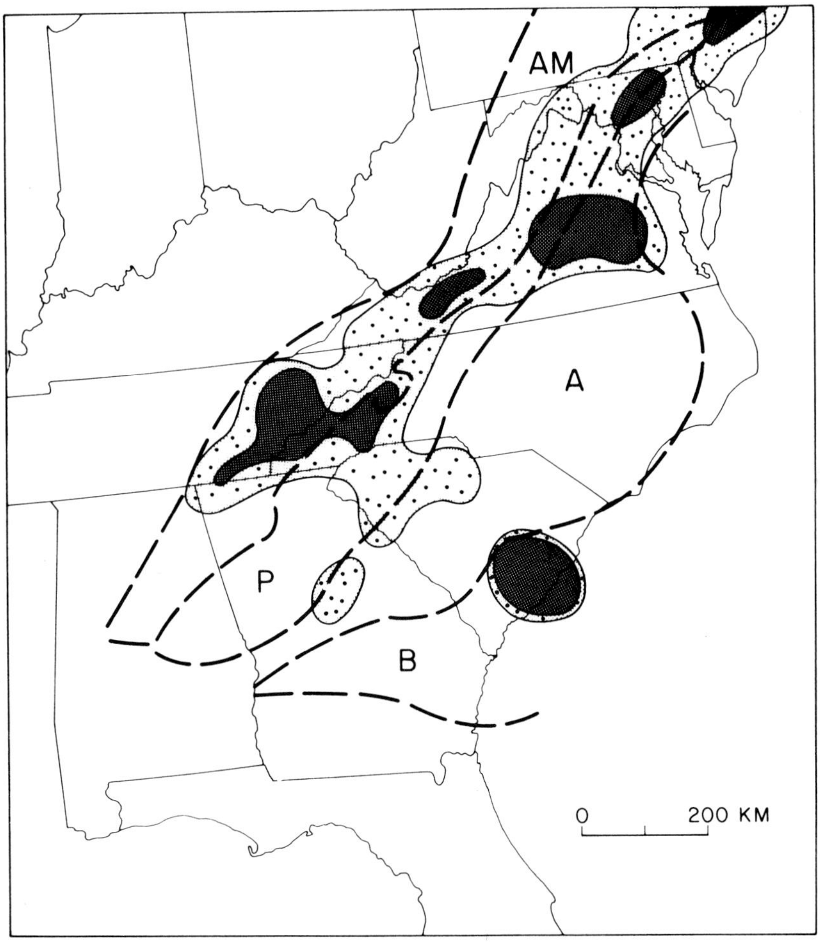

Fig. 4. Historical seismicity (1800–1972) and suspect terrane boundaries for the southeastern United States. Stippled areas indicate four or more earthquakes per 10^4 km^2, dark grey areas indicate sixteen or more earthquakes per 10^4 km^2. Dashed lines show approximate suspect terrane boundaries: Appalachian miogeocline (*AM*), Piedmont (*P*), Avalon (*A*), and Brunswick (*B*) (after Wheeler and Bollinger, 1984).

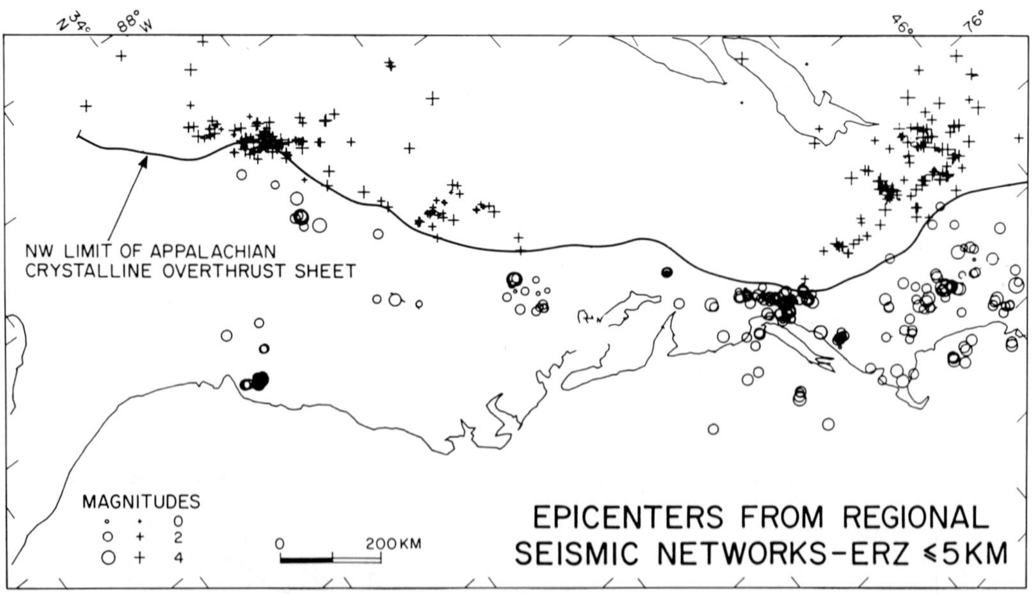

Fig. 5. Earthquake epicenters from regional seismic networks along the Appalachians. Thick line is the northwest limit of crystalline rocks in the Appalachian allochthon. Earthquakes located in the cratonic basement below the detachment (crosses) and earthquakes located in the allochthonous crystalline slab above the detachment (circles) delineate two northeast-southwest trending seismic zones (after Seeber and Armbruster, 1988).

1987; Seeber and Armbruster, 1988) and a nearly perpendicular set trending northwest (Sbar and Sykes, 1973; Bollinger, 1973a; Fletcher et al., 1978; Sykes, 1978). Causal explanations for these trends are less easily categorized, but most rely on (re)activation of pre-existing zones of crustal weakness or unhealed faults of Mesozoic age.

Northeast trending structural features (Fig. 4) have been invoked to explain inferred northeasternly seismic trends. Along the Atlantic margin early Paleozoic sutures separate suspect terranes that were accreted to the proto-North Atlantic margin (Williams and Hatcher, 1982). These sutures, acting as pre-existing zones of weakness or as zones of stress amplification causes by lithologic contrast, have been correlated with seismicity (Wheeler and Bollinger, 1984; Hatcher et al., 1987). A discontinuous zone of seismicity consisting of two distinct, northeast trending belts paralleling the Appalachian Front (Fig. 5), has been identified by Seeber and Armbruster (1988). Seismicity is confined to the Precambrian basement west of the Front, but east of the Front seismicity appears to be confined to the allochthonous, crystalline slab above the Appalachian master decollement. The structure along the Appalachian Front may control the spatial distribution of seismicity by juxtaposing contrasting lithologies and thus causing stress concentrations (McKeown, 1978; Seeber and Coles, 1984) in which case the Appalachian Front would separate distinct seismic areas without being a seismic source itself (Bollinger et al., 1985). Alternately, the Appalachian décollement could be active and possibly seismogenic (Seeber and Armbruster, 1981).

A thorough study of intraplate seismicity, alkaline magmatism, and continental fragmentation led Sykes (1978) to convincingly propose continentward extensions of northwest trending, oceanic, fracture zones (Fig. 6) as sources for some of the intraplate seismicity in the eastern United States and Canada. Seismic activity in the Charleston, S.C. and central Virginia seismic zones appears to be landward of the Blake Spur and Norfolk fracture zones,

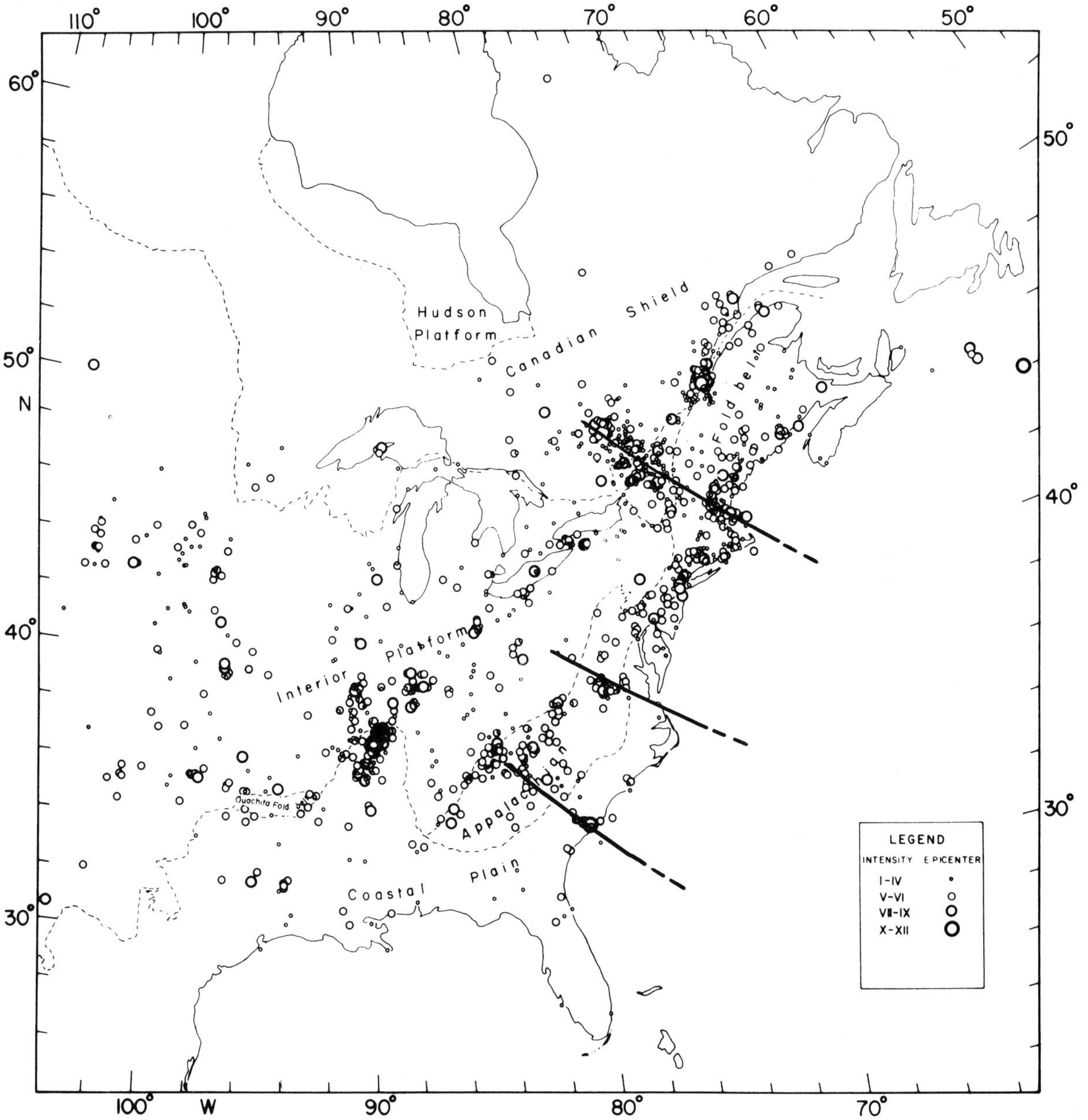

Fig. 6. Historical seismicity (1534–1971) and continentward projects of the Blake Spur and Norfolk fracture zones. A small circle passing through the New England seamount chain and extending into New England is drawn about the center of rotation for the Mesozoic movement of North America with respect to Africa (after Sykes, 1978). See Fig. 1 for location of fracture zones in oceanic crust.

while the Boston–Cape Ann and western Quebec–Ottawa seismic zones fall along a small circle extension of the New England seamount chain. However, more recent work on the location and continentward extension of these oceanic fractures (Klitgord et al., 1983) suggests a less rigorous correlation with these seismic zones (Seeber and Armbruster, 1988).

It is apparent that these causal inferences cannot individually explain all of the historical seismicity, but they do account for some of it. It is appropriate to ask then, if our knowledge of historical seismicity in eastern North America is consistent with neotectonic deformation as inferred from crustal deformation data.

Vertical neotectonism

General attributes

An extensive body of geologic data for the eastern United States has produced one of the most detailed reconstructions of crustal deformation along a passive continental margin. Significant attributes of this geologic framework that constrain the style and rate of neotectonism include: the basin and arch architecture along the continental margin (Fig. 1) and the complex pattern of basin filling; generally northeast trending Cretaceous and Cenozoic reverse faults along the inner Coastal Plain; the distribution, elevation, and age of Coastal Plain and continental shelf terraces, shorelines, and associated marine deposits; the controversial nature of extensive (in some cases Appalachian wide) plantation surfaces and associated variation in denudation and sediment delivery rates; numeric models of crustal response to mass transfer and thermal history; an expanding body of surficial, geomorphic data from fluvial, hillslope, and pedogenic systems associated with some historical seismic zones; and short-term tidal gauge records and topographic leveling surveys. The rate and spatial distribution of neotectonic deformation along the passive continental has been variously apportioned to several fundamental causes including: isostatic adjustments to ice volume (Walcott, 1972a), water loading of the continental shelf (Bloom, 1967; Clark and Bloom, 1979; Cronin et al., 1981), and continental denudation (Ahnert, 1970; Cronin et al., 1981; Pavich, 1985); peripheral bulge movements in response to ice volume (Clark et al., 1978; Dillon and Oldale, 1978; Newman et al., 1980; Peltier, 1986); crustal flexure from sediment loading (Walcott, 1972b, Beaumont, 1978, 1979; Newman et al., 1980; Cronin, 1981; Steckler et al., 1988); thermally induced expansion and contraction (Steckler and Watts, 1978; Sawyer et al., 1982; Sawyer, 1988); geoidal deformation due to gravitational fluctuations (Whitcomb, 1976) and mass redistribution from earth rotational variations which can significantly affect the sea-level datum (Morner, 1980); and stress from plate tectonic sources (Brown et al., 1972; Sykes, 1978; Prowell, 1988, among others).

Neotectonism is apparently dominated by two distinct domains. In the northeastern United States local and rapidly decreasing, short-term deformation is occurring in response to glacial induced isostasy and peripheral bulge migration. This is superimposed along the passive continental margin on a long-term, slower deformation caused by sediment loading in depositional basins, isostatic uplift from continental denudation, water loading of the shelf, and stress from plate tectonic sources. These processes have produced a complex pattern of neotectonic deformation. The most extensive data for determining rates and spatial distribution of neotectonism come from coastal terraces, shorelines, tidal gauge records and releveling surveys along the Coastal Plain and as such offers an appropriate starting point.

Atlantic margin

To calculate a vertical crustal velocity or uplift rate the age of a deposit and its elevation with respect to a datum must be known. For the Atlantic margin, sealevel is commonly chosen as the datum, but because sealevel has fluctuated through time, the elevation of sealevel at the time of deposition must be known. Uplift rate is then calculated from the equation:

$$U=\frac{H+D-SL_{p}}{A} \qquad (1)$$

where U is the vertical crustal velocity, H is the elevation of the deposit above modern sealevel,

D is the water depth of the deposit at the time of deposition, SL_p is the eustatic sealevel (positive for sealevel elevations above modern), and A is the age of the deposit. Of these parameters H can be measured with a high degree of accuracy relative to the others. D is less well constrained and depends upon facies reconstructions, and A, if radiometrically determined, is known within statistical limits. SL_p is usually the least well constrained and is dependent upon ice volume changes and sea-floor spreading rates. Unfortunately, local relative sealevel curves and eustatic sealevel curves disagree on the timing, magnitude and rate of post-Miocene and especially late Pleistocene through Holocene sealevel fluctuation for eastern North America (most recently summarized in Fletcher, 1988). Estimates of maximum, late Wisconsinan sealevel lowering range from 50 to 130 m (Fig. 7). Furthermore, age-estimates of the late Wisconsinan low-stand can vary by more than 5000 years (Fig. 7). These large variations in predicted sealevel elevations have seriously complicated attempts to calculate late Pleistocene and Holocene absolute crustal movements using eqn. (1).

Post-Miocene marine terraces and shorelines along the Coastal Plain in the southeastern United States that have been correlated with eustatic sealevel maxima (Blackwelder, 1981a; Cronin et al., 1981) suffer the same problem in estimating rates of vertical crustal

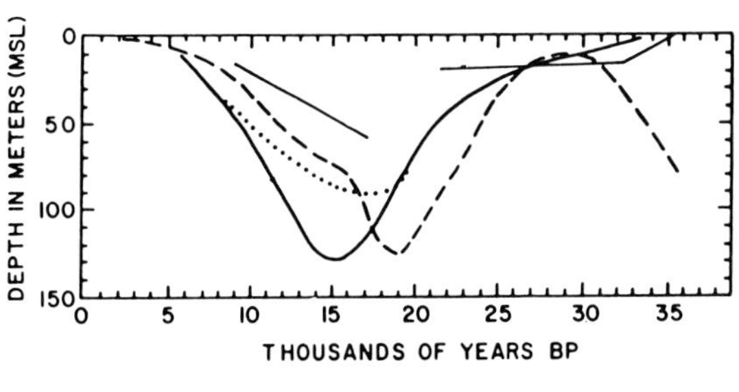

Fig. 7. Post 35 ka sealevel curves based on radiocarbon dates from the western Atlantic and Gulf of Mexico coasts. Data from Milliman and Emery (1968) thick solid curve, Curray (1965) dashed curve, Dillon and Oldale (1978) dotted curve, and Blackwelder et al. (1979) thin solid curve (after Fletcher, 1988).

deformation using eqn. (1). For example, early studies considered either periodic, continental uplift (Kerr, 1875; Shattuck, 1906) or change in eustatic sealevel (Cooke, 1930, 1931) to be the cause of the emergent Coastal Plain terraces. Although more refined data for ice volumes (Braun, this volume), sea-floor spreading rates (Pitman, 1978), and the timing of worldwide sealevel changes have improved estimates for eustatic sealevel curves (Vail and Hardenbol, 1978; Cronin, 1981; Haq et al., 1987), vertical deformation rates may still vary up to nearly an order magnitude depending upon assumed position of eustatic sealevel (Cronin, 1981).

However, given these limitations post-Miocene marine shoreline data together with historic tidal trends and releveling surveys consistently identify several major deformational features along the Atlantic margin (Fig. 1). These features include: an uplift anomaly in the vicinity of Cape Fear, N.C.; a northward tilting of the Coastal Plain north of Cape Fear with maximum long-term subsidence in the vicinity of the Salisbury Embayment; a complex pattern of late Pleistocene through Holocene uplift and subsidence in the northeast; and seaward tilting along the entire Coastal Plain. Although the sense of motion is generally quite consistent among the different types of data, rates of deformation (Table 1) vary over several orders of magnitude with the highest rates calculated from releveling data.

Vertical crustal velocities calculated from marine shoreline and terrace deposits for the Cape Fear area are $\sim 10^{-2}$ mm yr^{-1} relative to modern sealevel, southeast Georgia, or the Salisbury Embayment for the time interval 5 Ma–200 ka (Winkler and Howard, 1977; Blackwelder, 1981; Cronin, 1981; Markewich, 1985). There is some indication for a post 200 ka increase in uplift rate to $\sim 10^{-1}$ mm yr^{-1} (Cronin et al., 1981a; Markewich, 1985). Approaching 1 mm yr^{-1} in the Holocene for the area south of Cape Fear relative to the Salisbury Embayment as suggested from basal, transgressive peats

TABLE 1

Rates of neotectonic, vertical, crustal deformation for the eastern United States

Data type	Vertical crustal velocity (mm yr^{-1})	Source	Qualification[1]
Atlantic margin			
Cape Fear Arch Area			
Marine, tidal, and estuarine deposits and shorelines	2.0×10^{-2}	Winkler and Howard, 1977, fig. 3[2]	Post 3 Ma, relative to southeast Ga.
	7.0×10^{-3}	Winkler and Howard, 1977, fig. 3[2]	Post 1 Ma, relative to southeast Ga.
	6.0×10^{-2}	Winkler and Howard, 1977, fig. 3[2]	Post 100 ka, relative to southeast Ga.
	1.3×10^{-2}	Blackwelder, 1981a	Post 5 Ma, Cape Fear area
	2.0×10^{-2}	Markewich, 1985	3–1.6 Ma, Cape Fear area
	9.0×10^{-2}	Markewich, 1985	Post 200 ka, Cape Fear area
	1.0×10^{-2}–3.0×10^{-2}	Cronin, 1981	Post 3–5 Ma, Cape Fear area
	1.0×10^{-1}–2.0×10^{-1}	Cronin et al., 1981	Post 200 ka, Charleston, S.C. and southeast Va.
	2.0×10^{-1}–4.5×10^{-1}	Cronin et al., 1981	Post 100 ka, north of Cape Fear Arch
	1.0	Cinquemani et al., 1982	Post 8 ka, Savannah, Ga/Charleston relative to N.Y.C./Salisbury Embayment
Tidal gauge	1.7	Fadaie and Brown, 1984	Post 50 yr, New Bern, N.C. relative to Norfolk, Va.
Topographic releveling	9.6	Fadaie and Brown, 1984	Post 50 yr, New Bern, N.C. relative to Norfolk, Va.
	10.0	Balazs, 1974	Post 50 yr, New Bern, N.C. relative to Norfolk, Va.
Cape Fear Arch to New England			
Marine, tidal, and estuarine deposits and shorelines	−2.3	Newman et al., 1980	Post 6 ka, N.Y.C. relative to S.C.
	2.0	Newman et al., 1980	Post 12 ka, N.Y.C. to Salisbury Embayment,
	0.0		assumes 12 ka sealevel 50 m below present
	−10.0	Blackwelder, 1981b	9–12 ka, N.Y.C. relative to S.C.
	−3.8	Dillon and Oldale, 1978	13–15 ka, continental shelf off N.Y.C.
	−1.7	Dillon and Oldale, 1978	9–13 ka, continental shelf off N.Y.C.
		Dillon and Oldale, 1978	0–9 ka, continental shelf off N.Y.C.
Tidal gauge	−2.5	Walcott, 1972a	1940–1962, Chesapeake Bay, assumes 1.5 mm yr^{-1} eustatic sealevel rise
Tidal gauge and Topographic releveling	−2.0	Holdahl and Morrison, 1974	1930–1970, average on Delmarva Peninsula, assumes 1 mm yr^{-1} eustatic sealevel rise
Coast-parallel regional tilt[3] (north side down)			
Marine, tidal and estuarine deposits and shorelines	3.0×10^{-3}	Fletcher, 1988	10 ka, N.Y.C. to Savannah, Ga.
	1.0×10^{-3}	Fletcher, 1988	5 ka, N.Y.C. to Savannah, Ga.
	7.0×10^{-4}	Fletcher, 1988	2.5 ka N.Y.C. to Savannah, Ga.
Continental shelf break	3.0×10^{-3}	Fadaie and Brown, 1984, fig. 1[2]	Post 18 ka, N.Y.C. to Charleston S.C.
Tidal gauge	1.3×10^{-3}	Brown and Oliver, 1976	1940–1962, Del. to Savannah, Ga.
Topographic releveling	3.0×10^{-2}	Brown and Oliver, 1976	~1930–1972, Del. to Savannah, Ga.
Coast perpendicular regional tilt[3] (coast side down)			
Erosion rate	$3.5\text{-}10^{-4}$	Hack, 1980	Modern, Potomac River watershed
Tilt on strata	1.7×10^{-4}	Hack, 1980	100–30 Ma, Washington, D.C. area
Marine shorelines	1.0×10^{-4}–2.0×10^{-4}	Cronin, 1981	Post 3–5 Ma, landward of Carolina Trough
	3.0×10^{-2}	Newman et al., 1987	13 ka, New England (Champlain Sea datum)
	2.5×10^{-2}	Newman et al., 1987	Post 6 ka, Lower Hudson River estuary
	9.3×10^{-2}	Hack, 1980	12 ka, New England (Champlain Sea datum)
Topographic releveling	6.4×10^{-2}	Koteff, 1984	Post 14 ka, New England
	1.0×10^{-2}	Brown and Oliver, 1976, figs. 7,8[2]	~1930–1970, Appalachian Plateau to Coastal Plain at Pennsylvania and Carolinas
Seismic zones			
Charlevoix			
Topographic releveling	7.0	Frost and Lilley, 1966	1920–1964, Lake St. John relative to central Charlevoix

TABLE 1 (*continued*)

Data type	Vertical crustal velocity (mm yr^{-1})	Source	Qualification[1]
Passamaquoddy Bay			
Marine, tidal and estuarine deposits and shorelines	−1.2	Reilinger, 1987	Post 10 ka, Eastport, Maine, relative to southwest Maine from data in Thompson et al., (1983)
	−1.0 to −2.0	Reilinger, 1987	Post 2.5 ka, near Eastport Maine; from data in Belknap et al., (1987)
Isostatic model	−1.0	Peltier, 1986	Modern, coastal Maine
Tidal gauge	−1.3 to −1.8	Reilinger, 1987	1930–1985. Eastport Maine, assumes 1.0–1.5 mm yr^{-1} sealevel rise
Topographic releveling	−1.0 to −2.0	Reilinger, 1987	~1933–1978, Eastport, Maine, relative to inland Maine
	−9.0	Anderson et al., 1984	~1927–1966, Eastport, Maine, relative to inland Maine; from data in Tyler and Ladd (1980)
Adirondack Mountains			
Topography releveling	2.2–2.8	Isachsen, 1975	1955–1973, relative to Saratoga Spring, N.Y.
Topographic releveling[3]	3.0×10^{-2}	Hack, 1980	1955–1973, Lake Champlain to Ticonderoga
Terrace profile[3]	5.0×10^{-2}	Treadwell and Knuepfer, 1988	Postglacial, northeast flank of Adirondack Mountains
Topographic releveling[3]	6.0×10^{-2}	Treadwell and Knuepfer, 1988	1955–1973, East Branch Ausable River
Ramapo			
Marine, tidal and estuarine deposits	-5.0×10^{-1}	Newman et al., 1987	Post 2 ka, within Ramapo fault zone
Giles County			
Fluvial terrace	3.0×10^{-2}–2.0×10^{-1}	Mills, 1986	Post 1 Ma, New River, Va., rate depends on age estimate for terrace
Charleston			
Topographic releveling[3]	3.0×10^{-2}	Poley and Talwani, 1986, fig. 13[2]	1955–1979, Augusta, Ga. to Savannah, Ga.
Topographic releveling	5.0	Balazs, 1974	1932–1963, Charleston, S.C., relative to N.Y.C.
Tidal gauge	−2.3	Hicks, 1978, table 1[2]	1922–1975, assumes 1.5 mm yr^{-1} eustatic sealevel rise
Tidal gauge and topographic releveling	−2.5	Holdahl and Morrison, 1974	1955–1973, assumes a 1.0 mm yr^{-1} eustatic sealevel rise
Topographic releveling	−2.0	Lyttle et al., 1979	1963–1974, down-to-the-coast movement across Stono and Clubhouse Crossroads structures west of Charleston
Marine deposits	-3.0×10^{-3}	Lyttle et al., 1979, fig 1[2]	Post 5 Ma, down-to-the-coast relief on Miocene unit
Southern Appalachian			
Topographic releveling	4.0	Citron and Brown, 1979	1935–1967, across Brevard fault zone, Asheville, N.C.
Inner Coastal Plain			
Fault offset	3.0×10^{-4}–1.5×10^{-3}	Prowell, 1988	Post 110 Ma, average for reverse faults
	1.0×10^{-3}	Wentworth and Mergner-Keefer, 1983	Post 110 Ma, average for reverse faults
Regional average			
Denudation model	2.5×10^{-2}	Pavich, 1985	Post 100 Ma, mid-Atlantic Piedmont, isostatic compensation
Erosion rates	1×10^{-2}	Hack, 1980, table 3[2]	Regional uplift equals average short- and long-term regional erosion rates
	2.0×10^{-2}	Ahnert, 1970	Modern erosion rates, Delaware River Basin, isostatic compensation
	-2.0×10^{-2}	Ahnert, 1970	Modern erosion rates; Juniate River Basin, isostatic compensation

[1]Qualifications list time interval for which rate is calculated, specific location, and datum where appropriate.
[2]Rate is calculated from data in the reference cited.
[3]Tilt rate is given in mm km^{-1} yr^{-1}.

(Cinquemani et al., 1982) and 1.7 mm yr^{-1} from historic tidal trends (Fadaie and Brown, 1984). However, for the same historic period as the tidal data, releveling surveys indicate uplift of the Cape Fear area relative to the Salisbury Embayment (Fig. 8) approaching 10 mm yr^{-1} (Balazs, 1974; Fadaie and Brown, 1984), an order of magnitude higher than the Holocene and historical tidal data. This is also several orders of magnitude higher than longer-term geologic rates, suggesting systematic leveling errors in the topographic surveys (Fadaie and Brown, 1984).

Surficial geomorphology and subsurface stratigraphy further identify the Cape Fear area as a neotectonic anomaly. Fluvial patterns of Piedmont and Coastal Plain rivers change markedly north and south of the Cape Fear River (Hack, 1982). The Cape Fear River, located just south of the Cape Fear Arch, has shifted southwestward during the past 3 Ma apparently in response to uplift (possibly fault controlled) along the Cape Fear Arch (Markewich, 1985). The seaward bulge on structure contours drawn on top of pre-Upper Jurassic rocks (Gohn, 1988) along the North–South Carolina border is consistent with tectonic uplift along the Cape Fear Arch.

Southward tilt of the Coastal Plain from the Cape Fear Arch into the Southeast Georgia Embayment (Fig. 1) is indicated by post 3 Ma marine terraces (Fig. 9) and the landward bulge on pre-Upper Jurassic structure contours (Gohn, 1988). Rates of subsidence (Table 1) vary markedly from $\sim 3 \times 10^{-3}$ mm yr^{-1}, for relative relief on post 5 Ma marine units (Lyttle et al., 1979) to ~ 2 mm yr^{-1} for historic tidal and releveling data (Holdahl and Morrison, 1974; Hicks, 1978; Lyttle et al., 1979).

Subsidence of the Coastal Plain and shelf in the mid-Atlantic region relative to the Cape Fear area at ~ 2 mm yr^{-1} (Table 1) is indicated by post 15 ka marine shorelines and basal peats (Newman et al., 1980). The northward tilt is, in part, attributed to absolute subsidence in the Salisbury Embayment approaching 2 mm yr^{-1} from tidal (Walcott, 1972a) and releveling data (Holdahl and Morrison, 1974) and a rapidly decreasing rate of post 15 ka subsidence on the continental shelf off New York City (Dillon and Oldale, 1978). However, in a study of the 12–9 ka shoreline, Blackwelder (1981b) concluded that there was no relative tectonic movement between New York City and South Carolina.

The rapid decrease in rate of deformation and the controversy about the amount of deforma-

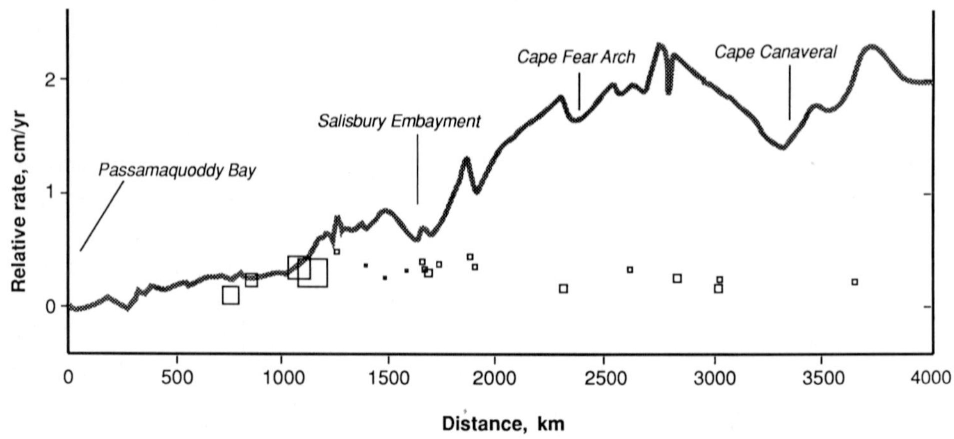

Fig. 8. Plot of apparent vertical crustal velocities as inferred from precise leveling surveys (crosses) and tidal gauge records (squares) along the Atlantic coast. Size of square equals two standard deviations of movement estimate. Leveling curve equalized to sea-level estimate at Portland, Maine ($X = 144$). Sealevel assumed to be stationary. Distance along the x-axis is measured from Calais, Maine (after Brown et al., 1980). See Fig. 1 for general location of the survey.

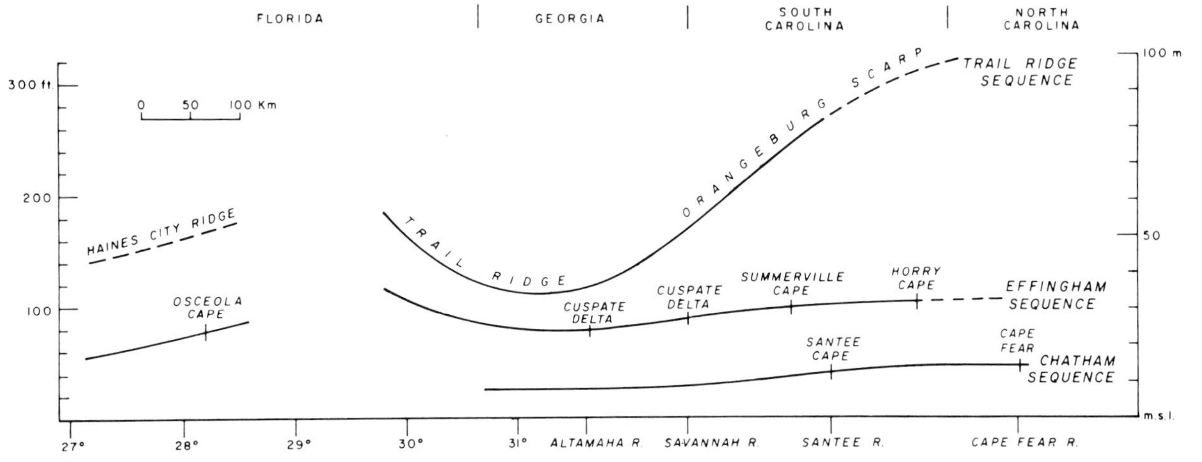

Fig. 9. Tilting along the southeastern United States Coastal Plain as inferred from shoreline sequences that indicate maximum height of transgression. See Winkler and Howard (1977) for tentative age assignments (after Winkler and Howard, 1977).

tion north of the Cape Fear Arch region result from the chosen sealevel curve and the complex, but predictable, northward migration of the glacially induced, peripheral bulge (Clark et al., 1978; Dillon and Oldale, 1978; Newman et al. 1980; Clark, 1981). Paleogeoidal isobase maps (Marcus and Newman, 1983) have been used to construct submergence velocity curves for the last 13 ka for the eastern United States (Fletcher, 1988). These curves (Fig. 10) clearly indicate the northward migration of maximum subsidence and a general decrease in submergence velocity over the last 10 ka.

The difficulty of calculating absolute uplift magnitude and rate for an isolated point on the ground and the attendant need to know eustatic sealevel elevation can be generally avoided by comparing present elevation differences along a previously horizontal datum. Coast-parallel tilt rates calculated in this manner indicate consistent down-to-the north regional tilting from the Cape Fear area to the Salisbury Embayment area. An average tilt rate of $\sim 10^{-3}$ mm km^{-1} yr^{-1} (Table 1) is indicated for post 18 ka marine shoreline (Fletcher, 1988), continental shelf topography (Fadaie and Brown, 1984), and historic tidal data (Brown and Oliver, 1976). However, topographic releveling surveys yield a value of $\sim 10^{-2}$ mm km^{-1} yr^{-1}, an order of magnitude higher. Coast-perpendicular tilt rates (coastside down) range from $\sim 10^{-4}$ mm km^{-1} yr^{-1} for geologically long-term intervals (Hack, 1980; Cronin, 1981) to $\sim 10^{-2}$ mm km^{-1} yr^{-1} for post glacial deformation in New England (Hack, 1980; Koteff, 1984; Newman et al., 1987) and topographic releveling surveys (Brown and Oliver, 1976). However, Hack (1980) calculated a modern tilt rate from erosion rates for the Potomic River watershed of $\sim 10^{-4}$ mm km^{-1} yr^{-1}, two orders of magnitude lower than releveling data (Table 1).

Inner Coastal Plain reverse faults

A significant data set for constraining Appalachian neotectonism is the high angle, reverse faults along the inner Coastal Plain from New Jersey to Georgia. Fault inventories and histories are excellently summarized in Wentworth and Mergner-Keefer (1983) and Prowell (1983, 1988). Detailed studies are available for the Stafford fault system, northeastern Va. (Mixon and Newell, 1977), the subsurface Brandywine-fault system, southwestern Md. (Jacobeen, 1972) which is possibly related to

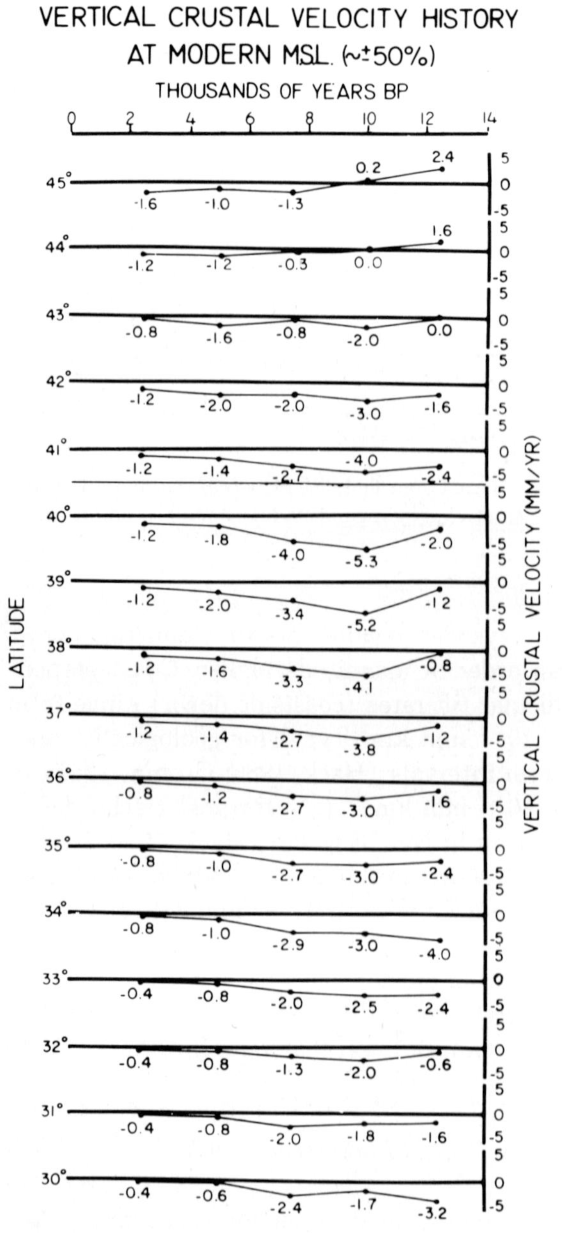

Fig. 10. Estimates of eastern United States Coastal Plain vertical velocities (±50%) for post 13 ka deformation. Datum is modern mean sealevel. For reference, latitude is located on Fig. 1 (after Fletcher, 1988).

the Stafford system, the Belair fault zone, eastern Ga. (Prowell and O'Connor, 1978), and the Cooke fault (Behrendt et al., 1981) and the offshore Helena Banks fault (Behrendt et al., 1983) near Charleston, S.C.

The fault systems are complex zones of closely spaced en-echelon shear planes generally located close to the Fall Line. Individual fault traces that range up to 1 km apart may extend for up to 10 km along strike. Individual faults are grouped in fault zones that extend for 25–100 km along strike (Prowell, 1988) and may be parts of more extensive fault systems (Mixon and Newell, 1977; Wentworth and Mergner-Keefer, 1983). Displacement along these dominantly northeast striking (Wentworth and Mergner-Keefer, 1983, Fig. 1), northwest or southeast dipping (40°–85°) faults is primarily dip-slip, reverse motion, producing horsts and grabens (Prowell, 1988). Some strike-slip motion is possible but is not considered to be dominant (Prowell, 1988), although outcrop exposure and lateral facies variations make identification difficult. However, the exact sense of motion is significant because it should help constrain the post 110 Ma paleo-stress regime.

Individual fault zone displacements are generally tens of meters, with a maximum cumulative offset of as much as 80 m (Fig. 11). Where available, stratigraphic data generally indicate greater offset on older units or progressive movement. Vertical slip rates ranges from 3.0×10^{-4} to 1.5×10^{-3} mm yr^{-1} over the last 110 Ma, averaging 5×10^{-4} mm yr^{-1} (Prowell, 1988) to 1×10^{-3} mm yr^{-1} (Wentworth and Mergner-Keefer, 1983). Poorly defined time lines and relatively small fault offset limit evidence for late Cenozoic fault movement. However, reverse faulting involving crystalline basement along the Mountain Run fault zone, central Va. (Pavlides et al., 1983) offsets possible Pleistocene colluvium by 1.5 m. Reverse faulting on fluvial terraces on Rock Creek near Washington, D.C. (Darton, 1950) along an extension of the Stafford fault system may indicate post-Pliocene movement (Mixon and Newell, 1977). Interestingly, longitudinal profiles of streams on the west side of the Stafford fault system are consistently steeper and more convex than those on the east side, in agreement with the sense of motion on the fault sys-

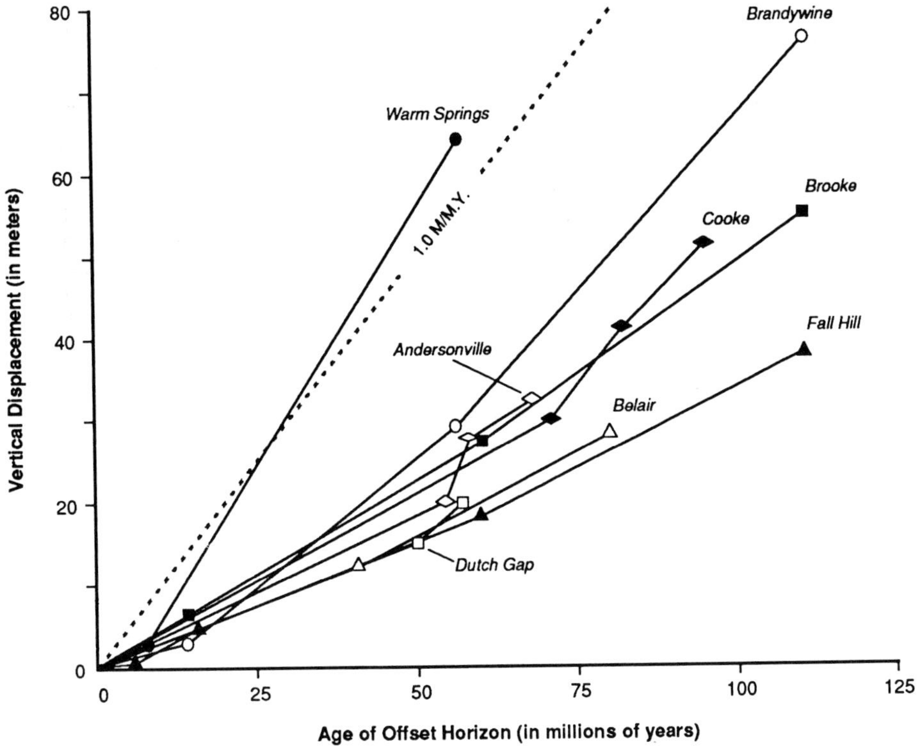

Fig. 11. Cumulative displacement of faults along the Atlantic coastal margin showing age of offset horizon versus greatest vertical displacement of horizon (after Prowell, 1988; data for the Cooke fault is from Wentworth and Mergner-Keefer, 1983, fig. 2).

tem (Mayer and Wentworth, 1983).

Reverse faulting in the Coastal Plain has been attributed to reactivation of pre-existing faults and discontinuities, especially Mesozoic normal faults (Mixon and Newell, 1977; Wentworth and Mergner-Keefer, 1983; Prowell, 1988). The reverse sense of motion indicates a compressional stress field since the Late Cretaceous. They are a significant body of data that have been used to infer the existence of a northwest–southeast oriented, maximum horizontal compression stress province for the Atlantic margin (Fig. 2), as well as the source of some historical seismicity.

Other seismic zones

There is a significant and growing body of evidence indicating neotectonic deformation for some seismic zones (Table 1). There is compelling evidence for coastal subsidence near the Charleston seismic zone from topographic releveling surveys (Holdahl and Morrison, 1974; Lyttle et al., 1979; Poley and Talwani, 1986), tidal gauge data (Hicks, 1978), and structural contours on subsurface formations (Lyttle et al., 1979), although subsidence rate varies over several orders of magnitude from 3×10^{-3} mm yr^{-1} to 2.5 mm yr^{-1}. A nontectonic component of subsidence from subsurface water withdrawal may cause overestimation of subsidence from releveling data (Poley and Talwani, 1986). Conversely, Bollinger (1973b) in an analysis of

Meade's (1971) leveling data suggested little relationship between vertical crustal velocities and seismicity in the Charleston area. Rhea (1989) combined gravity and magnetic surveys, seismic investigations, general topography, and convex inflections of river profiles to identify a 400 km^2 uplifted area 2.5 km northwest of Charleston that is consistent with the Charleston meizoseismal area and inferred motion along the Cooke fault.

Releveling surveys across the Brevard fault zone in North Carolina within the southern Appalachian seismic zone indicate an uplift rate of 4 mm yr^{-1} just west of the Brevard zone relative to Morristown, Tenn. The area of anomalous uplift is in proximity to the largest Bouguer gravity low in eastern North America and is associated with high values of the stream gradient index (Citron and Brown, 1979) although Hack (1973) suggested a lithologic rather than a neotectonic cause for the stream gradient index anomaly.

Several interesting geomorphic features are associated with the Giles County, Va. seismic zone. Offset fluvial terraces along the New River where it crosses the seismic zone indicate relative (northwest-side-up) uplift ranging from 3×10^{-2} to 2×10^{-1} mm yr^{-1} over the past 1 Ma (Mills, 1986). Uplift estimates depend upon terrace correlation and inferred age. Prehistoric, large rockslides, although not indicating deformation, have been mapped in the vicinity of the Giles County seismic zone and may be evidence for recurring late Cenozoic seismicity in the area, (Schultz, 1986; Southworth, 1987).

Topographic releveling surveys across the Adirondack Mountains indicate tilting rates of $\sim5\times10^{-2}$ mm km^{-1} yr^{-1} (Isachsen, 1975; Hack, 1980; Treadwell and Knuepfer, 1988), consistent with measured divergence of post glacial terraces on the East Branch Ausable River (Table 1). These rates are similar in magnitude to coast-perpendicular tilt rates on post-glacial marine shorelines in New England.

Extensive research has been conducted on the post-glacial subsidence in the Passamaquoddy Bay seismic zone. Subsidence rates range from 1.0 to 2.0 mm yr^{-1} for post 10 ka marine deposits, tidal trends, and corrected releveling surveys (Reilinger, 1987). These subsidence rates are consistent with those predicted from isostatic modeling for coastal Maine (Peltier, 1986), and are the same order of magnitude as those for the coastal areas in the Charleston seismic zone, the Salisbury Embayment, and the Ramapo fault zone (Table 1).

Discussion

Aside from glacially induced deformation, long-term neotectonism within the Atlantic drainage is caused by sediment loading (Cronin, 1981; Steckler et al., 1988), isostatic compensation of continental denudation (Ahnert, 1970; Pavich, 1985), and variations in the intraplate stress field (Brown et al., 1972; Gohn, 1988). The Atlantic Coastal Plain, north of Florida, lies landward of the hinge zone separating rapidly subsided, rift-stage crust from slowly subsided, continental crust. Compositional anisotropy, thermal history and fracture systems of the continental and rift-stage crust have produced differences in flexural strength and rates of thermal contraction. To a large degree, these factors have controlled the location of depocenters along the continental margin and the resulting crustal deformation due to sediment loading. The deepest sedimentary basins, i.e. the Baltimore Canyon Trough (Poag and Sevon, this volume) and Caroline Trough (Klitgord et al., 1988), are developed on the rift-stage crust (Klitgord et al., 1988). Lithospheric response to sediment loading in regions adjacent to these sedimentary basins, for example the Coastal Plain and Piedmont, can be predicted from various numeric, lithosphere flexure models (Walcott, 1972b; Beaumont, 1978, 1979). These models generally predict a peripheral bulge (uplift) equal to approximately 4–8% of the total basin subsidence within a distance of 100–400 km of the basin. The range of values depends upon the specific earth rheology model. The lithosphere is, in essence, bending or flexuring in response to the vertically imposed sedimentary load.

Cronin (1981) used this type of model to adequately explain uplifted and warped marine shorelines along the Carolina and Georgia coast landward of the Carolina Trough (Fig. 1). Predicted post-middle Miocene uplift of the peripheral bulge ranges from approximately 50 to 100 m, bracketing the observed 60 m of post-Pliocene uplift of the Orangeburg Scarp (Cronin, 1981). Lithospheric flexure in response to sediment loading of the Baltimore Canyon Trough may explain uplift in parts of the source areas for the Baltimore Canyon Trough sediments (Poag and Sevon, this volume). It may also explain uplift in the Piedmont (Hack, 1982) between the Potomac and Delaware Rivers that may exceed 100 m of post 5 Ma(?) uplift (Campbell, 1929). However, the evidence (deformed fluvial terraces) for the actual magnitude of warping has been questioned (Stose, 1930).

Regional-scale isostatic uplift inland of the Coastal Plain due to continental denudation has also been postulated. Uplift rates range from -2×10^{-2} mm yr^{-1} for the Juniata River basin (Ahnert, 1970) to $\sim2\times10^{-2}$ mm yr^{-1} for the Delaware River Basin (Ahnert, 1970) and the mid-Atlantic Piedmont (Pavich, 1985). These rates depend upon the assumed relationship between relief, erosion rate, and uplift (Ahnert, 1970) or a steady-state model of weathering, erosion and uplift (Pavich, 1985). Neither the isostatic model nor the lithospheric flexure model have adequately addressed the temporal nature of uplift. Is uplift periodic, thus helping to explain the Appalachian plantation surfaces as suggested by Davis (1889, see Morisawa, this volume for a review), Melhorn and Edgar (1975), or Poag and Sevon (this volume)? Is periodic uplift consistent with the steady-state model for a long-term balance between weathering, denudation, and uplift as suggested by Pavich (1985, this volume)? And importantly, what is the time lag between denudation, sediment loading, and uplift. These are significant questions that will be a fruitful area for future research.

Vertical crustal velocities calculated from historical tidal and releveling data are consistently one to two orders of magnitude higher than vertical crustal velocities calculated from early Quaternary or Pliocene data sets for the same areas. It has been inferred that either the historic data, especially the releveling surveys, contains systematic errors or that vertical crustal velocities have, in fact, increased in historic time. However, it is possible that the deformation rates calculated from these diverse data sets are compatible. In an analysis of the effects of measured time interval on the calculated rate of geomorphic and tectonic processes, Gardner et al. (1987) showed that a 4 or 5 order of magnitude difference in the measured time interval can account for approximately one order of magnitude difference in the measured process rate. Because surficial geologic processes are typically unsteady and nonuniform, rates of surficial processes measured over increasingly longer time intervals can incorporate longer intervals of process inactivity, thus producing an apparently slower rate. The measured time interval, δt, for the historic data sets is $\sim10^1$ yr and for the early Quaternary and Pliocene it is $\sim10^6$ yr. The five orders of magnitude difference in δt will produce a one order of magnitude correction for the two types of data, suggesting a much closer agreement. This indicates that the historical data, especially the tidal gauge data, is consistant with the longer-term geologic data. It further suggests that deformation along the Atlantic continental margin is periodic because the slower, long-term rates are incorporating periods of tectonic inactivity as inferred from the Appalachian planation surfaces (Melhorn and Edgar, 1975; Poag and Sevon, this volume). It also indicates that the present is the beginning of a period of active deformation.

Uplift rates calculated from lithospheric flexure and isostatic compensation are one to two orders of magnitude more than uplift rates calculated for reverse faults along the inner Coastal Plain. This indicates, as expected, that not all deformation is taken up along those faults (Pavich, this volume) and/or that all post-100 Ma reverse faults have not been identified.

It is apparent that all neotectonism (excluding glacially induced and hydroisostatic) may not be attributable to sediment loading of the Atlantic margin basins or isostatic uplift from continental denudation. The presence of inner Coastal Plain, high angle reverse faults indicate vertical crustal deformation involving a compressional stress field from intraplate tectonic sources (Wentworth and Mergner-Keefer, 1983; Gohn, 1988; Prowell, 1988) that may have existed since the Early Cretaceous. Brown et al. (1972) proposed a wrench-fault tectonic model to explain the lateral and vertical geometry of sedimentary units in the Coastal Plain. The dominant features of the three-phase model include a phase of northeast trending fault block anticlines and half grabens, a second phase of northwest trending anticlines and synclines and a third phase of northwest trending anticlines and north trending synclines (Brown et al., 1972, fig. 2). Periodic rotation of the compressional stress field from plate tectonic sources, not unlike that proposed by De Boer et al. (1988), cause a change in the phase of deformation and differential vertical movement along the strike-slip and tensional fault-bounded blocks. This model may, in part, explain the geometry of Cretaceous and Cenozoic Coastal Plain deposits in the Mid-Atlantic region (Newell, 1985; Olsson et al., 1988). It is also possible that rotation of the compressional stress field may periodically activate the reverse faults along the inner Coastal Plain, accounting for a significant component of the observed neotectonic deformation.

Summary

An extensive body of geologic data significantly constrains the style and rate of neotectonic deformation for the Appalachians and Atlantic passive continental margin. These data include the modern state of stress, historical seismicity, surface and subsurface stratigraphy, numeric models of crustal deformation, surficial geomorphic systems, and historical, precise leveling and tidal gauge records. The modern stress field is characterized by a northeast–southwest oriented maximum horizontal compression caused by far-field plate tectonic sources. The existence of a northwest–southeast oriented, maximum horizontal compression for the Atlantic margin, east of the Appalachian Front is currently being debated. Further work on the timing and sense of motion on Cretaceous and Cenozoic faults in the Coastal Plain together with more data from earthquake focal mechanisms, hydrofractures and well bore breakouts should help to resolve the problem.

Analysis of historical, seismic data define numerous, discrete seismic zones. Proposed causes for the seismicity are numerous, including reactivation of Paleozoic and Mesozoic structures, continentward extension of oceanic fracture systems, stress amplification along boundaries between contrasting lithologic materials (especially mafic igneous bodies) and transient increases in hydraulic head. It is apparent that as a group these mechanisms can account for most historic seismicity, but the cause(s) of individual seismic events and even some seismic zones, are still debated.

There are two major domains of neotectonism in the eastern United States. The northeastern United States are dominated by isostatic uplift and northward migration of peripheral bulge collapse in response to deglaciation. This locally rapid, but decreasing deformation rate is superimposed along the Atlantic margin on slower, long-term deformation. There, most deformation is attributed to lithospheric flexuring in response to sediment loading in major depocenters, isostatic deformation in response to continental denudation and water loading of the continental margin, and stress from far-field plate tectonic sources. These multiple processes, operating on an anisotropic lithosphere with different thermal histories, have resulted in a complex, deformation pattern.

Significant deformational features that are consistently identified by diverse data include the uplift anomaly near Cape Fear, N.C., northward and southward tilting of the Coastal Plain into the Salisbury and Southeast Georgia Embayments, respectively, seaward tilting of the Coastal Plain, and a complex pattern of post-glacial uplift and later subsidence in the northeast.

Estimates of vertical crustal velocities can vary over several orders of magnitude with Holocene deposits and historical, tidal and releveling data consistently giving the highest values. Systematic leveling error and/or subsurface water withdrawal may seriously affect the rapid deformation rates calculated from historical data. Yet, correction for differences in the time interval over which deformation rates are calculated, indicate that the historical rates and long-term rates are more consistent than originally thought. It has been suggested that crustal deformation has been episodic since the opening of the Atlantic and it is quite possible that the more rapid deformation determined from historical data indicate another phase of accelerated neotectonism. However, these rapid rates, if continued, will soon become evident in the landscape.

References

Aggarwal, Y.P. and Sykes, L.R., 1978. Earthquakes, faults, and nuclear power plants in southern New York and northern New Jersey. Science, 200: 425–429.

Ahnert, F., 1970. Functional relationships between denudation, relief, and uplift in large mid-latitude drainage basins. Am. J. Sci., 268: 243–263.

American Geological Institute, 1977. Glossary of Geology. Am. Geol. Inst., Falls Church, VA, 805 pp.

Anderson, W.A., Kelley, J.T., Thompson, W.B., Borns, H.W., Sanger, D., Smith, D.C., Tyler, D.A., Anderson, R.S., Bridges, A.E., Crossen, K.J., Ladd, J.W., Anderson, B.G. and Lee, F.T., 1984. Crustal warping in coastal Maine. Geology, 12: 677–680.

Balazs, E.I., 1974. Vertical crustal movements on the middle Atlantic Coastal Plain as indicated by precise leveling. Geol. Soc. Am., Abstr. Programs, 6(2): 73 (abstr.).

Barosh, P.J., 1981. Seismicity and tectonics of the Passamaquoddy Bay area, Maine. Geol. Soc. Am., Abstr. Programs, 13(3): 122 (abstr.).

Basham, P.W., Weichert, D.J. and Berry, M.J., 1979. Regional assessment of seismic risk in eastern Canada. Seismol. Soc. Am., Bull. 69: 1567–1602.

Beaumont, C., 1978. The evolution of sedimentary basins on a viscoelastic lithosphere: theories and examples. R. Astron. Soc. Geophys. J., 55: 471–497.

Beaumont, C., 1979. On rheological zonation of the lithosphere during flexure. Tectonophysics, 59: 347–365.

Behrendt, J.C., Hamilton, R.M., Ackerman, H.D. and Henry, V.J., 1981. Cenozoic faulting in the vicinity of Charleston, South Carolina, 1886 earthquake. Geology, 9: 117–122.

Behrendt, J.C., Hamilton, R.M., Ackerman, H.D. and Henry, V.J., 1983. Marine multichannel seismic-reflection evidence for Cenozoic faulting and deep crustal structure near Charleston, South Carolina. U.S. Geol. Surv., Prof. Pap., 131-J: J1–J29.

Belknap, D.F., Borns, H.W., Jr., Kelley, J.T. and Stuckenrath, R., 1987. Late Quaternary sealevel changes in Maine. In: D. Nummedal, O.H. Pilkey and J.D. Howard (Editors), Sea-level Fluctuation and Coastal Evolution. Soc. Econ. Paleontol. and Mineral., Spec. Publ., 41: 71–85.

Blackwelder, B.W., 1981a. Late Cenozoic marine deposition in the United States Atlantic Coastal Plain related to tectonism and global climate. Palaeogeogr. Palaeoclimatol. Palaeoecol., 34: 87–114.

Blackwelder, B.W., 1981b. Late Wisconsinan and Holocene tectonic stability of the United States mid-Atlantic coastal region. Geology, 8: 534–537,

Blackwelder, B.W., Pilkey, O.H. and Howard, J.D., 1979. Late Wisconsinan sea levels on the southeast U.S. Atlantic shelf based on in-place shoreline indicators. Science. 204: 618–620.

Block, J.W., Clement, R.C., Lew, L.R. and deBoer, J., 1979. Recent thrust faulting in southeastern Connecticut. Geology, 7: 79–82.

Bloom, A.L., 1967. Pleistocene shorelines: a new test of isostasy. Geol. Soc. Am., Bull. 78: 1477–1494.

Bollinger, G.A., 1973a. Seismicity of the southeastern United States. Seismol. Soc. Am., Bull., 63: 1785–1808.

Bollinger, G.A., 1973b. Seismicity and crustal uplift in the southeastern United States. Am. J. Sci., 273A: 396–408.

Bollinger, G.A., 1983. Speculations on the nature of seismicity at Charleston, South Carolina. U.S. Geol. Surv., Prof. Pap., 1313-T: T1–T11.

Bollinger, G.A. and Wheeler, R.L., 1983. The Giles County, Virginia, seismic zone. Science, 219: 1063–1065.

Bollinger, G.A., Chapman, M.C., Sibol, M.S. and Costain, J.K., 1985. An analysis of earthquake focal depths in the southeastern U.S. Geophys. Res. Lett., 12: 735-788.

Braun, D.D., 1989. Glacial and periglacial erosion of the Appalachians. In: T.W. Gardner and W.D. Sevon (Editors), Appalachian Geomorphology. Geomorphology, 2: 233-256 (this volume).

Brown, L.D. and Oliver, J.E., 1976. Vertical crustal movements from leveling data and their relation to geologic structure in the eastern United States. Rev. Geophys. Space Phys., 14: 13-35.

Brown, P.M., Miller, J.A. and Swain, F.M., 1972. Structural and stratigraphic framework, and spatial distribution of permeability of the Atlantic coastal plain, North Carolina to New York. U.S. Geol. Surv., Prof. Pap., 796, 79 pp.

Brown, L.D., Reilinger, R.E. and Citron, G.P., 1980. Recent vertical crustal movements in the US: evidence from precise leveling. In: N. Morner (Editor), Earth Rheology, Isostasy, and Eustasy. Wiley, New York, N.Y., pp. 389-405.

Campbell, M.R., 1929. Late geologic deformation of the Appalachian Piedmont as determined by river gravels. Natl. Acad. Sci., Proc., 15: 156-161.

Cinquemani, L.J., Newman, W.S., Sperling, J.A., Marcus, L.F. and Pardi, R.R., 1982. Holocene sea level changes and vertical movements along the east coast of the United States: a preliminary report. In: D. Colquhoun (Editor), Holocene Sea Level Fluctuations, Magnitudes and Causes, Proc. 1981 INQUA Meet., Columbia, S.C., pp.13-33.

Citron, G.P. and Brown, L.D., 1979. Recent vertical crustal movements from precise leveling surveys on the Blue Ridge and Piedmont provinces, North Carolina and Georgia. Tectonophysics, 52: 223-238.

Clark, J.A., 1981. Comment on Late Wisconsin and Holocene tectonic stability of the United States mid-Atlantic coastal region. Geology, 9: 438-439.

Clark, J.A. and Bloom, A.L., 1979. Hydroisostasy and Holocene emergence of South America, In: K. Surguis et al. (Editors), International Symposium on Coastal Evolution in the Quarternary. San Paulo, 1978, pp. 41-60.

Clark, J.A., Farrell, W.E. and Peltier, W.R., 1978. Global changes in postglacial sea level: a numerical calculation. Quat. Res. 9: 265-287.

Condie, K.C., 1982. Plate Tectonics and Crustal Evolution. Pergamon, New York, N.Y., 310 pp.

Cooke, C.W., 1930. Correlation of coastal terraces. J. Geol., 38: 577-589.

Cooke, C.W., 1931. Seven coastal terraces in the southeastern states. Washington Acad. Sci., J., 21: 503-513.

Costain, J.K., Bollinger, G.A. and Speer, J.A., 1987. Hydroseismicity - a hypothesis for the role of water in the generation of intraplate seismicity. Geology, 15: 618-621.

Cronin, T.M., 1981. Rates and possible causes of neotectonic vertical crustal movements of the emerged southeastern United States Atlantic Coastal Plain. Geol. Soc., Am., Bull., 92: 812-833.

Cronin, T.M., Szabo, B.J., Ager, T.A., Hazel, J.E. and Owens, J.P., 1981. Quaternary climates and sea levels of the U.S. Atlantic Coastal Plain. Science, 211: 233-240.

Curray, J.R., 1965. Late Quaternary history, continental shelves of the United States. In: H.E. Wright and D.G. Frey (Editors), The Quaternary of the United States, Princeton University Press, Princeton, N.J., pp. 723-735.

Darton, N.H., 1950. Configuration of the bedrock surface of the District of Columbia and vicinity. U.S. Geol. Surv., Prof. Pap., 217, 42 pp.

Davis, W.M., 1889. The rivers and valleys of Pennsylvania. Natl. Geol. Mag., 1: 183-253.

Davis, W.M., 1902. Base-level, and peneplain. J. Geology, 10: 77-111.

Davis, W.M., 1932. Piedmont benchlands and primarrumpfe. Geol. Soc. Am., Bull., 43: 399-440.

De Boer, J.Z., McHone, J.G., Puffer, J.H., Ragland, P.C. and Whittington, D., 1988. Mesozoic and Cenozoic magmatism. In: R.E. Sheridan and J.A. Grow (Editors), The Atlantic Continental Margin. Geol. Soc. Am., The Geology of North America, I-2: 217-241.

Dewey, J.E. and Gordon, D.W., 1984. Map showing recomputed hypocenters of earthquakes in the eastern and central United States and adjacent Canada, 1925-1980. U.S. Geol. Surv., Misc. Field Stud. Map, MF-1699, Scale 1:2,500,000.

Dillon, W.P. and Oldale, R.N., 1978. Late Quaternary sea-level curve: reinterpretation based on glaciotectonic influence. Geology, 6: 56-60.

Doornkamp, J.C., 1986. Geomorphological approaches to the study of neotectonics. Geol. Soc. London J., 143: 335-342.

Ebel, J.E., 1985. A study of Seismicity and Tectonics in New England. U.S. Nucl. Regul. Comm. Rep. CR-4354, Washington, D.C., 88 pp.

Ebel, J.E., Vudler, V. and Celata, M., 1982. The 1981 earthquake swarm near Moodus. Connecticut. Geophys. Res. Lett., 9: 397-400.

Fadaie, K. and Brown, L., 1984. Contemporary tilting in the central Atlantic Coastal Plain. J. Geol., 92: 547-559.

Fletcher, C.H., 1988. Holocene sea level history and neotectonics of the United State mid-Atlantic region: applications and corrections. J. Geol., 96: 323-337.

Fletcher, J.B., Sbar, M.L. and Sykes, L.R., 1978. Seismic trends and travel-time residuals in eastern North America and their tectonic implications. Geol. Soc. Am., Bull., 89: 1656-1676.

Frost, N.H. and Lilley, J.E., 1966. Crustal movement in the Lac St. Jean area, Quebec. Can. Surv., XX: 292-299.

Gardner, T.W., Jorgensen, D.W., Shuman, C. and Lemieux, C.R., 1987. Geomorphic and tectonic process rates: effects of measured time interval. Geology, 15: 259-261.

Gephart, J.W. and Forsyth, D.D., 1985. On the state of stress in New England as determined from earthquake focal mechanisms. Geology, 13: 70-72.

Gohn, G.S., 1988. Late Mesozoic and early Cenozoic geology of the Atlantic Coastal Plain: North Carolina to Florida. In: R.E. Sheridan and J.A. Grow (Editors), The Atlantic Continental Margin, U.S., Geol. Soc. Am., The Geology of North America, I-2: 107–130.

Grow, J.A. and Sheridan, R.E., 1988. U.S. Atlantic continental margin; a typical Atlantic-type or passive continental margin. In: R.E. Sheridan and J.A. Grow (Editors), The Atlantic Continental Margin, Geol. Soc. Am. The Geology of North America, I-2: 1–7.

Hack, J.T., 1973. Drainage adjustment in the Appalachians. In: M. Morisawa (Editor), Fluvial Geomorphology, State Univ. New York, Binghamton, pp. 51–69.

Hack, J.T., 1980. Rock control and tectonism - their importance in shaping the Appalachian Highlands. U.S. Geol. Surv., Prof. Pap., 1126-B: B1–B17.

Hack, J.T., 1982. Physiographic divisions and differential uplift in the Piedmont and Blue Ridge. U.S. Geol. Surv., Prof. Pap., 1265, 49 pp.

Hadley, J.B. and Devine, J.F., 1974. Seismotectonic Map of the Eastern United States. U.S. Geol. Surv., Misc. Field Stud. Map, MF-620, Scale 1:5,000,000.

Hancock, P.L. and Williams, G.D., 1986. Neotectonics. Geol. Soc. London, J., 143: 325–326.

Hansen, H.J., 1988. Buried rift basin underlying coastal plain sediments, central Delmarva Peninsula, Maryland. Geology, 16: 779–782.

Haq, B.U., Hardenbol, J. and Vail, P.R., 1987. Chronology of fluctuating sea levels since the Triassic. Science, 235: 1156–1167.

Hatcher, R.D., Fietz, I. and Litehiser, J.J., 1987. Crustal subdivision of the eastern and central United States and a seismic boundary hypothesis for eastern seismicity. Geology, 15: 528–532.

Hicks, S.D., 1978. An average geopotential sea level series for the United States. J. Geophys. Res., 83C: 1377–1379.

Higgins, M.W., Fietz, I. and Fisher, G.W., 1974. Interpretation of aeromagnetic anomalies bearing on the origin of upper Chesapeake Bay and river course changes in the central Atlantic seaboard region: speculations. Geology, 2: 73–76.

Holdahl, S.R. and Morrison, N.L., 1974. Regional investigations of vertical crustal movements in the U.S., using precise relevelings and mareograph data. Tectonophysics, 23: 373–390.

Isachsen, Y.W., 1975. Possible evidence for contemporary doming of the Adirondack Mountains, New York, and suggested implications for regional tectonics and seismicity. Tectonophysics, 29: 169–181.

Jacobeen, F.H., 1972. Seismic evidence for high angle reverse faulting in the Coastal Plain of Prince Georges and Charles County. Maryland. Maryland Geol. Surv., Inf. Circ., 13: 21 pp.

Kane, M.F., 1977. Correlation of major eastern earthquake centers with mafic/ultramafic basement masses. U.S. Geol. Surv., Prof. Pap., 1029-0: 199–204.

Kerr, W.C., 1875. Report of the geological survey of North Carolina. Physical Geography Resume and Economic Geology, Vol. 1, Raleigh, 325 pp.

Klitgord, K.D., Dillon, W.P. and Popenoe, P., 1983. Mesozoic tectonics of the southeastern United States Coastal Plain and continental margin, U.S. Geol. Surv., Prof. Pap., 1313-P: P1–P15.

Klitgord, K.D., Hutchinson, D.R. and Shouten, H., 1988. U.S. Atlantic continental margin; structure and tectonic framework. In: R.E. Sheridan and J.A. Grows (Editors), The Atlantic Continental Margin. Geol. Soc. Am., The Geology of North America, I-2: 19–55.

Koteff, C., 1984. Postglacial uplift in northeastern United States. In: Summary of Technical Reports V.XVIII. U.S. Geol. Surv., Open File Rep., 84-628, pp. 61.

Lyttle, P.T., Gohn, G.S., Higgins, B.B. and Wright, D.S., 1979. Vertical crustal movements in the Charleston, South Carolina–Savannah, Georgia area. Tectonophysics, 52: 183–189.

Marcus, L.F. and Newman, W.S., 1983. Hominid migrations and the eustatic sea level paradym. In: P.M. Masters and M.C. Flemming (Editors), Quaternary Coast Lines and Marine Archeology, Academic Press, London, pp. 63–85.

Markewich, H.W., 1985. Geomorphic evidence for Pliocene–Pleistocene uplift in the area of the Cape Fear Arch, North Carolina. In: M. Morisawa and J. Hack (Editors), Tectonic Geomorphology. Proc. Vol. of the 15th Annu. Geomorphol. Ser., Binghamton, New York, 1984. Allen and Unwin, London, pp. 279–297.

Mayer, L. and Wentworth, C.M., 1983. Geomorphic differences east and west of the afford fault system, northeastern Virginia. Geol. Soc. Am., Abstr. Programs, 15: 56 (abstr.).

McKeown, F.A., 1978. Hypothesis: many earthquakes in the central and southeastern United States are causally related to mafic intrusive bodies. U.S. Geol. Surv., J. Res., 6: 41–50.

Meade, B.K., 1971. Report of the sub-commission on recent crustal movements in North America. Paper presented to the 15th Gen. Assem. Int. Union Geodyn. Geophys., Int. Assoc. Geodyn., Brussels.

Melhorn, W.N. and Edgar, D.E., 1975. The case for episodic, continental-scale erosion surfaces: a tentative geodynamic model. In: W.N. Melhorn and R.C. Flemal (Editors), Theories of Landform Development. Proc. Vol. 6th Geomorphol. Symp. Ser., Binghamton, N.Y., pp. 243–276.

Milliman, J.D. and Emery, K.P., 1968. Sea-levels during the past 35,000 years. Science, 162: 1121–1123.

Mills, H.H., 1986. Possible differential uplift of New River terraces in southwestern Virginia. Neotectonics, 1: 75–86.

Mixon, R.B. and Newell, W.L., 1977. Stafford fault system: structures documenting Cretaceous and Tertiary deformation along the Fall Line in northeastern Virginia. Geology, 5: 437–440.

Morisawa, M., 1989. Rivers and valleys of Pennsylvania, revisited. In: T.W. Gardner and W.D. Sevon (Editors), Appalachian Geomorphology. Geomorphology, 2: 1–22 (this volume).

Morner, N.-A., 1980. Earth Rheology, Isostasy, and Eustasy. Wiley, New York, N.Y., 599 pp.

Muncey, J.W. and Bollinger, G.A., 1985. Focal mechanism analysis for Virginia earthquakes. Seismol. Soc. Am., Bull., 75: 1613–1636.

Newell, W.L., 1985. Architecture of the Rappahannock estuary-neotectonics in Virginia. In: M. Morisawa and J. Hack (Editors), Tectonic Geomorphology. Proc. Vol. 15th Annu. Geomorphol. Ser., Binghamton. N.Y., Allen and Unwin, London, pp. 322–342.

Newman, W.S., Cinquemani, L.J., Pardi, R.R. and Marcus, L.F., 1980. Holocene deleveling of the United States' East Coast. In: N. Morner (Editor), Earth Rheology, Isostasy and Eustacy. Wiley, New York, N.Y., pp. 449–463.

Newman, W.S., Cinquemani, L.J., Sperling, J.A., Marcus, L.F. and Pardi, R.R., 1987. Holocene neotectonics and the Ramapo fault zone sea-level anomaly: a study of varying marine transgression rates in the lower Hudson estuary, New York and New Jersey. In: D. Nummedal, O.H. Pilkey and J.D. Howard (Editors), Sea-Level Fluctuations and Coastal Evolution. Soc. Econ. Paleontol. Miner., Spec. Publ., 41: 97–111.

Olsson, R.K., Gibson, T.G., Hansen, H.J. and Owens, J.P., 1988. Geology of the northern Atlantic coastal plain: Long Island to Virginia. In: R.E. Sheridan and J.A. Grow (Editors), The Atlantic Continental Margin. Geol. Soc. Am., The Geology of North America, I-2: 87–107.

Pavich, M.J., 1985. Appalachian Piedmont morphogenesis: weathering, erosion, and Cenozoic uplift. In: M. Morisawa and J. Hack (Editors), Tectonic Geomorphology. Proc. Vol. 15th Annu. Geomorphol. Ser., Binghamton, N.Y., Allen and Unwin, London, pp. 299–319.

Pavich, M.J. 1989. Regolith residence time and the concept of surface age of the Piedmont "peneplain". In: T.W. Gardner and W.D. Sevon (Editors), Appalachian Geomorphology. Geomorphology, 2: 181–196 (this volume).

Pavlides, L., Bobyarchick, A.B., Newell, W.L. and Pavich, M.J., 1983. Late Cenozoic faulting along the Mountain Run fault zone, central Virginia Piedmont. Geol. Soc. Am., Abstr. Programs, 15(2): 55 (abstr.).

Peltier, W.R., 1986. Deglaciation-induced vertical motion of the North American continent and transient lower mantle rheology. J. Geophys. Res., 91: 9099–9123.

Pitman, W.C., III, 1978. Relationship between eustasy and stratigraphic sequences of passive margins. Geol. Soc. Am., Bull., 89: 1389–1403.

Pitman, W.C. III and Talwani, M., 1972. Sea floor spreading in the north Atlantic. Geol. Soc. Am., Bull., 83: 619–646.

Plumb, R.A. and Cox, J.W., 1987. Stress directions in eastern North America determined to 4.5 km from borehole elongation measurements. J. Geophys. Res., 92(B): 4805–4816.

Poag, C.W. and Valentine, P.C., 1988. Mesozoic and Cenozoic stratigraphy of the United States Atlantic continental shelf and slope. In: R.E. Sheridan and J.A. Grow (Editors), The Atlantic Continental Margin. Geol. Soc. Am., The Geology of North America, I-2: 67–85.

Poag, C.W. and Sevon, W.D., 1989. A record of Appalachian denudation in post-rift Mesozoic and Cenozoic sedimentary deposits of the U.S. middle Atlantic continental margin. In: T.W. Gardner and W.D. Sevon (Editors), Appalachian Geomorphology. Geomorphology, 2: 119–157 (this volume).

Poley, C.M. and Talwani, P., 1986. Recent vertical crustal movements near Charleston, South Carolina. J. Geophys. Res., 91(B): 9056–9066.

Prowell, D.C., 1983. Index of faults of Cretaceous and Cenozoic age in eastern United States. U.S. Geol. Surv., Misc. Field Stud. Map, MF-1269, Scale 1:2,500,000.

Prowell, D.C., 1988. Cretaceous and Cenozoic tectonism on the Atlantic coastal margin. In: R.E. Sheridan and J.A. Grown (Editors), The Atlantic Continental Margin. Geol. Soc. Am., The Geology of North America, I-2: 557–564.

Prowell, D.C. and O'Connor, B.J., 1978. Belair fault zone: evidence of Tertiary fault displacement in eastern Georgia. Geology, 6: 681–684.

Quittmeyer, R.C., Stratton, C.T., Mrotek, K.A. and Houlday, M., 1985. Possible implications of recent microearthquakes in southern New York State. Earthquakes Notes, 56: 35–42.

Ratcliffe, N.M., 1971. The Ramapo fault system in New York and adjacent northern New Jersey: a case of tectonic heredity. Geol. Soc. Am., Bull., 82: 125–142.

Reilinger, R., 1987. Reanalysis of crustal warping in coastal Maine. Geology, 15: 958–961.

Rhea, S., 1989. Evidence of uplift near Charleston, South Carolina. Geology, 17: 311–315.

Sawyer, D.S., 1988. Thermal evolution. In: R.E. Sheridan and J.A. Grow (Editors), The Atlantic Continental Margin. Geol. Soc. Am., The Geology of North America, I-2: 417–428.

Sawyer, D.S., Swift, B.A., Toksz, M.N. and Sclater, J.G., 1982. Thermal evolution of the Baltimore Canyon Trough and Georges Bank Basin. In: J.S. Watkins and C.L. Drake (Editors), Studies in Continental Margin Geology. Am. Assoc. Pet. Geol., Mem., 34: 743–764.

Sbar, M.L. and Sykes, L.R., 1973. Contemporary compressive stress and seismicity in eastern North America: an example of intra-plate tectonics. Geol. Soc. Am.,Bull., 84: 1861–1882.

Sbar, M.L., Jordan, R.R., Stephens, C.D., Pickett, T.E., Woodruff, K.D. and Sammis, C.G., 1975. The Delaware–New Jersey earthquake of February 28, 1973. Seismol. Soc. Am., Bull., 65: 85–92.

Schultz, A.P., 1986. Ancient, giant rockslides, Sinking Creek Mountain, southern Appalachians, Virginia. Geology, 14: 11-14.

Seeber, L. and Armbruster, J., 1981. The 1886 Charleston, South Carolina earthquake and the Appalachian detachment. J. Geophys. Res., 86: 7874-7894.

Seeber, L. and Armbruster, J., 1985. Spatio-temporal distribution of seismicity along the Appalachians and the case for detachment reactivation. Earthquake Notes, 56: 66 (abstr.).

Seeber, L. and Armbruster, J.C., 1988. Seismicity along the Atlantic seaboard of the U.S.; intraplate neotectonics and earthquake hazard. In: R.E. Sheridan and J.A. Grow (Editors), The Atlantic Continental Margin. Geol. Soc. Am., The Geology of North America, I-2: 565-582.

Seeber, L. and Coles, K.S., 1984. Seismicity in the central Adirondacks with emphasis on the Gooknow, October 7, 1983, epicential zone and to geology. In: D.B. Potter (Editor), Field Trip Guidebook, 56th Annu. Meet., New York State Geol. Assoc., Hamilton College, pp. 334-352.

Shattuck, G.B., 1906. The Pliocene and Pleistocene deposits of Maryland. Md. Geol. Surv., Pliocene and Pleistocene, pp. 21-137.

Smith, W.E.T., 1962. Earthquakes of eastern Canada and adjacent areas 1534-1927. Dominion Observ. Publ. No. 26, Ottawa, Ont., pp. 271-301.

Smith, W.E.T., 1966. Earthquakes of eastern Canada and adjacent areas 1928-1959. Dominion Observ., Publ. No. 32, Ottawa, Ont., pp. 87-121.

Southworth, C.S., 1987. Prehistoric giant rockslides of the Appalachian Valley and Ridge Province: types and characteristics. U.S. Geol. Surv., Circ. 1008: 24-31.

Steckler, M.S. and Watts, A.B., 1978. Subsidence of the Atlantic-type continental margin off New York. Earth Planet. Sci. Lett., 42: 1-3.

Steckler, M.S., Watts, A.B. and Thorne, J.A., 1988. Subsidence and basin modeling at the U.S. Atlantic passive margin. In: R.E. Sheridan and J.A. Grow (Editors), The Atlantic Continental Margin. Geol. Soc. Am., The Geology of North America, I-2: 399-416.

Stockar, D.V., 1986. Contemporary tectonics of the Lancaster, Pennsylvania Seismic Zone. M.Sc. Pap., Pennsylvania State University, University Park, 226 pp.

Stose, G.W., 1930. Is the Bryn Mawr peneplain a warped surface. Am. J. Sci., 219: 177-184.

Summerfield, M.A., 1987. Neotectonics and landform genesis. Progr. Phys. Geogr., 11: 384-397.

Sykes, L.R., 1978. Intraplate seismicity, reactivation of preexisting zones of weakness, alkaline magmatism, and other tectonism postdating continental fragmentation. Rev. Geophys. Space Phys., 16: 621-688.

Talwani, P., 1982. Internally consistent pattern of seismicity near Charleston, South Carolina. Geology, 10: 654-658.

Thompson, W.B., Crosser, K.J., Borns, H.W. and Anderson, B.G., 1983. Glacial-marine deltas and late Pleistocene-Holocene crustal movements in southern Maine. Maine Geol. Surv., Open File Rep., 83-3, 18 pp.

Treadwell, C.J. and Knuepfer, P.L.K., 1988. Deformation of Holocene river terrace profiles northeast Adirondack Mountains, New York. Geol. Soc. Am., Abstr. Programs, 20(1): 77 (abstr.).

Tyler, D.A. and Ladd, J.W., 1980. Vertical crustal movement in Maine. Maine Geol. Surv., Open File Rep., 80-34, 53 pp.

Vail, P.R. and Hardenbol, J., 1979. Sea level change during the Tertiary. Oceanus, 22: 71-79.

Vita-Finzi, C., 1986. Recent Earth Movements: An Introduction to Neotectonics. Academic Press, London, 226 pp.

Voight, B., 1969. Evolution of North Atlantic Ocean: relevance of rock-pressure measurements. In: M. Kay (Editor), North Atlantic: Geology and Continental Drift. Am. Assoc. Pet. Geol., Mem., 12: 955-962.

Walcott, R.I., 1972a. Late Quaternary vertical movements in eastern North America: quantitative evidence of glacio-isostatic rebound. Rev. Geophys. Space Phys., 10: 849-884.

Walcott, R.I., 1972b. Gravity, flexure, and the growth of sedimentary basins at a continental edge. Geol. Soc. Am., Bull., 83: 1845-1848.

Wentworth, C.M. and Mergner-Keefer, M., 1983. Regenerate faults of small Cenozoic offset-probable earthquake sources in the southeastern United States. U.S. Geol. Surv., Prof. Pap., 1313S: S1-S20.

Wheeler, R.L. and Bollinger, G.A., 1984. Seismicity and suspect terranes in the southeastern United States. Geology, 12: 323-326.

Whitcomb, J.H., 1976. New vertical geodesy. Jour. Geophys. Res., 81: 4937-4946.

Williams, H. and Hatcher, R.D., 1982. Suspect terrains and accretionary history of the Appalachian origin. Geology, 10: 530-536.

Winkler, C.D. and Howard, J.D., 1977. Correlation of tectonically deformed shorelines on the southern Atlantic coastal plain. Geology, 5: 123-127.

Woollard, G.P., 1958. Areas of tectonic activity in the United States as indicated by earthquake epicenters. Am. Geophys. Union, Trans., 39: 1135-1150.

Woollard, G.P., 1965. Tectonic activity in North America as indicated by earthquakes. In: P.J. Hart (Editor), The Earth's Crust and Upper Mantle. Am. Geophys. Union, Washington, D.C., Geophys. Monogr. Ser., 10: 125-133.

Yang, J.P. and Aggarwal, Y.P., 1981. Seismotectonics of northeastern United States and adjacent Canada. Jour. Geophys. Res., 86(B): 4981-4998.

York, J.E. and Oliver, J.E., 1976. Cretaceous and Cenozoic faulting in eastern North America. Geol. Soc. Am., Bull., 87: 1105-1114.

Zoback, M.L. and Zoback, M., 1980. State of stress in the conterminous United States. Jour. Geophys. Res., 85(B): 6113-6156.

Zoback, M.L. and Zoback, M.D., in press. Tectonic stress field of the continental United States. In: L. Pakiser and W. Mooney (Editors), Geophysical Framework of the Continental United States, Geol. Soc. Am., Mem., 172.

Geomorphology, 2 (1989) 99-118
Elsevier Science Publishers B.V., Amsterdam — Printed in The Netherlands

Climate Variations and the Appalachians from the Late Paleozoic to the Present: Results from Model Simulations

ERIC J. BARRON

Earth System Science Center, Pennsylvania State University, University Park, PA 16802 (U.S.A.)

(Received May 1, 1989; accepted after revision June 8, 1989)

Abstract

Barron, E.J., 1989. Climate variations and the Appalachians from the late Paleozoic to the Present: Results from model simulations. In: T.W. Gardner and W.D. Sevon (Editors), Appalachian Geomorphology, 2: 99-118.

A series of climate model simulations are utilized to examine the climate evolution of the Appalachians from the Carboniferous to the Present through consideration of large scale changes in paleogeography and potential variations in carbon dioxide on long time scales. The model simulations predict considerable variations in temperature and precipitation through time. In addition, a number of shorter time scale phenomena may influence the climatic record of the Appalachians. In particular Milankovitch orbital variations can produce rhythmic climate and weathering. Particular geographic configurations since the Carboniferous tend to promote sensitivity to orbital variations. Although far from comprehensive, the model simulations provide independent predictions of climate which can be utilized to investigate the evolution of the Appalachians deductively.

Introduction

The geomorphic evolution of the Appalachians and the associated evolution of sedimentary basins involve a complex interaction between tectonics, climate and drainage evolution. The evolution of the Appalachians can be addressed in two ways. First, the stratigraphic and depositional history in adjacent basins, combined with regional climatic evidence, can be utilized to reconstruct the history of the Appalachians. In this case, these data are used to separate the relative importance of climate and tectonism and to infer a consistent geomorphic evolution. As with other complex, coupled nonlinear systems, problems occur because the observations are almost always insufficient to derive a comprehensive picture which includes all the major variables. In addition, co-evolution of climate and tectonics and complex interactions between these two factors make clear separation of their relative significance or their specific roles difficult.

Second, geologic systems can be addressed from physical laws or principles (a model). In this case, the denudation rates and depositional history are predicted from a simulated climatic and tectonic evolution. The observations of stratigraphic and depositional history then become an independent test or verification of the capability of the model, and hopefully of the level of knowledge concerning climate and global tectonics. Unfortunately, this second approach is lacking because physically based models of the components of the climatic-topography–global tectonics system are presently inadequate.

However, efforts to simulate Phanerozoic climate evolution from physical principles have advanced considerably over the last decade. These model efforts provide some of the first independent predictions of climate which can be utilized to investigate the evolution of the Appalachians deductively. Although preliminary, the results provide a number of interesting insights.

The results described below are derived primarily from General Circulation Models (GCMs) of the atmosphere, with additional information derived from energy balance climate models and qualitative models of the general circulation. First, the long time scale climate evolution, primarily in response to changing geography and changing atmospheric composition, is considered. Second, short term climate variations, with orbital variations as the key example, are addressed as a possible additional influence during particular climatic regimes or for particular geographic configurations. Finally, the limitations of the model results are addressed.

Long time scale climate variations

The variations in the Earth's climate (Fig. 1) on long time scales (greater than 10 million years) have been attributed to a variety of terrestrial and extraterrestrial factors. Changes in geography and changes in atmospheric composition (primarily carbon dioxide concentration) are generally considered to be the two most important forcing factors for long term changes in climate. Since the acceptance of plate tectonics, changes in continental positions, and the distribution, elevation and size of land masses have been frequently cited as explanations of climate change (e.g. Crowell and Frakes, 1970; Donn and Shaw, 1977; Beaty, 1978; Tarling, 1978). More recently, Berner et al. (1983) demonstrated, using geochemical models, the potential for carbon dioxide concentrations to

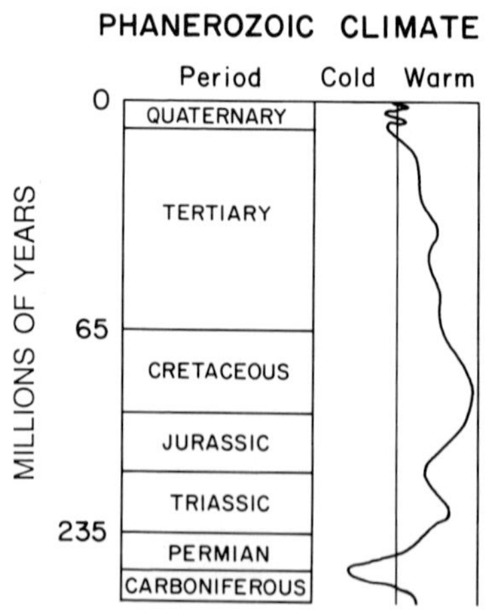

Fig. 1. Phanerozoic temperature history modified from Frakes (1979).

vary substantially (more than ten times present day levels) as a function of the rates of plate tectonic processes (CO_2 input from volcanoes) and the rates of chemical weathering of continental crust (dependent of precipitation, temperature and continental area). The importance of these two climatic forcing factors governs the following discussion of the long term climate variations from the Late Paleozoic to the present.

Geography and climate

The climatic response to changes in geography is complex. Land–sea distribution and the distribution and height of mountains influence the general circulation of the atmosphere and ocean through changes in the balances of energy, moisture and angular momentum. Regionally, climates are influenced by the nature of global circulation, but also by such factors as topography and proximity to moisture sources or continentality. Changes far afield from a region (e.g. mountain building on another continent) can produce large climate changes at a particular site even if the elevation and latitude of the site remains static because changes in atmospheric boundary conditions can influence the global circulation. Conversely, the large-

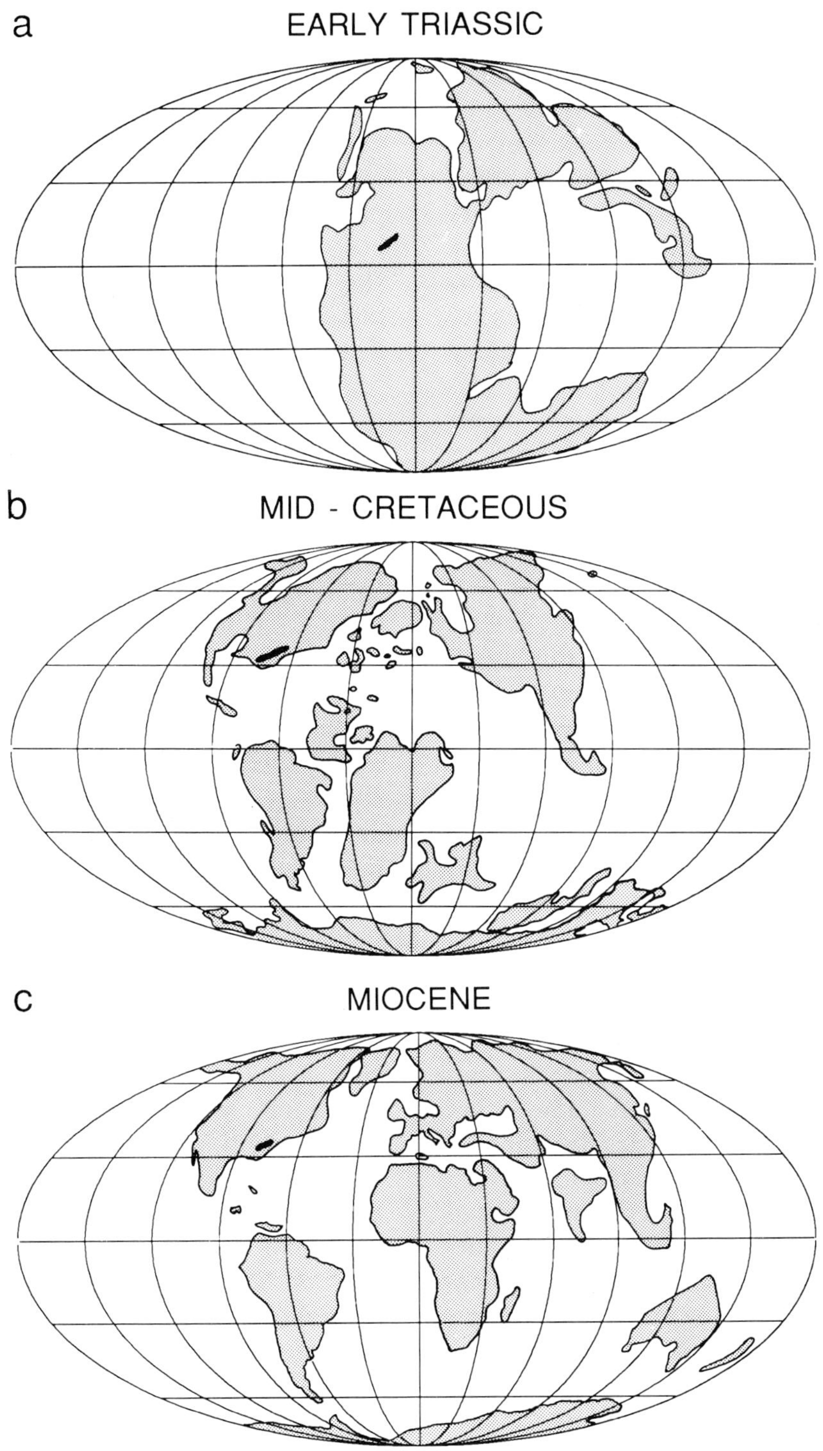

Fig. 2. The major paleogeographic regimes since the Carboniferous: (a) Early Triassic, after Ziegler et al. (1983); (b) mid-Cretaceous, after Barron et al. (1981); and, (c) Miocene, after Barron et al. (1981). Shading indicates regions above sea level. The Appalachians are indicated in black.

scale forcing factors may not change dramatically and the features of the general circulation may remain static, yet a relatively small latitudinal change at a specific site may lead to a large regional climate change because of the proximity of the site to a climatic transition zone. Consequently, the climatic history of individual regions is difficult to extrapolate from general rules and each time period must by viewed as an independent climatic regime with a potentially unique set of global and regional climatic controls. Despite the importance of this point, the discussion which follows is of necessity highly general. Therefore, the conclusions presented here will have many exceptions.

The global paleogeography from the Carboniferous to the present can be divided into three major regimes with regard to the evolution of the Appalachians (Fig. 2). First, from the Carboniferous to the Jurassic, the major continents are assembled into a supercontinent, Pangaea, and the Appalachians are a part of the central megacontinent at a tropical to subtropical latitude (Ziegler et al., 1979). From the Late Carboniferous to the Early Jurassic the Appalachians moved from a position of approximately 10–15° South to a position centered approximately on 15° North. According to the reconstructions of Ziegler et al. (1979) and given by Ziegler et al. (1983), the mountains are oriented northeast to southwest throughout this interval. Starting in the Jurassic the ocean enters the northern most portion of the rift between Africa and North America, bringing an oceanic region close to the Appalachians.

The opening of the North Atlantic, dated from 180 to 165 million years ago (e.g. see survey by Barron et al., 1981), places a large moisture source adjacent to the mountains and initiates the formation of the second major type of paleogeographic configuration – dominated by the large zonal Tethys Ocean within the subtropics (see e.g. the global paleogeographic maps given by Barron, 1987).

Third, the continued opening of the North and South Atlantic Oceans throughout the Cenozoic and the closure of the east–west Tethyan gateways results in a longitudinal ocean, connecting both poles and marked by east–west barriers. During this time period the Appalachians move from a subtropical position to a mid-latitude position. These three distinctive paleogeographic regimes which dominate the geographically-controlled climate evolution of the Appalachians guide the following discussion.

Pangaean climate

In the last decade, the global-scale Pangaean climate has been investigated using four different model methodologies. Parrish (1982), Parrish et al. (1982) and Parrish and Curtis (1982) applied general principles of atmospheric circulation to assess qualitatively pressure patterns, winds and rainfall for a series of paleogeographic reconstructions throughout the Phanerozoic. Crowley et al. (1987) has applied a two-dimensional energy balance climate model (vertically integrated) to examine the seasonal temperature cycle on the supercontinent-sized land mass of Gondwanaland during the Paleozoic. Kutzbach and Gallimore (1989) were the first to apply a fully-resolved GCM of the atmosphere to study Pangaean climate quantitatively. These model experiments specifically address the geography of Pangaea for the late Permian and Triassic. Finally, Hay et al. (1989, in press) completed GCM experiments for a series of idealized pole-to-pole supercontinents with variable topography. The sensitivity experiments by Hay and others have specific implications for the role of topographic expression as a control of Pangaean climate. Each of the above studies contributes new and different information which is a basis for interpreting the Pangaean climate record of the Appalachians.

The qualitative climate models developed by Parrish and colleagues are based upon superimposing the zonal climate (easterlies and westerlies) on paleogeographic reconstructions and then taking into consideration how the continents disrupt the zonal circulation (e.g. monsoons). The disruption of the zonal circulation occurs because of the different thermal properties of land and sea, and because seasonality and continentality influence the nature and distribution of high pressure and low pressure cells.

Parrish (1982) reconstructed the atmospheric circulation for the Late Carboniferous and Late Permian (Fig. 3), with the Appalachians as part of central Pangaea at a paleolatitude ranging from 15°S to the equator. During the Late Carboniferous, the Appalachian region is reconstructed with low pressure during the southern hemisphere summer, and as a transition area from high pressure in the south to low pressure in the north during the southern hemisphere winter. The uplift of air in low pressure regions should be associated with precipitation, while regions of dominantly high pressure are likely to be arid.

During the Permian, the shape of the tropical ocean to the east of Pangaea becomes more wedge-like, and Parrish (1982) interprets this condition as appropriate to disrupt the zonal circulation and promote monsoonal climates

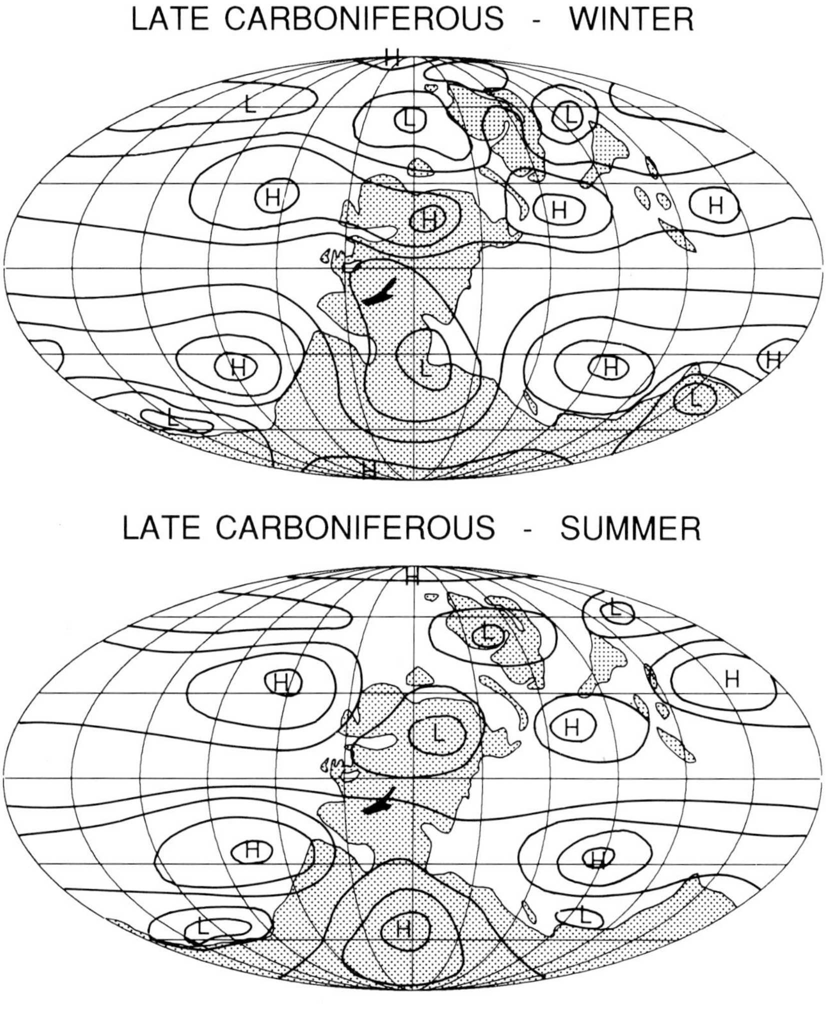

Fig. 3. Late Carboniferous and Late Permian atmospheric circulation indicating high (*H*) and low (*L*) pressure, after Parrish (1982). The Appalachians are indicated in black.

LATE PERMIAN - WINTER

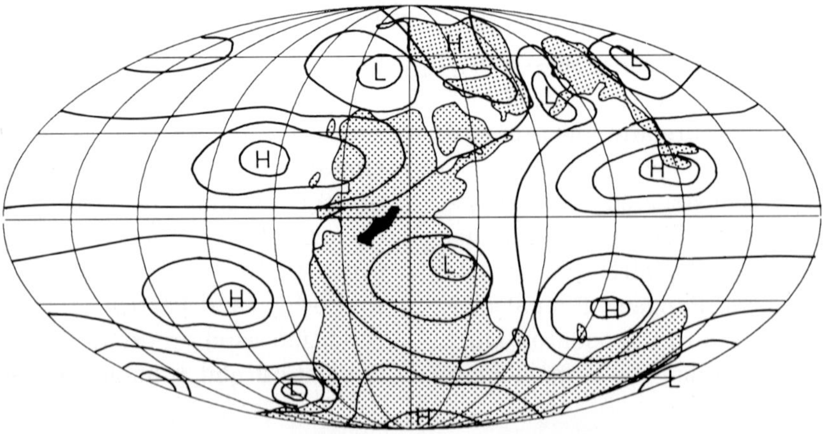

LATE PERMIAN - SUMMER

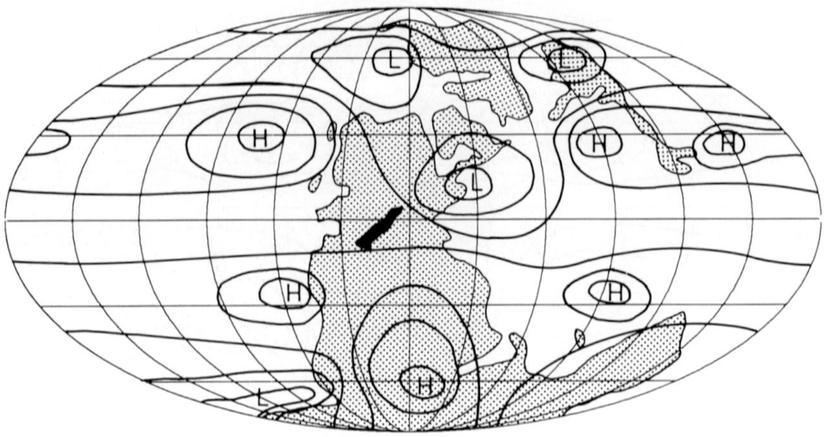

Fig. 3 (continued).

seasonally on both margins of the wedge-shaped ocean. In this case, the tropical low-pressure system with associated high rainfall moves seasonally from the northern and southern margins of the wedge-shaped ocean , skirting the eastern margins of the Appalachians. The rainfall is likely to be largely restricted to the continental margins, but may bring seasonally wet conditions to the eastern margin of the Appalachians during the southern hemisphere summer.

According to Parrish et al. (1982) and Parrish and Curtis (1982), the northward movement of the Appalachians through the Early Mesozoic accentuates this Permian pattern and places the Appalachians in a distinctly subtropical arid environment (Fig. 4), largely unaffected by the tropical low-pressure systems. Parrish and Curtis (1982) continue to suggest that the greatest potential for moisture supply to the Appalachians remains from easterly flow.

The contribution by Crowley et al. (1987) employs a seasonal energy balance climate model with realistic resolution of land–sea distribution to address the amplitude of the seasonal cycle of temperature in continental interiors. The most important point is that the temperatures in the interiors of the large con-

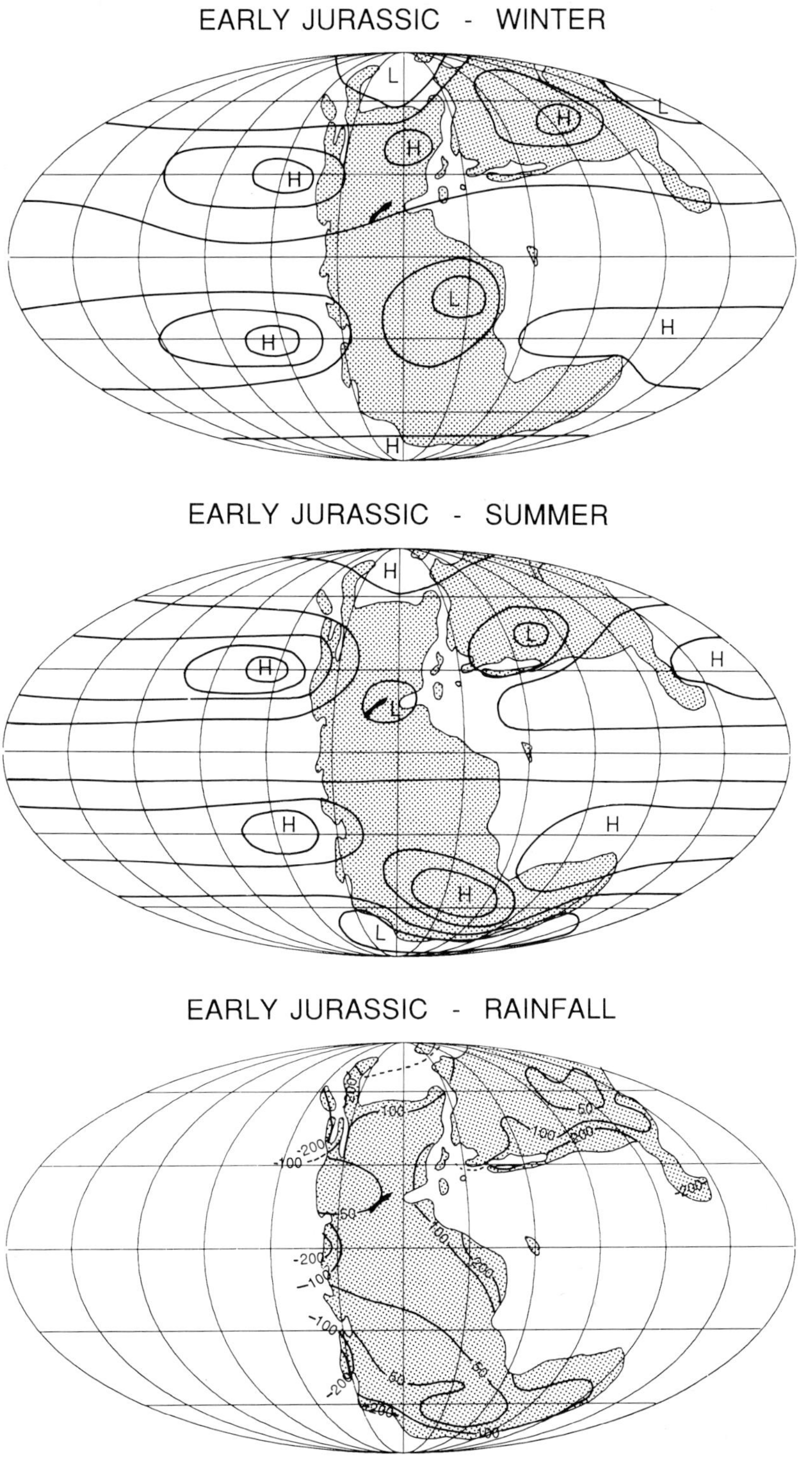

Fig. 4. Early Jurassic atmospheric circulation from Parrish et al. (1982) and rainfall patterns (< 50 = low; 50–100 = moderately low; 100–200 = moderately high; > 200 = high rainfall) after Parrish and Curtis (1982). The Appalachians are indicated in black.

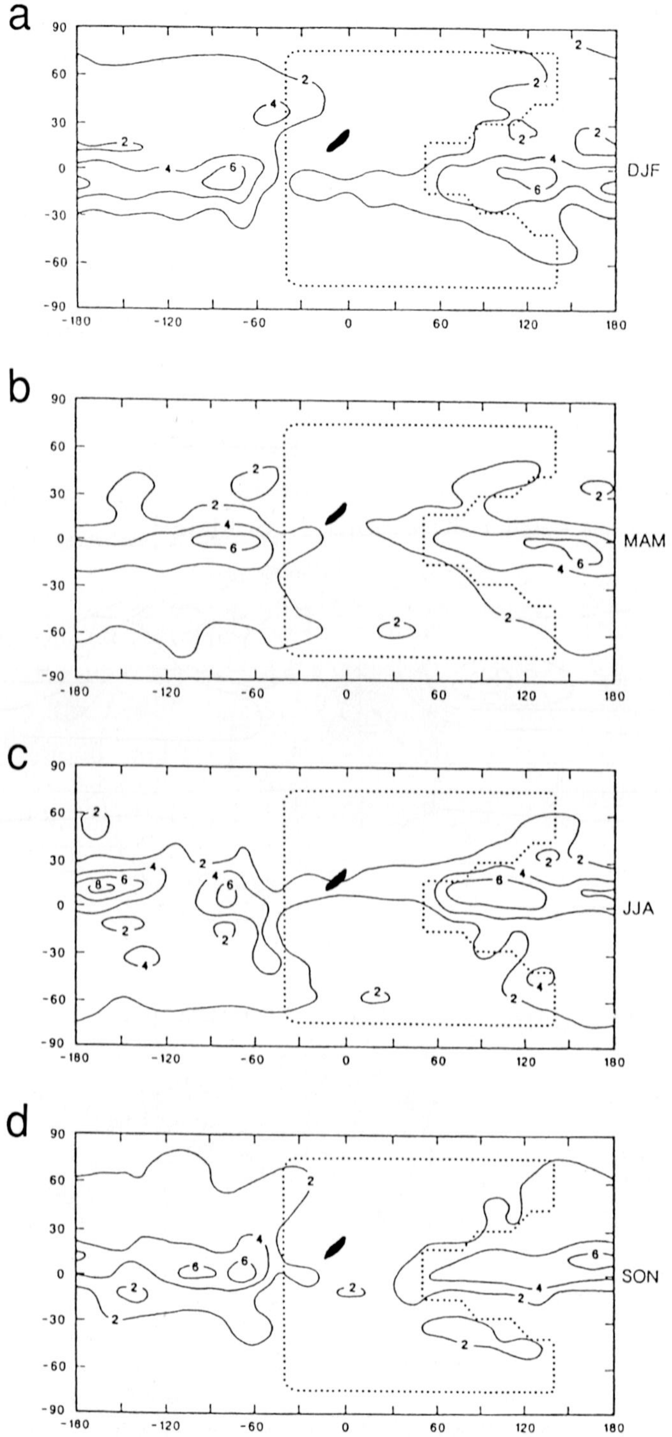

Fig. 5. Precipitation rate (mm d^{-1}) for the four seasons with Pangaea geography (position of the Appalachians are indicated in black although see text for a description of the model topography) from Kutzbach and Gallimore (1989).

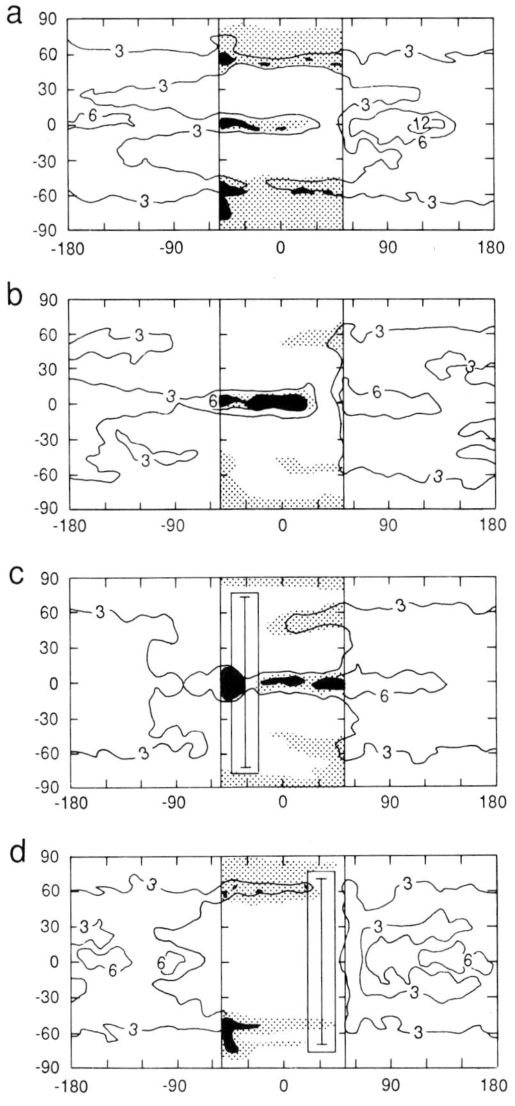

Fig. 6. Precipitation rate (mm d^{-1}) for four idealized pole to pole continents (the land area is centered in the map projection): (a) average relief, 750 m; (b) high plateau, 1500 m; (c) average relief, 750 m with western margin 3 km mountain belt; (d) average relief, 750 m with eastern margin 3 km mountain belt. Areas of runoff are in black. Regions of soil moisture > 5 cm are stippled.

tinents are primarily governed by thermal inertia arguments. Land has a small heat capacity and therefore the temperature reflects the amount of incoming solar energy. In short, larger continents will closely track the seasonal range in solar insolation. In the case of the low-latitude Appalachians, the contribution by Crowley et al. (1987) would indicate quite high summer temperatures, with a relatively small seasonal range, and generally arid conditions in continental interiors.

The GCM experiments of Kutzbach and Gallimore (1989) address the Late Permian and Triassic case specifically using a low-resolution (11.6 by 11.25° latitude–longitude grid) climate model with a coupled mixed layer ocean of 50 m depth. The model experiments use an idealized geography which is symmetric about the equator and includes the wedge-shaped reentrant of the Tethys Sea between Laurasia and Gondwanaland (Fig. 5). Kutzbach and Gallimore (1989) performed experiments for a flat sea level continent and a flat 1 km high continent. The results of these two experiments were qualitatively similar.

In the simulations of Kutzbach and Gallimore (1989) the central tropical and subtropical region of Pangaea during the Late Permian and Triassic has a predicted seasonal temperature range between 2 and 20°C (depending on latitude) with maximum temperatures greater than 35°C. The annually-averaged precipitation minus evaporation is negative throughout the Appalachian region. The predicted, annually-averaged precipitation for the Appalachian region is less than 2 mm d^{-1}, and is mostly less than 1.5 mm d^{-1}. Precipitation exceeding 2 mm d^{-1} occurs only during model June, July and August and is associated with southeasterly flow and the northern hemisphere summer monsoon.

Essentially these GCM experiments suggest that tropical conditions were quite warm and arid, except along the east and west coasts. Seasonally high, continental precipitation occurs only at mid-latitudes.

Hay et al. (1989, in press) performed mean annual experiments using a hemispherically symmetric wedge-shaped pole-to-pole continent (a highly idealized megacontinent) with the Community Climate Model, a spectral General Circulation Model of the Atmosphere from the National Center for Atmospheric Research.

The model includes an energy balance ocean which lacks heat storage or heat transport. Four experiments (Fig. 6) were performed: (1) low relief, 750 m average elevation; (2) high plateau, 1500 m average elevation; (3) case (1) with a 3 km mountain range to the west; and (4) case (1) with a 3 km mountain range to the east. All of the model experiments produced significantly more arid conditions than exist today. High overall topographic expression and mountains on the western margin promoted tropical precipitation from the west. Eastern mountains produced the most arid tropical continental interior, but increased rainfall on the eastern margin.

In summary the climate model results suggest that with the possible exception of the Late Carboniferous, the Pangaean geography should promote extreme tropical and subtropical aridity. At most, the Appalachians would have one season of moderate precipitation when the atmospheric flow is from the east or southeast. The extreme aridity is largely confirmed by observations from evaporites, coals, bauxites, cherts and calcrete (Van Houten, 1982; Frakes, 1979).

Tethyan climate

A series of GCM experiments and qualitative climate models have also been utilized to address changes in climate as a function of geography for the Cretaceous. The northern hemisphere geography during this time period is dominated by Tethys, a zonal subtropical ocean bordered to the north by a north America–Greenland–Eurasia continent. Tethys is particularly large and zonal during the high sea level periods of the middle and late Cretaceous (Fig. 2b). The decrease in size and decrease in zonality and continuity of the Tethys Ocean is one of the most distinctive changes in geography from the Cretaceous to the present day.

Mid-Cretaceous climate has been simulated using several GCMs for fixed January and July insolation (Barron and Washington, 1982), mean annual insolation (Barron and Washington, 1984) and for the full annual cycle (Barron et al., 1989, in press). The role of Tethyan geography in governing Appalachian climate is clear from this series of experiments.

First, the zonal subtropical nature of Tethys promotes geographically-forced regions of high continental rainfall on both northern and southern borders of Tethys. In early grid-point GCM experiments (Barron and Washington, 1982), the position of Tethys controlled the position of the Intertropical Convergence Zone and the mid-latitude precipitation maxima. During summer, high mid-latitude precipitation was broadly distributed over the northern hemisphere continents. During winter, land–sea thermal contrasts controlled, and apparently fixed, the position of high and low pressure systems along the northern continental margin of Tethys, promoting high winter precipitation. More recent annual cycle experiments (Fig. 7) also illustrate the degree to which high precipitation is focussed on the northern and southern borders of the subtropical Tethys Ocean. The position of centers of high rainfall are clearly associated with the Tethyan margin.

In short, the condition of a subtropical ocean promotes high rainfall over large areas of the northern hemisphere continent. In particular, the Appalachians, as part of the northern margin of Tethys, have a predicted high precipitation (approaching modern tropical values) during the winter. Barron et al. (1989, in press) attributes the large rainfall to mid-Cretaceous warmth, to the role of Tethys as a source of moisture and to the role of a zonal land–ocean boundary with associated thermal contrasts as a control on the nature of circulation patterns. Essentially, the land–sea configuration is controlling the position of the mid-latitude precipitation belt. Global warmth and the large subtropical moisture source influence the rate of precipitation. The qualitative climate model reconstructions for the Cenomanian by Parrish (1982) and Parrish and Curtis (1982) also suggest high rainfall on the margins of Tethys based

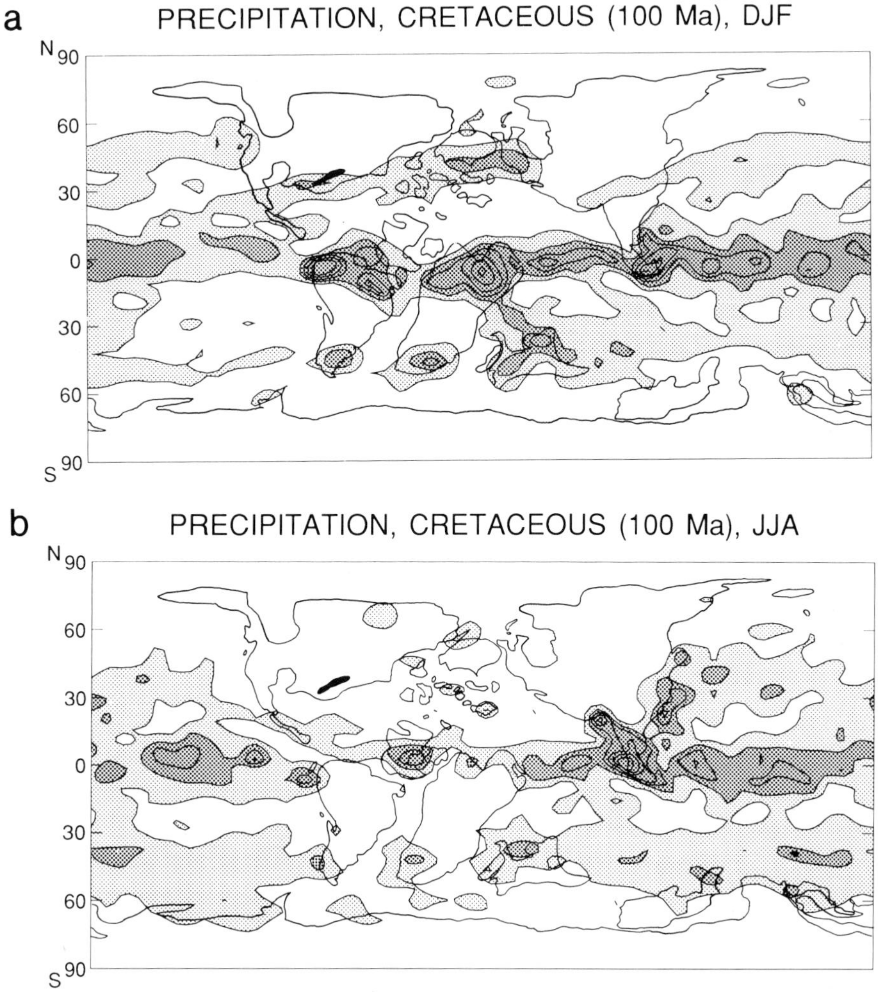

Fig. 7. The distribution of precipitation (mm d^{-1}) from a full annual cycle simulation with Cretaceous geography (~ 100 Ma) averaged over (a) DJF = December, January and February and (b) JJA = June, July and August. Light stipple is 3-6 mm d^{-1} and heavy stipple is ≥ 6 mm d^{-1}. The location of the Appalachians is indicated in black.

on the suggested position of high and low pressure systems.

These predictions indicate that the Cretaceous record from the northern hemisphere mid-latitude continents should reflect unusually high precipitation rates. The evidence clearly supports the model results. The northern hemisphere continents are characterized by abundant laterites (Khain and Ronov, 1960). Widespread bauxites with other indications of humid, seasonally wet conditions occur over large areas of North America and Europe (Hallam, 1985). Cretaceous kaolin deposits in central Georgia and South Carolina are the largest kaolin resource in the United States (Tourtelot, 1983). Many deep sea drilling sites in the North Atlantic are characterized by abundant Cretaceous kaolinite and smectites (Chamley, 1979) indicative of warm and humid conditions everywhere surrounding the North Atlantic. The Cretaceous is also a time of extensive coal formation, and over 50% of the Cretaceous coals of North America are of mid-Cretaceous age (Beeson, 1984). The climate model results suggest that the great extent of Cretaceous coals, bauxites and kaolinite reflect the optimal geographic distribution of land and sea and climatic warmth which promoted high continental precipitation.

The Cretaceous is frequently considered as the warmest time period in Earth history which can be confidently reconstructed (see review by Barron, 1983). Equability and lack of seasonality in temperature is usually considered as a characteristic of Cretaceous warmth. However, in model simulations by Barron and Washington (1982) and Schneider et al. (1985), continental interiors at high to mid-latitudes experienced subfreezing temperatures in winter. Basically, the small heat capacity of the continental surface and the limited capability of the atmosphere to advect heat into continental interiors, result in the interior of large continents being governed largely by the amount of incoming solar energy (see also results by Crowley et al., 1986). Consequently, in winter at mid to high latitudes, the climate models predict subfreezing conditions with winter warmth largely restricted to the continental margins. Apparently this model result is not at odds with the observations, as almost all sites indicative of Cretaceous warmth plot on continental margins when paleogeographic reconstructions are utilized (Barron and Washington, 1982).

Transition to modern geography

From the Late Cretaceous to the present the Appalachians remain close to 40°N in latitude and the primary large scale changes in geography consist of a general decrease in global eustatic sea level (with exceptions such as notable sea level rises in the Eocene and Miocene), and the decrease in size, zonality and continuity of Tethys with the development of a longitudinal North Atlantic Ocean.

The qualitative climate model reconstructions of Parrish and Curtis (1982) suggest continued high rainfall on the eastern North American margin with an easterly or southeasterly source. High rainfall is generally restricted to the margin except during the Eocene, at which time much of North America is predicted to have high rainfall.

Barron (1985), Barron et al. (1989, in press) and Barron and Frakes (1989, in press) described precipitation results for mean annual general circulation model experiments for the Campanian, Paleocene, Eocene, Miocene and Present. First, these experiments illustrate that continental precipitation rates on land surfaces poleward of Tethys are a function of the area (% ocean in the subtropics) of the Tethys Ocean (Fig. 8). The relationship between higher continental precipitation rates and larger subtropical ocean areas supports the previously stated argument that land–sea distribution influences continental precipitation rates by virtue of modifying the source area of moisture. This relationship suggests a general decline in precipitation rates though the Cenozoic for the Appalachians, as modified by changes in sea level. Higher sea level (Eocene, Miocene) stands should tend to promote somewhat higher precipitation, if all other factors are equal.

A second significant aspect of this series of experiments is the degree to which the model simulations are characterized by regions of highly focussed precipitation on the margins of

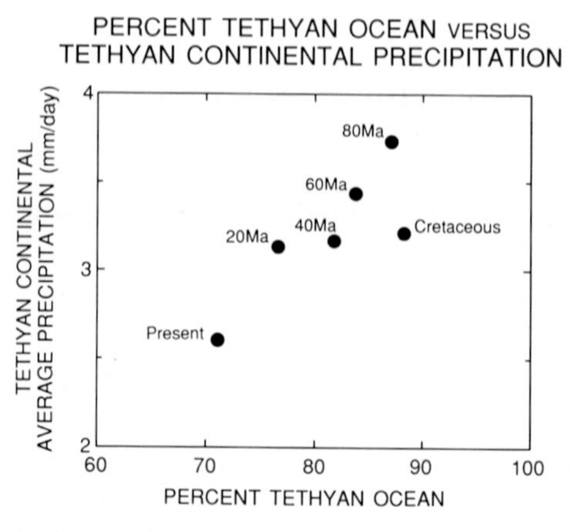

Fig. 8. A model comparison of the relative area of the Tethyan Ocean or Subtropical ocean (%) with the rate of continental precipitation (mm d^{-1}) poleward of Tethys for the present day, 20 Ma (Miocene), 40 Ma (Eocene), 60 Ma (Paleocene), 80 Ma (Campanian), and 100 Ma (Cretaceous).

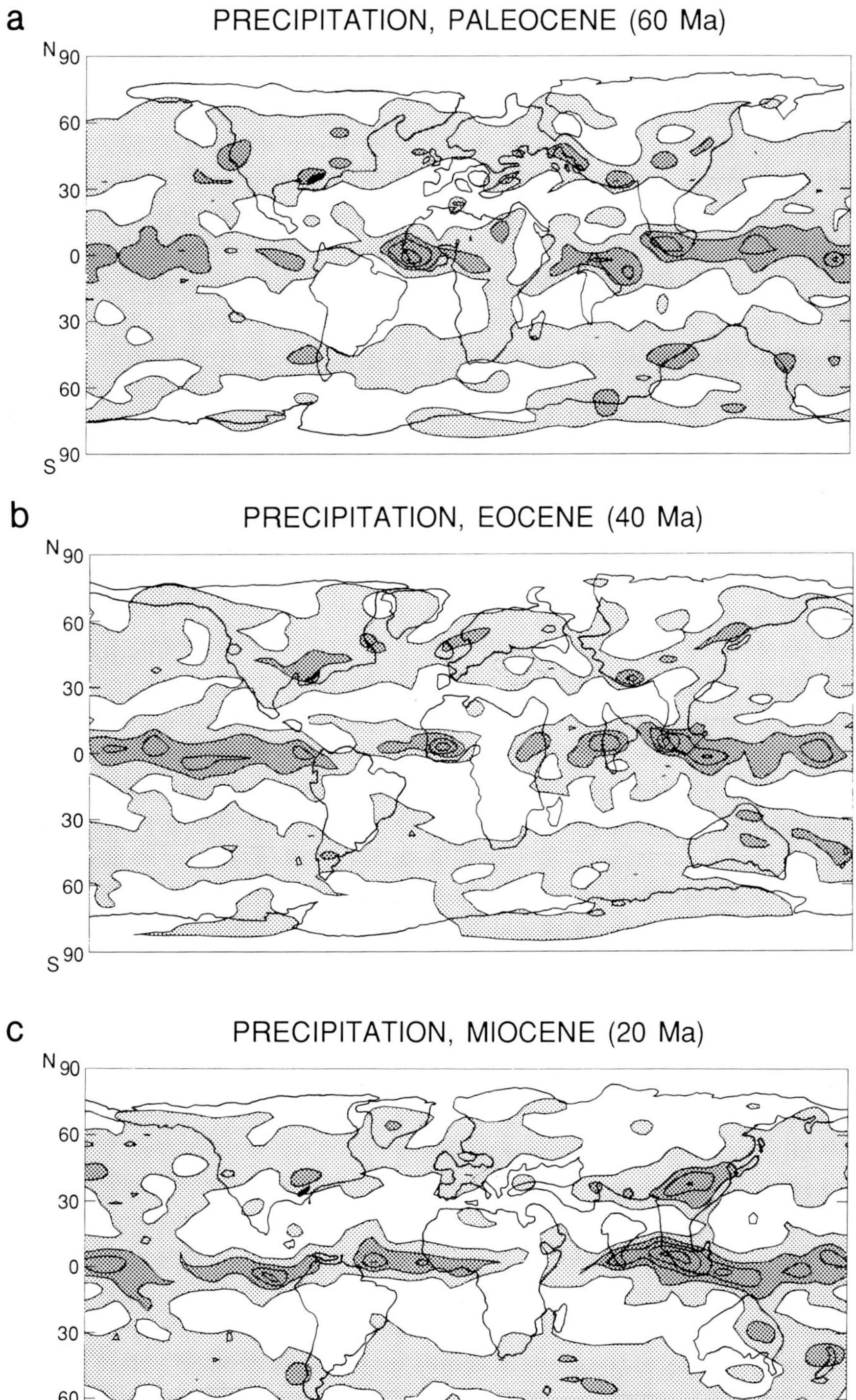

Fig. 9. Precipitation fields (mm d^{-1}) from mean annual simulations for: (a) Paleocene (60 Ma), (b) Eocene (40 Ma), and (c) Miocene (20 Ma). Light stipple is 3-6 mm d^{-1} and heavy stipple is ≥ 6 mm d^{-1}. The Appalachians are indicated in black.

the Tethys Ocean. As Tethys becomes smaller and less continuous, the mid-latitude precipitation maxima become less well defined and less focussed on the continental margin. In particular, from the Miocene to the present the mid-latitude precipitation maxima are less continuous and the least focussed on the continental margin (Fig. 9). In all cases, the Appalachians are a region of higher mid-latitude precipitation compared to the other mid-latitude regions. The precipitation over the Appalachians, as predicted by the model, is the best developed for the Eocene of all the Cenozoic simulations.

The climatic record from the Cretaceous to the present is dominated by the transition between the warmth of the Cretaceous to the widespread glaciation of the Late Cenozoic (Fig. 1). This global cooling is documented by a variety of data (see the review by Frakes, 1979). Despite a number of hypotheses to explain the global cooling trend of the Cenozoic (Donn and Shaw, 1977; Beaty, 1978; Barron et al., 1980; Berggren, 1982; Barron, 1985; Crowley et al., 1986), the cause of the global cooling trend has not been convincingly demonstrated, although several hypotheses concerning paleogeography, topography and changes in atmospheric composition remain valid. The model simulations described above do not reflect a global cooling trend, suggesting that the paleogeographic variables do not provide an explanation and/or the mean annual simulations are inadequate for this purpose (Barron, 1985).

Barron et al. (1989, in press) demonstrates through a series of experiments that the intensity of the hydrologic cycle as indicated by the global rate of precipitation and evaporation is related to planetary warmth, suggesting that the global cooling (although not simulated) may result in a decreased intensity of the global hydrologic cycle. This decrease may accentuate the Cenozoic decline in northern hemisphere continental precipitation which is illustrated in Fig. 8.

Again, the general results from the model simulations are confirmed by observations from North America. The data suggest that a well-defined Mesozoic mid to high latitude rainy belt persisted into the Paleocene and Eocene, with the floral data suggesting that the Eocene of North America was characterized by increased warmth and humidity (Frakes, 1979). Much of the Late Cenozoic changes are dominated by the global cooling and polar glaciation.

Carbon dioxide and climate

Carbon dioxide variations have been proposed as a major influence on climate during Earth history. Budyko and Ronov (1979) proposed changes in atmospheric carbon dioxide as an explanation of the icehouse–greenhouse climates during the Phanerozoic. Berner et al. (1983) and Lasaga et al. (1985) used geochemical mass balance arguments to suggest that carbon dioxide was as much as ten times present day values during the Cretaceous, based on rates of seafloor spreading, inferred rates of volcanism, and calculated rates of chemical weathering of the continents. In the geochemical model, CO_2 levels declined to the present day values, but with somewhat higher values predicted for the Eocene. Paleozoic predictions of CO_2 levels based on the volumes of specific rock types (Budyko and Ronov, 1979) and utilizing carbonate sedimentology (Sandberg, 1985) suggest higher CO_2 concentrations during the Carboniferous and lower concentrations from the Late Permian to the Jurassic.

The climate response to changes in atmospheric carbon dioxide has been addressed by a broad range of sensitivity experiments with GCMs (e.g. Schlesinger and Mitchell, 1987) because of the prospects of human-induced global warming due to the burning of fossil fuels. In addition, Barron and Washington (1985) completed a mid-Cretaceous mean annual GCM experiment with four times present day carbon dioxide concentration. Because of the different thermal characteristics of land and sea and the number of ways in which surface conditions influence the atmospheric circulation, the climate response to carbon dioxide is likely to be geography dependent, and consequently, generalizations from modern and Cretaceous experiments may have limited application. However, the climate response in modern simulations and in the one Cretaceous experiment have some similar features.

All the model experiments (Schlesinger and Mitchell, 1987) illustrate an enhanced hydrologic cycle with higher concentrations of carbon dioxide. For purposes of illustration, the Cretaceous experiment by Barron and Washington (1985) can be compared with present day experiments utilizing the same GCM (Washington and Meehl, 1983; 1984). In both cases, the largest increases in precipitation occurred in the tropics, in association with the Intertropical Convergence Zone (Fig. 10). This increase is accentuated over the tropical ocean. In addition, the central, mid-latitudinal continental regions tended to become more arid. The precipitation was enhanced at higher latitudes, poleward of the mid-latitude maxima in the present day CO_2 concentration cases.

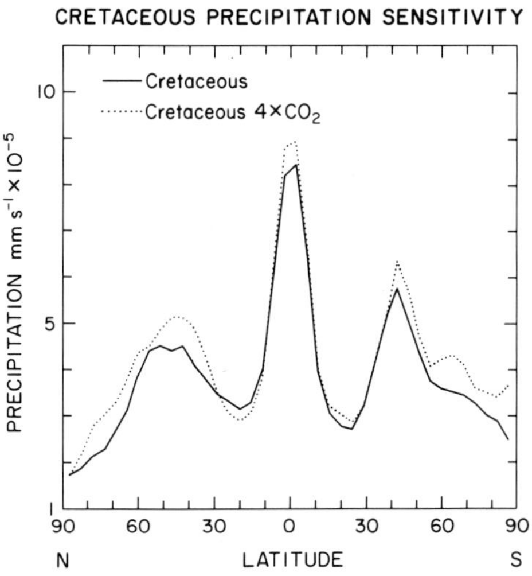

Fig. 10. Comparison of zonally averaged precipitation (mm $s^{-1} \times 10^{-5}$) with respect to latitude for mean annual simulations of the Cretaceous ($\sim$100 Ma) and Cretaceous with four times present CO_2 levels.

Short time scale climate variations

The climatic record of the Phanerozoic is the product of a number of forcing factors which operate over tens of millions of years, but is also a response to forcing factors which operate over tens of thousands or hundreds of thousands of years or is in response to specific events. The role of specific events, such as opening or closing oceanic gateways, episodic volcanism, mountain building, and changes in deepwater formation, have not been addressed with climate models. Consequently, their role in governing the climate evolution of the Appalachians would be largely speculative.

However, one shorter time scale climatic forcing factor, Milankovitch orbital variations, has received considerable attention with climate models. Since original variations modulate the amplitude of seasonal cycle of temperature and precipitation, this forcing factor may be of significance for reconstructing the climatic and geomorphologic evolution of the Appalachians.

Variations in the Earth's orbital parameters (obliquity or tilt, eccentricity, and time of perihelion) results in temporal and latitudinal changes in solar insolation. As an illustration, if the obliquity or tilt is large, if the eccentricity is large, and if the earth is closest to the sun in summer (perihelion in summer) then the amplitude of the seasonal cycle is large. For example, 9000 years BP the time of perihelion was 30 July (compared to January 3 today) and the obliquity was 24.23° (compared to 23.45° today), resulting in an increased amplitude of the northern hemisphere seasonal cycle of insolation of 14%.

Kutzbach (1981) and Kutzbach and Guetter (1984) pioneered the efforts which demonstrated that these changes accentuated the Indian-Asia monsoon 9000 years BP. The explanation is straightforward. The oceanic response to a change in the amplitude of the seasonal cycle is minimal because of the ocean's large heat capacity. In contrast, the continental regions have little heat capacity. Consequently, a 7% increase in summer insolation may result in a continental temperature increase of nearly 5°C. The increased thermal contrast between land and sea than drives a stronger monsoon. Kutzbach and Guetter (1984; 1986) illustrate a 30–35% increase in precipitation over the monsoonal area of Asia for 9000 years BP.

Two points, in this regard, are worth focussing on. First, the orbital parameters are strongly periodic (eccentricity; ~100,000 years, obliquity; ~41,000 years, and perihelion, ~19,000; 23,000 years). Consequently, the changes in precipitation and temperature will be rhythmic. Simple oscillation cycles have been recognized as rhythmic sedimentation during many different periods of Earth history (e.g. Schwarzacher and Fisher, 1982). In particular, many Cretaceous sequences from the epicontinental and pelagic realms (Fischer, 1980; Arthur et al., 1985; Barron et al., 1985) have been related to Milankovitch orbital periodicities. Apparently, orbital variations of this type should be characteristic of the history of the Appalachians.

Second, any climatic phenomena which are associated with changes in land–sea thermal contrast will be influenced by changes in the seasonal cycle of solar insolation due to orbital variation. The following are proposed as factors which will be affected by orbital variations: (1) Winter storm distribution and intensity is governed by the juxtaposition of warm (oceanic) and cool (continental) air masses. Increased amplitude of the seasonal cycle of insolation will therefore enhance winter storm distributions. (2) Any site of geographically controlled precipitation (e.g. northern margin of Tethys in the Cretaceous and Permo–Triassic as given in Figs. 4 and 7 will be enhanced or modulated by orbital variations following the arguments given above. (3) Changes in freshwater influx into basins or marginal seas may modulate basin stratification or circulation, and hence sediment character. Depending on the connections with the open ocean, changes in marginal sea characteristics may influence the global ocean circulation. (4) If the site of interest is adjacent to a climatic transition zone (location of atmospheric convergences or divergences) small shifts in surface pressure patterns may result in marked local changes in prevailing winds and moisture availability. (5) Changes in land–sea thermal contrasts which result in significant changes in coastal winds may modulate basin productivity and hence the sediment character.

This "list" of possible climatic responses to orbital variations may have importance for the evolution of the Appalachians in three cases. First, as described earlier, the seasonal precipitation associated with a southeasterly flow or easterly flow during the Permo–Triassic was associated with the northern hemisphere summer monsoon. The strength of the summer monsoon should be modulated by orbital variations, resulting in rhythmic wetter versus drier climatic conditions. Second, the precipitation on the northern margin of Tethys during the Cretaceous is clearly geographically forced and related to land–sea thermal contrasts. In a set of sensitivity experiments to examine the importance of orbital variations during the Cretaceous, Glancy et al. (1986; 1989, in press) demonstrated an 8–12 mm d^{-1} increase in precipitation on the northern margin of Tethys (including the Appalachians) during the winter in the case of a larger amplitude of the seasonal cycle of insolation. Third, with the development of the North Atlantic and northern hemisphere glaciation, the Appalachians should be influenced increasingly by winter storms. The role of orbital variations in modulating the intensity of these storms has not been evaluated.

In summary, particular geographic conditions tend to promote rhythmic climatic change over periods of tens of thousands of years. Particular time periods in Appalachian evolution, because of the distribution of continents, should be influenced by substantial climatic change on these time scales.

Summary of model predicted climate changes in the Appalachians

In general the above studies provide a consistent view, with the following key points:

(1) The Late Carboniferous has the greatest potential to have seasonal high rainfall over the Appalachians of any of the Pangaean reconstructions.

(2) The continental configurations for the Permian to Early Jurassic would promote extreme tropical and subtropical aridity unlikely to have a positive annual precipitation minus evaporation balance, and would promote warm temperatures in comparison with the present day. Moderate precipitation, with an easterly source, may characterize one season of the year when the flow is easterly or southeasterly, but most high precipitation at low latitudes will be confined to the margins of the early Tethys Ocean.

(3) The extreme continental aridity predicted by models for the Permian to the Early Jurassic seems largely confirmed by the observations. Frakes (1979) reconstructs a tropical arid zone based on the distributions of evaporites, coals, reefs, bauxites and cherts. The occurrence of calcite also suggests a low-latitude belt of dryness (Van Houten, 1982).

(4) In the late Mesozoic and early Tertiary, the condition of a zonal, subtropical Tethys to the south of the Appalachians promotes intense, highly focussed rainfall on the continental margin including the Appalachians. Precipitation is predicted to be most extreme in winter. Continental interiors may have experienced subfreezing conditions. The continental margin (eastern margin of the Appalachians) was likely to have experienced mild seasonality.

(5) Times of higher sea level stands such as the middle and late Cretaceous for Tethyan geographies probably were also times of higher Appalachian precipitation.

(6) The predicted high precipitation for the Cretaceous is, in general, confirmed based on the distribution of extensive laterites, kaolins and coals.

(7) The model simulations indicate that the decrease in size, zonality and continuity of Tethys through the Tertiary result in a decrease in northern hemisphere precipitation. In addition, the regions of highly focussed continental margin precipitation become less continuous.

(8) The Appalachians remain a region of focussed precipitation throughout the Cenozoic in the model simulations although in general the rates decline through time.

(9) The global cooling trend may accentuate the Cenozoic decline in precipitation described in points (7) and (8).

(10) High stands in sea level (Eocene, Miocene) should tend to promote higher precipitation in comparison with the Cenozoic trends.

(11) Higher concentrations of carbon dioxide, hypothesized for the Carboniferous, Cretaceous and Eocene (Budyko and Ronov, 1979; Berner et al., 1983) may promote (a) higher tropical rainfall, (b) continental interior aridity at mid-latitudes, and (c) increased precipitation poleward of the mid-latitude precipitation maxima.

(12) Many short time scale events and factors may have influenced Appalachian climates, but these factors are largely untested.

(13) Periodic climatic forcing such as orbital variations, are likely to be important climatic modulators for specific geographic conditions. In particular, rhythmic climatic forcing may have been important during the Permo–Triassic, the mid-Cretaceous and during Cenozoic periods of particularly intense, geographically-focussed precipitation (for example, the Eocene).

Discussion

The climate predictions described here are largely a first-order view of climatic change computed from basic physical principles in response to particular forcing factors (geography, carbon dioxide and orbital variations). The results provide a number of insights into climate sensitivity and represent a prediction or reconstruction which is independent of regional observations of Appalachian climate. In large measure, the results are confirmed by general paleoclimatic data, although no regional comparison has been undertaken. The strength of the results stems from a firm foundation in physical principles, and the fact that the predicted climate was derived from a global approach.

The weaknesses of the predictions are equally important. First, it is likely that all of the potentially important forcing factors have not been identified for any time period in Earth history. Second, the predictions for the Appalachians are from coarse resolution models, and consequently many regional climate changes (e.g. along the axis of the Appalachians) have not been considered. Third, the importance of episodes of Appalachian uplift and the role of topography in influencing regional climates has been ignored. Fourth, climate models have a number of limitations. For example, the role of the oceans is only crudely incorporated in modern, state-of-the-art climate models, and the potential role of the ocean in climate change is largely untested. In addition, precipitation, which is characterized by high spatial and temporal resolution, is difficult to simulate in coarse resolution climate models. In general, the large-scale features of precipitation and the areas of geographically-forced precipitation are well simulated. Convective precipitation is poorly simulated. Finally, the Phanerozoic climate history of glaciations and warm time periods is yet to be adequately explained. Consequently, much of the climate sensitivity or important climatic forcing factors may yet be discovered.

Despite the limitations, the climate models appear to yield a first-order view of climate evolution which can be verified by observations. This point may reflect the degree to which geography controls the nature of the atmospheric circulation. Most importantly, the model results suggest large and well-defined climate changes as a function of geography, carbon dioxide concentration and orbital variations which can be independently compared with the geologic record. The proposed changes in climate are large, ranging from extreme aridity to highly focussed rainfall, and from periods susceptible to rhythmic variations to periods much less susceptible. These changes impact the rate of denudation, and certainly support the concept that the history of the Appalachians is likely to be a complex product of tectonics and climate. The differences are large enough to question interpretations attributed solely to one factor, such as uplift (e.g. Poag and Sevon, this volume).

Physically-based models represent a building block for a tectonics-climate-morphology model of the Appalachians. The coupling of climate models with topographic evolution and the evolution of independent, physically-based predictions with the geologic record represents the frontier areas which will provide major new advances.

References

Arthur, M.A., Dean, W.E., Bottjer, D.J., and Scholle, P.A., 1985. Rhythmic bedding in Mesozoic–Cenozoic pelagic carbonate sequences: the primary and diagenetic origin of Milankovitch-like cycles. In: A. Berger, J. Imbrie, J. Hays, G. Kukla and B. Saltzman, (Editors), Milankovitch and Climate. Reidel, Dordrecht, pp. 375–401.

Barron, E.J., 1983. A warm, equable Cretaceous: The nature of the problem. Earth Sci. Rev., 19: 305–338.

Barron, E.J., 1985. Explanations of the Tertiary global cooling trend. Palaeogeogr. Palaeoclimatol. Palaeoecol., 50: 45–61.

Barron, E.J., 1987. Global Cretaceous Paleogeography – International Geologic Correlation Program project 191. Palaeogeogr. Palaeoclimatol. Palaeoecol., 59: 207–214.

Barron, E.J. and Frakes, L.A., 1989. Climate model evidence for variable continental precipitation and its significance for the phosphate problem. In: W.C. Burnett and Riggs (Editors), Phosphate Deposits of the World – Vol. 3. Cambridge University Press (in press).

Barron, E.J. and Washington, W.M., 1982. Cretaceous climate: a comparison of atmospheric simulations with the geologic record. Palaeogeogr. Palaeoclimatol. Palaeoecol., 40: 103–133.

Barron, E.J. and Washington, W.M., 1984. The role of geographic variables in explaining paleoclimates: Results from Cretaceous climate model sensitivity studies. J. Geophys. Res., 89: 1267-1279.

Barron, E.J. and Washington, W.M., 1985. Warm Cretaceous climates: High atmospheric CO_2 as a plausible mechanism. In: E.T. Sundquist and W.S. Broecker (Editors), The Carbon Cycle and Atmospheric CO_2: Natural Variations Archean to Present. Am. Geophys. Union, Washington, D.C., pp. 546-553.

Barron, E.J., Sloan, J.L., II and Harrison, C.G.A., 1980. Potential significance of land-sea distribution and surface albedo variations as a climatic forcing factor: 180 m.y. to the present. Palaeogeogr. Palaeoclimatol. Palaeoecol., 30: 17-40.

Barron, E.J., Harrison, C.G.A., Sloan, J.L., II and Hay, W.W., 1981. Paleogeography, 180 million years ago to the present. Eclogae Geol. Helv., 74: 443-470.

Barron, E.J., Arthur, M.A. and Kauffman, E.G., 1985. Cretaceous rhythmic bedding sequences: A plausible link between orbital variations and climate. Earth Planet. Sci. Lett., 72: 327-340.

Barron, E.J., Hay, W.W. and Thompson, S.L., 1989. The hydrologic cycle: A major variable during Earth history. Global Planet. Change (in press).

Beaty, C., 1978. The causes of glaciation. Am. Sci., 66: 452-459.

Beeson, D.C., 1984. Relative significance of tectonics, sea level fluctuations, and paleoclimate to Cretaceous coal distribution in North America. Natl. Center Atmos. Res. Coop. Thesis, University of Colorado, 83, 210 pp.

Berggren, W., 1982. Role of ocean gateways in climatic change. In: Climate in Earth History, Studies in Geophysics. National Academy Press, Washington, D.C., pp. 118-125.

Berner, R.A., Lasaga, A.C. and Garrels, R.M., 1983. The carbonate-silicate geochemical cycle and its effect on atmospheric carbon dioxide over the last 100 million years. Am. J. Sci., 283: 641-683.

Budyko, M.I. and Ronov, A.B., 1979. Chemical evolution of the atmosphere in the Phanerozoic. Geochem. Int., pp. 1-9 (Engl. Transl.)

Chamley, H., 1979. North Atlantic clay sedimentation and paleoenvironment since the Late Jurassic. In: M. Talwani, W. Hay and W.B.F. Ryan, (Editors), Deep Drilling Results in the Atlantic Ocean: Continental Margins and Paleoenvironment. Am. Phys. Union, Washington, D.C., Ewing, Ser. 3, pp. 342-361.

Crowell, J. and Frakes, L.A., 1970. Phanerozoic glaciation and the causes of the ice ages. Am. J. Sci., 268: 193-224.

Crowley, T.J., Short, D.A., Mengel, J.G. and North, G.R., 1986. Role of seasonality in the evolution of climate during the last 100 million years. Science, 231: 579-584.

Crowley, T.J., Mengel, J.G. and Short. D.A., 1987. Gondwanaland's seasonal cycle. Nature, 329: 803-807.

Donn, W. and Shaw, D., 1977. Model of climate evolution based on continental drift and polar wandering. Geol. Soc. Am., Bull., 88: 390-396.

Fischer, A.G., 1980. Gilbert-bedding rhythms and geochronology. In: E.I. Yochelson, (Editor), The Scientific Ideas of G.K. Gilbert. Geol. Soc. Am., Spec. Pap., 183: 93-104.

Frakes, L.A., 1979. Climates Throughout Geologic Time. Elsevier, S New York, N.Y. 310 pp.

Glancy, T.J., Barron, E.J. and Arthur, M.A., 1986. An initial study of the sensitivity of modeled Cretaceous climate to cyclical insolation forcing. Palaeoceanography, 1(4): 523-537.

Glancy, T.J., Arthur, M.A., Barron, E.J. and Kauffman, E.G., 1989. The model paleoclimate of the North American Cretaceous epicontinental sea. In: V.W.G. Caldwell and E.G. Kauffman (Editors), Evolution of the Cretaceous Western Interior Basin. Geol. Assoc. Can., Spec. Pap. (in press).

Hallam, A., 1985. A review of Mesozoic climates. J. Geol. Soc., 142: 433-445.

Hay, W.W., Barron, E.J. and Thompson, S.L., 1989. Global circulation experiments on idealized geographies. Lyell volume. Geol. Soc. London (in press).

Khain, V.E. and Ronov, A.B., 1960. World paleogeography and lithological associations of the Mesozoic Era. Rep. 21st Int. Geol. Congr., 12: 152-164.

Kutzbach, J.E., 1981. Monsoon climate of the early Holocene: Climatic experiment using the Earth's orbital parameters for 9000 years ago. Science, 214: 59-61.

Kutzbach, J.E. and Gallimore, R.G., 1989. Pangaean climates: Megamonsoons of the Megacontinent. J. Geophys. Res., No. D3, 94: 3341-3357.

Kutzbach, J.E. and Guetter, P.J., 1984. The sensitivity of monsoon climates to orbital parameter changes for 9000 BP: Experiments with the NCAR General Circulation Model. In: A. Berger, J. Imbrie, J. Hays, G. Kukla and B. Saltzman (Editors), Milankovitch and Climate. Reidel, Dordrecht, pp. 801-820.

Kutzbach, J.E. and Guetter, P.J., 1986. The influence of changing orbital parameters and surface boundary conditions on climate simulations for the past 18,000 years. J. Atmos. Sci., No. 16, 43: 1726-1759.

Lasaga, A.C., Berner, R.A. and Garrels, R.M., 1985. An improved geochemical model of atmospheric CO_2 fluctuations over the past 100 million years. In: E.T. Sundquist and W.S. Broecker (Editors), The Carbon Cycle and Atmospheric CO_2: Natural Variations Archean to Present. Am. Geophys. Union, Geophys. Monogr., 32, Washington, D.C., pp. 397-411.

Parrish, J.T., 1982. Upwelling and petroleum source beds, with reference to the Paleozoic. Am. Assoc. Pet. Geol., Bull., 66: 750-774.

Parrish, J.T. and Curtis, R.L., 1982. Atmospheric circulation, upwelling, and organic-rich rocks in the Mesozoic and Cenozoic Eras. Palaeogeogr. Palaeoclimatol. Palaeoecol., 40: 31-66.

Parrish, J.T., Ziegler, A.M. and Scotese, C.R., 1982. Rainfall patterns and the distribution of coals and evaporites in the Mesozoic and Cenozoic. Palaeogeogr. Palaeoclimatol. Palaeoecol., 40: 67-101.

Poag, C.W. and Sevon, W.D., 1989. A record of Appalachian denudation in post-rift Mesozoic and Cenozoic sedimentary deposits of the U.S. Middle Atlantic continental margin. In: T.W. Gardner and W.D. Sevon (Editors), Appalachian Geomorphology. Geomorphology, 2: 119-157 (this volume).

Sandberg, P.A., 1985. Nonskeletal aragonite and p_{CO^2} in the Phanerozoic and Proterozoic. In: E.T. Sundquist and W.S. Broecker (Editors), The Carbon Cycle and Atmospheric CO_2: Natural Variations Archean to Present. Am. Geophys. Union, Geophys. Monogr., 32, Washington, D.C., pp. 585-594.

Schlesinger, M.E. and Mitchell, J.F.B., 1987. Climate model simulations of the equilibrium climatic response to increased carbon dioxide. Rev. Geophys., 25 (4): 760-798.

Schneider, S.H., Thompson, S.L. and Barron, E.J., 1985. Mid-Cretaceous, continental surface temperatures: Are high CO_2 concentrations needed to simulate above freezing winter conditions. In: E.T. Sundquist and W.S. Broecker (Editors), The Carbon Cycle and Atmospheric CO_2: Natural Variations Archean to Present. Am. Geophys. Union, Geophys. Monogr., 32, Washington, D.C., pp. 554-559.

Schwarzacher, W. and Fischer, A.G., 1982. Limestone-shale bedding. In: G. Einsele and A. Seilacher (Editors), Cyclic and Event Stratification. Springer, New York, N.Y., pp. 72-95.

Tarling, D.H., 1978. The geological-geophysical framework of ice ages. In: J. Gribben (Editor), Climatic Change. Cambridge University Press, Cambridge, pp. 3-24.

Tourtelot, H.A., 1983. Continental aluminous weathering sequences and their climatic implications in the United States. In: T.M. Cronin, W.F. Cannon and R.Z. Poore (Editors), Paleoclimate and Mineral Deposits. U.S. Geol. Surv., Circ., 822: 1-5.

Van Houten, F.B., 1982. Ancient soils and ancient climates. In: Climate in Earth History. National Academy Press, Washington, D.C., pp. 112-117.

Washington, W.M. and Meehl, G.A., 1983. General circulation model experiments on the climatic effects due to a doubling and quadrupling of carbon dioxide concentration. J. Geophys. Res., 88: 6600-6610.

Washington, W.M. and Meehl, G.A., 1984. Seasonal cycle experiment on the climate sensitivity due to a doubling of CO_2 with an atmospheric general circulation model coupled to a simple mixed-layer ocean model. J. Geophys. Res., D6, 89:9475-9503.

Ziegler, A.M., Scotese, C.R.., McKerrow, W.S., Johnson, M.E. and Bambach, R.K., 1979. Paleogeography. Annu. Rev. Earth Planet. Sci., 7: 473-502.

Ziegler, A.M., Scotese, C.R. and Barrett, S.F., 1983. Mesozoic and Cenozoic paleogeographic maps. In: P. Brosche and J. Sundermann, (Editors), Tidal Friction and the Earth's Rotation, II. Springer Berlin, pp. 240-252.

Geomorphology, 2 (1989) 119–157
Elsevier Science Publishers B.V., Amsterdam — Printed in The Netherlands

A Record of Appalachian Denudation in Postrift Mesozoic and Cenozoic Sedimentary Deposits of the U.S. Middle Atlantic Continental Margin

C. WYLIE POAG and WILLIAM D. SEVON

U.S. Geological Survey, Woods Hole, MA 02543 (U.S.A.)
Pennsylvania Bureau of Topographic and Geologic Survey, Harrisburg, PA 17105 (U.S.A.)

(Received March 16, 1989; accepted after revision May 1, 1989)

Abstract

Poag, C.W. and Sevon, W.D., 1989. A record of Appalachian denudation in postrift Mesozoic and Cenozoic sedimentary deposits of the U.S. Middle Atlantic continental margin. In: T.W. Gardner and W.D. Sevon (Editors), Appalachian Geomorphology. Geomorphology, 2: 119–157.

The complex interplay between source-terrain uplift, basin subsidence, paleoclimatic shifts, and sea-level change, left an extensive sedimentary record in the contiguous offshore basins of the U.S. middle Atlantic margin (Salisbury Embayment, Baltimore Canyon Trough, and Hatteras Basin). Isopach maps of 23 postrift (Lower Jurassic to Quaternary) allostratigraphic units, coupled with a revised stratigraphic framework, reveal that tectonism, by regulating sediment supply (accumulation rate), dominated the interplay of forcing mechanisms. Tectonic pulses are evidenced by abruptly accelerated sediment accumulation, marked latitudinal shifts in the location of depocenters, and regional changes in lithofacies. Relatively rapid tectonic subsidence during the Early and Middle Jurassic history of the basins may have enhanced sediment accumulation rates. Beginning in the Late Jurassic, however, subsidence rates decreased significantly, though occasional short pulses of subsidence may have effected relative sea-level rises. Sea-level change heavily influenced the distribution and redistribution of sediments one they reached the basins, and paleoclimate regulated the relative abundance of carbonates and evaporites in the basins.

We conclude that source terrains of the central Appalachian Highlands were tectonically uplifted, intensely weathered, and rapidly eroded three times since the Late Triassic: (1) Early to Middle Jurassic (Aalenian to Callovian); (2) mid-Early Cretaceous (Barremian); and (3) Late Cenozoic (Middle Miocene). Intervals of tectonic quiescence following these three tectonic pulses provided conditions suitable for the formation of regional erosion surfaces, geomorphic features commonly reported to characterize the central Appalachian Highlands. This series of three, irregularly spaced, tectonic/quiescent cycles does not, however, match the traditional four-cycle concept of post-Triassic Appalachian "peneplanation".

Introduction

In the past few years numerous authors have attempted to assess the relative effects of diverse forcing mechanisms on the stratigraphy, depositional and erosional history, and geomorphic features of the U.S. Atlantic offshore basins and their bordering highland source terrains (Fig. 1). The primary forcing mechanisms considered have been tectonic and isostatic uplift and subsidence (Gibson, 1970; Owens, 1970; Watts and Steckler, 1979; Denny, 1982; Hack, 1982; Heller et al., 1982; Sawyer et al., 1982; Watts and Thorne, 1984; Pavich, 1985) sea-level change (Vail et al., 1977; Pitman, 1979; Poag and Schlee, 1984; Watts and Thorne, 1984; Pitman and Golovchenko, 1988), and paleoclimatic fluctuations (Sevon, 1985; Poag, 1987a;

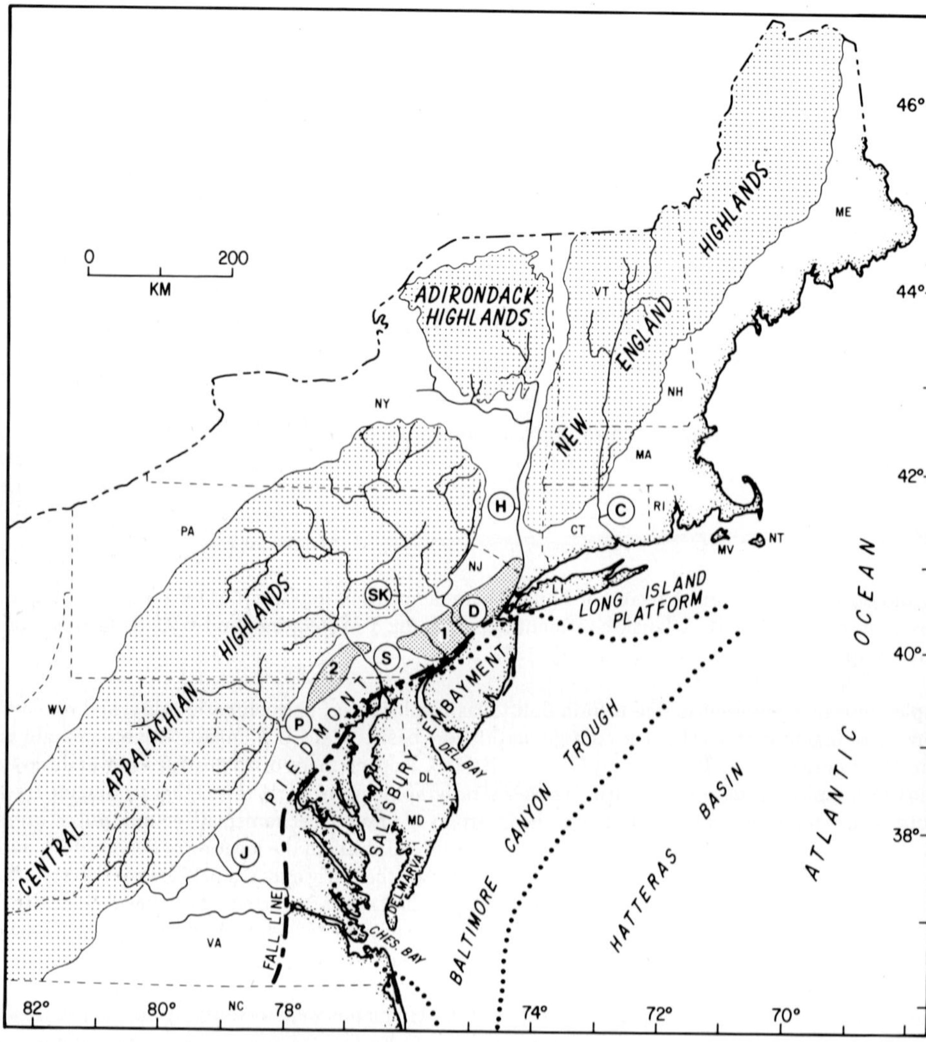

Fig. 1. Location map of the U.S. middle Atlantic continental margin. *MV*=Martha's Vineyard; *NT*=Nantucket; *LI*=Long Island; *Del. Bay*= Delaware Bay; *Ches. Bay*=Chesapeake Bay; *C*=Connecticut River; *H*=Hudson River; *D*=Delaware River; *SK*=Schuylkill River; *S*=Susquehanna River; *P*=Potomac River; *J*=James River; *1*= Newark Basin; *2*=Gettysburg Basin.

O'Leary, 1988). Particularly contentious have been arguments regarding the role of episodic tectonism during the late geologic history (Mesozoic–Cenozoic) of the central Appalachians (Rodgers, 1967; Ahnert, 1970; Owens, 1970; Pierce, 1966; Hack, 1975, 1982; Melhorn and Edgar, 1980; Pavich, 1985; Sevon, 1985) and the presence, origin, and age of broad, regional erosional surfaces ("peneplains").

Hack (1982) argued, for example, that major eastward-flowing streams that drain the central Appalachians (such as the Potomac and Susquehanna rivers; Fig. 1) were affected by differential movement at the Fall Line (convex longitudinal profiles, increased coastal gradients) in the Tertiary. He believed that this differential movement was caused by tectonic uplift of the Piedmont and central Appalachian Highlands (Fig. 1). Hack cited Ahnert's (1970) analysis of denudation versus relief to argue that uplift was tectonic rather than isostatic. Ahnert had calculated that any isostatically uplifted terrain would be reduced to 10% of its original mean relief at the end of a 30-Myr period. On the basis of this calculation, Hack showed that the Blue Ridge segment of the Appalachians, whose mean relief is now 300 m, would have had an impossible relief of 30,000 m in the late Paleocene, if isostacy were the only Tertiary uplift mechanism. Hack (1980), Gibson (1970), and Owens (1970) all inferred that depositional cycles and variable sediment accumulation rates in the Atlantic Coastal Plain and adjacent offshore basins evidenced episodic tectonic uplift in the central Appalachians.

Mathews (1975) used early estimates of Atlantic offshore sediment volumes (e.g., Emery et al.,1970) to speculate on denudation and erosion surfaces of the central Appalachians. He suggested that the particularly large volume of Miocene sediments might indicate a significant period of subaerial erosion in the uplands of New Jersey and Pennsylvania (Schooley and Kittatinny mountains) during the Miocene. Hack (1982) used more recent offshore data to reach the same conclusion. In addition, rejuvenated landforms in the central Appalachians (Hack, 1982), the southwestward diversion of the Potomac River (Schlee, 1957), and a series of en echelon faults west of the Potomac estuary (Mixon and Newell, 1977) have been cited as evidence of middle Tertiary (probably Miocene) tectonism.

Pavich (1985), on the other hand, believed that rapid tectonic uplift was infrequent in the post-Triassic history of the central Appalachians. He cited unpublished fission-track data of Durrant (1978) to conclude that the Piedmont region was uplifted 2–3 km during the Cenozoic (an estimate that fits the sediment-volume data of Mathews, 1975), but he attributed the uplift to gradual isostatic rebound responding to extensive sediment removal. Pavich presented a steady-state model (balancing weathering, erosion, and uplift rates) to show that mass removal and uplift could be accomplished with little elevation change, if the uplift were chiefly isostatic.

To the north, middle Cenozoic uplift of the Adirondack Highlands (Fig. 1) is indicated by geodetic leveling data compiled by Gable and Hatton (1983). These authors concluded that the Adirondack Dome has been uplifted ~2,000 m since the beginning of the late Miocene. Also, estimates of erosion rates compared to current highland elevations suggest that the White Mountains of New Hampshire and other parts of New England also may have been uplifted during the Miocene (Denny, 1982). O'Leary (1988), in addition, has suggested that a notable abundance of rock fragments reported by Kaye (1983) in middle to upper Miocene strata of southern Massachusetts might reflect regional uplift in the New England Appalachians (Fig. 1).

Evidence that tectonism (faulting, uplift) has been active during the last few million years has been documented in the Atlantic Coastal Plain and in the Adirondack Mountains in several recent papers (Isachsen, 1975; Cronin, 1981; Newell, 1985).

Sevon (1985), taking a different approach, argued that variable paleoclimate dominated Appalachian denudation, and thereby also produced the greatly varying depositional rates recorded in the flanking basins. Manspeizer (1985), too, emphasized paleoclimate and demonstrated its crucial role in producing arkosic and evaporitic deposits of the newly formed offshore basins during the Late Triassic and Early Jurassic. Mountain and Tucholke (1985) and O'Leary (1988) also recognized the influence of paleoclimate in their analyses of offshore deposits, but they made no attempt to evaluate its relative impact on depositional or erosional patterns.

Most geologists studying the postrift (Early Jurassic–present) depositional history of the offshore basins (particularly the Baltimore Canyon Trough) have stressed the importance of such phenomena as basin thermotectonism (Steckler and Watts, 1978; Watts and Steckler, 1979; Heller et al., 1982; Sawyer et al., 1982), eustatic sea-level cycles (Vail et al.,1977; Pitman, 1979; Watts, 1982; Poag and Schlee, 1984; Watts and Thorne, 1984; Poag, 1987a; Greenlee et al., 1988; Pitman and Golovchenko, 1988), abyssal current systems (Mountain and Tucholke, 1985), carbonate compensation depth (Mountain and Tucholke, 1985), and sediment gravity flows (Poag, 1987a; Van Hinte et al., 1987; O'Leary, 1988). But these authors have largely ignored the role of hinterland tectonism and paleoclimate, which regulate the supply of sediment reaching the basins.

Our goal is to integrate these divergent data, interpretations, and speculations with new stratigraphic and depositional data from the offshore region. We hope to create a coherent synthesis that relates the principal episodes of postrift depositon to four forcing mechanisms, namely source-terrain uplift, basin subsidence, paleoclimate, and sea level. We will evaluate the relative importance of each forcing mechanism by successively examining the genetic implications of three aspects of the sedimentary record: (1) volumetric sediment-accumulation rates; (2) location of sediment dispersal routes and depocenters; and (3) regional lithofacies patterns.

Methods

Our chief basis for interpreting the depositional episodes of the study area is a series of 23 isopach maps (scale = 1:1,000,000; see Appendix figs. 1–23) drawn from a network of numerous outcrops, 88 boreholes, and ~10,000 line-km of multichannel seismic reflection profiles on the coastal plain and on the continental shelf, slope, and rise (Fig. 2). The maps allow us to identify principal depocenters, to delineate sediment dispersal routes and distribution patterns, to estimate offshore sediment volumes, and to determine the dominant sources of terrigenous detritus.

We used a volumetric method to calculate sediment accumulation rates. First, in order to derive sediment thickness values from the two-way travel-time displayed on the seismic profiles, we used depth conversions published by Grow et al. (1979, 1988), Schlee et al., (1979), Ewing (1984), and Poag (1985; 1987a), as well as several unpublished ones. From these we extrapolated thickness values to unconverted profiles. Approximate sediment volumes were then calculated by treating each mapped unit as a series of stacked tabular slices. The volume of each 100-m thick slice was calculated individually and the results summed to produce the total volume for the mapped unit. From those units which contain significant volumes of carbonate deposits, we subtracted the carbonate values to obtain the volume of siliciclastics alone. This method yields rough volume estimates whose values are imprecise, but are sufficient for the types of analyses and interpretations we have applied.

A more common way to estimate sediment accumulation rates, especially in areas of relatively uniform deposition, like the deep sea, is to measure thickness of depositional units at a particular site (borehole or seismic reflection profile) and to calculate the rate in linear units (centimeters) per time interval (thousand years). When dealing with a broad, depositionally complex continental margin, however, we find that the column of sediments for any given depositional unit varies greatly in thickness from place to place (within a single basin and from basin to basin) along the margin. The volumetric method of calculating sediment accumulation rates is the more accurate under such circumstances.

The values obtained for accumulation rates, whether based on volume or thickness, are directly dependent upon the time scale used to determine the duration of deposition for each mapped unit. We have used the time scale recently adopted by the Geological Society of America for the Decade of North American Geology (DNAG) series (see Vogt and Tucholke, 1986). We compared our results with calculations using the time scale of Haq et al. (1987), and found no significant departures in calculated rates, except for the Aalenian interval. The

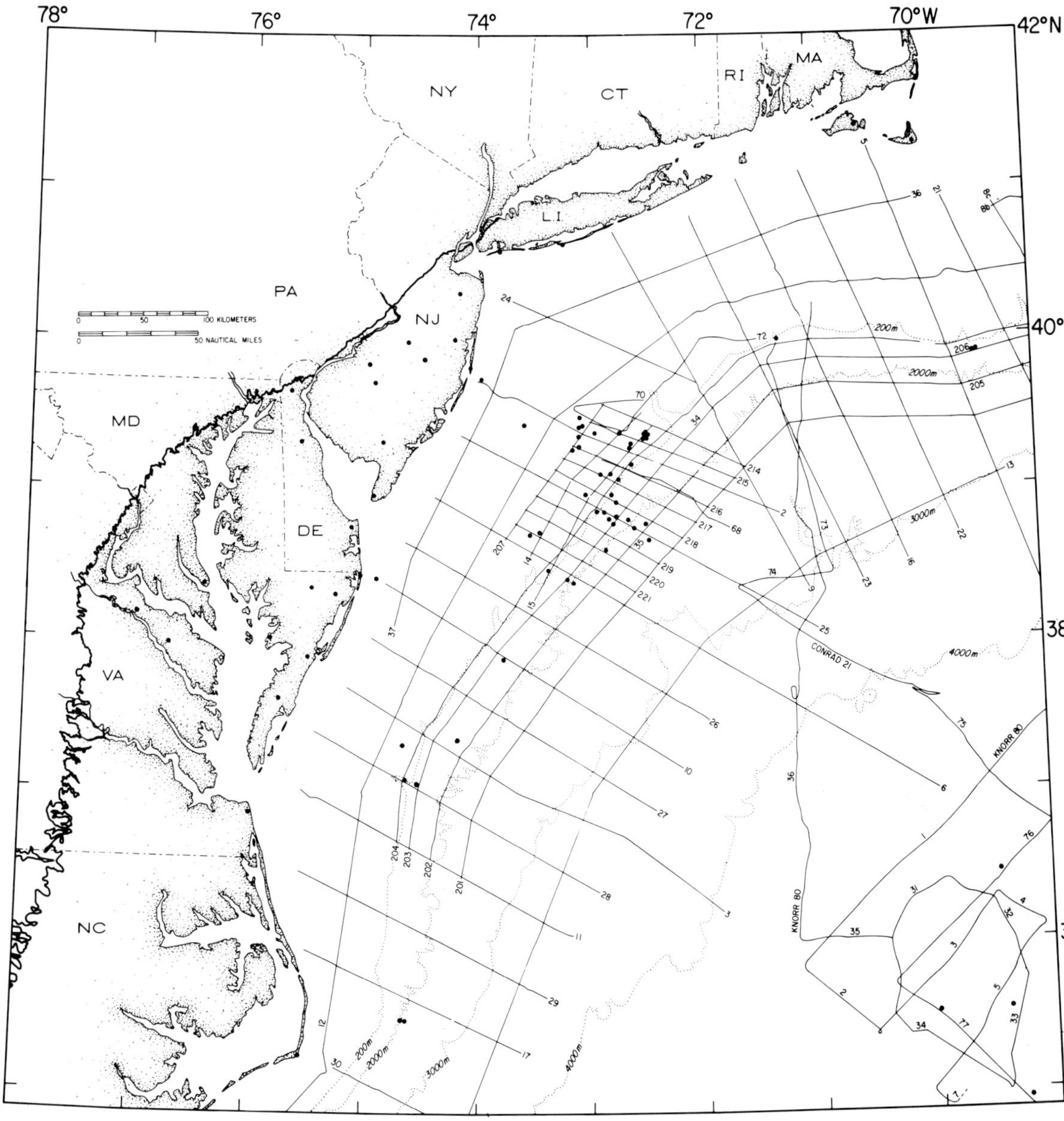

Fig. 2. Location of boreholes and seismic reflection profiles incorporated in this study.

Aalenian time span given by Haq and others is twice that of the DNAG scale, so the accumulation rate is reduced by half.

The accuracy of rate calculations also depends intimately, of course, on which geological ages are represented by the depositional units mapped and whether or not the units represent deposition during the entire span of geological time assigned to a given age interval. The depositional units we mapped are generally those recognized by Poag (1987a), which have been dated by biochronological methods (planktonic and benthic foraminifera, radiolarians, calcareous nannofossils, and palynomorphs). Because the mapped units are allostratigraphic units (Poag, 1987b), the upper and lower boundary of each is a widespread unconformity. As a result, each mapped unit represents an incomplete portion of the sedimentary prism that originally constituted each unit. We cannot know exactly how much time is represented by the bounding unconformities at any one spot, except at outcrops and boreholes. Furthermore, the duration of the hiatus varies from place to place. To partially compensate for this condition, we arbitrarily assumed that on average, at least 1 Myr of hiatus is represented by each allostratigraphic boundary.

The oldest biostratigraphic unit drilled to date in the study area is of Oxfordian age, so older units had to be identified by seimostratigraphic methods and correlated with the cyclic stratigraphic framework proposed by Vail et al. (1977) and modified most recently by Haq et al. (1987, 1988). Thus the stratigraphic age of the Callovian and older units is not as precisely determined as that of the Oxfordian and younger units. The maximum possible age of the oldest postrift unit is controversial, but most workers agree that sea-floor spreading (and postrift deposition) began approximately 165–190 Ma (see Klitgord and Schouten, 1986). Our current seimostratigraphic analysis, along with revised borehole data from the Georges Bank Basin (Ascoli, 1988), suggest that a 190–187 Ma date (late Toarcian–early Aalenian) is probably most realistic, a conclusion also reached by Jansa (1986) and Benson and Doyle (1988). (Note that this is a markedly different opinion than the earlier seimostratigraphic inferences of Poag, 1985; 198a.) Furthermore, the Toarcian–Aalenian age agrees with a radiometric data of 183±12 Ma (Jansa, 1986) for evaporites presumed to have precipitated in the Baltimore Canyon Trough during the transition from rifting to sea-floor spreading.

Another inaccuracy in our calculations is the result of the compaction of the sedimentary column through roughly 190 Myr of its history, which reduced the original volume. Because the lithology (porosity) and depth of burial of each mapped unit varies considerably over the study area, we have not attempted to decompact the units; thus our accumulation rates are minimum rates.

Finally, through we speak generally of sea-level cycles in terms of alternating eustatic rise and fall, we agree with Pitman (1979), Pitman and Golovchenko (1988), and Christie-Blick et al. (in press), that in some cases, the changes we infer may be only relative shifts, which resulted from variable rates of sea-level fall imposed on a continuously (but variably) subsiding basin. Whatever the eustatic relationship or its ultimate cause may have been, we can document a succession of *relative* rises and falls, which initiated transgressive and regressive depositional sequences.

Accumulation rates

The temporal succession of sediment accumulation rates (Fig. 3) can be divided into five main phases of deposition: Phase I (Aalenian–Callovian), an early postrift phase characterized by variable, but high accumulation rates (mean = 16,000 km^3 Myr^{-1}); Phase II (Oxfordian–Turonian), in which accumulation rates fluctuated, but were generally an order of magnitude lower than in Phase I (mean = 6,000 km^3 Myr^{-1}); Phase III (Coniacian–Maestrichtian), in which rates initially increased again by an order of magnitude, but declined by half toward the end (mean = 14,000 km^3 Myr^{-1}); Phase IV (Paleocene–early Miocene), in which accumulation rates dropped by an order of magnitude to reach the lowest values calculated (mean = 2000 km^3 Myr^{-1}; and Phase V (middle Miocene–present), a final phase, featuring a 20-fold increase to the highest rates of all (mean = 28,000 km^3 Myr^{-1}).

To help interpret this history of variable sediment accumulation, we turn to Galloway (1989a), whose method of sedimentary basin analysis compares the relative influence of three forcing mechanisms (sediment supply, subsidence history, and sea level) on sequences of basin fill. He found that high sediment accumulation rates, particularly when initiated ab-

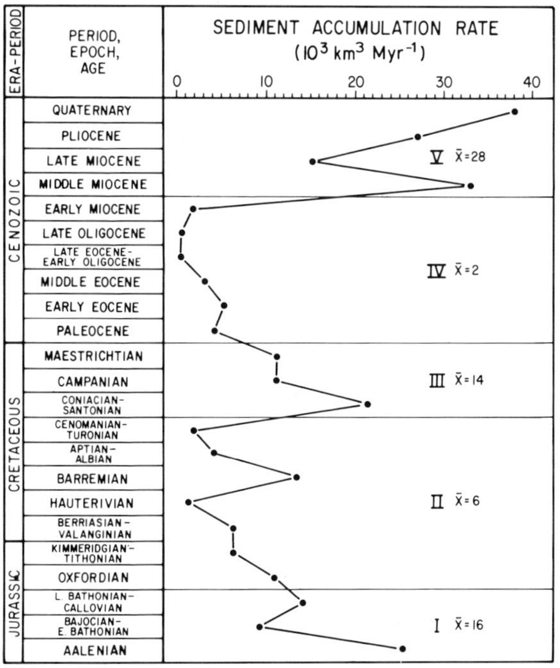

Fig. 3. Sediment accumulation rates for the Salisbury Embayment, Baltimore Canyon Trough, and Hatteras Basin. Rates calculated as volume of sediment (km^3) per million years, using the DNAG time scale (Vogt and Tucholke, 1986; see Fig. 9 for DNAG time scale). Roman numerals indicate five phases of deposition defined according to accumulation rates. $\bar{x}$ = mean sediment accumulation rate for each depositional phase.

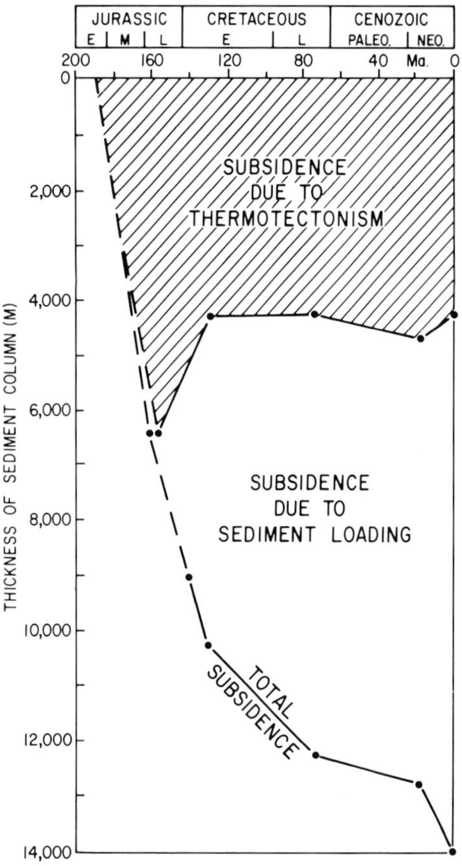

Fig. 4. Postrift subsidence history of the Baltimore Canyon Trough. The hachured field indicates the relative portion of total subsidence due to thermotectonism; the blank field indicates the relative amount of total subsidence due to sediment loading. We have extended (by linear extrapolation) Watts and Steckler's (1979) model of the COST B-2 well to include the early postrift history (dashed boundary lines) of basin subsidence (140-187 Ma). *E* = Early; *M* = Middle; *L* = Late; *Paleo.* = Paleogene; *Neo.* = Neogene; Quaternary cannot be distinguished at this scale. Time scale is given in mega-annums (millions of years ago).

ruptly, are strong evidence that sediment supply (reflecting source-terrain uplift and paleoclimate) dominated the basin-fill process. The high accumulation rates of Phase I (Early–Middle Jurassic) appear to indicate such a situation. Independent evidence of source-terrain uplift during Phase I is provided by geophysical models of this margin.

According to current theoretical models, (Steckler and Watts, 1978; Watts and Steckler, 1979; Sawyer et al., 1982; Watts, 1982; Sawyer, 1985; Steckler et al., 1988), the first ~40–60 Myr of postrift development of the Baltimore Canyon Trough was dominated by rapid lithospheric contraction (subsidence) caused by asthenospheric cooling following rifting (Fig. 4). The flexural style of subsidence, believed appropriate to this margin, was accompanied by partly compensating uplift landward of the basin's main hinge line (a 30–50 km zone of rapid downwarping located a few kilometers seaward of the present shoreline; Watts and Thorne, 1984; Watts and Swift, 1988). Paleoclimatic and sea-level change appear, on the other hand, to have had relatively minor influence on sediment accumulation rates during Phase I. Jurassic paleoclimatic interpretations applicable to Phase I are necessarily broad because few specific data are available for the Appalachian region. We rely mainly on the presence of climate-sensitive deposits (evaporites, carbonates, coal), relative global rainfall estimates (Parrish et al., 1982), and paleogeographic considerations (Manspeizer, 1985). Local paleobotanical data, which allow more accurate paleoclimate interpretations of our specific area of study, are available, however, for the Late Cretaceous and younger deposits.

In the Early Jurassic, the nascent Baltimore Canyon Trough lay between latitudes 15° and 20°N, was initially only ~45 km wide (Fig. 5A), and also was closed at the southwest end (Jansa, 1986), which restricted marine circulation. Early Jurassic climate was warm and arid or semi-arid (Parrish et al., 1982; Manspeizer, 1985), and these conditions undoubtedly enhanced the production of large volumes of mechanically weathered detritus for eventual transfer to the offshore basins. Without the high source-terrain relief produced by tectonic uplift, however, this detritus probably would not have been rapidly delivered to the offshore basins.

The Baltimore Canyon Trough widened during the Middle Jurassic (Fig. 5B), and alternating wet and dry periods were probable for the Appalachian area, considering the amounts of Jurassic coal formed in various parts of the Baltimore Canyon Trough (Poag, 1985; Jansa, 1986) and elsewhere in Pangea (Frakes, 1979; Parrish et al., 1982; Hallam, 1984).

Sea-level change may have influenced Early and Middle Jurassic sediment accumulation also, but to a relatively minor degree. The long-term sea-level curve of Haq et al., (1987), for example, places sea level generally lower during depositional Phase I than at any other time in the postrift history of the region. This low position may have steepened coastal stream gradients and perhaps accelerated basinal deposition somewhat. We conclude, however, that the bordering tectonic uplift, which accompanied rapid thermotectonic subsidence of the Baltimore Canyon Trough, would have produced sufficient source-to-basin relief to supply terrigenous detritus in large volumes, regardless of the specific prevailing sea level.

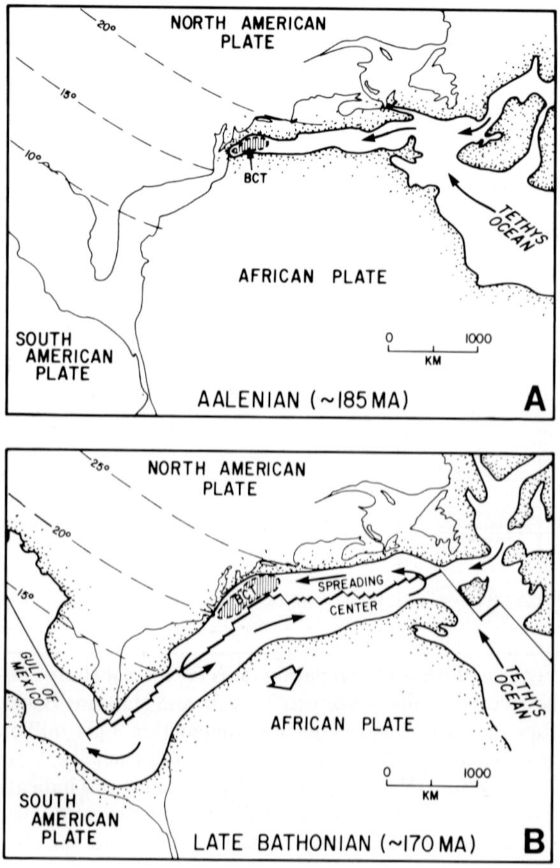

Fig. 5. Early Jurassic and Middle Jurassic reconstructions of the North Atlantic region, showing the changing paleogeographic and paleoceanographic relationships pertinent to deposition in the Baltimore Canyon Trough (BCT). (A) Reconstruction at some time during early Aalenian deposition, near the onset of sea-floor spreading. (B) Reconstruction approximately at the end of Bathonian deposition. Small arrows indicate surface circulation as interpreted by Jansa (1986); large arrow indicates motion of African plate relative to the North American plate (Klitgord and Schouten, 1986); paleogeography modified from Jansa and Tucholke (1986); paleolatitudes from Firstbrook et al. (1979) and Manzpeizer (1985).

Depositional Phase II (Late Jurassic–middle Cretaceous) began more or less coincident with the termination of rapid thermotectonic subsidence of the Baltimore Canyon Trough and its coeval source-terrain uplift (Fig. 4). Phase II also coincided with significant changes in the paleoclimate of eastern North America. A ~39 Myr stabilization of highland source terrains, following nearly 25 Myr of uplift, is suggested by the 60% reduction in average sediment accumulation rage (Fig. 3). This stability was interrupted, however, by a brief (~5 Myr) tectonic episode (accelerated deposition) in the Barremian. Heller et al. (1982) noted a pulse of increased basin subsidence during the Aptian–Albian interval of Phase II, which they attributed to sediment loading. Our data show, however, that sediment accumulation slowed during the Aptian–Albian interval, which precludes sediment loading as the forcing mechanism for increased subsidence. Thus a brief episode of thermotectonic subsidence apparently took place at this time, but it did not increase the rate of sediment accumulation.

During Phase II, the continental climate of the Late Jurassic and Early Cretaceous was remarkable because of its widespread warmth and humidity (tropical to subtropical conditions at mid-latitudes; Frakes, 1979; Parrish et al., 1982). This paleoclimate must have had a considerable influence on Appalachian denudation rates by promoting chemical weathering, thereby reducing the supply of siliciclastic sediment available for transport to the basins. The postulated increase in rainfall did not, however, accelerate the delivery of terrigenous detritus to the offshore basins.

The effects of long-term sea level may be reflected in the variability of accumulation rates during Phase II. Long-term sea level (Haq et al., 1987) was lowest during intervals of most rapid accumulation and rose to maxima during the two intervals when accumulation rates were lowest (late Hauterivian and Cenomanian–Turonian). The second-order (supercycle) fluctuations of sea-level (Haq et al., 1987, 1988), however, had negligible effect on sediment accumulation rates. Of the eight notable short-term sea-level falls (supercycle boundaries) assignable to Phase II, not one is associated with an increase in sediment accumulation.

During the early part of depositional Phase III (Coniacian–Santonian), siliciclastic deposition in the offshore basins abruptly accelerated by a factor of 10 (Fig. 3). This Late Cretaceous terrigenous increase has been noted in cores as far as 1,000 km from the Late Cretaceous shelf edge (Graciansky et al.,1987). By this time basin subsidence was controlled mainly by sediment loading, and thus was primarily an effect, not a cause of rapid accumulation (Steckler and Watts, 1978; Poag, 1985). Heller et al. (1982), for example, noted a brief pulse of increased basin subsidence during the Coniacian–Santonian interval, but attributed it to sediment loading.

The sedimentary record of Phase III in the Salisbury Embayment contains an abundance of fossilized leaves, whose physiognomy Wolfe and Upchurch (1987) analysed (leaf-margin analysis) for paleoclimatic implications. They found that the middle Atlantic coastal region had a generally subhumid, equable, megathermal paleoclimate (little seasonality in either temperature of precipitation) during Phase III. This paleoclimate may have caused somewhat more intense chemical weathering than during Phase II. Parrish et al. (1982) postulated that Late Cretaceous rainfall had increased relative to Early Cretaceous rates in the Appalachian region, and this increase may have enhanced rapid delivery of siliciclastic detritus to the basins.

Sea level was generally high in the Late Cretaceous (Haq et al., 1987), though declining from the middle Cretaceous maximum. No significant effect of either long-term or short-term (supercycle) sea-level change on the rate of sediment supply is apparent in our data. The abrupt acceleration and the magnitude of Phase III deposition convinces us that renewed uplift of highland source terrains was its chief forcing mechanism.

Depositional Phase IV (Paleocene–early Miocene) saw a marked 35-fold decline to unusually low rates of sediment accumulation, and includes the lowest rates recorded in the study area (Fig. 3). This nearly 50-Myr interval of subdued deposition indicates that source terrains were stabilized, and perhaps low-lying (see Cleaves, this volume), after the intense erosion of Phase III. This reduced sedimentation also coincided with a significant change in global and regional climate. In the early Paleocene, temperature and rainfall increased and a megathermal rain-forest ecosystem evolved in the coastal region of eastern North America (Wolfe and Upchurch, 1987). A temperature maximum in the Eocene is indicated by both continental (Wolfe, 1978; Fredericksen, 1984) and marine (Miller et al., 1987) records. Tropical conditions along the coast were supported, in part, by warm, northward-flowing waters of the proto-Gulf Stream (Pinet and Popenoe, 1985; Popenoe, 1985). Chemical weathering would have dominated under the extremes of the rain-forest climate.

A major decrease in global temperature and equability took place in the later part of Phase IV (early Oligocene), when the first known Cenozoic continental ice sheets began to form in the Antarctic (Wolfe, 1978; Miller et al., 1987). We see no evidence, however, that this paleoclimatic shift affected sediment accumulation rates in the offshore basins of the study area.

The early Tertiary was a time of increasingly high sea levels (Haq et al., 1987), and the highest Cenozoic sea levels noted along the Atlantic margin occurred in the Eocene (Charletta, 1980; Poag, 1985; Olsson et al.,1988; Poore and Bybell, 1988). Thus the diminished sediment accumulation rates of Phase IV also could be attributed, in part, to reduced coastal stream gradients caused when high sea levels flooded the former coastal plain. The relative height of sea level (depth of water over the continental shelf) may have been augmented in the Baltimore Canyon Trough during the Eocene by a pulse of increased basin subsidence noted by Heller et al., (1982). This brief subsidence pulse, however, had no discernible affect on sediment accumulation rates (in fact, the rate *decreased*).

As for short-term sea-level change, Haq et al. (1987, 1988) postulated four significant sea-level falls (supercycle boundaries) assignable to Phase IV, including the widely discussed middle Oligocene fall. These authors indicated that sea level fell as much as 200 m during the middle Oligocene, but our record contains no evidence of accelerated sedimentation during this or any other part of Phase IV.

Depositional Phase V (middle Miocene–present) is characterized by an abrupt 20-fold increase in terrigenous sediment accumulating in the middle Atlantic basins (Fig. 3). Average accumulation rate was significantly greater than in any other post-rift depositional Phase and is taken as a signal of renewed source–terrain uplift. A short pause in uplift took place, however, in the late Miocene. Heller et al. (1982) inferred a Middle Miocene pulse in thermotectonic subsidence of the Baltimore Canyon Trough, which also may have contributed to the accelerated accumulation rate. But the affect of this subsidence on deposition rates, if any, is not readily distinguishable from that of source-terrain uplift.

Phase V also experienced dramatic changes in continental and marine peleoclimates, beginning in the middle Miocene. A pronounced drop in temperature (Pollack, 1982; Savin, 1977; Miller et al., 1987), accompanied by accelerated continental ice buildup and changes in annual precipitation, resulted in a humid, subtropical Miocene paleoclimate bordering the Baltimore Canyon Trough. This paleoclimatic shift is recorded in the composition of palynomorph floras (Fredericksen, 1984). This subtropical regime initiated a period of intense mechanical weathering and rapid erosion, whose products accumulated in unusually thick sedimentary prism offshore. Climatic deterioration culminated in the Pleistocene, as the North American ice sheet spread over the northern source terrains of the study area and contributed voluminous glaciofluvial and glaciomarine sediments to the offshore basins.

Long-term sea level continued to fall during Phase V and reached its lowest position since the Jurassic, but it is unlikely that this gradual decline had much affect on the dramatic flood of terrigenous sediments that took place. The repetitious short-term sea-level fluctuations of the Quaternary, however, were undoubtedly very important in increasing sediment supply to the Hatteras Basin.

Shifting dispersal routes and depocenters

Included in the record of episodic deposition in the Baltimore Canyon Trough and Hatteras Basin is evidence of temporal shifts in the primary routes by which sediments were dispersed to the basins and in patterns of depocenter distribution within the basins (Fig. 6), factors which are clearly expressed by the contours of our isopach maps (Appendix figs. 1–23). We use these shifts to infer changing relationships (relative contribution of siliciclastic detritus) among the three highland source terrains.

The isopach patterns (Appendix) indicate that the main seaward-flowing drainage systems of the highland source terrains developed early in the Mesozoic, and have endured to the present (see also Hack, 1982; Sevon, 1985; Cleaves, this volume; Morisawa, this volume). Independent evidence of this is seen in the geology of the Appalachians. For example, clasts of the Triassic Hammer Creek Formation (southeastern Pennsylvania), which originated north of the present position of Blue Mountain (MacLachan, oral commun., 1988), indicate that the Schuylkill River drainage opened from the north into the Newark Basin (Fig. 1) early in the Triassic. The Schuylkill River supplied sediment to the Newark Basin throughout the remainder of the Triassic and presumably extended eastward in the Early Jurassic to eventually join the ancient Delaware River system in the Miocene (Appendix fig. 20).

Similarly, the Gettysburg Formation of the

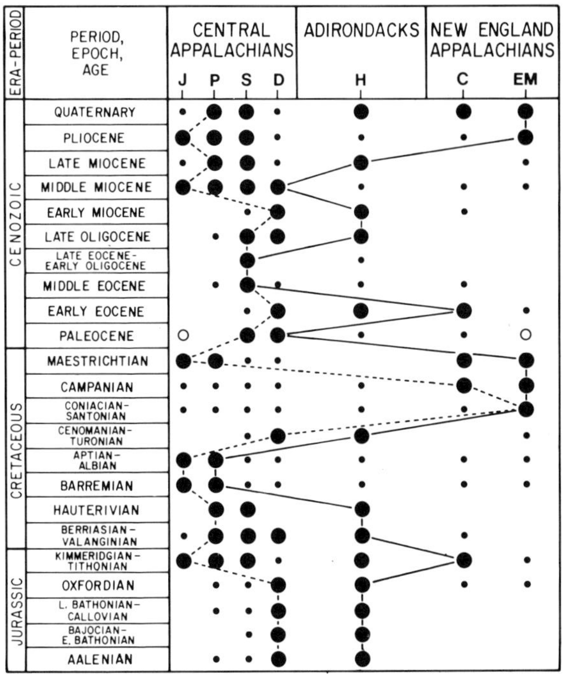

Fig. 6. Dominant terrigenous source terrains and sediment dispersal routes for each mapped depositional unit. J=ancient James River; P=ancient Potomac River; S=ancient Susquehanna River; D=ancient Delaware River; H=ancient Hudson River; C=ancient Connecticut River; EM=ancient river(s) of eastern Massachusetts. Large solid dots indicate primary dispersal routes; small solid dots indicate secondary dispersal routes; open circles indicate possible dispersal routes. Solid line connects northernmost primary dispersal route for each mapped sedimentary unit; dashed line connects southernmost primary dispersal route for each mapped sedimentary unit. Note the temporal latitudinal shifts of dominant source terrains and primary dispersal routes.

Gettysburg Basin (Fig. 1; Berg et al., 1980) contains sedimentary evidence that Susquehanna River drainage was initiated somewhat later in the Triassic. The Triassic Susquehanna system does not appear to have supplied as much sediment as the Triassic Schuylkill (MacLachan, oral commun., 1988), so the Susquehanna's development as a *major* drainage system probably occurred later in the Jurassic (Fig. 6).

We are not aware of any field evidence suggesting that the Hudson or Potomac rivers existed in the Triassic, but Lovegreen (1974) found evidence of Early Cretaceous Hudson River deposits in cores drilled along its present lower course. Depocenter evidence (Appendix fig. 1) indicates, however, that an ancient Hudson River was established by Aalenian time. The isopach maps also indicate that the Connecticut River was established as a major sediment delivery system in Late Jurassic time.

Once these major streams were notched into the resistant rocks (such as the Tuscarora and Pocono formations for the Susquehanna, Schuylkill, and Delaware rivers), they were trapped, and being the major drainage systems, were not likely objects for stream capture. Our isopach maps (Appendix figs. 1–23) indicate, furthermore, that during most of Mesozoic and Cenozoic time, the major rivers draining the central Appalachian Highlands exited the Piedmont at the Fall Line, near where they exit today (Fig. 1; see Cleaves, this volume). The southwestward diversion of these river systems (James, Potomac, Susquehanna, Schuylkill, and Delaware rivers) at the inner edge of the present coastal plain apparently was initiated in the Neogene (middle Miocene; Appendix fig. 20; Owens and Minard, 1979) and generally did not affect sediment distribution patterns through most of the postrift history of this region. Thus to facilitate discussion we apply modern river names to the dominant ancient drainage systems associated with Mesozoic and Cenozoic depocenters of the middle Atlantic margin. For simplicity we include the Schuylkill drainage as part of the Delaware River system.

Let us now examine the genetic implications of the sediment distribution patterns displayed by each isopach map. The primary sediment dispersal systems feeding the Baltimore Canyon Trough during the Aalenian interval, as indicated by the isopach contours, were the ancient Delaware River (draining the central Appalachian Highlands) and the ancient Hudson River (draining the Adirondack Highlands; Fig. 6; Appendix fig. 1). The principal offshore depocenter for the Aalenian period (>3000 m thick) occupied the continental slope seaward of what now is New Jersey. Additional sediment was dispersed through secondary drainage systems, the ancient Susquehanna and Potomac rivers. Similar relationships were maintained during the Bajocian–early Bathonian interval (as the basin continued to widen), except that the potomac sediment supply diminished (Appendix fig. 2).

During the Middle Jurassic (late Bathonian–Callovian), the offshore depocenters began to spread out along the continental slope and stretched from what now is Virginia to Long Island, seaward of a narrow (20–30 km wide) carbonate bank that occupied the outer shelf (Appendix fig. 3). The largest depocenter (>3000 m thick, along the continental slope) continued to be fed primarily through the ancient Delaware and Hudson rivers, but additional small deep-water depocenters (also ~3000 m thick) developed in front of secondary dispersal routes to the southwest (ancient Susquehanna and Potomac rivers; Fig. 6).

During the Late Jurassic (Oxfordian–Tithonian), offshore siliciclastic depocenters steadily enlarged (500–>700 m thick) and elongate, fan-like prisms began to extend hundred of kilometers across the continental rise in the Hatteras Basin (Appendix figs. 4,5). At the same time, distinct shallow-water depocenters (500–1000 m thick) developed on the continental shelf, as source-terrain uplift and basin subsidence rates diminished. In addition, a series of long, narrow depocenters (20-30 km wide; >1000 m thick) were formed by carbonate production along the shelf edge. Sediment still arrived in abundance from the ancient Hudson and Delaware river dispersal routes, but dispersal from the central Appalachian Highlands included the ancient James, Potomac, and Susquehanna rivers. In addition, the New England Highlands began to furnish moderate to large volumes of sediment through the ancient Connecticut River and an undetermined river (or rivers) in eastern Massachusetts (Fig. 6; Appendix figs. 4,5).

During the Early Cretaceous (Berriasian–Valanginian), shelf deposition diminished (though several distinct depocenters remained), whereas thick, fan-like sedimentary prisms (700–>2000 m thick) developed in the Hatteras Basin seaward of the ancient Potomac, Susquehanna, Delaware, and Hudson drainage systems (Appendix fig. 6; see also Cleaves, this volume). The lack of large depocenters associated with the ancient Connecticut and eastern Massachusetts rivers, however, indicates a significant reduction in sediments supplied by the New England Highlands (Fig. 6).

Hauterivian deposits are thin, relatively fine-grained, often calcareous (low sediment accumulation rate; Fig. 3), and essentially limited to the continental shelf and slope (Appendix fig. 7). Small depocenters (>100 m thick) developed seaward of the ancient Potomac, Susquehanna, and Hudson rivers.

A significant change in sediment provenance and dispersal routes took place in the latter part of the Early Cretaceous (Barremian and Aptian–Albian; Fig. 6). During the Barremian (Appendix fig. 8), a broad depocenter (300–>400 m thick), fed principally by the ancient James and Potomac rivers, occupied the outer continental shelf between what now is the Delmarva Peninsula (Fig. 1) and southern New Jersey. Lesser depocenters developed off northern New Jersey and eastern Massachusetts. Corresponding slope aprons (300–>600 m thick) and submarine fan complexes (100–300 m thick) developed in the Hatteras Basin, although much of the middle continental rise is barren of Barremian deposits.

During the Aptian–Albian interval, the ancient James and Potomac rivers added more than 700 m of sediment thickness to the largest Barremian shelf depocenter (Fig. 6; Appendix fig. 9). To the north, the two smaller shelf depocenters thickened by 200–300 m, and additional small depocenters (300–500 m thick) developed in front of the ancient Hudson and Connecticut rivers. Aptian–Albian deposition in the Hatteras Basin, however, was much reduced compared to that of the Barremian.

During the Cenomanian–Turonian interval (earliest Late Cretaceous), sediment accumulation faltered again (Fig. 3), and shelf depocenters dominated (Appendix fig. 10). The principal depocenter (100–>200 m thick) developed off New Jersey (ancient Delaware River) and Long Island (ancient Hudson River), as northern source terrains (Adirondack Highlands) began to compete with the central Appalachian Highlands (Fig. 6).

The Cretaceous ended (Coniacian–Maestrichtian) with a switch to dominantly northern source terrains (New England Appalachians via the ancient Connecticut River and the undetermined river(s) in eastern Massachusetts (Fig. 6; Appendix figs. 11–13). Most of these sediments were dumped on the continental slope and inner continental rise of the Hatteras Basin (depocenters 500–>1000 m thick), along the base of the Long Island Platform. Smith (1980) also noted this Late Cretaceous shift to northern source terrains in his analysis of heavy mineral suites in the COST B-2 well, on the New Jersey continental shelf. In the Maestrichtian, however, sediment supply from the central Appalachian Highlands began to revive, and the ancient James and Potomac rivers again became primary dispersal routes (Appendix fig. 13).

In the early Cenozoic (Paleocene), the central Appalachian Highlands once again became the main terrigenous source terrain. The ancient Susquehanna and Delaware rivers were the primary sediment dispersal routes (Fig. 6) and built a small, elongate prism (>100 m thick) on the middle shelf seaward of New Jersey (Appendix fig. 14). The rest of the continental shelf is virtually bare of Paleocene sediments. Another unusual aspect of Paleocene deposition is that the two largest and thickest depocenters (200–>300 m thick) developed on the continental slope and rise, seaward of broad swaths where Paleocene shelf sediments are missing. Much of the Paleocene section in these two deep-water depocenters, therefore, may have been derived from erosion of the continental shelf or upper continental slope, rather than from highland source terrains. The possibility remains, however, that the ancient James River and eastern Massachusetts rivers also contributed significantly to these abyssal depocenters. Narrow, elongate Paleocene depocenters (> 200 m thick) along the base of the continental slope comprise chiefly fine-grained pelagic carbonates.

With the exception of the early Eocene (Appendix fig. 15), the ancient Susquehanna, Delaware, and Hudson rivers continued to dominate sediment dispersal through the early Miocene (Fig. 6; Appendix figs. 16–9), and built successive middle-to-outer-shelf depocenters seaward of the middle Atlantic states. In contrast, depocenters in the Hatteras Basin were small and widely scattered (or too thin to identify) and comprised chiefly pelagic carbonate deposits.

During the notable pulse of middle Miocene deposition, the central Appalachian source terrain remained dominant, and all four ancient Appalachian dispersal routes contributed large volumes of sediment (Fig. 6; Appendix fig. 20). But significant contributions of terrigenous debris came from the other two bordering highland regions as well, through all the main dispersal routes. Massive siliciclastic prisms (> 1300 m thick) accumulated in deltas along the shelf edge, but even thicker ones (> 1600 m thick) formed huge submarine fan complexes, which spread out across the outer continental rise in the Hatteras Basin.

During the late Miocene, the central Appalachian Highlands (ancient Potomac, Susquehanna, and Delaware dispersal routes) and the Adirondack Highlands (Fig. 6) contributed significant volumes of sediment, which created depocenters (> 800 m thick) on the outer continental rise. Vigorous abyssal currents redistributed some of these deposits to form the crescentic Chesapeake Drift (Appendix fig. 21; Mountain and Tucholke, 1985; Tucholke and Mountain, 1986).

During the Pliocene, the New England Appalachians again began to shed large volumes through the eastern Massachusetts river(s), and the ancient James River joined the Potomac, Susquehanna, and Delaware as primary western dispersal routes (Fig. 6; Appendix fig. 22).

Finally, in the Quaternary, the offshore basins drew major sediment contributions from all highland source terrains, primarily through the ancient Potomac–Susquehanna, Delaware, Hudson, Connecticut, and eastern Massachusetts rivers (Fig. 6; Appendix fig. 23). Throughout the Pliocene and Quaternary intervals, the main depocenters (400– > 800 m thick) were concentrated on the continental slope and inner rise of the Hatteras Basin.

The succession of isopach maps (Appendix figs. 1–23) certifies explicitly that depocenters in the Baltimore Canyon Trough and Hatteras Basin shifted through geologic time in two principal ways: (1) latitudinal shifts from south to north (and vice versa; Fig. 6); and (2) bathymetric shifts from continental shelf to continental rise (and vice versa; Fig. 7). It is obvious that most of the major latitudinal shifts, such as that from James and Potomac-fed depocenters in the Barremian–Albian interval (Appendix figs. 8,9), to the Connecticut and eastern Massachusetts-fed depocenters in the Coniacian–Campanian interval (Appendix figs. 11, 12), were not related to paleoclimatic change, because, as far as we can determine, the entire study area was subject to the same general paleoclimatic regime during this period. Neither would sea-level changes have brought about this pattern of latitudinal depocenter shifts. The most likely reason for most of the latitudinal shifts, then, is changing differential uplift among the three dominant highland source terrains (central Appalachians, Adirondacks, New England Appalachians; Figs. 1, 6). Winker (1982) and Galloway (1989b) reached similar conclusions regarding latitudinal shifts in Cenozoic depocenters of the Gulf Coast. Central Appalachian source terrains were tectonically (topographically) dominant or co-dominant during most of the postrift history of the offshore depocenters (Fig. 6), but New England sources dominated during much of the Late Cretaceous (Coniacian–Campanian). The Adirondack Highlands never dominated alone, but were co-dominant during the Jurassic and Early Cretaceous (Aalenian–Hauterivian) and throughout much of the Cenozoic.

The one significant exception to the covariance of latitudinal depocenter shifts and differential highland uplift is the Quaternary deposition. The enormous Quaternary sediment volumes could be attributed almost entirely to North American continental glaciation, which was active in New England, New York, New Jersey, and Pennsylvania. (Ice sheets produced moraines as far south as Long Island and the Massachusetts islands of Martha's Vineyard and Nantucket [Fig. 1; Borns, 1973; Delcourt and Delcourt, 1981; Oldale, 1982].)

The bathymetric type of depocenter shift (Fig. 7), such as the shift from dominant outer shelf-slope-rise deposition in the Maestrichtian and early Paleogene (Appendix figs. 13, 14), to dominant middle shelf deposition in the middle Eocene–early Miocene (Appendix figs. 16–19), appears to be unrelated to differential uplift of the highland source terrains, but related rather to changes in sea level. The Maestrichtian to Eocene shift, for example, coincides with a long-term sea-level rise. A short-term sea-level at the end of the Cretaceous fall (and perhaps, in part, a bolide-related tsunami (Bourgeois et al., 1988) or vigorous shelf current systems (Olsson and Wise, 1987), caused widespread shelf erosion. This condition contrasts with dominant shelf deposition in the Eocene brought about by a long-term sea-level maximum (Poag, 1985; Haq et al., 1987; Olsson et al.,1988). Likewise, the gradual seaward progression of shelf depocenters during the mid to late Tertiary (Appendix figs. 16–22) can be related, in part, to gradually falling long-term sea level (Haq et al., 1987).

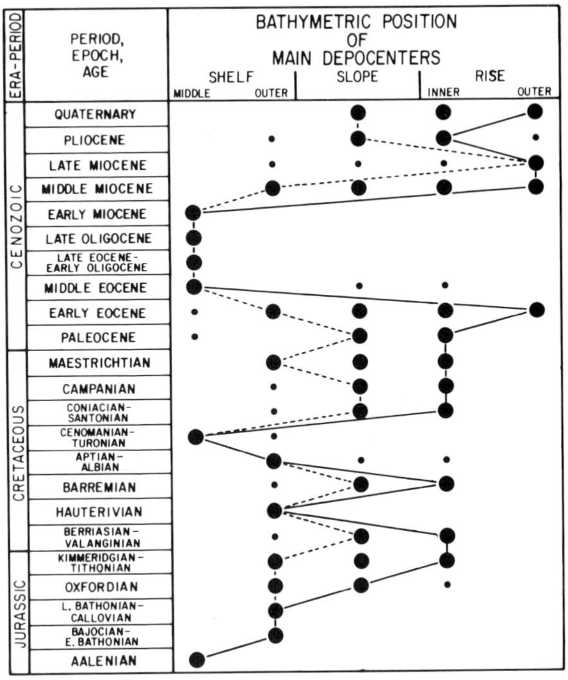

Fig. 7. Bathymetric position of main offshore depocenters for each mapped depositional unit. Large solid dots indicate dominant depocenters (thickest prism of sediments); small solid dots indicate secondary depocenters. Solid line tracks bathymetric shifts of main deepest-water depocenters through time; dashed line tracks bathymetric shifts of main shallowest-water depocenters through time.

The middle Miocene–Quaternary dominance of bathyal and and abyssal depocenters (Fig. 7) is partly attributable also to intense erosion of the continental shelf during intermittent sea-level lows, especially during the rapid fluctuations of the Pleistocene. In fact, nearly every deep-water depocenter we mapped contains several depositional subsequences, some of which are low-stand deposits that accumulated during the erosional hiatuses that intersperse and separate the allostratigraphic units.

Lithofacies

During postrift deposition, several long-term changes took place in the general composition of sediments reaching the Salisbury Embayment, Baltimore Canyon Trough, and Hatteras Basin (Fig. 8). During the early part of depositional Phase I (Aalenian), the narrow (~45 km), low-latitude, restricted basin (Fig. 5A), typified by semi-arid paleoclimate, was the site of evaporitic deposition (Fig. 8; Grow, 1980; Jansa, 1986; Poag, 1987a). Circulation in the ever-widening basin quickly improved, however, to produce normal marine waters for the next ~20 Myr (late Aalenian–Callovian; Fig. 5B), and during this time terrigenous siliciclastics appear to have been the dominant sediment type over most of the continental shelf (as determined from seismic facies analysis; Schlee, 1981; Poag, 1985; Schlee et al., 1988). This lithofacies indicates that mechanical weathering dominated production of siliciclastic detritus in the highland source terrains during this period, a circumstance compatible with

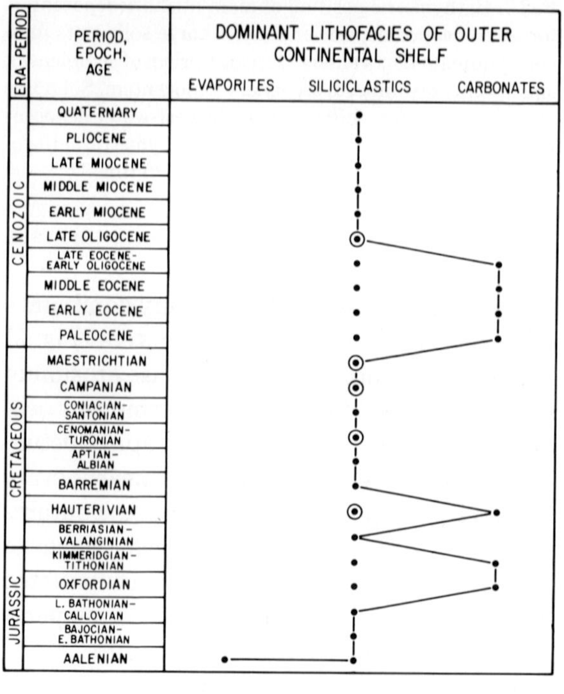

Fig. 8. Temporal changes in dominant lithofacies of the outer continental shelf of the Baltimore Canyon Trough. Open circles indicate that siliciclastics are dominantly fine-grained (silt, clay). Line connecting dots helps emphasize intervals when evaporites or carbonates were significant lithic components.

warm, alternating wet and dry paleoclimate, low sea levels. relatively rapid thermotectonic basin subsidence, and elevated source terrains.

By the start of early Phase II deposition (Oxfordian), however, (and perhaps somewhat earlier) shallow-water limestones had formed a narrow (~20–30 km) band across the outer continental shelf of the Baltimore Canyon Trough (Fig. 8; Jansa, 1981, 1986; Poag, 1987a; Schlee et al., 1988). Carbonate deposition, in dominating the outer shelf (including shelf-edge reefs) and parts of the continental slope and rise (Hollister et al., 1972) for the next ~20 Myr (Oxfordian–Tithonian), attests to a tropical or subtropical coastal climate and pervasive chemical weathering in the highland source terrains. Paleoclimatic warming and increased precipitation, rising sea level, slowed basin subsidence, and stabilized source terrains, all probably contributed to the initiation and maintenance of this carbonate regime.

During most of the remaining ~55 Myr of Phase II (Berriasian–Turonian), however, coarse, siliciclastic sediments again dominated continental shelf deposition (Fig. 8); they buried the shelf-edge carbonate platform and spilled voluminously into the Hatteras Basin (e.g., Van Hinte et al., 1987; Graciansky et al., 1987; Poag et al., 1988). The absence of an accompanying change in sediment accumulation rate, however, indicates that tectonism in the source terrains did not trigger this siliciclastic pulse. The continued presence of warm, humid, non-monsoonal paleoclimates likewise precludes any significant long-term paleoclimatic effect on basin lithofacies. Long-term sea-level changes, similarly, were too slow to drown healthy carbonate platforms (Schlager, 1981; Hallock and Schlager, 1986). On the other hand, carbonate banks and reefs are built and maintained under a tenuous combination of environmental properties (e.g., light availability, temperature, salinity, dissolved oxygen, nutrient levels, turbidity, bioerosion, wave energy, and water depth). Short-term (even annual) disturbances in one or more of such properties could damage and eventually destroy carbonate-producing communities. Several authors have recently suggested a reef-drowning scenario that involves short-term sea-level changes (Hallock and Schlager, 1986; Scott, 1988; Macintyre, 1989). In this scenario, sea level falls and exposes the continental shelf to subaerial weathering and soil formation. Carbonate-producing communities that occupied the shelf edge migrate downward to compatible habitats on the upper slope. This activity is followed by a rapid sea-level rise, which floods the shelf and erodes the nutrient-rich soil to form a turbid, hypernutrient water column. The turbidity and excessive nutrients retard repopulation of the shelf edge by the displaced reef organisms, until they are submerged below the critical depth from which they could "catch up" with sea-level rise. Stratigraphic and petrologic evidence indicates that this is probably what happened to the Late Jurassic reefs of the middle Atlantic shelf. Seimostratigraphic analysis shows that a widespread erosional unconformity separates the Kimmeridgian–Tithonian carbonates from the Berriasian–Valanginian siliciclastics. Petrologic analysis of limestone cores from the reef show evidence of freshwater diagenesis (Eliuk et al., 1986; Meyer, 1986; G. Edson, written commun., 1988), evidence which indicates that the reef was emergent, or nearly so. Furthermore, the reef community contains dominantly corals, algae, and stromatoporoids (Karlo, 1986; G. Edson, written commun. 1988), all of which are zooxanthellate organisms that rely on photosynthesis for their life processes. They are particularly susceptible to the lethal effects of turbid, hypernutrient waters.

Siliciclastic dominance in the Early Cretaceous was relinquished briefly, however, during the Hauterivian and Cenomanian–Turonian intervals, when elevated sea levels and stabilized source terrains allowed chiefly carbonates and fine-grained siliciclastic detritus to blanket the outer continental shelf.

Depositional Phase III (Coniacian–Maestrichtian) also was characterized by dominant siliciclastic sedimentation on the continental shelf, slope, and rise (Fig. 8), but the generally finer grained sequences of the Campanian and Maestrichtian could indicate slowed uplift of hinterland source terrains, and (or) higher sea levels. Haq et al. (1987) showed that sea level was significantly higher in the Campanian than during the preceding Coniacian–Santonian interval.

During the succeeding ~40 Myr (most of Phase IV; Paleocene–early Oligocene), a carbonate regime was re-established on the outer continental shelf and in the Hatteras Basin (Fig. 8; Poag, 1987a). This carbonate buildup can be attributed, in large part, to the reduction in siliciclastic sediments brought to the basins as a result of sustained tectonic quiescence in the source terrains. The carbonate increase also can be attributed, in part, to the global paleoclimatic change that brought a rain-forest ecosystem and increased chemical weathering to the middle Atlantic continental margin. At the same time, high sea levels allowed the coarser terrigenous sediment to be trapped in large depocenters on the inner and middle shelf.

The remaining part of depositional Phase IV (late Oligocene–early Miocene) and the ~25 Myr of Phase V (middle Miocene–present) are characterized by dominantly siliciclastic deposition in the middle Atlantic offshore basins (Fig. 8). This interval corresponds to a period of renewed source-terrain uplift, deteriorating global climates (cooling), intensified latitudinal thermal gradients, and, eventually, widespread continental glaciation. These conditions reinforced mechanical weathering and intensified erosion of the source terrains, as long-term sea level continued its downward course (Haq et al., 1987).

Summary

Using stratigraphic and depositional data, we have sorted out some of the individual tectonic, paleoclimatic, and sea-level signals contained in the sedimentary record of the Salisbury Em-

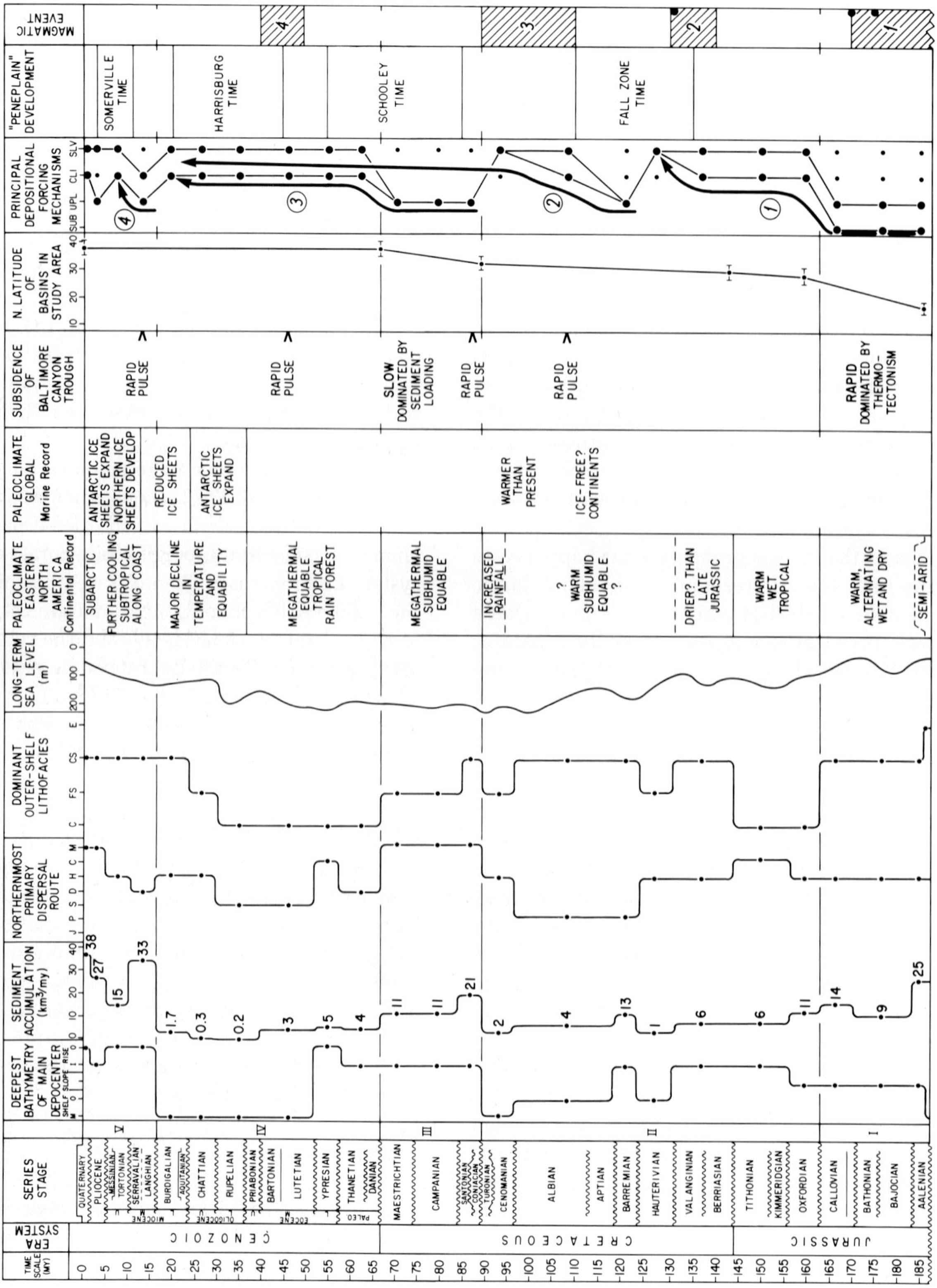

1 TIME SCALE (MY) ERA SYSTEM
2 SERIES STAGE
3
4 DEEPEST BATHYMETRY OF MAIN DEPOCENTER
5 SEDIMENT ACCUMULATION (km³/my)
6 NORTHERNMOST PRIMARY DISPERSAL ROUTE
7 DOMINANT OUTER-SHELF LITHOFACIES
8 LONG-TERM SEA LEVEL (m)
9 PALEOCLIMATE EASTERN NORTH AMERICA Continental Record
10 PALEOCLIMATE GLOBAL Marine Record
11 SUBSIDENCE OF BALTIMORE CANYON TROUGH
12 N. LATITUDE OF BASINS IN STUDY AREA
13 PRINCIPAL DEPOSITIONAL FORCING MECHANISMS
14 "PENEPLAIN" DEVELOPMENT
15 MAGMATIC EVENT
CENOZOIC
CRETACEOUS
JURASSIC
SOMERVILLE TIME
HARRISBURG TIME
SCHOOLEY TIME
FALL ZONE TIME
RAPID PULSE
SLOW DOMINATED BY SEDIMENT LOADING
RAPID DOMINATED BY THERMO-TECTONISM
ANTARCTIC ICE SHEETS EXPAND NORTHERN ICE SHEETS DEVELOP
REDUCED ICE SHEETS
ANTARCTIC ICE SHEETS EXPAND
WARMER THAN PRESENT
ICE-FREE? CONTINENTS
SUBARCTIC
FURTHER COOLING, SUBTROPICAL ALONG COAST
MAJOR DECLINE IN TEMPERATURE AND EQUABILITY
MEGATHERMAL EQUABLE TROPICAL RAIN FOREST
MEGATHERMAL SUBHUMID EQUABLE
INCREASED RAINFALL
? WARM SUBHUMID EQUABLE ?
DRIER? THAN LATE JURASSIC
WARM WET TROPICAL
WARM, ALTERNATING WET AND DRY
SEMI-ARID

bayment, Baltimore Canyon Trough, and Hatteras Basin, and have related them to four primary forcing mechanisms, as summarized graphically in Fig. 9.

The rapid accumulation of dominantly coarse siliciclastic sediments during depositional Phase I of the Early Jurassic (Aalenian–Callovian) took place as the narrow, low-latitude Baltimore Canyon Trough underwent rapid thermotectonic subsidence, accompanied by uplift of highland source terrains, during a period of low sea level. Isopach patterns indicate that the ancient Delaware and Hudson rivers, draining the central Appalachian and Adirondack highlands, dominated sediment dispersal routes at first. The warm climate must have been dominantly arid or semi-arid, judging from the presence of evaporitic strata near the base of the sedimentary succession. The combination of great highland-to-basin relief and intense mechanical weathering maximized the production of coarse terrigenous detritus in the central Appalachians, which was delivered rapidly to the bordering offshore basins because of high source-to-basin relief.

Depositional Phase II (Late Jurassic–middle Cretaceous), characterized by fluctuating, moderate to low sediment-accumulation rates, provides a more complex record than Phase I. Basin subsidence had slowed significantly by this time, and was controlled mainly by sediment loading rather than thermotectonism. Thus subsidence probably was a minor factor in producing the depositional sequences of Phase II and succeeding phases. On the other hand, differential highland uplift, paleoclimatic shifts, and sea-level change continued to be major forcing mechanisms. A broad carbonate shelf-edge and barrier reef developed in the Late Jurassic (Oxfordian–Tithonian) as a response, in part, to a tropical coastal paleoclimate, to increased rainfall and chemical weathering in the now stabilized highlands, and, in part, to a sea level that was higher than during preceding Phase I.

A sudden return to dominantly siliciclastic deposition during the Early Cretaceous (Berriasian–Valanginian) appears, in light of the generally low accumulation rates, the absence of a major latitudinal shift in depocenters, and the absence of significant paleoclimatic change, to have been initiated by a short-term sea-level

Fig. 9. Summary chart correlating deposition-related events of the study area. Column 1 shows the DNAG time scale (Vogt and Tucholke, 1986). Column 2 shows the allostratigraphic units mapped herein. Column 3 shows the five depositional phases determined from volumetric sediment-accumulation rates (defined in the section "Accumulation Rates"). Column 4 shows temporal shifts in dominant deep-water depocenters. I=inner; M=middle; O=outer. Column 5 shows temporal shifts in volumetric sediment-accumulation rates. Column 6 shows temporal shifts in the position of the northernmost primary sediment dispersal routes for each mapped unit. J=James River; P=Potomac River; S=Susquehanna River; D=Delaware River; H=Hudson River; C=Connecticut River; M=eastern Massachusetts river(s). Column 7 shows temporal shifts in dominant sediment composition (lithofacies) of the outer continental shelf. C=carbonate; FS=fine-grained siliciclastics CS=coarse-grained siliciclastics; E=evaporites. Column 8 shows temporal shifts in long-term sea level according to Haq et al. (1987). Column 9 shows the continental record of paleoclimate for eastern North America according to Wolfe and Upchurch (1987). Column 10 shows general global paleoclimate derived from the marine stable-isotope record according to Miller et al. (1987). Column 11 shows temporal changes in subsidence rates and subsidence mechanisms for the Baltimore Canyon Trough. Column 12 shows increasingly northern paleolatitudes of the Baltimore Canyon Trough as sea-floor spreading separated North America and Africa. Column 13 shows the principal depositional forcing mechanisms for each mapped depositional unit as inferred from the data in column 4–12. SUB=tectonic subsidence; UPL=tectonic uplift; CLI=paleoclimate; SLV=long-term sea level. Large solid dots indicate dominant forcing mechanisms; small solid dots indicate secondary forcing mechanisms. Tectonic/quiescent Cycles (1–4) are shown by the numbered heavy arrows. Cycles 1,2,4 apply to central Appalachians; Cycle 3 applies to New England Appalachians. Column 14 shows age ranges assigned by Melhorn and Edgar (1980) to the commonly cited Appalachian "peneplain" surfaces. Column 15 shows the stratigraphic position of significant magmatic events in eastern North America (hachured) according to De Boer et al. (1988). Solid circles indicate times of notable change in the direction of Atlantic plate motion (sea floor spreading; Klitgord and Schouten, 1986).

drop in the late Tithonian to early Berriasian. This drop was followed by a rapid sea-level rise, which drowned the shelf-edge Jurassic reefs.

The further influence of sea-level rise (or of slowed rates of sea-level fall, see Pitman, 1979; Pitman and Golovchenko, 1988) is manifest by the bathymetric shift of major depocenters from the continental rise to the continental shelf during the Hauterivian and Aptian–Albian intervals. Latitudinal depocenter shifts during the middle of Phase II (Hauterivian, Barremian, and Aptian–Albian) and the marked increase in Barremian accumulation rates were responses to differential tectonic uplift among the three highland source terrains.

Depositional Phase III (late Cretaceous) was characterized by significantly accelerated accumulation of siliciclastic detritus and a switch to dominantly northern source terrains and depocenters. These characteristics suggest renewed uplift of the New England Highlands. The coarser deposits might also suggest intensified mechanical weathering, but no particularly significant change in paleoclimate has been proposed for this interval. Sea-level effects are evident, however, in the sediment textures of Phase III deposits. Coarse-grained strata dominated the Coaniacian–Santonian outer shelf during a sea-level fall, and fine-grained strata (clays, silts) characterized higher sea levels on the Campanian and Maestrichtian shelves.

The dominance of deep-water depocenters, which also characterizes Phase III, normally would be taken as an indication of low sea level, but in this case, sea level was high, so another explanation must be sought. The continental shelf bordering the New England source terrains during Phase III rested on the Long Island Platform, a structural high, which resisted subsidence under sediment load. The shelf, therefore, was essentially bypassed by the increasing volume of sediment, in spite of the prevailing high sea level.

The Cenozoic was a time of frequent change and strong contrast among the properties we have analyzed. Depositional Phase IV (Paleocene–Early Miocene) began with an order-of-magnitude reduction in sediment accumulation rate and a switch to dominantly carbonate deposition on the outer continental shelf of the Baltimore Canyon Trough and throughout the Hatteras Basin (Fig. 9). These changes reflect, in part, the warm, tropical–subtropical paleoclimate and rain-forest ecosystem that developed during the Paleocene and Eocene. Although these basins had reached their approximate present latitude by this time (Fig. 9), elevated global temperature and warming effects of the north-flowing proto-Gulf Stream maintained the tropical coastal paleoclimate.

Source-terrain dominance shifted from the north to the west again in Phase IV, although the ancient Potomac, Susquehanna, and Delaware dispersal systems carried severely reduced sediment volumes from the central Appalachians. Physiographic relief between the source terrains and the offshore basins was apparently at an all-time low. The combination of the highest sea levels and lowest accumulation rates fo the Cenozoic caused terrigenous detritus to be trapped in large middle-shelf depocenters during Phase IV, especially during the middle eocene and late Oligocene. Deep-water depocenters of the Hatteras Basin were generally small, sparse, and comprised chiefly carbonates, though widespread erosion of the shelf and slope during the Paleocene and early Eocene appears to have nourished a few broad, abyssal depocenters.

The late part of Phase IV, however, (late Oligocene and early Miocene), saw a mild increase in accumulation rates and a return to dominantly coarse-grained siliciclastic deposition on the continental shelf of the Baltimore Canyon Trough, and fine-grained siliciclastics in the Hatteras Basin (Fig. 8). This change probably reflects, in part, somewhat cooler (subtropical) and dryer paleoclimate, which increased mechanical weathering. It also reflects an abrupt drop in long-term sea level.

Depositional Phase V (middle Miocene–present) is notable for an abrupt acceleration to exceptionally high sediment accumulation rates, for the coolest paleoclimates (including subarctic climates of the Pleistocene), for the lowest sea levels since the Jurassic, for rapid switching of dominant source terrains and primary dispersal routes, and for the dominance of huge abyssal depocenters from which thick sediment blankets spread to the deepest parts of the Hatteras Basin. The abruptness and magnitude of the middle Miocene and Pliocene depositional shifts strongly suggest rejuvenated tectonic uplift, first in the central Appalachians, and then in the New England Highlands. Rapidly fluctuating (glacial–interglacial) paleoclimate and sea level probably were the dominant forcing mechanisms for Quaternary deposition.

Conclusions

Our analysis of the data at hand convinces us that sediment supply, too often neglected in studies of the Atlantic basins, nearly always played a dominant role in determining the composition, architecture, and distribution of the offshore sedimentary sequences. Furthermore, of the two forcing mechanisms that controlled sediment supply (tectonism and paleoclimate), tectonism clearly was paramount. Relative tectonic uplift among the highland source terrains (and, early on, thermotectonic subsidence of the basins) directly determined the dispersal routes, latitudinal location of depocenters and sediment accumulation rates throughout most of postrift history. Episodic uplift and accelerated deposition took place repeatedly in pulses lasting 5–24 Myr, without regard for prevailing paleoclimate or the position of long-term or short-term sea level. In contrast, the four brief pulses of increased basin subsidence noted by Heller et al. (1982) had no consistent discernible affect on sediment accumulation rates.

In spite of Scumm's (1977) and Galloway's (1989b) claim that paleoclimatic change should stimulate a surge in sediment supply to depocenters, the Atlantic basins record only minimal effects of paleoclimate on sediment accumulation rates, other than during the Pleistocene (though undocumented, short-term, paleoclimatic changes may have enhanced highland erosion rates or sediment delivery at times). As a case in point, the change from megathermal tropical rain-forest conditions in the Paleocene and Eocene, to cooler, less equable paleoclimates in the Oligocene and early Miocene, constitutes the second most significant paleoclimatic shift in the Cenozoic history of this margin (the "terminal Eocene event" of Wolfe, 1978). Yet this "event" was accompanied by a barely measurable change in the sediment accumulation rate. There was, however, another kind of depositional change associated with this "event" – a notable reduction in the relative amount of carbonate deposition. The carbonate reduction, in fact, exemplifies what we believe to have been the chief effect of paleoclimate (aside from Pleistocene glaciation) on Atlantic margin deposition. That effect was to regulate the amount of evaporitic and carbonate deposition relative to siliciclastic detritus reaching the offshore basins. The regulatory effect was strongest when highland source terrains were stabilized, as during Phase IV (Paleocene–early Miocene) and mid-Phase II (Hauterivian), but could be cancelled by renewed uplift, as in later Phase II (Barremian).

Sea-level change had practically no effect on hinterland events that could affect sediment supply. The modest (100–200 m) vertical range commonly cited for Mesozoic and Cenozoic sea level shifts, is insufficient to significantly change the source to basin relief (Galloway, 1989a), even on a short-term scale (5–10 Myr). The role of sea-level change appears to have been chiefly to distribute (and redistribute) sediments, once they reached the basins. Sea level acted in two main ways. First, long-term (and short-term) sea-level change effected shifts in the bathymetric position of depocenters, particularly by allowing sediment to build up on the continental shelf during sea-level highs [e.g., late Phase II (Cenomanian–Turonian) and Phase IV (Paleocene–early Miocene)]. This effect could be overridden, however, by sediment surges brought about by highland tectonism [e.g., early Phase III (Coniacian–Santonian)]. Second, short-term sea level changes were particularly effective in redistributing sediments from place to place within the basins, especially by exposing large areas (or all) of the continental shelf and upper continental slope to erosion and mass sediment displacement (forming allostratigraphic sequence boundaries) during low stands and subsequent marine transgressions (Mitchum, 1985). Schumm (1977) and Galloway (1989a), among others, have pointed out, however, that in some cases a widened coastal plain produced by sea-level fall might reduce the average stream gradients enough to retard erosion and ensuing deposition rates.

Consequent of the relationships cited above, we support earlier workers (Anhert, 1970; Gibson, 1970; Owens, 1970; Mathews, 1975; Denny, 1982; Hack, 1982; Gable and Hatton, 1983; O'Leary, 1988), who concluded that highland source terrains were episodically uplifted during the post-Triassic history of the U.S. Atlantic margin. The inference of cyclic tectonic pulses followed by quiescence (or epeirogenesis, e.g., Pavich, 1985), leads us to consider the phenomenon of broad regional erosion surfaces, known widely in Appalachian literature as "peneplains". Melhorn and Edgar (1980) reviewed the principal hypotheses and arguments regarding "peneplains". They then synthesized modern records (although many are already outdated) of sea-floor spreading rates, continent-scale marine transgressions and regressions, North Atlantic deep-sea sediment accumulation records, and regional Atlantic Coastal Plain unconformities to derive tectonic/epeirogenic cycles for the U.S. east coast and Appalachian area. Their purpose was to distinguish intervals when net erosion could persist long enough to develop subdued, low-relief (denudational) landscapes ("peneplains").

The four intervals of net erosion they (among others) inferred for Appalachian "peneplanation" are shown in Fig. 9, column 14 (Fall Zone Time, etc.). Underlying their inference is the assumption that episodes of tectonic uplift were followed by intervals of extended tectonic stability, or quiescence, in which erosion continued to progress. They recognized, however, that isostatic adjustments might interrupt erosion cycles and cause multiple or warped erosion surfaces.

When viewed from the perspective of tectonic/quiescent cycles and Appalachian "peneplanation", the offshore depositional record can be seen to represent four unequal (1 short, 2 intermediate, 1 long) cycles of tectonic uplift (rapid sediment accumulation) followed by quiescence (reduced sediment accumulation; Fig. 9, columns 5, 13). One of these tectonic cycles (Cycle 3), however, is not related to events in the central Appalachian Highlands. Cycle 3 is, instead, the result of uplift in the New England Appalachians.

In Cycle 1 (central Appalachians), the initial tectonic pulse, though fluctuating, lasted ~24 Myr (equivalent to depositional Phase I). The subsequent stable, or quiescent interval lasted ~39 Myr (early part of depositional Phase II) and culminated in Hauterivian time, which is correlative with the early part of Fall Zone time of the Appalachian "peneplain" succession (Fig. 9, column 14).

The initial tectonic pulse of Cycle 2 (central Appalachians) lasted only ~5 Myr (middle of depositional Phase II; Barremian). Subsequent quiescence in the central Appalachians, however, lasted ~100 Myr (to the end of depositional Phase II), overlapping Cycle 3 and culminating in the early Miocene. The late part of Fall Zone time and all of Schooley time and Harrisburg time are encompassed by this unusually long quiescent interval.

The initial tectonic pulse of Cycle 3 (New England Appalachians) lasted ~22 Myr (depositional Phase III; Coniacian–Maestrichtian), followed by ~50 Myr of quiescence (depositional Phase IV), and culminated in the early Miocene. No "peneplain" development in the central Appalachians is related to Cycle 3.

Initial tectonic uplift in Cycle 4 (central Appalachians) lasted only ~7 Myr (early part of depositional Phase V; middle Miocene) and was followed by a ~5 Myr quiescent interval (spanning the late Miocene), which correlates with Somerville time. The beginning of a fifth cycle may be indicated by uplift in the Pliocene, but the complications of Pleistocene glaciation obscure the pattern.

De Boer et al. (1988) reviewed magmatic events in eastern North America that could be related to the four tectonic cycles we perceive. These authors recognized six magmatic events on the basis of radiometric dates, relative age information (from cross cutting field relationships), paleomagnetism, and geochemistry. Five of the six magmatic events took place in post-Triassic time, and their stratigraphic positions are compared in Fig. 9 (column 15) to the succession of tectonic cycles we infer (column 13). [We treat the oldest two magmatic events as a single event (Event 1) because their respective age ranges are contiguous (190 ± 10 Ma and 175 ± 5 Ma).] Figure 9 shows that magmatic Event 1 is stratigraphically coincident with the uplift pulse of tectonic Cycle 1. Magmatic Event 2 precedes the uplift pulse of tectonic Cycle 2 by ~5 Myr, and magmatic Event 3 is contiguous with the uplift pulse of tectonic Cycle 3. Magmatic Event 4, however, is not associated with any of the tectonic pulses we recognize.

Given the imprecise dates for both the magmatic events and tectonic cycles, we conclude that these phenomena are probably correlative, and that magmatic events 1–3 are additional tangible evidence of the cyclic uplift of middle Atlantic source terrains. The stratigraphic offset between magmatic Events 2–3 and tectonic Cycles 2–3 may indicate that, in some cases, the abrupt acceleration of sediment accumulation in the basins may lag behind the initiation of source-terrain uplift by several million years.

De Boer et al. (1988) related the episodic magmatism to periods of divergent subhorizontal stress in the earth's crust, caused by changes in the direction of sea floor spreading, reactivation of transform faults, or proximity to hot spots. A recent re-examination of changes in Atlantic plate motions by Klitgord and Schouten (1986) indicates that significant changes in plate motion are correlative with (Fig. 9, column 15), but not exclusive to, magmatic Events 1 and 2.

Obviously, this irregular, overlapping series of tectonic cycles is incompatible with the traditional concept of four post-Triassic, Davisian geomorphic cycles in the central Appalachians (Fig. 9, column 14), whose merits have been so vehemently debated by many previous authors (e.g., Ritter, 1988). The inferred tectonic cycles do, however, meet the fundamental requirements of Melhorn and Edgar's (1980) "peneplanation" model, in providing a potential framework for persistent, long-term, widespread intervals of denudation.

In our view, the sedimentary record of the Salisbury Embayment, Baltimore Canyon Trough, and Hatteras Basin reflects, in a definitive way, how tectonism, paleoclimate, and sea level interacted to determine the provenance, dispersal routes, accumulation rates, lithofacies, depocenter locations, and stratigraphic framework of the basin fill. Although tectonism (sediment supply) generally dominated depositional patterns, one must remember that the forcing mechanisms we have cited always worked in concert with each other and with myriad additional environmental processes. If one were to examine these supplementary processes, and choose finer time slices than those we analyzed, additional complexities could be expected to alter and, perhaps, to obscure some aspects of the simplified scenarios we have presented.

Appendix

Isopach maps of allostratigraphic units in the Salisbury Embayment, Baltimore Canyon Trough, and Hatteras Basin. Contours in hundreds of meters; e.g., 5=500 m, 10=1000 m. Principal depocenters are marked by shaded pattern. Diagonal hachures indicate areas where mapped units are either absent or are too thin to be resolved on the multichannel seismic reflection profiles. Solid arrows indicate primary sediment dispersal routes; dashed arrows indicate secondary sediment dispersal routes; dotted arrows indicate possible sediment dispersal routes; heavy dashed line indicates approximate position of shelf edge during deposition of mapped unit. *J*=ancient James River; *P*=ancient Potomac River; *S*=ancient Susquehanna River; *D*ancient Delaware River; *H*=ancient Hudson River; *C*=ancient Connecticut River; *EM*=ancient river(s) of eastern Massachusetts. On fig. 21, *CD*=Chesapeake Drift. All maps are simplified for presentation at this reduced size.

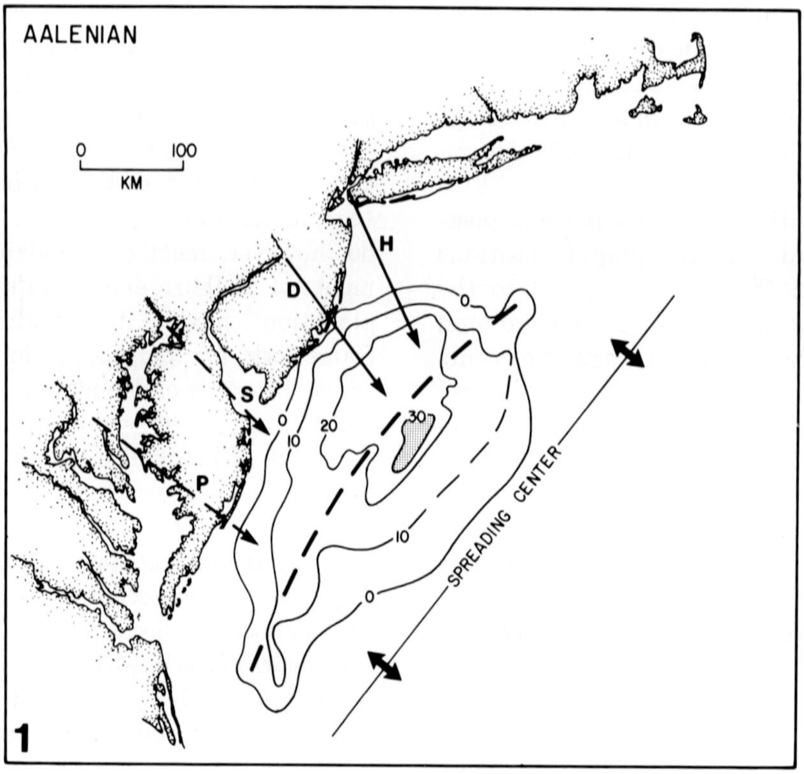

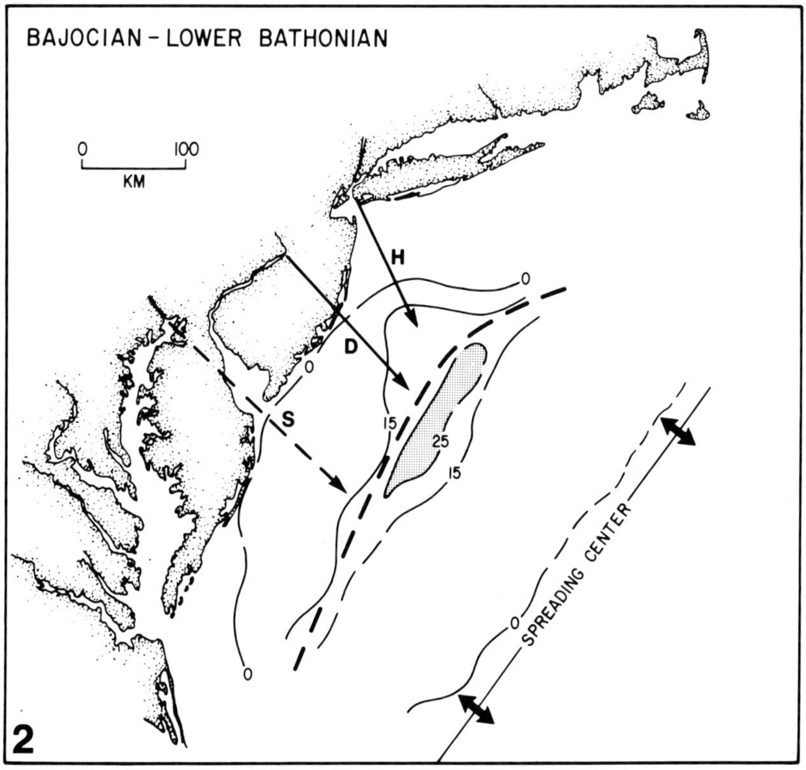
BAJOCIAN - LOWER BATHONIAN
0 100
KM
H
D
S
0
15
25
15
0
SPREADING CENTER
0
2

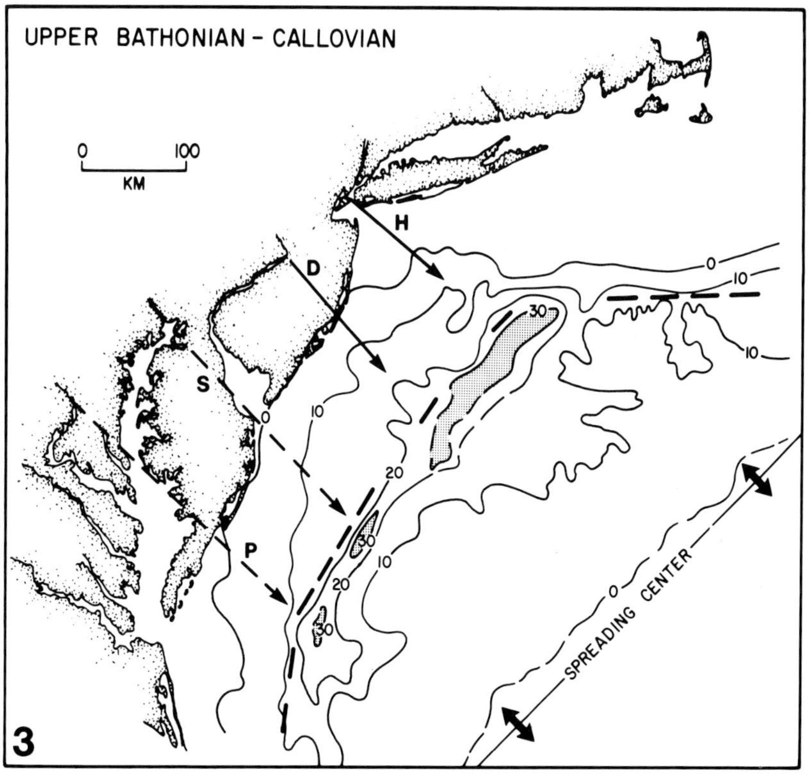
UPPER BATHONIAN - CALLOVIAN
0 100
KM
H
D
S
P
0
10
20
30
10
20
30
30
0
10
10
0
SPREADING CENTER
3

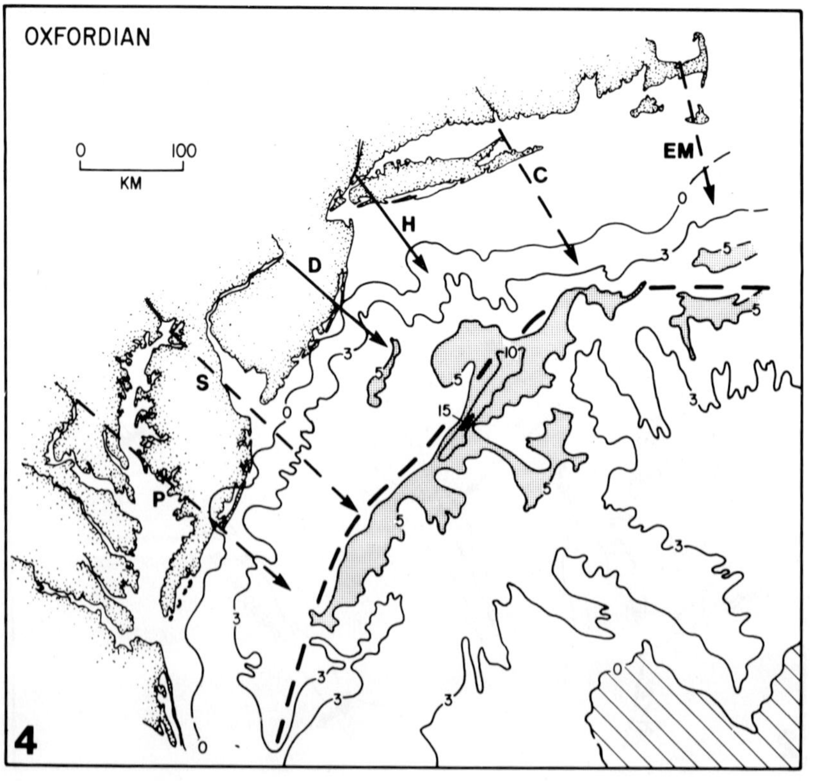
OXFORDIAN
0
100
KM
EM
C
H
D
S
P
4

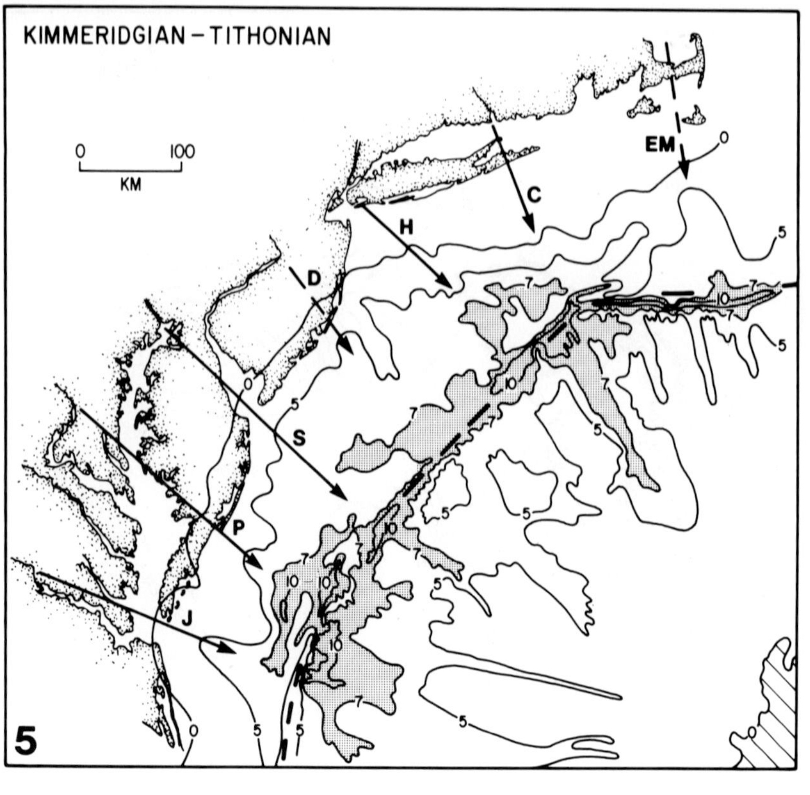
KIMMERIDGIAN – TITHONIAN
0
100
KM
EM
C
H
D
S
P
J
5

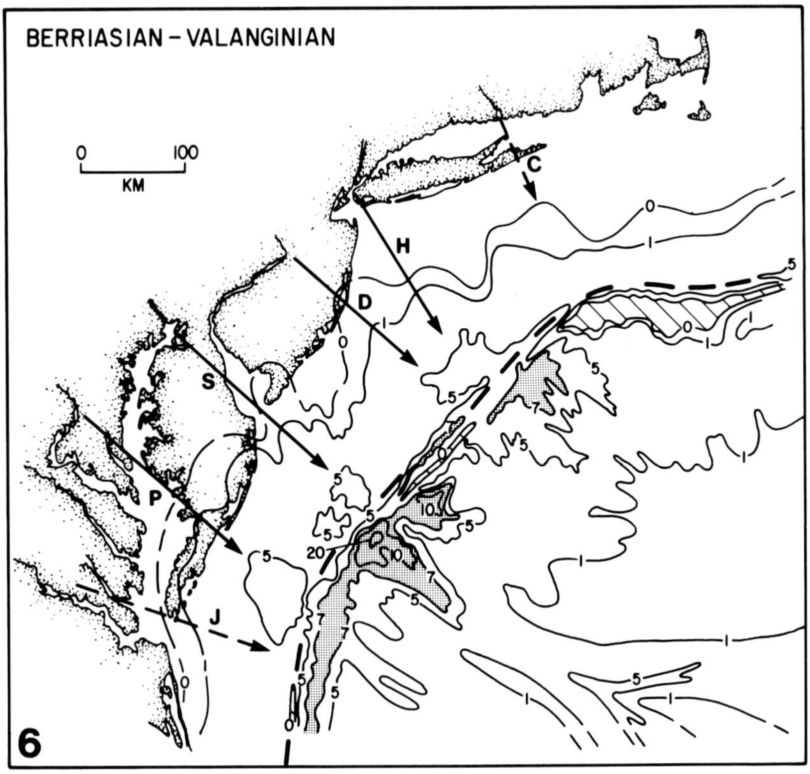
BERRIASIAN – VALANGINIAN
0
100
KM
C
H
D
S
P
J
6

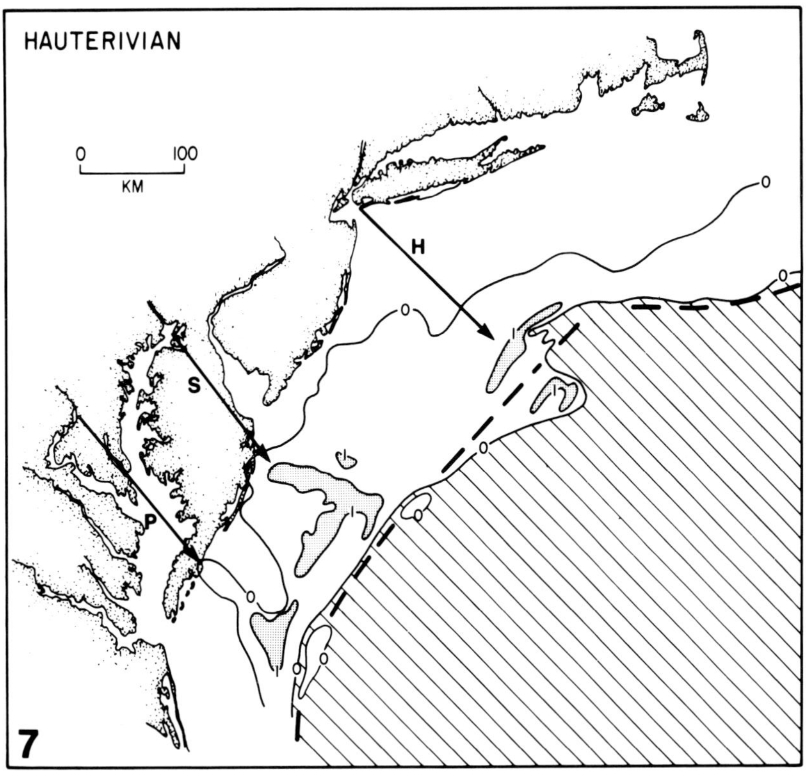
HAUTERIVIAN
0
100
KM
H
S
P
7

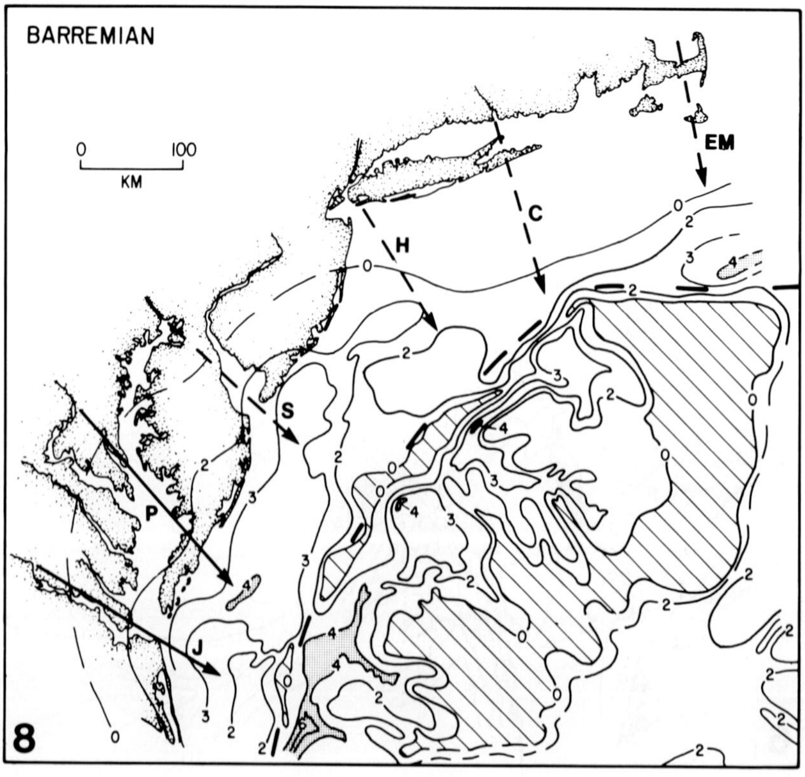
BARREMIAN
0 100
KM
EM
C
H
S
P
J
8

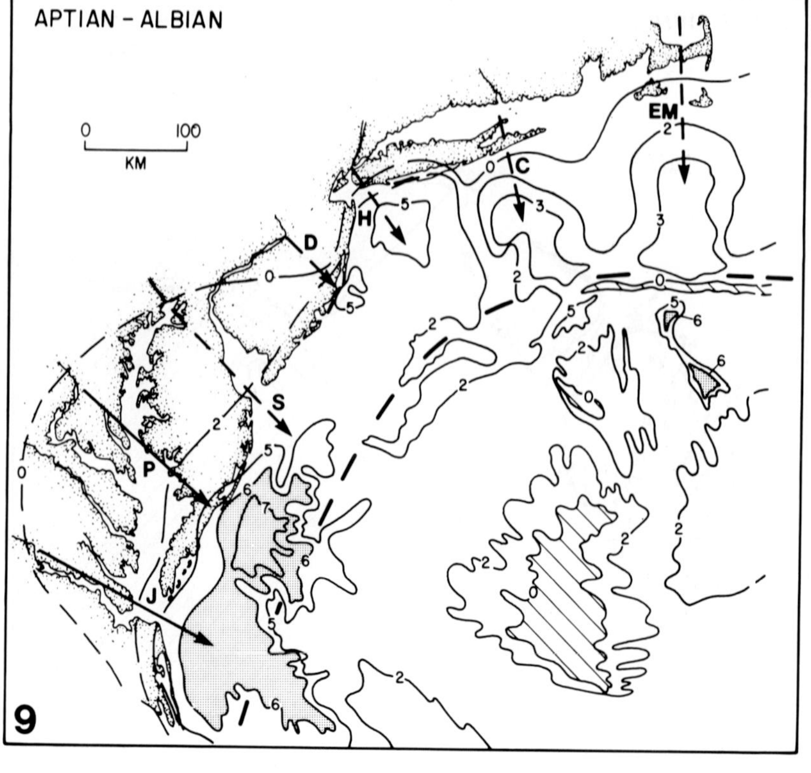
APTIAN - ALBIAN
0 100
KM
EM
C
H
D
S
P
J
9

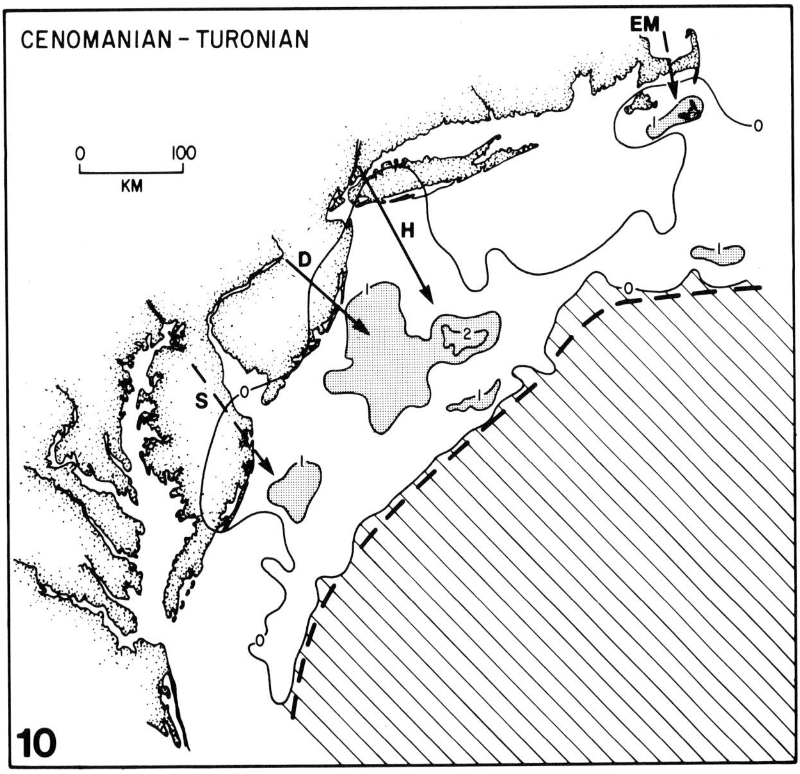
CENOMANIAN - TURONIAN
0
100
KM
EM
H
D
S
10

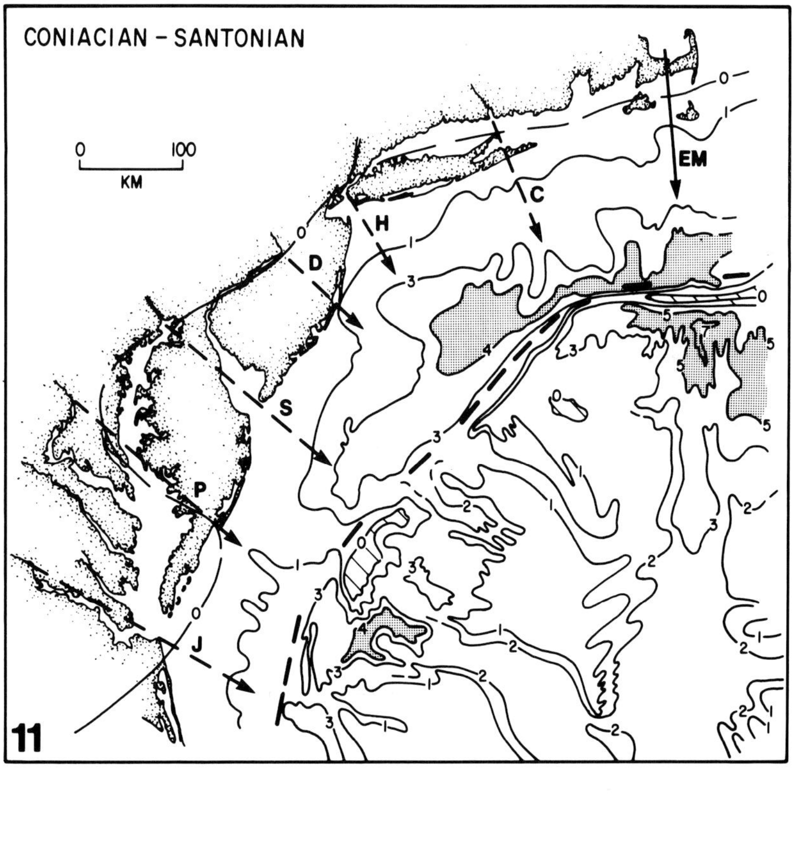
CONIACIAN - SANTONIAN
0
100
KM
EM
C
H
D
S
P
J
11

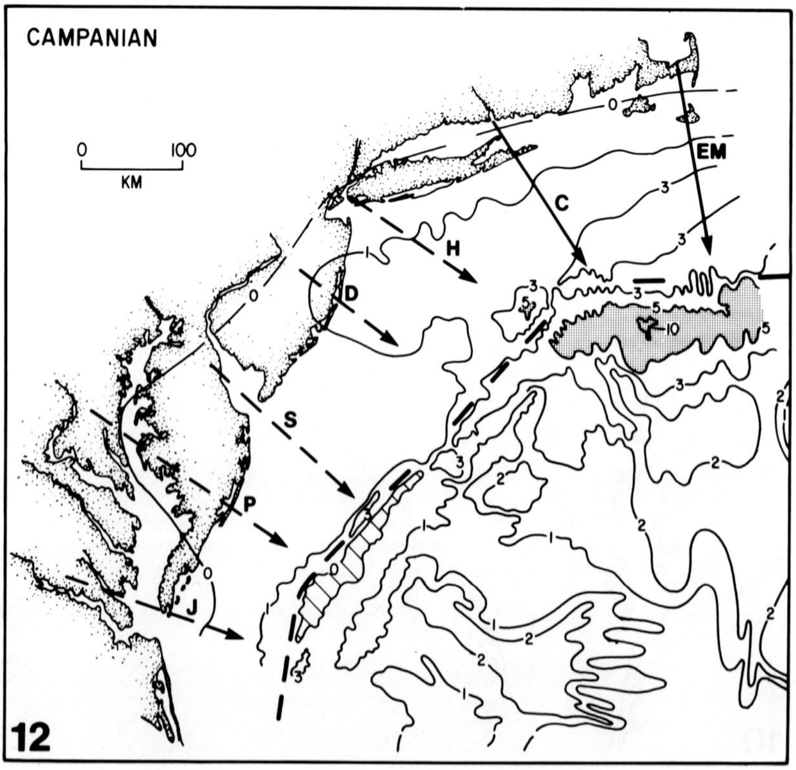
CAMPANIAN
0 100
KM
EM
C
H
D
S
P
J
12

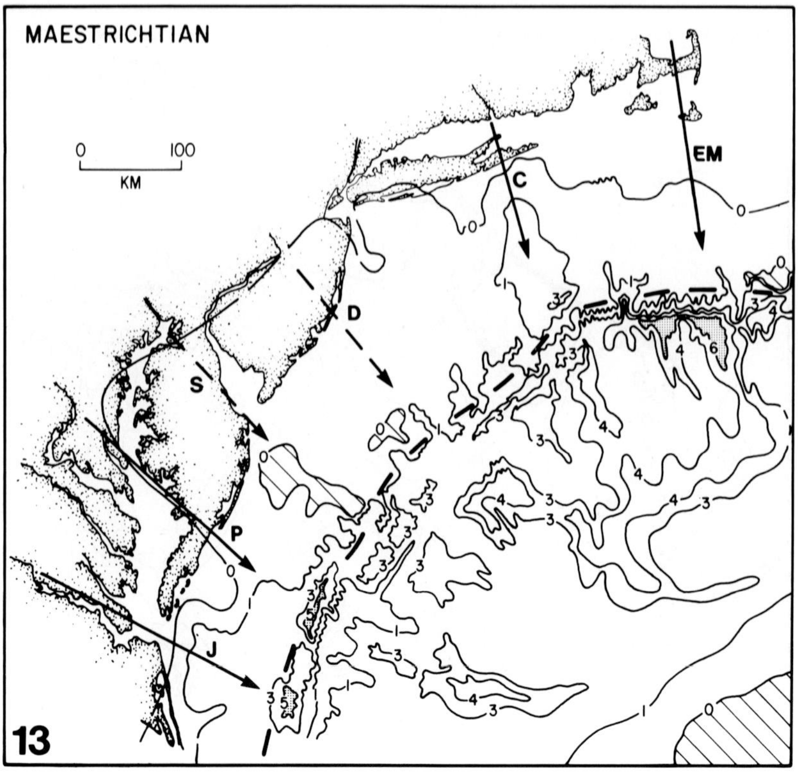
MAESTRICHTIAN
0 100
KM
EM
C
D
S
P
J
13

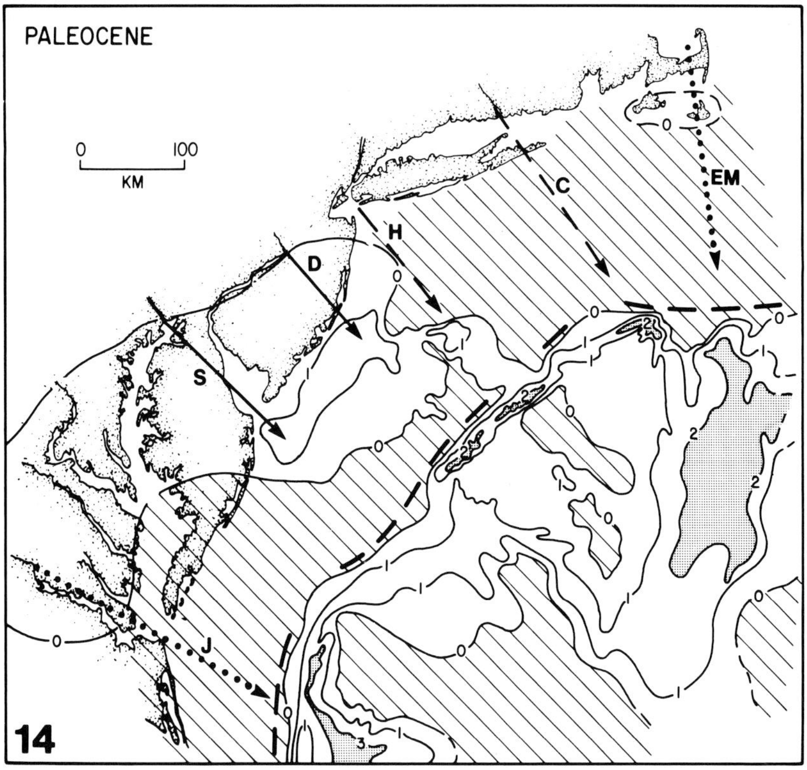
PALEOCENE
0 100
KM
C
EM
H
D
S
J
14

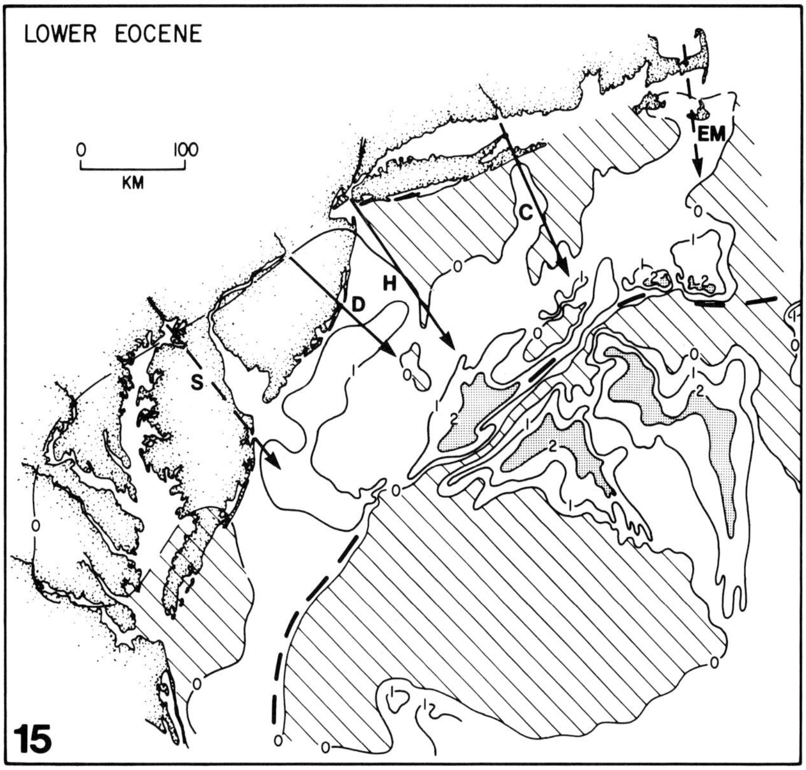
LOWER EOCENE
0 100
KM
EM
C
H
D
S
15

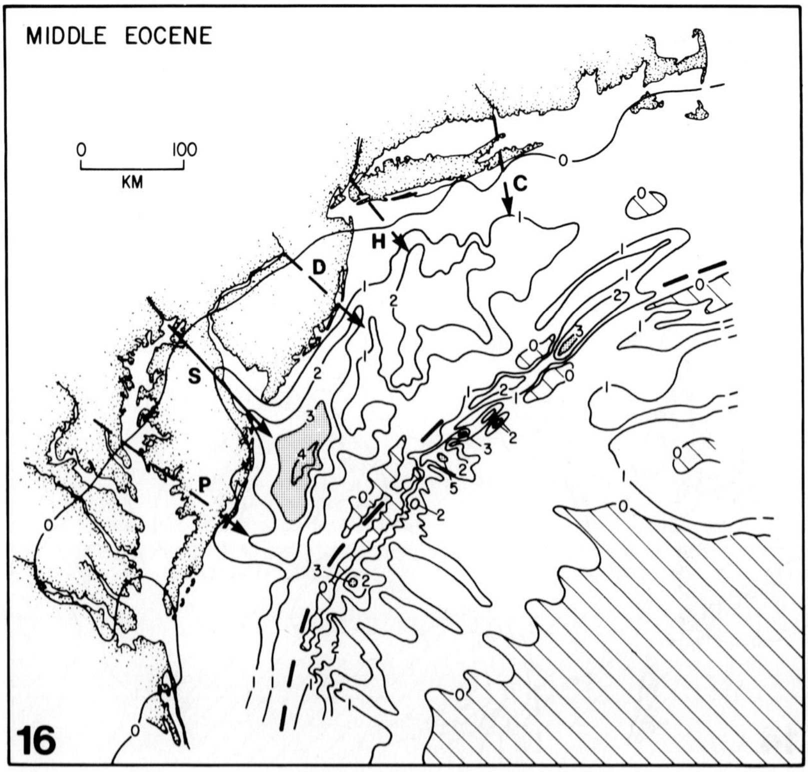
MIDDLE EOCENE
0 100
KM
C
H
D
S
P
16

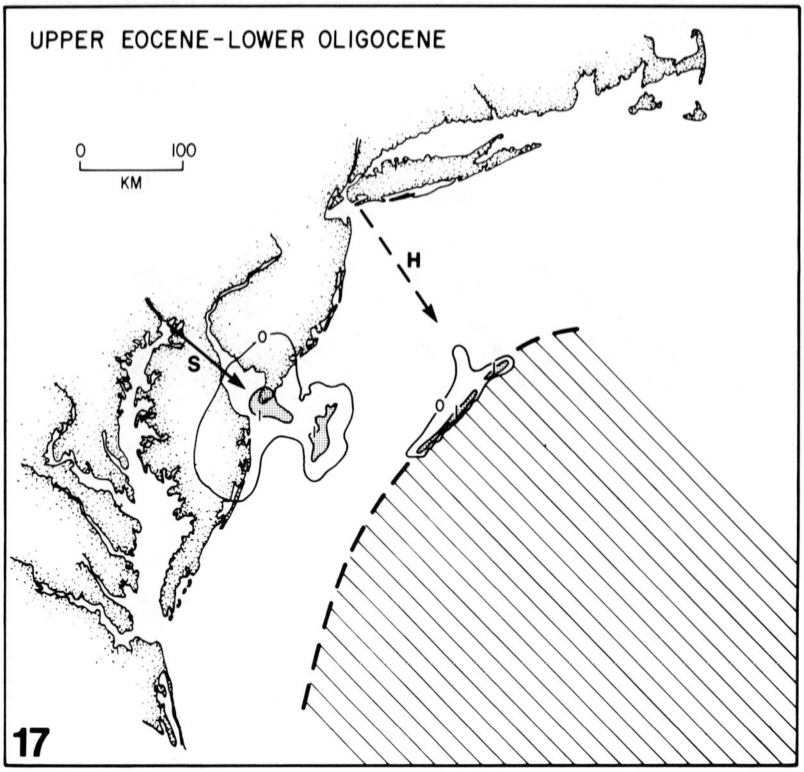
UPPER EOCENE-LOWER OLIGOCENE
0 100
KM
H
S
17

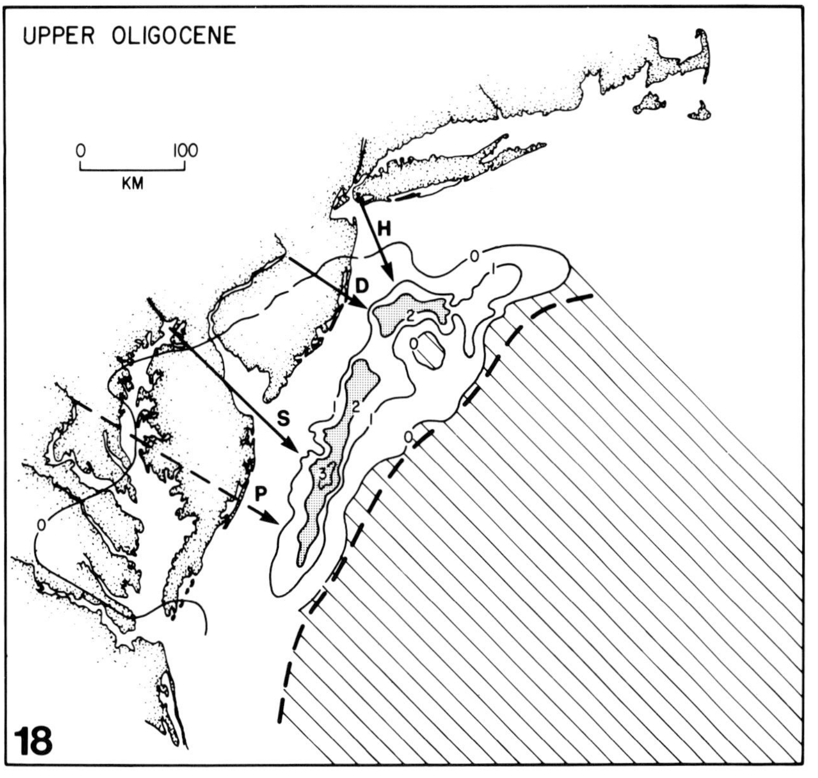
UPPER OLIGOCENE
0 100
KM
H
D
S
P
18

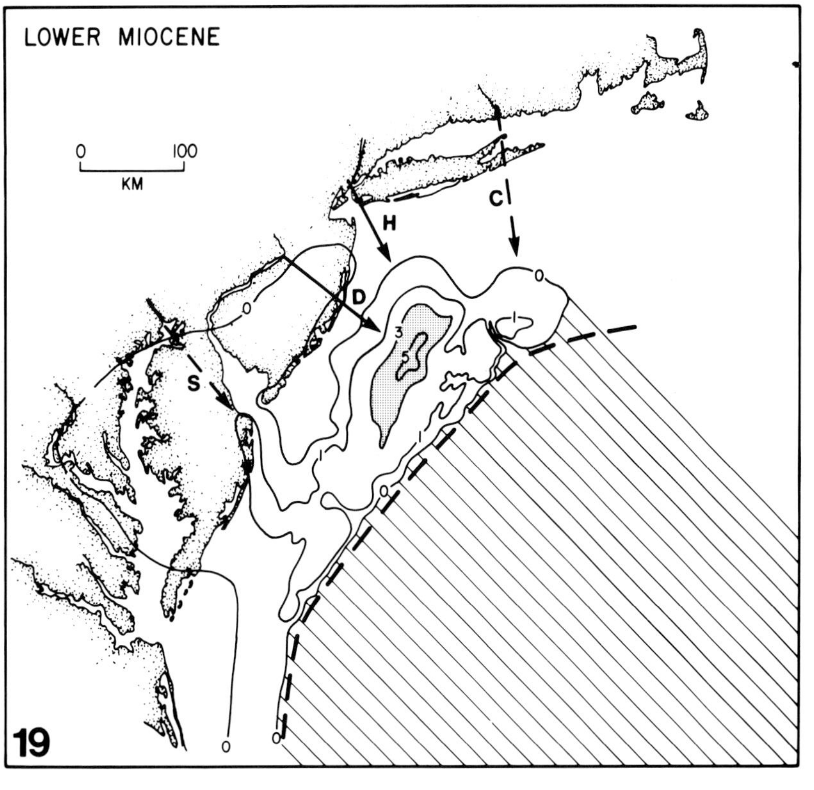
LOWER MIOCENE
0 100
KM
H
C
D
S
19

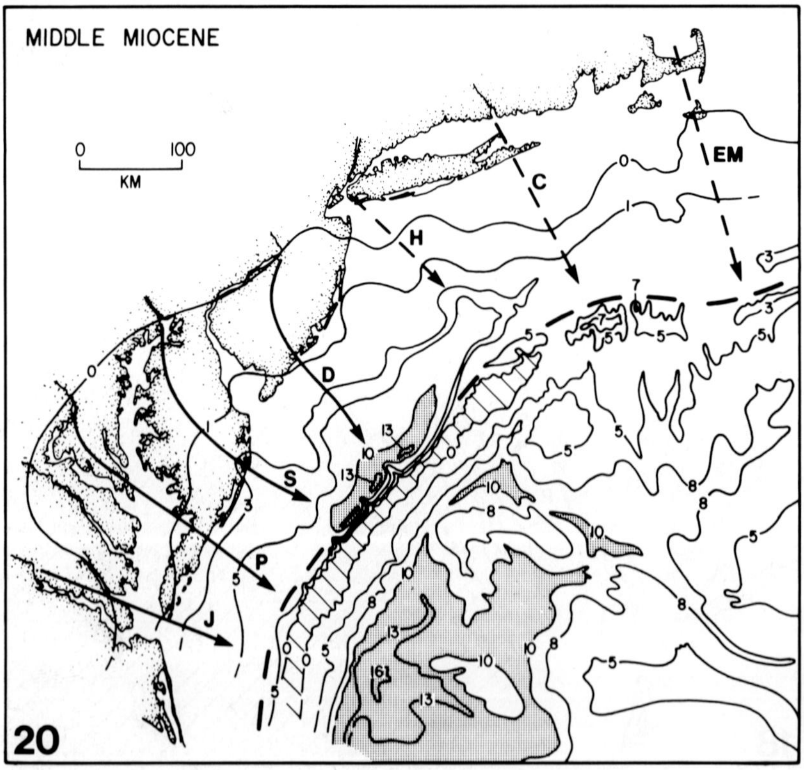
MIDDLE MIOCENE
0 100
KM
EM
C
H
D
S
P
J
20

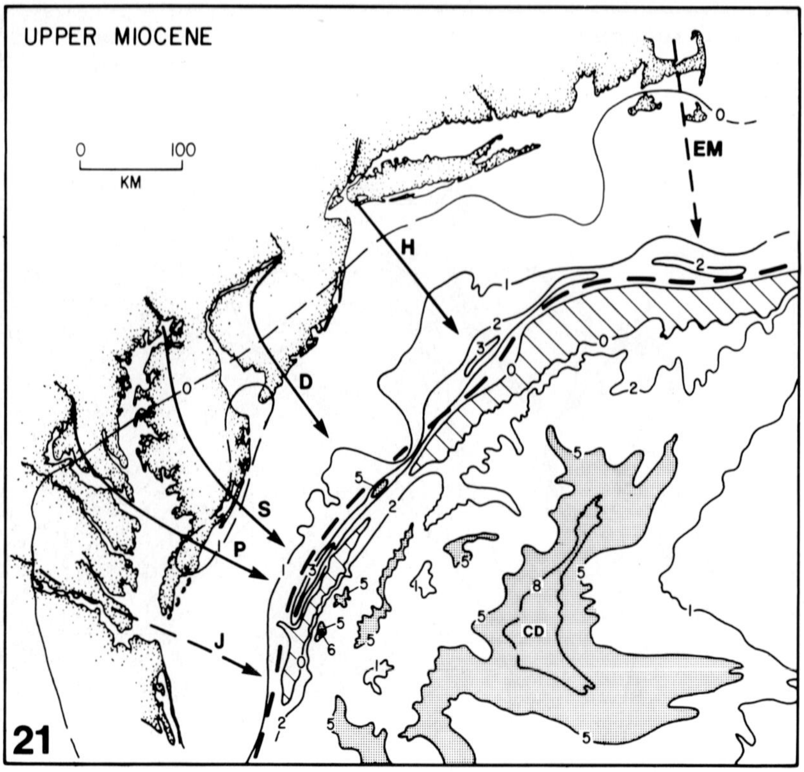
UPPER MIOCENE
0 100
KM
EM
H
D
S
P
J
CD
21

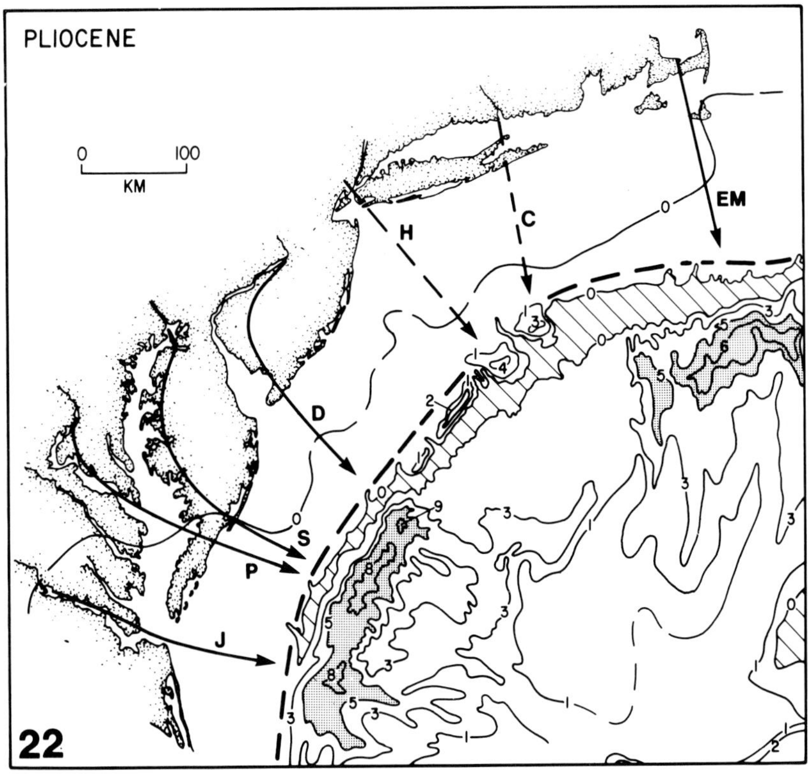
PLIOCENE
0 100
KM
EM
C
H
D
S
P
J
22

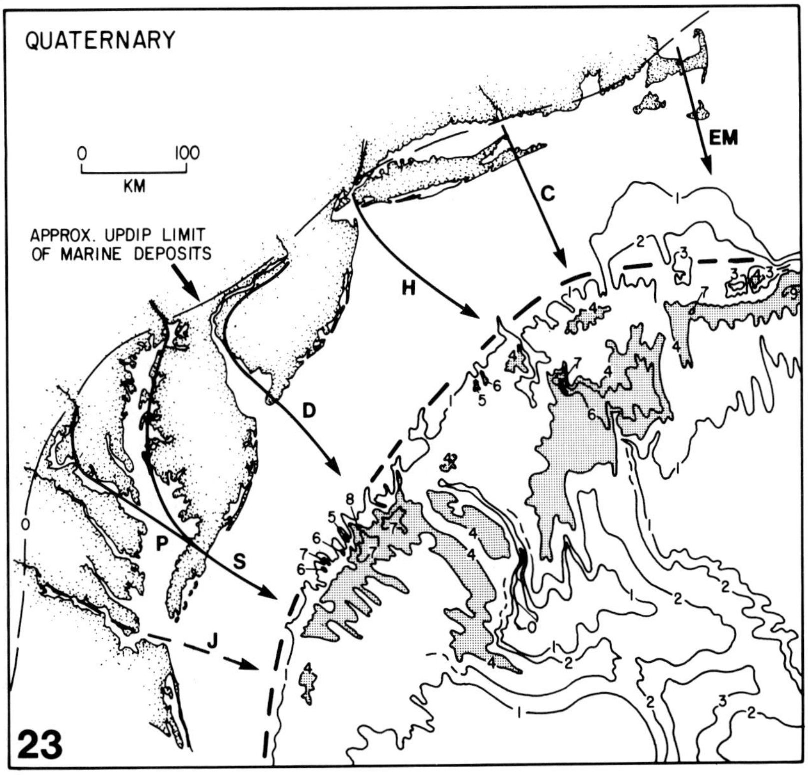
QUATERNARY
0 100
KM
APPROX. UPDIP LIMIT
OF MARINE DEPOSITS
EM
C
H
D
P
S
J
23

References

Ahnert, F., 1970. Functional relationships between denudation, relief, and uplift in large mid-latitude drainage basins. Am. J. Sci., 268: 243-263.

Ascoli, P., 1988. Mesozoic-Cenozoic foraminiferal, ostracod, and calpionellid zonation of the N.W. Atlantic margin of North America: Georges Bank, Scotian basins, and the N.E. Grand Banks (Jeanne d'Arc-Carson basins and Flemish Pass). Biostratigraphic correlation of 51 wells. Geol. Surv. Can. Open File No. 1791, 41 pp.

Benson, R.N. and Doyle, R.G., 1988. Early Mesozoic rift basins and the development of the United States middle Atlantic continental margin. In: W. Manspeizer (Editor), Triassic-Jurassic Rifting. Elsevier, Amsterdam, pp. 99-127.

Berg, T.M. et al., 1980. Geologic map of Pennsylvania. Pennsylvania Geological Survey, 4th ser., Map 1.

Borns, W.H., Jr., 1973. Late Wisconsin fluctuations of the Laurentide ice sheet in southern and eastern New England. In: R.F. Black, R.P. Goldthwait, and H.B. Willman (Editors), The Wisconsinan Stage. Geol. Soc. Am., Mem. 136, pp. 37-45.

Bourgeois, J., Hansen, T.A., Wiberg, P.K. and Kauffman, E.G., 1988. A tsunami deposit at the Cretaceous-Tertiary boundary in Texas. Science, 241, 567-570.

Charletta, A.C., 1980. Eocene benthic foraminiferal paleoecology and paleobathymetry of the New Jersey continental margin. Ph.D. Thesis, Rutgers University, 83 pp.

Christie-Blick, N., Mountain, G.S. and Miller, K.G., in press. Seismic stratigraphic record of sea level change. In: National Research Council, Sea Level Change: Studies in Geophysics. Natl. Acad. Sci.

Cleaves, E.T., 1989. Piedmont landscapes from the Permian to the Holocene. In: T.W. Gardner and W.D. Sevon (Editors), Appalachian Geomorphology. Geomorphology, 2: 159-179 (this volume).

Cronin, T.M., 1981. Rates and possible causes of neotectonic vertical crustal movements of the submerged southeastern Atlantic Coastal Plain. Geol. Soc. Am. Bull., 92: p. 812.

De Boer, J.Z., McHone, J.G., Puffer, J.H., Ragland, P.C. and Whittington, D., 1988. Mesozoic and Cenozoic magmatism. In: R.E. Sheridan and J.A. Grow (Editors), The Geology of North America, Volume I-2. The Atlantic Continental Margin, U.S. Geol. Soc. Am., Boulder, Colo., pp. 217-241.

Delcourt, P.A. and Delcourt, H.R., 1981. Vegetation maps for North America; 40,000 yr. B.P. to the present. In: R.C. Romans (Editor), Geobotany II. Plenum, New York, N.Y., pp. 123-165.

Denny, C.S., 1982. The geomorphology of New England. U.S. Geol. Surv., Prof. Pap., 1208, 18 pp.

Durrant, J.M., 1978. Structural and metamorphic history of the Eastern Virginia Piedmont province near Richmond, Va. M. Thesis, The Ohio State University, Columbus, Ohio.

Eliuk, L.S., Cearly, S.C. and Levesque, R., 1986. West Atlantic Mesozoic carbonates: Comparison of Baltimore Canyon and offshore Nova Scotian basins. Am. Assoc. Pet. Geol., Bull., 70: 586-587.

Emery, K.O., Uchupi, E., Phillips, J.D., Bowin, C.O., Bunce, E.T. and Knott, S.T.,1970. Continental rise off eastern North America. Am. Assoc. Pet. Geol., Bull., 54: 44-108.

Ewing, J.I., 1984. Sections along USGS Line 27, Baltimore Canyon Trough. In: J.I. Ewing and P.D. Rabinowitz (Editors), Eastern North American Continental Margin and Adjacent Ocean Floor, 34° to 41°N and 68° to 78°W. Ocean Margin Drilling Program. Regional Atlas Series. Mar. Sci. Int., Woods Hole. Mass., p. 30.

Firstbrook, P.L., Funnell, B.M., Hurley, A.M. and Smith, A.G., 1979. Paleoceanic Reconstructions. Deep Sea Drilling Project, Scripps Inst. Oceanogr., La Jolla, Calif., 41 pp.

Frakes, L.A., 1979. Climates Throughout Geologic Time. Elsevier, Amsterdam, 310 pp.

Fredericksen, N.O., 1984. Stratigraphic, paleoclimatic, and paleobiogeographic significance of Tertiary sporomorphs from Massachusetts. U.S. Geol. Surv. Prof. Pap. 1308, 25 pp.

Galloway, W.E., 1989a. Genetic stratigraphic sequences in basin analysis I: Architecture and genesis of flooding-surface bounded depositional units. Am. Assoc. Pet. Geol., Bull., 73: 125-142.

Galloway, W.E., 1989b. Genetic stratigraphic sequences in basin analysis II: Application to northwest Gulf of Mexico Cenozoic basin. Am. Assoc. Pet. Geol., Bull., 73: 143-154.

Gable, D.J. and Hatton, T., 1983. Maps of vertical crustal movements in the conterminous United States over the last 10 million years. U.S. Geol. Surv. Misc. Invest. Map I-1315, 2 sheets, scale 1:5,000,000.

Gibson, T.G., 1970. Late Mesozoic-Cenozoic tectonic aspects of the Atlantic coastal margin. Geol. Soc. Am., Bull., 81: 1813-1822.

Graciansky, P.C. de, Brosse, E., Deroo, G., Herbin, J.-P., Montadert, L., Muller, C., Sigal, J. and Schaaf, A., 1987. Organic-rich sediments and paleoenvironmental reconstructions of the Cretaceous North Atlantic. In: J. Brooks and A.J. Fleet (Editors), Marine Petroleum Source Rocks. Geol. Soc. Spec. Publ. 26, 317-344.

Greenlee, S.M., Schroeder, F.W. and Vail, P.R., 1988. Seismic stratigraphic and geohistory analysis of Tertiary strata from the continental shelf off New Jersey; Calculation of eustatic fluctuations from stratigraphic data. In: R.E. Sheridan and J.A. Grow (Editors), The Geology of North America, Volume I-2. The Atlantic Continental Margin, U.S. Geol. Soc. Am. Boulder, Colo., pp. 437-444.

Grow, J.A., 1980. Deep structure and evolution of the Baltimore Canyon Trough in the vicinity of the COST No.

B-3 well. In: P.A. Scholle (Editor), Geological Studies of the COST No. B-3 Well, United Stated Mid-Atlantic Continental Slope Area. U.S. Geol. Surv., Circ., 833: 117-126.

Grow, J.A., Mattick, R.E. and Schlee, J.S., 1979. Multichannel seismic depth sections and interval velocities over outer continental shelf and upper continental slope between Cape Hatteras and Cape Cod. In: J.S. Watkins, L. Montadert and P.W. Dickerson (Editors), Geological and Geophysical Investigations of Continental Margins. Am. Assoc. Pet. Geol., Mem., 29: 65-84.

Grow, J.A., Klitgord, K.D., Schlee, J.S. and Dillon, W.P., 1988. Representative seismic profiles of U.S. Atlantic continental margin. In: R.E. Sheridan and J.A. Grow (Editors), The Geology of North America, Volume I-2, The Atlantic Continental Margin, U.S. Geol. Soc. Am., Boulder, Colo., Plate 4.

Hack, J.T., 1975. Dynamic equilibrium and landscape evolution. In: W.N. Melhorn and R.C. Edgar (Editors), Theories of Landform Development. Publications in Geomorphology, Binghamton, N.Y., pp. 87-102.

Hack, J.T., 1980. Rock control and tectonism - Their importance in shaping the Appalachian Highlands. U.S. Geol. Surv., Prof. Pap., 1126-B, B1-B17.

Hack, J.T., 1982. Physiographic divisions and differential uplift in the Piedmont and Blue Ridge. U.S. Geol. Surv., Prof. Pap., 1265, 49 pp.

Hallam, A., 1984. The Jurassic climate. In: National Research Council, Climate in Earth History. National Academy Press, Washington, D.C., pp. 159-163.

Hallock, P. and Schlager, W., 1986. Nutrient excess and the demise of coral reefs and carbonate platforms. Palaios, 1: 389-398.

Haq, B.U., Hardenbol, J. and Vail, P.R., 1987. Chronology of fluctuating sea levels since the Triassic. Science, 235: 1156-1167.

Haq, B.U., Hardenbol, J. and Vail, P.R., 1988. The new chronostratigraphic basis of Cenozoic and Mesozoic sea level cycles. In: C.A. Ross and D. Haman (Editors), Timing and Depositional History of Eustatic Sequences: Constraints on Seismic Stratigraphy. Cushman Found. Foraminiferal Res., Spec. Publ. 24: 7-13.

Heller, P.L., Wentworth, C.M. and Poag, C.W., 1982. Episodic post-rift subsidence of the U.S. Atlantic continental margin. Geol. Soc. Am., Bull., 93: 379-390.

Hollister, C.D., Ewing, J.I., Habib, D., Hathaway, J.C., Lancelot, Y., Luterbacher, H.P., Paulus, F.J., Poag, C.W., Wilcoxon, J.A. and Worstell, P., 1972. Initial Report of the Deep Sea Drilling Project, Volume 11. U.S. Gov. Print. Off., Washington, D.C., 1077 pp.

Isachsen, Y.W., 1975. Possible evidence for contemporary doming of the Adirondack Mountains, New York, and suggested implications for regional tectonics and seismicity. Tectonophysics, 29: 161-181.

Jansa, L.F., 1981. Mesozoic carbonate platforms and banks of the eastern North American margin. Mar. Geol., 44: 97-117.

Jansa, L.F., 1986. Paleoceonography and evolution of the North Atlantic Ocean basin during the Jurassic. In: P.R. Vogt and B.E. Tucholke (Editors), The Geology of North America, Volume M. The Western North Atlantic Region. Geol. Soc. Am., Boulder, Colo., pp. 603-616.

Jansa, L.M. and Tucholke, B.E., 1986. Paleogeography: Triassic to Late Cretaceous. In: P.R. Vogt and B.E. Tucholke (Editors), The Geology of North America, Volume M. The Western North Atlantic Region. Geol. Soc. Am., Boulder, Colo., Plate 9.

Karlo, J.F., 1986. Results of exploration in Mesozoic shelf-edge carbonates, Baltimore Canyon Basin. Am. Assoc. Pet. Geol., Bull., 70: 605-606.

Kaye, C.A., 1983. The autochthonous and allochthonous coastal plain deposits of Martha's Vineyard and the Marshfield-Scituate area, southeastern Massachusetts. In: Atlantic Coastal Plain Geological Association Field Trip Guidebook. Del. Geol. Surv., Newark, Del., 34 pp.

Klitgord, K.D. and Schouten, H., 1986. Plate kinematics of the Central Atlantic. In: P.R. Vogt and B.E. Tucholke (Editors), The Geology of North America, Volume M. The Western North Atlantic Region. Geol. Soc. Am., Boulder, Colo., pp. 351-378.

Lovegreen, J.R., 1974. Paleodrainage history of the Hudson estuary. M. Thesis, Columbia University, New York, N.Y.

Macintyre, I.G., 1989. Modern coral reefs of Western Atlantic: New geological perspective. Am. Assoc. Pet. Geol., Bull., 72: 1360-1369.

Manspeizer, W., 1985. Early Mesozoic history of the Atlantic passive margin. In: C.W. Poag (Editor), Geologic Evolution of the United States Atlantic Margin. Van Nostrand Reinhold, New York, N.Y., pp. 1-23.

Mathews, W.H., 1975. Cenozoic erosion and erosion surfaces of eastern North America. Am. J. Sci., 275: 818-824.

Melhorn, W.N. and Edgar, D.E., 1980. The case for episodic, continental-scale erosion surfaces: A tentative geodynamic model. In: W.N. Melhorn and R.C. Flemal (Editors), Theories of Landform Development. G. Allen and Unwin, London, pp. 243-276.

Meyer, F.O., 1986. Facies specificity of megaporosity in Mesozoic shelf-edge carbonates Baltimore Canyon Basin. Am. Assoc. Pet. Geol., Bull., 70: p. 621.

Miller, K.G., Fairbanks, R.G. and Mountain, G.S., 1987. Tertiary oxygen isotope synthesis, sea level history, and continental margin erosion. Paleoceanography, 2: 1-19.

Mitchum, R.M., Jr., 1985. Seismic stratigraphic expression of submarine fans. In: O.R. Berg and D.G. Woolverton (Editors), Seismic Stratigraphy II: An Integrated Approach to Hydrocarbon Exploration. Am. Assoc. Pet. Geol., Mem., 39: 117-138.

Mixon, R.B. and Newell, W.L., 1977. Stafford fault system - Structures documenting Cretaceous and Tertiary deformation along the Fall Line in northeastern Virginia. Geology, 5: 437-441.

Morisawa, M., 1989. Origins of Appalachian drainages. In: T.W. Gardner and W.D. Sevon (Editors), Appalachian Geomorphology. Geomorphology 2: 1-22 (this volume).

Mountain, G.S. and Tucholke, B.E., 1985. Mesozoic and Cenozoic Geology of the U.S. continental slope and rise. In: C.W. Poag (Editor), Geologic Evolution of the United States Atlantic Margin. Van Nostrand Reinhold, New York, N.Y., pp. 293-341.

Newell, W.L., 1985. Architecture of the Rappahannock estuary - neotectonics in Virginia. In: M. Morisawa and J.T. Hack (Editors), Tectonic Geomorphology. Allen and Unwin, Boston. Mass., pp. 321-342.

Oldale, R.N., 1982. Pleistocene stratigraphy of Nantucket, Martha's Vineyard, the Elizabeth Islands, and Cape Cod, Massachusetts. In: G.J. Larson and B.D. Stone (Editors), Late Wisconsinian Glaciation of New England. Kendall/Hunt, Dubuque, Iowa, pp. 1-34.

O'Leary, D.W., 1988. Shallow stratigraphy of the New England continental margin. U.S. Geol. Surv. Bull., 1767, 40 pp.

Olsson, R.K. and Wise, S.W., Jr., 1987. Upper Maestrichtian to middle Eocene stratigraphy of the New Jersey slope and coastal plain. In: J.E. van Hinte, S.W. Wise, Jr., et al. Initial Reports of the Deep Sea Drilling Project. U.S. Gov. Print. Off., Washington, D.C., 93: 1343-1366.

Olsson, R.K., Melillo, A.J. and Schreiber, B.L., 1988. Miocene sea level events in the Maryland coastal plan and the offshore Baltimore Canyon Trough. In: C.A. Ross and D. Haman (Editors), Timing and Depositional History of Eustatic Sequences: Constraints on Seismic Stratigraphy. Cushman Found. Foraminiferal Res. Spec. Publ. 24: 85-97.

Owens, J.P., 1970. Post-Triassic tectonic movements in the central and southern Appalachians as recorded by sediments of the Atlantic Coastal Plain. In: G.W. Fisher and others (Editors), Studies of Appalachian Geology - Central and Southern. Interscience, New York, N.Y., pp. 417-427.

Owens, J.P. and Minard, J.P., 1979. Late Cenozoic sediments of the lower Delaware Valley and the northern Delmarva Peninsula, New Jersey, Pennsylvania, Delaware, and Maryland. U.S. Geol. Surv., Prof. Pap., 1067-D, 47 pp.

Parrish, J.T., Ziegler, A.M. and Scotese, C.R., 1982. Rainfall patterns and distribution of coals and evaporites in the Mesozoic and Cenozoic. Palaeogeogr. Palaeoclimatol., Palaeoecol., 40: 67-101.

Pavich, M.J., 1985. Appalachian piedmont morphogenesis: Weathering, erosion, and Cenozoic uplift. In: M. Morisawa and J.T. Hack (Editors), Tectonic Geomorphology. Allen and Unwin, Boston, Mass., pp. 299-319.

Pierce, K.L., 1966. Bedrock and surficial geology of the McConnellsburg Quadrangle, Pennsylvania. Pa. Geol. Surv., 4th ser., Atlas 109a, 111 pp.

Pinet, P.R. and Popenoe, P., 1985. A scenario of Mesozoic-Cenozoic ocean circulation over the Blake Plateau and its environs. Geol. Soc. Am., Bull., 96: 618-626.

Pitman, W.C., III, 1979. The effect of sea level changes on stratigraphic sequences at Atlantic margins. In: J.S. Watkins, L. Montadert and P.W. Dickerson (Editors), Geological and Geophysical Investigations of Continental Margins. Am. Assoc. Pet. Geol., Mem., 29: 453-460.

Pitman, W.C. and Golovchenko, X., 1988. Sea-level changes and their effect on the stratigraphy of Atlantic-type margins. In: R.E. Sheridan and J.A. Grow (Editors), The Geology of North America, Vol. I-2. The Atlantic Continental Margin, U.S. Geol. Soc. Am., Boulder, Colo., pp. 429-436.

Poag, C.W., 1980. Foraminiferal stratigraphy, paleoenvironments and depositional cycles in the outer Baltimore Canyon Trough. In: P.A. Scholle (Editor), Geological Studies of the COST No. B-3 Well, United States Mid-Atlantic Continental Slope Area, U.S. Geol. Surv., Circ., 833: 44-65.

Poag, C.W., 1985. Depositional history and stratigraphic reference section for central Baltimore Canyon Trough. In: C.W. Poag (Editor), Geological Evolution of the United States Atlantic Margin. Van Nostrand Reinhold, New York, N.Y., pp. 217-263.

Poag, C.W., 1987a. The New Jersey Transect: Stratigraphic framework and depositional history of a sediment-rich passive margin. In: C.W. Poag, A.B. Watts et al. Initial Reports of the Deep Sea Drilling Project, Volume 95. U.S. Gov. Print. Off., Washington, D.C., 817 pp.

Poag, C.W., 1987b. Allostratigraphy: Principles and practice on the U.S. Atlantic margin. Geol. Soc. Am., Abstr. Programs, 19: p. 807.

Poag, C.W. and Schlee, J.S., 1984. Depositional sequences and stratigraphic gaps on submerged United States Atlantic margin. In: J.S. Schlee (Editor), Interregional Unconformities and Hydrocarbon Accumulation. Am. Assoc. Pet. Geol., Mem., 36: 165-182.

Poag, C.W., Schlee, J.S., Swift, B.A., Ball, M.M. and Sheetz, L.L., 1988. Banks, reefs, deltas, and fans: Jurassic-Cretaceous transition in the Baltimore Canyon Trough. In: L.M.H. Carter (Editor), USGS Research on Energy Resources, 1988. Program Abstr., U.S. Geol. Surv., Circ., 1025: 48-49.

Pollack J.B., 1982. Solar, astronomical, and atmospheric effects on climate. In: National Research Council, Climate in Earth History. National Academy Press, Washington, D.C., pp. 68-76.

Poore, R.Z. and Bybell, L.M., 1988. Eocene to Miocene biostratigraphy of New Jersey Core ACGS No. 4: Implications for regional stratigraphy. U.S. Geol. Surv., Bull., 1829, 39 pp.

Popenoe, P., 1985. Cenozoic depositional and structural history of the North Carolina margin from seismic-stratigraphic analyses. In: C.W.Poag (Editor), Geologic Evolution of the United States Atlantic Margin. Van Nostrand Reinhold, New York, N.Y., pp. 125–187.

Ritter, D.F., 1988. Landscape analysis and the search for geomorphic unity. Geol. Soc. Am., Bull., 100: 160–171.

Rodgers, J., 1967. Chronology of tectonic movements in the Appalachian region of eastern North America. Am. J. Sci., 265: 408–427.

Savin, S.M., 1977. The history of the earth's surface temperature during the last 100 million years. Annu. Rev. Earth Planet. Sci., 5: 319–355.

Sawyer, D.S., 1985. Total tectonic subsidence; A parameter for distinguishing crust type at the U.S. Atlantic continental margin. J. Geophys. Res., 90: 7751–7769.

Sawyer, D.S., Toksoz, M.N., Sclater, J.G. and Swift, B.A., 1982. Thermal evolution of the Baltimore Canyon trough and Georges Bank basin. In: J.S. Watkins (Editor), Studies in Continental Margin Geology. Am. Assoc. Pet. Geol., Mem., 34: 743–764.

Schlager, W., 1981. The paradox of drowned reefs and carbonate platforms. Geol. Soc. Am., Bull., 92: 197–211.

Schlee, J.S., 1957. Upland gravels of southern Maryland. Geol. Soc. Am. Bull., 68: 1371–1410.

Schlee, 1981. Seismic stratigraphy of Baltimore Canyon Trough. Am. Assoc. Pet. Geol., Bull., 65: 26–53.

Schlee, J.S., Dillon, W.P. and Grow, J.A., 1979. Structure of the continental slope off eastern United States. In: L.J. Doyle and O.H. Piley (Editors), Geology of Continental Slopes. Soc. Econ. Paleontol. Mineral., Spec. Publ., 27: 95–117.

Schlee, J.S., Manspeizer, W. and Riggs, S.R., 1988. Paleoenvironments: Offshore Atlantic U.S. margin. In: R.E. Sheridan and J.A. Grow (Editors), The Geology of North America, Volume I-2, The Atlantic Continental Margin, U.S. Geol. Soc. Am., Boulder, Colo., pp. 365–385.

Schumm, S.A., 1977. The Fluvial System. Wiley, New York, N.Y., 338 pp.

Scott, R.W., 1988. Evolution of Late Jurassic and Early Cretaceous reef biotas. Palaios, 3: 184–193.

Sevon, W.D., 1985. Pennsylvania's polygenetic landscape. Harrisburg Area Geological Society Guidebook, 4th Annual Field Trip, 55 pp.

Smith, R.V., 1980. Provenance of mid-Atlantic continental margin sediments from the COST B-2 test well. M. Thesis, University of Delaware, Newark, Del., 182 pp.

Steckler, M.S. and Watts, A.B., 1978. Subsidence of the Atlantic-type continental margin off New York. Earth Plant. Sci. Lett., 42: 1–13.

Steckler, M.S., Watts, A.B. and Thorne, J.A., 1988. Subsidence and basin modeling at the U.S. Atlantic passive margin. In: R.E. Sheridan and J.A. Grow (Editors), The Geology of North America, Volume I-2. The Atlantic Continental Margin, U.S. Geol. Soc. Am., Boulder, Colo., pp. 399–416.

Tucholke, B.E. and Mountain, G.S., 1986. Tertiary paleoceanography of the western North Atlantic Ocean. In: P.R. Vogt and B.E. Tucholke (Editors), The Geology of North America, Volume M. The Western North Atlantic Region. Geol. Soc. Am., Boulder, Colo., pp. 631–650.

Vail, P.R., Mitchum, R.M. and Thomson, S., III, 1977. Seismic stratigraphy and global changes of sea level. Part Four: Global cycles of relative changes of sea level. In: C.E. Payton (Editor), Seismic Stratigraphy - Applications to Hydrocarbon Exploration. Am. Assoc. Pet. Geol., Mem., 36: 83–98.

Van Hinte, J.E., Wise, S.W., Jr., Biart, B.N.M., Covington, J.M., Dunn, D.A., Haggerty, J.A., Johns, M.W., Meyers, P.A., Moullade, M.R., Muza, J.P., Ogg, J.G., Okamura, M., Sarti, M. and von Rad, U., 1987. Initial Reports of the Deep Sea Drilling Project, Volume 93. U.S. Gov. Print. Off., Washington, D.C., 1423 pp.

Vogt, P.R. and Tucholke, B.E., (Editors), 1986. The Geology of North America, Volume M., The Western North Atlantic Region. Geol. Soc. Am., Boulder, Colo., 610 pp.

Watts, A.B., 1982. Tectonic subsidence, flexure, and global changes of sea level. Nature, 297: 469–474.

Watts, A.B. and Steckler, M.S., 1979. Subsidence and eustasy at the continental margin of eastern North America. Am. Geophys. Union, Maurice Ewing Symp. Ser. 3, pp. 218–234.

Watts, A.B. and Swift, B.A., 1988. Seismic stratigraphy, flexure, and gravity anomalies at Baltimore Canyon Trough, northeastern U.S. continental margin. Geol. Assoc. Can., Mineral. Soc. Can., Can. Soc. Pet. Geol. Joint Annu. Meet., Program Abstr., 13: p. A132.

Watts, A.B. and Thorne, J.A., 1984. Tectonics, global changes in sea level, and their relationship to stratigraphic sequences at the U.S. Atlantic continental margin. Mar. Pet. Geol., 1: 319–339.

Winker, C.D., 1982. Cenozoic shelf margins, northwestern Gulf of Mexico basin. Gulf Coast Assoc. Geol. Soc., Trans., 32: 427–448.

Wolfe, J.A., 1978. A paleobotanical interpretation of Tertiary climates in the northern hemisphere. Am. Sci., 66: 694–703.

Wolfe, J.A. and Upchurch, G.R., Jr., 1987. North American nonmarine climates and vegetation during the Late Cretaceous. Palaeogeogr., Palaeoclimatol., Palaeoecol., 61: 33–77.

Geomorphology, 2 (1989) 159–179
Elsevier Science Publishers B.V., Amsterdam — Printed in The Netherlands

Appalachian Piedmont Landscapes from the Permian to the Holocene

EMERY T. CLEAVES

Maryland Geological Survey, 2300 St. Paul Street, Baltimore, MD 21218 (U.S.A.)

(Received December 20, 1988; accepted after revision May 30, 1989)

Abstract

Cleaves, E.T., 1989. Appalachian Piedmont landscapes from the Permian to the Holocene. In: T.W. Gardner and W.D. Sevon (Editors), Appalachian Geomorphology. Geomorphology, 2: 159–179.

Between the Potomac and Susquehanna Rivers and from the Blue Ridge to the Fall Zone, landscapes of the Piedmont are illustrated for times in the Holocene, Late Wisconsin, Early Miocene, Early Cretaceous, Late Triassic, and Permian. Landscape evolution took place in tectonic settings marked by major plate collisions (Permian), arching and rifting (Late Triassic) and development of the Atlantic passive margin by sea floor spreading (Early Cretaceous). Erosion proceeded concurrently with tectonic uplift and continued after cessation of major tectonic activity. Atlantic Outer Continental Shelf sediments record three major erosional periods: (1) Late Triassic–Early Jurassic; (2) Late Jurassic–Early Cretaceous; and (3) Middle Miocene–Holocene. The Middle Miocene–Holocene pulse is related to neotectonic activity and major climatic fluctuations.

In the Piedmont upland the Holocene landscape is interpreted as an upland surface of low relief undergoing dissection. Major rivers and streams are incised into a landscape on which the landforms show a delicate adjustment to rock lithologies. The Fall Zone has apparently evolved from a combination of warping, faulting, and differential erosion since Late Miocene. The periglacial environment of the Late Wisconsin (and earlier glacial epochs) resulted in increased physical erosion and reduced chemical weathering. Even with lowered saprolitization rates, geochemical modeling suggests that 80 m or more of saprolite may have formed since Late Miocene. This volume of saprolite suggests major erosion of upland surfaces and seemingly contradicts available field evidence. Greatly subdued relief characterized the Early Miocene time, near the end of a prolonged interval of tropical morphogenesis. The ancestral Susquehanna and Potomac Rivers occupied approximately their present locations. In Early Cretaceous time local relief may have been as much as 900 m, and a major axial river draining both the Piedmont and Appalachians flowed southeast past Baltimore. The Late Triassic landscape was influenced by rift basin development. Streams drained into a hydrologically closed basin: no through-flowing rivers seem to have been present. A limestone escarpment along the Blue Ridge may have existed as a consequence of a semi-arid climate. The Permian may have been a time of Himalayan-like mountains and mountain glaciers. Streams (and glaciers) generally flowed southwest and west.

Introduction

The paper is built around six landscape scenes selected to illustrate aspects of the evolution of the Piedmont landscape in the area from the Susquehanna River to the Potomac River and between the Blue Ridge and the Fall Zone (Fig. 1). Development of the Fall Zone from Late Miocene to Holocene time is considered within the context of the Holocene landscape. Within a Late Wisconsin setting, rates of saprolitization in a periglacial morphogenetic system are contrasted to those in a humid-temperate environment. Persistance of major rivers upon the Piedmont landscape is a theme of the Early Miocene landscape evaluation, and a concern for the Early Cretaceous landscape, as well. The semi-arid climate of Late Triassic results in a

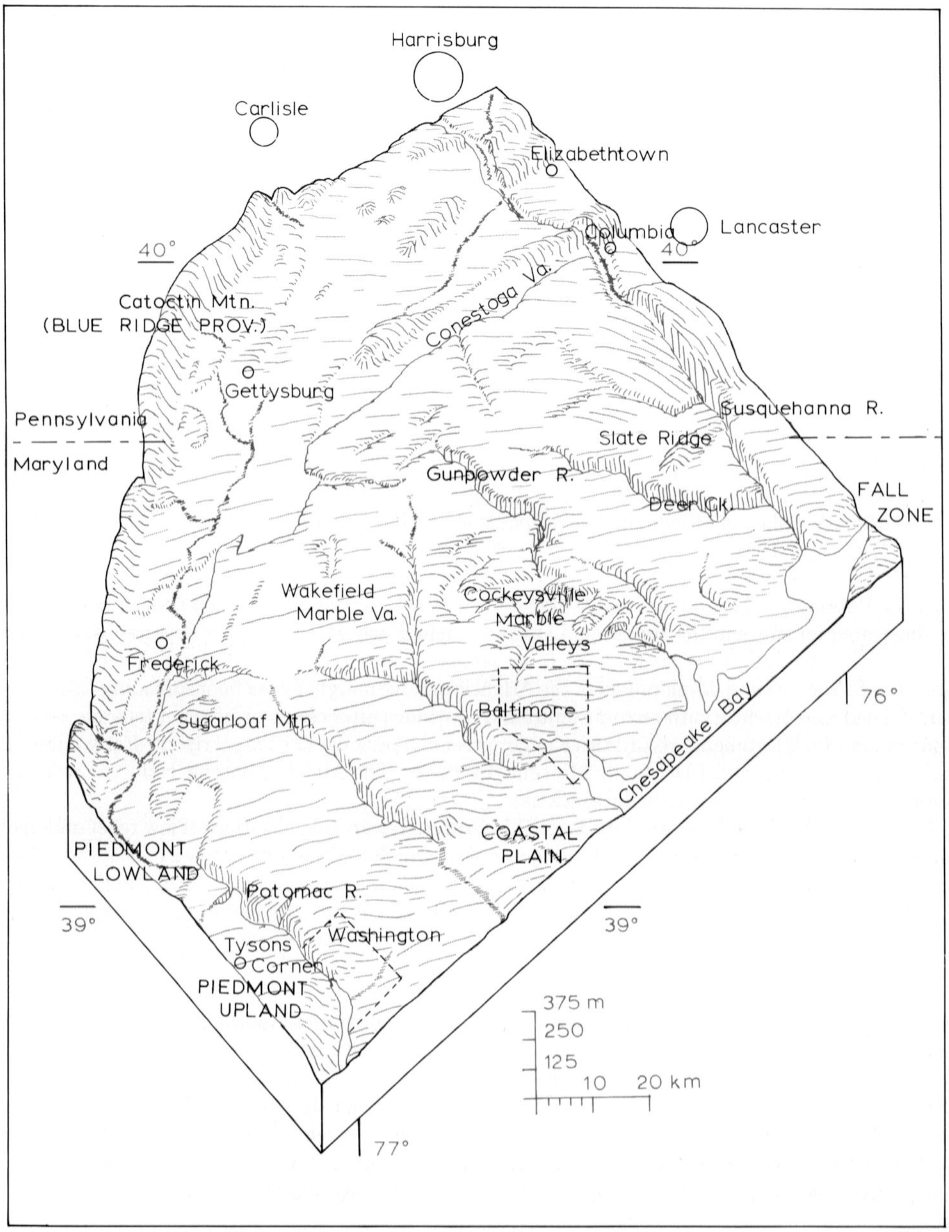

Fig. 1. Geographic location of landscape. Each landscape block diagram illustrates my interpretation of the appearance of approximately 14,250 km^2 part of the Maryland/Pennsylvania Piedmont. For reference purposes latitude and longitude are shown on each block diagram. Selected geographic features mentioned in the text are shown for purposes of orientation. The Holocene Landscape block diagram is used as the base map.

landscape interpretation startingly different from that of the Early Cretaceous. The mountainous land of the Permian is a speculative exercise growing out of an Alleghanian tectonic setting.

These landscapes, and the geomorphic systems which have shaped them, have been shaped within a broad framework of tectonic and climatic events. For example, a tropical climate operating from Late Cretaceous into Early

Miocene time supported a system of erosional, transportational and depositional processes dominated by chemical weathering (Cleaves, 1983, p. 47). Starting in Late Miocene time climatic change resulted in a shift from a chemically dominant geomorphic system to a physically dominant one. At the height of each Pleistocene glaciation a periglacial geomorphic systems appears to have dominated much of the middle Appalachian area. Consideration of the tectonic (and sedimentary) history of the Atlantic continental margin can be found in Klitgord et al. (1988), Manspeizer and Cousminer (1988), Poag (1979) and Poag and Sevon (this volume).

The paper is organized in an unorthodox way, proceeding from the present (Holocene Landscape) back into the past (Permian). In a very real sense, the present has been the key to the past. My interpretations of past landscapes are built around first describing the modern landscape, and then moving step by step back into the past. At each "backward" step various geologic data, interpretations, and speculations are utilized to construct a particular landscape.

Age dates for the various geologic epochs and ages used in the text are from Geological Society of America, DNAG, 1983 Geologic Time Scale.

Holocene landscape

The Holocene landscape of the Piedmont exhibits three major components, an upland surface of low relief which truncates rocks of varying lithologies, major rivers and streams occupying gorge-like steep-walled valleys incised into the upland surface, and a broad lowland adjacent to the Catoctin Mountain of the Blue Ridge Province (Fig. 2).

The upland surface (Piedmont Upland of Fenneman, 1938) truncates alumino-silicate rocks of many lithologies (Cleaves and Costa, 1979, p. 16) including Sams Creek Metabasalt and Ijamsville Phyllite in the western Piedmont and Loch Raven Schist, Baltimore Gabbro Complex, and Baltimore Gneiss in the eastern Piedmont. The alumino-silicate rocks are mantled by saprolite, a residual deposit with thicknesses related to landform and lithology (Cleaves and Costa, 1979, p. 7). A broad lowland, the Lancaster–Frederick Lowland of Fenneman (1938, p. 152), bounds the upland surface to the northwest. In the vicinity of Frederick the lowland is underlain by Cambro–Ordovician limestones and Mesozoic shales. Mesozoic shales and basalt of the Gettysburg Mesozoic basin underlie the lowland between the Susquehanna River and Gettysburg.

On both the Piedmont Lowland and Piedmont Upland there is a close relationship between landform and lithology. The Cockeysville and Wakefield Marbles display very distinctive broad-bottomed valleys. Valleys cut into schists and phyllites are commonly vee-shaped and steep-walled. The Baltimore Gneiss domes adjacent to the Cockeysville Marble Valleys are etched by erosion into a distinctive topography as a consequence of contrasting rock lithologies including gneiss, quartzite, marble, and schist. In the Lancaster-Frederick Lowland, limestones underlie the lower areas and Mesozoic shales crop out beneath higher parts of the Lowland. In many places the Lowland is bordered by ridges underlain by quartzite, e.g. Catoctin Mountain between the Potomac River and the Pennsylvania State Line. In the Gettysburg–Elizabethtown–Columbia area diabase flows occur in the shales, and the resulting topography has ridges and knobby hills underlain by diabase. In some areas monadnocks dominate the local landscape: Sugarloaf Mountain underlain by the Sugarloaf Quartzite, and Slate Ridge underlain by the Peach Bottom Slate.

The Susquehanna River and Potomac River flow northwest to southeast through the Piedmont. In the Upland the two rivers have cut gorge-like valleys. Rivers originating with the Piedmont Upland such as the Gunpowder River and Deer Creek flow in gorge-like steep-walled valleys throughout much of their length, but

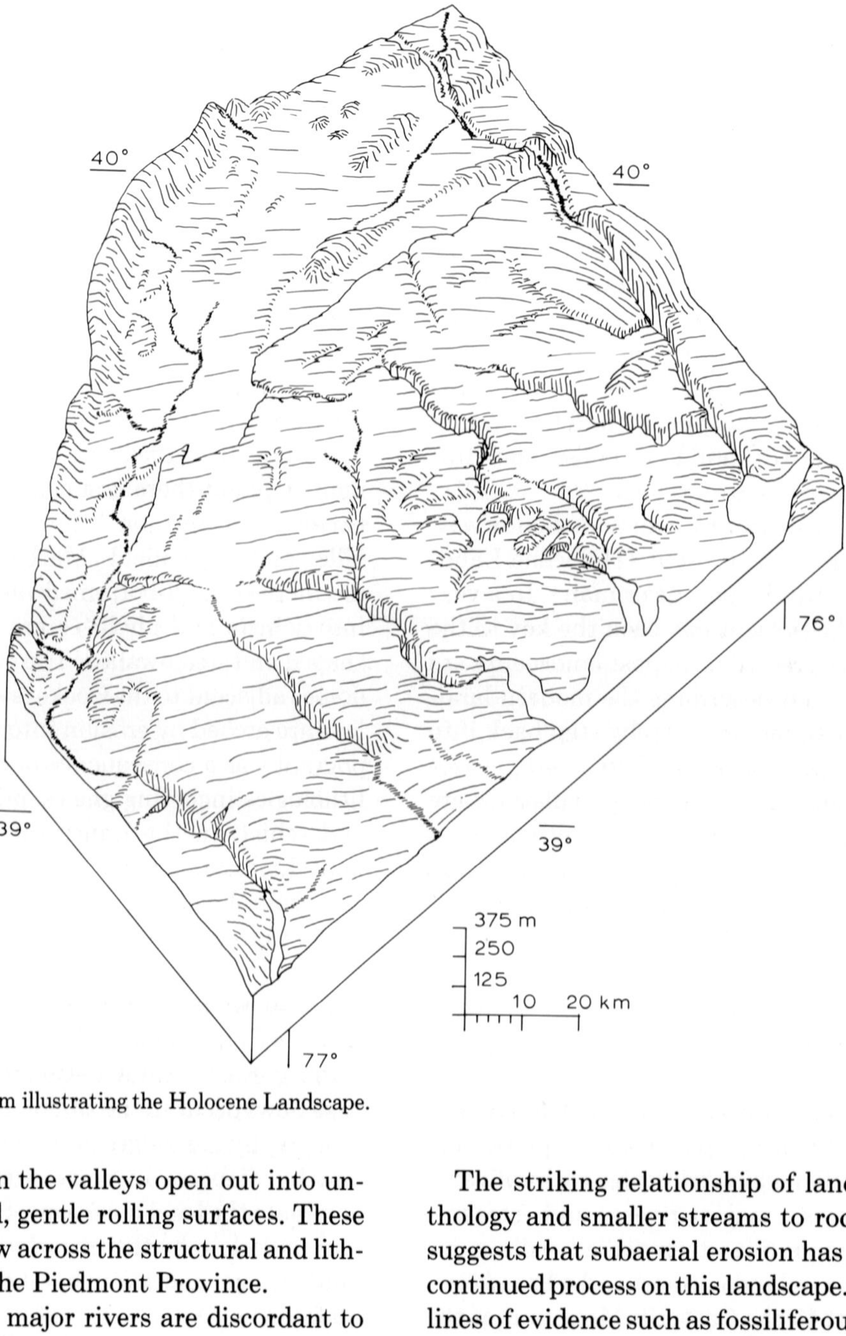

Fig. 2. Block diagram illustrating the Holocene Landscape.

near their origin the valleys open out into undissected, broad, gentle rolling surfaces. These major rivers flow across the structural and lithologic grain of the Piedmont Province.

Although the major rivers are discordant to structure and lithology, low-order streams, especially first and second-order ones, show a delicate adjustment to lithology and structure. In places jointing may control stream orientation, and in others thinly interdigitated layers of quartzite and schist determine stream orientation (Cleaves and Costa, 1979, p. 7).

The striking relationship of landform to lithology and smaller streams to rock structure suggests that subaerial erosion has been a long continued process on this landscape. From other lines of evidence such as fossiliferous lignite deposits of Cretaceous and Tertiary age on the Cambrian and Ordovician limestones in the "Great Valley", Hack (1980, pp. B8–9) came to the same conclusion.

The Piedmont Holocene landscape is bounded on the northwest by an escarpment along Catoctin Mountain (Blue Ridge Province) and to the southeast by the Fall Zone. The escarpment is considered to be the result of differential erosion (Hack, 1982, pp. 25–27). The modern relief along the escarpment probably has evolved since Late Miocene. See discussion below, and in "Miocene Landscape". Along Catoctin Mountain, the ridge crest is held up by the Weverton and Antietam Quartzites, and the adjacent valleys (Frederick Valley and Shenandoah or Great Valley) are underlain by limestones and shales. Along Blue Mountain (west of Catoctin Mountain and the Great Valley), the crest is underlain by the Tuscarora Sandstone. As long as such lithologies of contrasting resistance to weathering and erosion existed along the escarpment, this landform may have been present with subdued relief back into the Early Tertiary or Late Cretaceous.

The Fall Zone, in contrast, is interpreted as a feature which has evolved from Late Miocene to the Holocene due to warping and faulting as well as differential erosion. As a topographic feature the Fall Zone can be identified from the James River in Virginia north to the Delaware River. It constitutes a zone of transition from the Piedmont Uplands, underlain by crystalline rocks, to the Coastal Plain Province underlain by unconsolidated sediments.

Uplift of the Piedmont between the James and Delaware Rivers relative to the Coastal Plain has been well documented by Hack (1982, pp. 15–19) who cites as evidence the distinctive relief, plateau-like topography, deep narrow valleys of the Potomac and Susquehanna, and convexity of their profiles crossing the Piedmont. Faulting and warping along the Piedmont and Coastal Plain boundary has occurred from Early Cretaceous into Miocene and perhaps into the Holocene (Jacobeen, 1972; Mixon and Newell, 1977; Newell and Rader, 1982; and Newell, 1984).

The development of the modern Fall Zone, however, is probably restricted to the Late Miocene–Holocene interval. On the outer continental shelf a major pulse of terrigenenous sedimentation was initiated in Middle Miocene and continues into the Holocene (Poag and Sevon, this volume). However, the fluvial deposits laid down by the Potomac River southeast of Washington, D.C. unconformably overlie Chesapeake Group marine sediments of Middle Miocene, suggesting that major erosion in the Potomac River watershed was initiated somewhat later than the initial pulse recorded in the Baltimore Canyon Trough. Uplift of the area of the Piedmont shown in Fig. 1 and development of the Fall Zone also is indicated by the shift of the depositional basin located in the Salisbury Embayment in the Late Miocene southward into the Albermarle Embayment in Pliocene and Pleistocene time (Newell and Rader, 1982, pp. 14, 16; Newell, 1984, p. 327). Delineation of a Fall Zone beginning in the Late Miocene is dramatically highlighted by the diversion of the Potomac River from its original southeasterly course (when it flowed on Middle Miocene age sediments (Schlee, 1957) to a southwesterly course where the river is bracketed by the Stafford fault system to the northwest and Brandywine fault system to the southeast.

Late Wisconsin landscape

The Late Wisconsin landscape was basically similar in appearance to the Holocene landscape. However, geomorphic processes operating on the landscape were controlled by a periglacial morphogenetic system in contrast to the humid-temperate one of the Holocene landscape with its significant chemical weathering oriented processes (Cleaves et al., 1970, 1974). At least seven major glacial periods have occurred in the last 700,000 years (Bowen, 1979), and climatic instability extends back through the Pleistocene (1.6 Ma) into the Pliocene and probably into the Late Miocene. A glacial/interglacial cycle lasted approximately 120,000 years, at least with those documented for the past 700,000 years (Bowen, 1979). Glacials spanned perhaps 100,000 years and interglacials 10,000 to 20,000 years (Bowen, 1979, p. 63); consequently, periglacial morphogenetic regimes would appear to be considerably longer than humid-temperate periods. Although each morphogenetic system has contrasting weathering processes preparing material for eventual erosion, the switchover from one system to the other may be of greater importance in shaping in the landscape that their duration.

For 18,000 BP the general climate for the Piedmont area may be inferred from Kutzback and Wright (1985, fig. 11, Northeast area). The data are listed as changes (Δ) from the present for January and July. Temperature is listed as degrees Kelvin (K); precipitation (P) in percent; and precipitation minus evaporation (E) in percent: ΔK Jan, $-5/-10$; ΔP Jan, ~ 0; Δ (P–E) annual, $-10/-15$; ΔK Jul, $-5/-10$; ΔP Jul, $-10/-15$. These climatic parameters characterize a periglacial morphogenetic system, and suggest that frost and frozen ground penetrated to significantly greater depth than the current 45cm average; that considerable water was tied up in snow; that ground temperatures at the weathering front were substantially reduced — possibly in the same 5–10°K range indicated for air temperatures; and that plant activity was significantly modified.

Recharge of the groundwater was probably greatly reduced. Water table recharge occurs in late Fall, Winter and early Spring when plant activity is greatly reduced. However, with the ground frozen for much greater lengths of time under the influence of a periglacial climate, and precipitation accumulating as snow with few if any periods of thawing prior to Spring, recharge to the water table certainly was greatly reduced. Alternatively, thawing and consequent meltwater runoff must have produced stream runoffs of high magnitude. Much of the yearly surface water runoff probably took place in a relatively few weeks in the Spring. This kind of stream hydrology and groundwater recharge could be expected to result in enhanced fluvial erosion and reduced chemical weathering.

The periglacial environment in the Piedmont presumably was one in which mechanical processes dominated over chemical ones in removal of mass from the landscape. In contrast, Pavich and Newell, 1987, pp. 18, 20, suggest that periglacial processes were probably absent in much of the area. Besides enhanced fluvial erosion, strong frost action, intensive mass wasting, and eolian processes (especially in the Coastal Plain) functioned on terrain that was seasonally snow free. Although periglacial features cannot yet be conclusively demonstrated in the Maryland Piedmont, boulder streams and block fields document periglacial geomorphic processes in the Catoctin–Blue Ridge Mountain area (Godfrey, 1975, pp. 18–20). A jack-pine boreal forest occupied the Piedmont (Delacourt and Delacourt, 1981, p. 145) and may have significantly reduced the impact of frost wedging, congeliturbation, and related freeze-thaw processes. Alpine tundra extended south along the Appalachians to the Great Smokies (Delacourt and Delacourt, 1981). The landscape of the southern New Jersey Coastal Plain has been modified by frost wedges, congeliturbation, extensive mass movement, and eolian activity during periglacial climatic conditions (Newell et al., 1988). Colluvial deposits of probable Pleistocene age on side slopes and in gullies and low-order stream valleys have been documented in South Carolina, North Carolina, and Virginia (Whitehead and Barghorn, 1962, p. 367; Overstreet et al., 1968, pp. 31–36) and in Maryland (Cleaves and Costa, 1979, pp. 17–18).

Whereas periglacial processes have been of major importance during the Quaternary, removal of mass from the Piedmont landscape by isovolumetric chemical weathering of the crystalline rock has been a continuing process, although the intensity of weathering has varied. Distribution of residual deposits on the landscape suggests that fluvial erosion is dissecting an upland mantled by saprolite, but geochemical modeling indicates that large volumes of saprolite may have been generated since the Late Miocene (see discussion below). If the modeling is valid, it indicates substantial removal of saprolite from the upland must have taken place.

Field evidence indicates that the saprolite "mantle" is being dissected, and that fluvial erosion greatly exceeds chemical erosion. Overburden thicknesses in four quadrangles in the Baltimore area (Hereford, Cockeysville, Towson, White Marsh), some 520 km^2 in area, show that deep saprolite is present on alumino-silicate rock beneath uplands. Based upon data from 156 water wells, the thickness of saprolite ranged from 7 to 34 m, and averaged 17 m. In the four quadrangle area, saprolite is absent beneath major streams as well as many intermediate and small ones. Selected profiles further document the presence of the saprolite mantle and its dissection by streams (Cleaves and Costa, 1979; fig. 12). Modern rivers such as the Susquehanna and Potomac flow in rocky gorges through the Piedmont Upland, and major streams such as Gunpowder Falls and Deer Creek flow in steep-walled valleys incised into bedrock. Numerous quartz veins cut Piedmont crystalline rocks such as the Loch Raven Schist in the Towson and Cockeysville Quadrangles. Upland inspections of trenches cut for pipelines, of foundations for home sites, and detailed mapping of Pond Branch (Cleaves et al., 1970) indicate an absence of concentrations of residual gravel clasts of vein quartz. Such coarse-grained material could be expected to concentrate on upland surfaces as a residual rock pavement if significant removal of finer grained particulate matter was occurring, or if saprolite was being destroyed in significant amounts in the soil zone as hypothesized by Pavich (1986, p. 579). Lack of widespread residual vein quartz clast concentrates suggests other geomorphic agents must be operative of significant volumes of saprolite are being eroded.

Mass balance studies indicate short-term rates of saprolitization ranging from 38 m Ma^{-1} in the Smokey Mountains, Coweeta, N.C. (Velbel, 1985), to 2–10 m Ma^{-1}, Massanutten Mountain, Virginia (Afifi and Bricker, 1983), and 4 m Ma^{-1}, Occoquan, Virginia (Pavich, 1986). In the Piedmont crystallines near Baltimore a rate of 4.8 m Ma^{-1} has been calculated from the Pond Branch data of Cleaves et al. (1970). Each of these watershed mass balance studies suffer from similar shortcomings: they do not consider the impact of major climatological changes which have occurred during the Pliocene and Quaternary. As discussed below, major changes in water flow past the weathering front and temperature at the weathering front may have drastically influenced the rate of saprolitization. In addition the CO_2 (gas) content of the soil and saprolite in the vadose zone, which is in great part a biologic contribution, can also have a major affect on rate of saprolitization, but has not been considered in the mass balance studies.

Utilizing the Pond Branch lithologic, hydrologic, and weathering information (Cleaves et al., 1970), a saprolite geochemical weathering model was prepared to evaluate the relative importance of temperature, water flow, and CO_2 on the rates at which saprolite may be formed and the potential impact of a periglacial environment. Two conditions were simulated, the modern environment and the Late Wisconsin glacial environment. The weathering model utilizes a rock of composition quartz 50%, muscovite 30%, and plagioclase 30%, which mimics the Loch Raven Schist. The major assumptions are: equilibrium is achieved instantly; the only mineral reactions are albite altering to kaolinite, and kaolinite to gibbsite; the system equilibrates at a given pressure of CO_2 and then is isolated from the CO_2 source. Infinite dilution is assumed. Therefore the activity coefficients are approximately one; saprolite porosity is 50%.

Water moving into and through the soil reacts with soil CO_2 to form carbonic acid, which is considered to be the primary "weathering acid". The carbonic acid percolates through the saprolite and reacts with the schist at the weathering front. Saprolite is formed, and dissolved solids are removed in groundwater, which is discharged as base flow in adjacent streams.

The three parameters used in the model were determined as follows: Soil CO_2 concentrations in a soil beneath a forest seem to be about 10–20 times atmospheric CO_2 concentrations (O.P. Bricker, pers. commun., 1988), and depend upon the rate of plant decomposition, microbial and root respiration and the diffusion rate of CO_2 to the atmosphere. Groundwater moving past the weathering front was not directly measured; stream base flow served as a surrogate. At Pond Branch this was 15.97 cm yr^{-1}. This is rounded up to 17.78 cm yr^{-1} in the model. A temperature at the weathering front of 12°C was determined from temperature logs taken in water wells in Piedmont rocks (Nutter and Otton, 1969).

For the modern weathering environment these parameters indicate that saprolite forms at a rate of 25–48 m Ma^{-1} (Fig. 3a) depending upon the concentration of CO_2 in the soil atmosphere. The sensitivity of the model to CO_2 concentrations is paralleled by its sensitivity to water flow past the weathering front. For example, if water flow is halved to 8.89 cm yr^{-1} and CO_2 remains at 20× then saprolite formation decreases from 48 m Ma^{-1} to 22 m Ma^{-1}. The model is significantly less sensitive to temperature change as 28 m Ma^{-1} of saprolite is generated at 4°C with 8.89 cm yr^{-1} of water flow and 20×CO_2 (Fig. 3b) compared to 22 m Ma^{-1} at 12°C (Fig. 3a). Indeed, when only temperature changes, the weathering rate increases somewhat, as carbonic acid is more soluble in cooler than warmer waters. For the glacial environment I have assumed that water flow past the weathering front is halved, to 8.89 cm yr^{-1} as recharge to the groundwater would probably be greatly reduced in a periglacial environment (see discussion above). Similarly, major reductions in surface air temperature would be paralleled by significant decreases at the weathering front. For modeling purposes I have assumed an 8°C decrease. The same concentrations of CO_2 were used; I presently have no information to estimate what, if any, changes to anticipate. Presumably, plant growth (and vegetation decomposition) was reduced, suggesting a reduction in available soil CO_2.

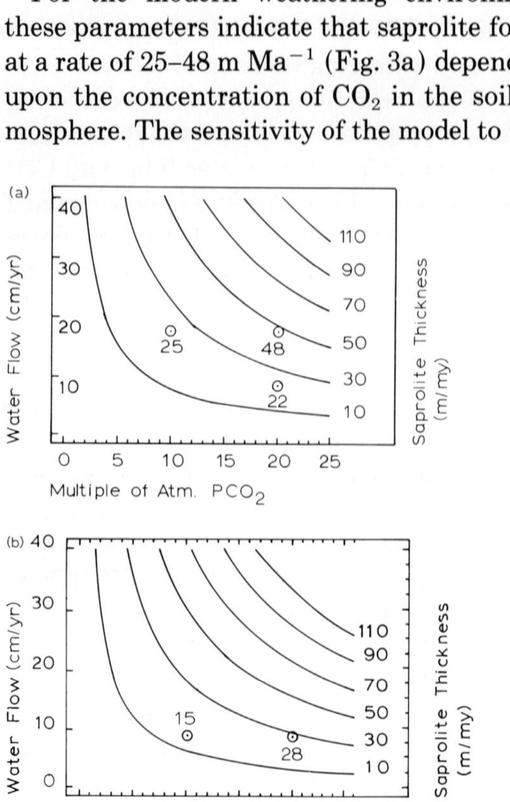

Fig. 3. Equilibrium weathering models for saprolite. (a) Modern environment, at 12°C; (b) Late Wisconsin glacial environment, at 4°C.

Despite the uncertainties, the "glacial model" (Fig. 3b) indicates major decreases in the rates of formation of saprolite. At P_{CO_2} of 10×, the rate decreases from 25 to 15 m Ma^{-1} and at P_{CO_2} of 20×, the rate decreases from 48 to 28 m Ma^{-1}.

Both modeling scenarios indicate that isovolumetric chemical weathering is very sensitive to concentration of soil CO_2 and to water flow past the weathering front. Weathering is much less sensitive to temperature, and actually increases with decreasing temperature. But in either scenario large volumes of saprolite are formed, ranging from a low of 15 m Ma^{-1} to a high of 48 m Ma^{-1}. Translated into the period from the end of the Late Miocene to the present (some 5.3 m Ma^{-1}), this means that saprolite ranging in thickness from 79.5 to 254.4 m could have been produced on the Piedmont landscape! Such calculations suggest significant erosion of upland surfaces, in the contradiction to the field evidence.

What needs to be done? Better information needs to be developed on soil CO_2 values, especially concentrations and rates of generation relative to soil temperatures and to forest cover. One of the major surprises of this study was the very great dependence of saprolitization on soil CO_2 concentrations.

Water flow past the weathering front needs to be investigated. It is as critical a parameter as soil CO_2. Groundwater recharge normally occurs when plants are inactive in late fall, winter and early spring. However, in a periglacial climate, how much recharge is retarded due to frozen ground and snow accumulation?

In addition to refinements of modeling parameters, geomorphic processes which may erode the upland surface need to be identified and investigated. None seems presently active, although the pedogenic mass loss concept of Pavich (1986) merits evaluation. Even hurricanes and major tropical storms such as "Agnes" (1972) left the forested uplands of the Piedmont unaffected.

The extent of the impact of Late Miocene to Holocene crustual warping and faulting on the landscape remains unclear. Assuming uplift of the Piedmont, has erosion of the landscape been mainly concentrated in cutting stream gorges and steep-walled valleys? Such a scenario, if valid, would better accord with Crickmay's hypothesis of unequal activity (Crickmay, 1975) rather than Hack's theory of dynamic equilibrium.

Alternatively, if the modeling results (considered above) are essentially valid, has the saprolite on the Piedmont Uplands been stripped off during one or more periglacial periods (by lateral corrosion of streams, headward migration of gullies [stream heads] and mass wasting on steeper slopes)? Or, has stripping been concentrated during periods of instability when the morphogenetic system was altered in response to major climate change?

Early Miocene landscape

The early Miocene landscape (Fig. 4) illustrates a scene near the end of a prolonged period of tropical morphogenetic regime (Cleaves, 1983, p. 47). The tropical climate supported erosional, transportational, and depositional processes in which chemical weathering agents dominated over physical weathering agents. Relief on the Piedmont is portrayed as subdued compared to the Holocene. Low hydrologic gradients resulted in reduced water flow past the weathering front, and saprolite thicknesses would have increased relatively slowly.

Subdued tectonic activity in the Appalachian interior from Late Cretaceous into the Early Miocene is marked by shifting depositional basins such as the Salisbury Embayment along the Atlantic Coastal Plain and by accompanying marine-estuarine sediment sequences (Owens, 1970; Owens and Gohn, 1985). Although the sediments remained chiefly clastic, their volume is greatly reduced from that of the Early Cretaceous fluvial deposits of the Potomac Group, and the gravels and coarse sands of the Early Cretaceous sediments change to fine sands, silts and clays in Late Cretaceous, Paleocene, and Eocene sediments. The moderate clastic input and subdued tectonic activity suggest that landscapes from Late Cretaceous into Early Miocene had subdued but not flat relief in the Maryland Piedmont and the adjacent Appalachians.

Although the Early Miocene landscape probably had subdued relief, many features shown in the Holocene model can still be recognized. The east flank of the Catoctin Mountains forms a subdued ridgeline, with relief perhaps one-half of that in the Holocene. The juxtaposition of quartzites and sandstones against limestones and shales, rocks of contrasting erosional resistance, maintained the erosional escarpment. Similar contrasts in erosional resistance resulted in the continued presence of limestone and marble valleys such as the Wakefield Valley, Conestoga lowlands, and the marble valleys

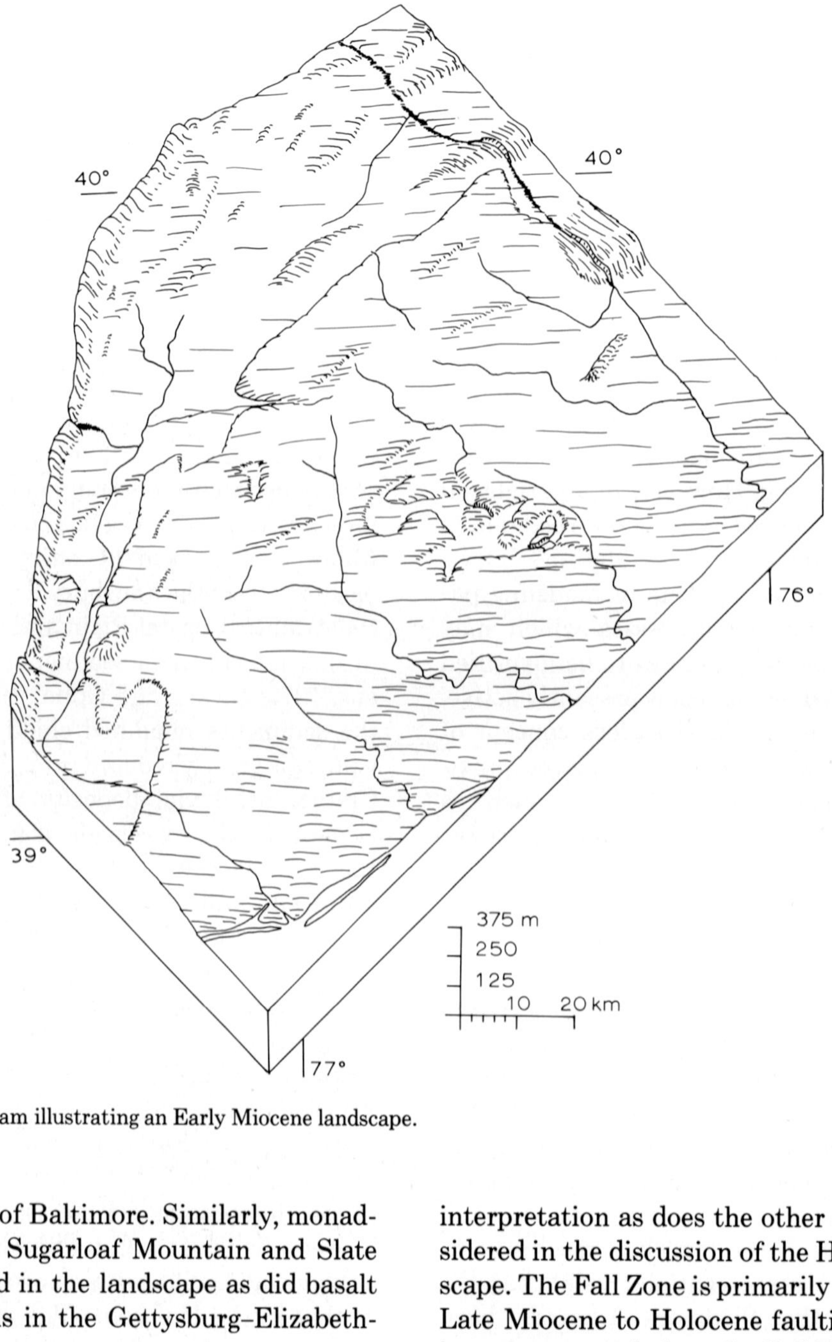

Fig. 4. Block diagram illustrating an Early Miocene landscape.

in the vicinity of Baltimore. Similarly, monadnocks such as Sugarloaf Mountain and Slate Ridge persisted in the landscape as did basalt ridges and hills in the Gettysburg–Elizabethtown areas.

On the other hand features such as the Fall Zone are absent. The presence of Potomac Group sediments of Cretaceous age and fluvial gravel deposits of apparent Late Miocene or Early Pliocene age capping ridges in the Fall Zone suggests, but does not compel, such an interpretation as does the other evidence considered in the discussion of the Holocene landscape. The Fall Zone is primarily a construct of Late Miocene to Holocene faulting and warping concentrated along the Fall Zone and inner Coastal Plain.

Major rivers such as the Potomac, Susquehanna and Gunpowder Falls are not shown incised into the landscape (Fig. 4). However, these rivers were apparently present on the landscape in approximately their present locations. Fluvial upland gravel and sands were deposited unconformably on marine Middle Miocene sediments (Schlee, 1957) in southern Maryland countryside by an ancestral Potomac. The ancestral Potomac River initially flowed east to southeastward prior to its diversion to its modern location (Schlee, 1957; Hack, 1982). The apex of the fluvial upland sands and gravels is the same as the present location of the Potomac River where it crosses from the Piedmont into the Coastal Plain. Gravel deposits of Tertiary age (possibly Miocene) at Tysons Corner, in the Virginia Piedmont, also provide evidence to an ancestral Potomac River. A recent discovery of fossil-bearing material in upland gravel deposits near Brandywine, Maryland, provides a tentative maximum age of Late Miocene of Early Pliocene (J. Brooks, Md. Geol. Survey and L. McCartan, U.S. Geol. Survey, pers. commun., 1988). These sedimentary deposits document the initiation or early phases of the current major pulse of terrigenous sedimentation and they imply the presence of an ancestral Potomac prior to Late Miocene and in approximately the same location as the modern river.

Similarly, upland gravel and sand deposits at the mouth of the Susquehanna River were laid down by an ancestral Susquehanna River, apparently in the Late Miocene (Conant, in prep.) The gravel clasts contain a high proportion of quartz and quartzite, some chert, and minor amounts of granite, gneiss, and schist (Conant, in prep.). This assemblage suggests that fluvial erosion was concentrated mainly in saprolite rather than underlying crystalline rocks and the chert clasts indicate that the Susquehanna extended back into the Valley and Ridge. No significant valley cutting in crystalline rocks appears to have take place prior to deposition of the upland gravels, but about 120 m of downcutting has occurred since the gravels were first deposited (Conant, in prep.).

Eastward from its mouth at the head of the Chesapeake Bay there is no sedimentary record of the Susquehanna River. It appears that an ancestral Delaware–Hudson River system, represented by the Pensauken Formation of Late Miocene age (Owens and Minard, 1979), has eroded Susquehanna River deposits that may have previously existed on the Delmarva Peninsula. However, at least three Pleistocene paleochannels of the Susquehanna River with a maximum possible age of 425–450 ka (Coleman and Mixon, in press) are preserved beneath Chesapeake Bay.

Incision by the Susquehanna River of an estimated 120 m into the landscape and by the Potomac River of possibly 150 m at Tyson's Corner (Hack, 1975) as well as a change in local relief along the Blue Ridge escarpment at Catoctin Mountain that I have postulated to be 90 m or more indicates significant erosion on the Piedmont upland landscape since the Early Miocene. What is the role of uplift of the Piedmont/Appalachians relative to the Coastal Plain in incision of the master streams of the region? Could the upland landscape morphology be a relict inherited from an Early Miocene surface of low relief and the Holocene upland surface actually have been denuded by 15–30 m or more? If the upland surface has been denuded, what processes accomplished the removal? How is the field evidence of dissection of an upland surface to be reconciled with the rates of saprolitization indicated by geochemical modeling? Has there been massive "stripping" of the saprolite during periglacial periods or at times of transition between periglacial and humid-temperate morphogenetic systems? To what extent does the soil zone on saprolite represent additional chemical weathering, destruction of saprolite and lowering of the upland surface? Answers to these questions should emerge from future investigations of Piedmont geology and geomorphology.

Early Cretaceous landscape

The Early Cretaceous landscape is reconstructed for a time during the Berriasian (144–138 Ma). The Berriasian occurs within a much longer interval marked by subsidence of the Atlantic passive margin and clastic in-filling of the Salisbury Embayment and the Baltimore Canyon Trough which began in Late Jurassic (Tithonian, 152–144 Ma) and continued through Early Cretaceous (Albian, 113–97.5 Ma) (Poag, 1985; Olson et al., 1988). This interval is characterized by uplift and erosion of the Piedmont.

The Salisbury Embayment is a major first-order basin formed in early evoluation of the Atlantic passive margin (Owens and Gohn, 1985). Approximately 50% of the Embayment beneath the emergent Coastal Plain is filled by clastic nonmarine fluvial and deltaic sediments of Tithonian to Albian age (Olson et al., 1988). The lowermost Waste Gate Formation consists of nonmarine alluvial sequences of arkosic sandstones and shales (Hansen, 1984). The overlying Potomac Group units (Sedimentation Sequence I, Owens, 1970; Depositional Sequence I, Owens and Gohn, 1985) represent fluvial and fluvial-deltaic environments. The Embayment was progressively filled from east to west by sediments eroded from the Piedmont and Appalachians. All together some 1615 m of sediment accumulated beneath Ocean City, Maryland and 460 m beneath Annapolis, Maryland (Hansen, 1984).

The stratigraphy and mineralogy of the sediments provide some indication of the prevailing paleoenvironments. Evaluation of subsurface Potomac Group fluvial-deltaic lithofacies in the Baltimore–Washington area allowed Hansen (1969) to document the presence of a major axial river which emerged from the Piedmont near Baltimore. The river was actively depositing sediments during the Barremian (Patuxent Formation) and Albian (Patapsco). Study of the outcropping Potomac Group by Glaser (1969) led him to conclude that the basal sediments in the Maryland–Delaware area were deposited by streams whose source areas were mainly deeply weathered crystalline rocks in the Maryland–Pennsylvania–Delaware Piedmont; and that the streams flowed eastward to southeastward with the headwaters of larger ones extending into the folded Appalachians. The distribution and relative exposure of Piedmont rocks was similar to that of the modern landscape. Local relief on the contact between Patuxent sediments and underlying saprolite/crystalline rock was as much as 45 m (Glaser, 1969, p. 74). Similar subdued relief, 15–30 m along a 10 km transect, on the Patuxent-crystalline rock contact, has been documented by Cleaves (1968, fig. 7).

This sketch of tectonics, stratigraphy, and sedimentation suggests the following attributes for the Early Cretaceous landscape (Fig. 5). Major erosion of the Piedmont/Appalachian landmass took place from the Tithonian through the Albian, but by Barremian time (outcropping Patuxent Formation) the Piedmont area of he Maryland–Pennsylvania–Delaware region was characterized by low relief which truncated noncarbonate rocks of different lithologies and a deep mantle of saprolite. Consequently, a Piedmont landscape set in the Berriasian is selected. This time period in the lowermost Early Cretaceous better portrays relief and landforms resulting from uplift caused by basin subsidence. Based upon the volume of Late Jurassic–Early Cretaceous clastic fill in the Salisbury Embayment beneath the emergent Coastal Plain (see above, and Hansen, 1984), I have somewhat arbitrarily chosen to replace 900 m of rock presumably eroded from the landscape. The distribution of rock lithologies in the Cretaceous landscape is generally similar to that of the Holocene, based upon geologic cross sections from available county geologic maps (e.g., Baltimore County (Crowley et al., 1976) and 7.5-min quadrangle geologic maps (e.g., Hereford Quadrangle; Muller, 1985).

The major drainage was towards the east and southeast. Subsidence of the continental margin and the development of second-order basin and arch features in Late Jurassic and Early Cretaceous time was accompanied by development of easterly and southeasterly drainage into the structural and topographic trough of the Salisbury Embayment. Based upon strati-

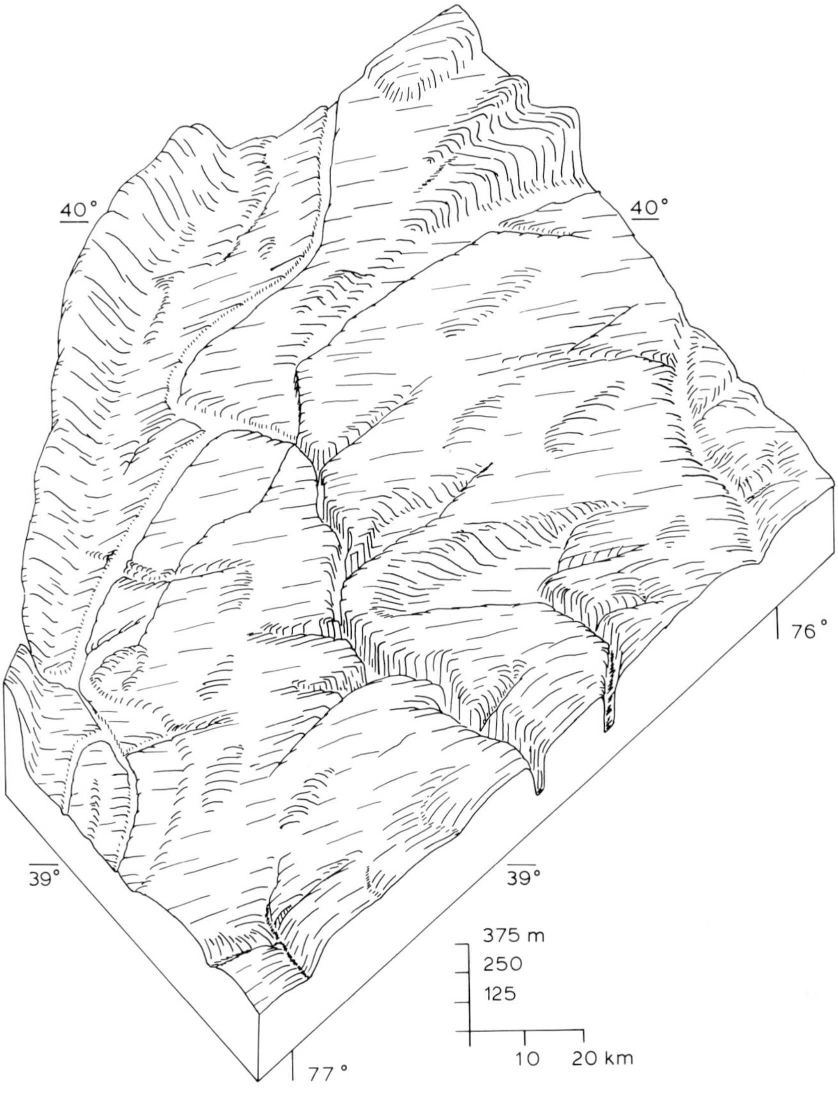

Fig. 5. Block diagram illustrating an Early Cretaceous landscape.

graphic and lithologic evidence from Patuxent sediments in the Maryland Coastal Plain (Hansen, 1984), a major axial river flowed southeast past Baltimore. These same Coastal Plain sediments show no ancestral Potomac or Susquehanna. Consequently, an ancestral Potomac and Susquehanna are merged together in the Cambro–Ordovician/Triassic lowland (located just east of the Catoctin/Blue Ridge) to form the Phoenix River. The path of these rivers on the Piedmont is hypothetical.

Alternatively, both the ancestral Potomac and ancestral Susquehanna are hypothesized to have cut through the Blue Ridge and Catoctin at the same locations that they presently do. Their persistence upon the landscape of the folded Appalachians is based upon Oberlander's (1984, pp. 174–178) hypothesis that the ancestral Susquehanna River was established and in place through the folded Appalachians by Early Cretaceous.

Oberlander's hypothesis requires some 4500 m of erosion (Oberlander, 1984, p. 178), of which all but 915 m may have been accomplished by Berriasian time (see above). Apparently the major volumes of rock eroded from the Piedmont (and by implication from the Appalachians) were removed early in the uplift/erosion cycle, but significant relief presumably still existed in Berriasian time. Local relief of 760–900 m is arbitrarily assumed in constructing the Early Cretaceous landscape.

One final landscape attribute concerns the drainage divide between local streams draining westward into the Gettysburg Lowland and those flowing eastward off of the Piedmont Upland. In the Holocene landscape this divide is called Parrs Ridge (see Fig. 1), and a similar divide is constructed on the Early Cretaceous landscape. I have done so because of the presence of the Gettysburg lowland underlain by limestones and shales along the northwest margin of the Piedmont. Local westward drainage from the Piedmont into the lowland is assumed to have been maintained from the time of development of the lowland as a rift basin in the Jurassic into the Early Cretaceous, and to have lasted up to the present. However, heavy minerals in the Patuxent Formation of the Maryland Coastal Plain indicate source areas in the adjacent Piedmont terrain (Glaser, 1969). This suggests that local rivers headed in the Piedmont east of the Blue Ridge, and presumably east of an ancestral Parrs Ridge divide, and flowed to the east and southeast.

Late Triassic landscape

A Late Triassic landscape is reconstructed for a time during the Norian (226–208 Ma). The Pangaean supercontinent, formed during the Late Carboniferous and Permian, had started to pull apart along pre-existing fracture zones, sutures, and transforms of the Variscan–Alleghanian orogenic belt. Prior to actual splitting apart and rifting there was a prolonged period of arching and uplift. Along the incipient rift zone Hay et al. (1981) estimated vertical uplift as much as 1.7 km and 720 km across occurring over a period of 30 Ma. This uplift combined with an average elevation of Pangaea of 1320 m (Hay et al., 1981) suggests elevations in excess of 3000 m along the axis of the arch. The rift stage was also characterized by faulting coupled with rapid deposition in elongate fault-governed troughs (Manspeizer and Cousminer, 1988). Igneous activity did not begin until Early Jurassic (Sutter, 1985, p. 113; Sutter, 1988, p. 194).

With development of the arch, northeast-trending, elongate half-grabens developed on the American plate (Manspeizer and Cousminer, 1988), and Newark-type basins formed west of a hinge zone, with the west and north margins having en echelon normal faults, and the south and east margins having a sedimentary onlap.

Somewhat more specific parameters can be developed for the Culpeper and Gettysburg basins (Glaeser, 1966; Faill, 1973; Lindholm et al., 1979; Lindholm, 1979). For the Gettysburg basin, Faill and Glaeser indicates a two-step process of basin development: Symmetrical basin subsidence and infilling followed by igneous intrusion and major tilting. Lindholm, however, considers sedimentation coincided with half-graben development in the Culpeper basin. In the Hartford basin, LeTourneau (1985) shows the influence of syndepositional tectonic activity on distribution of sedimentary deposits. The preponderance of evidence favors a modified half-graben model, with the north and west side being faulted and I have used this model in constructing the Triassic landscape (Fig. 6).

Sedimentological evidence suggests a semi-arid climate. Numerous limestone conglomerates which occur along both western and eastern basin margin (such as Leesburg Limestone Conglomerate; Lindholm, 1979) suggest a semi-

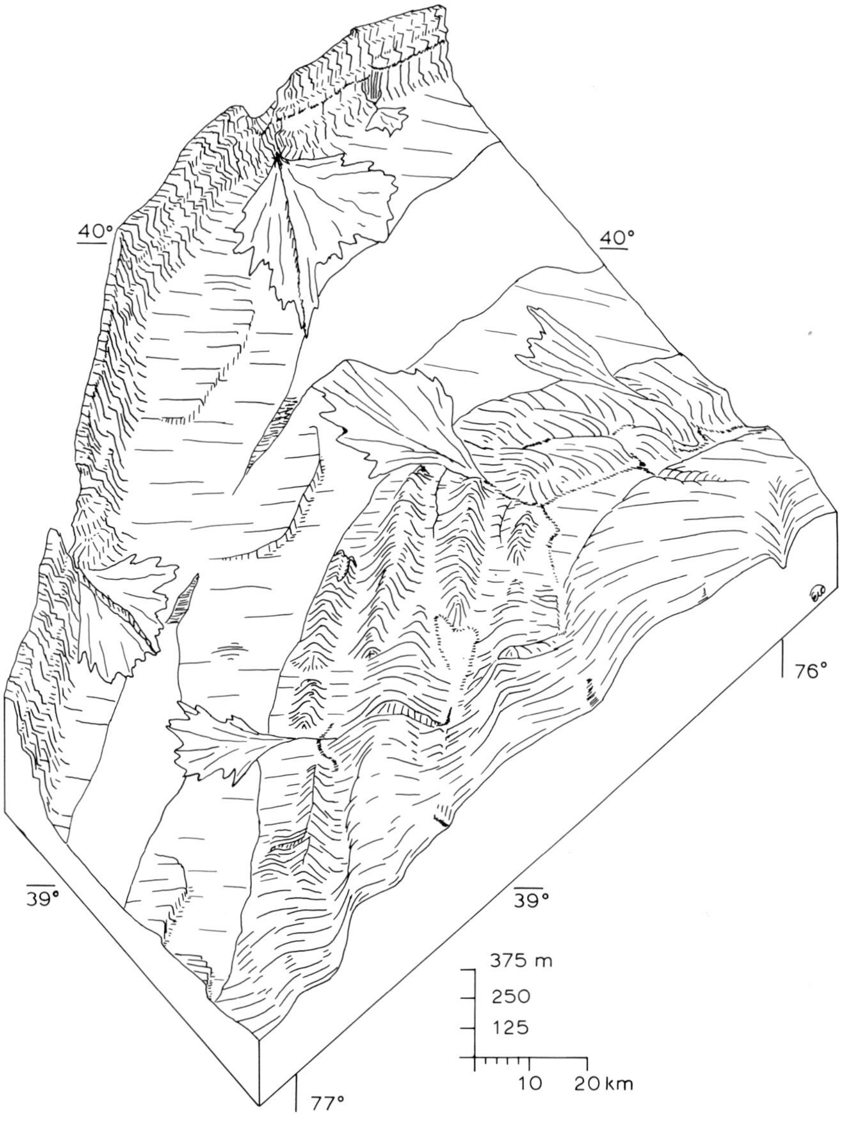

Fig. 6. Block diagram illustrating a Late Triassic landscape.

arid climate in the adjacent source area. Massive mudstones of the Newark Supergroup contain numerous features, such as mudcracks, evaporite mineral casts, and carbonate nodules interpreted as caliche nodules, which indicate that the deposits probably formed under semi-arid climatic conditions (Smoot and Olson, 1985). Basal pebble and boulder conglomerates containing schist, gneiss, granite, greenstone, and basalt clasts, and plagioclase-rich arkosic sandstones suggest an environment with subdued chemical weathering (Glaeser, 1966; Lindholm, 1979). The Culpeper–Gettysburg basins probably underwent climatic oscillations between semi-arid and humid conditions like that described in the Hartford basin for the

Lower Jurassic Portland Formation in Connecticut (McDonald and LeTourneau, 1988).

Streams bringing sediments into the trough headed in local source areas adjacent to the basin (Glaeser, 1966; Faill, 1973; Lindholm et al., 1979; Hentz, 1985). A major river apparently discharged into Gettysburg basin on the northwest margin in the vicinity of the modern Schuylkill River (Hammer Creek Formation) (Faill, 1973). Large lakes periodically occupied the basins, and at other times extensive playas were probably present. The basin presumably was hydrologically closed; there is no present evidence indicating rivers flowing southeastward or eastward away from it. The Piedmont area apparently was located west of the central arch of the uplift, implying that major through-flowing rivers should drain towards the west. However, development of the Gettysburg basins (and others, as well) appears to have disrupted pre-basin drainage patterns, and resulted in stream flow being channelled into periodically closed basins.

The landscape cartoon (Fig. 6) depicts a semi-arid interval. A limestone escarpment borders the northwestern margin of a half-graben rift valley in which two shallow lakes are located. Within the basin its surface is modified by localized fault scarps and alluvial fans and debris flows. The Piedmont terrain to the south and east of the basin is one of subdued hills. Where schists and phyllites occur, the landscape has a linear grain; massive, granite-like rock results in a domal aspect to the land. Streams drain into the basin, except along the southeastern edge – these streams flow towards the Taylorsville basin located southeast of the area of Fig. 6.

Permian landscape

The mountainous landscape in the Permian (Fig. 7) was a consequence of the Alleghanian orogeny which resulted from the convergence and collision of the Gondwana and Laurentia during the Late Carboniferous and Permian Periods. The final assembly of the supercontinent Pangaea was accompanied by major thrust faulting and folding (in the Valley and Ridge of Tennessee, 140 km of northwest transport; the Piedmont terrane was displaced 175 km northwestward) as well as metamorphism and deformation associated with plutonic activity (Secor et al., 1986; Klitgord et al., 1988). The Permo–Triassic Appalachian Mountains may have attained an average relief of 3.5–4.5 km (Slingerland and Furlong, this volume). The mountain belt generated by this orogenic activity shed sediments into the continental interior in the Mississippian and Pennsylvanian (Potter and Pryor, 1961; Secor et al., 1986, fig. 2) but Permian age sediments are absent. The Alleghanian orogenic activity may well have resulted in Himalayan-like mountain systems (or Andean-like system; Slingerland and Furlong, this volume), and voluminous clastic sediments presumably were derived from their erosion. Where is the sediment of Permian age which must have been eroded from the Appalachian Mountain belt? Minor volumes are preserved in the Dunkard Series sediments, but the depositional sites of the bulk of the material remain unidentified.

The Permian landscape (Fig. 7) is modeled after the Himalayan physiography, in part because both mountain systems resulted from plate collisions, and both mountain systems lie within major continental clocks. The block diagram covers approximately 14,250 km^2, and the local relief ranges from 610 to 1130 m. The local relief should be superimposed on a general average elevation which could have ranged from 1310 m, the average elevation of Pangaea (Hay, 1981), to 4025 m, the lowermost elevations in the valleys of Mt. Everest area (Mt. Everest, Map Supplement, National Geographic Magazine, 1988). Major valleys drain towards the south and west, in contrast to the interior basin drainage of Late Triassic and the east and southeast flow of streams from Late Jurassic to Holocene. Given the continental interior setting and the probability of high mountains,

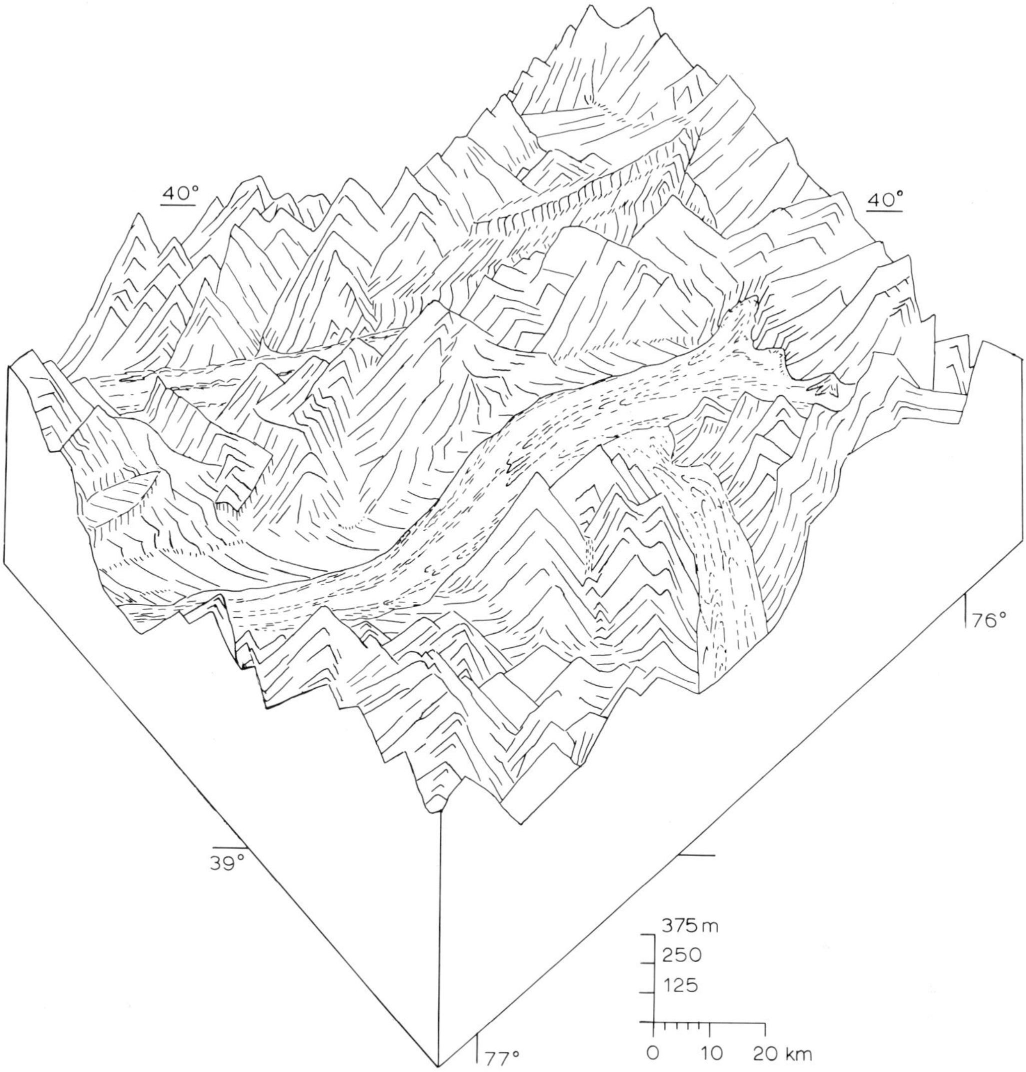

Fig. 7. Block diagram illustrating a Permian landscape.

mountain glaciers have been included on the landscape. Gondwana glaciation lasting from Late Carboniferous into Early Permian has been documented in the southern hemisphere (Martin, 1981), but glaciation in the Appalachians during this time is speculative.

Conclusions/speculations

Consideration of the 14,250 km^2 of the Maryland and Pennsylvania Piedmont at six different periods of time from the Permian to the Holocene demonstrates that at least three cycles of tectonism/erosion have affected the Piedmont landscape. Each tectonic episode appears to have resulted in a distinctive landscape. The Alleghanian orogeny resulted in a Himalayan-like mountain system. The rupturing and splitting apart of Pangaea created rift basin-type topography. Subsequent sea floor spreading and uplift of the Appalachians (the passive margin phase) resulted in the Appalachian fold belt and complex metamorphic hill terrain of the Piedmont. Since the Late Miocene, the area has been undergoing a neotectonic/glacial cycle which has produced a major pulse of clastic sedimentation. The Appalachian fold belt and the Piedmont upland surface are being dissected and the Early Miocene landscape of subdued relief is evolving into one in which landforms of contrasting lithologies are arising in bolder relief.

The Piedmont of Maryland and Pennsylvania has been an area of positive relief subject to subaerial weathering and erosion since at least the start of Permian time, some 286 million years ago. The bulk of the erosion/denudation of the Piedmont and Appalachians predates the Berriasian (and probably the Tithonian). Approximately 900 m of rock appears to have been eroded from Berriasian to Holocene, but in excess of 6000 m of rock is estimated to have been removed from the folded Appalachians between the start of Late Jurassic time through the Holocene. The Appalachian Mountain system generated by the Alleghanian orogeny was removed by erosion during Permian to Early of Middle Jurassic time. Conceivably, the volume of clastic sediment generated by this erosional period more than equalled the material which has accumulated in the rift basins, and Atlantic coastal plain basins.

The region has undergone significant rearrangement of its drainage network. Major river systems flowing southward into the continental interior of America and way from the Appalachian Mountain system have been documented for Mississippian and Pennsylvanian time. Eastward flowing rivers draining the continental interior of Pangaea presumably existed but remain undocumented. Major drainage networks cannot be documented in the Permian due to a paucity of Permian clastic deposits. Late Triassic–Early Jurassic rifting disrupted previously existing Pangaean drainage systems. The rift basins were apparently hydrologically closed systems. The present easterly to southeasterly direction of major rivers was probably established in Late Jurassic, and certainly existed by Early Cretaceous time based upon the sedimentary record in the Salisbury Embayment.

Any direct impact of vegetation upon land-forming processes it not apparent from the evidence at hand. In a way this is surprising because land plants greatly influence the chemical composition of surface and groundwater and strongly impact fluvial/erosional processes. Evolution of angiosperms in the Early Cretaceous, and their rhizomatous growth must have effectively reduced stream bank erosion (Tiffney, 1985). However, the sedimentological record, and the Early Cretaceous and Early Miocene landscape reconstructions, reflect tectonic and climatic impacts. Vegetational influences are not evident and apparently are many orders of magnitude less than tectonic and climatic impacts. In the Oligocene and Miocene grasses diversified and spread over the landscape (Tiffney, 1985). Grasses are especially effective in retarding erosion; yet, ironically, the Late Miocene–Holocene has been a period of greatly increased erosion compared to the Late Cretaceous–Early Miocene period. Although vegetation dies not seem to directly influence major land-forming processes, the strong effect soil CO_2 has on the rate of saprolization emphasizes the need to better evaluate the role of plants in geomorphic processes.

The Holocene and Late Wisconsin landscapes highlight the influence of climate upon landscape evolution. Throughout the Quaternary periglacial processes have alternated with humid-temperate processes in the Piedmont. Their impact combined with neotectonic activity has resulted in incision of major rivers and streams by fluvial erosion to form long, deep valleys, mass-wasting (creep and tree throw) on steep-walled valley sides, and slope undercutting (corrosion) by streams. On the interfluves and upland areas precipitation infiltrates through the soil and the saprolite and reacts with the bedrock. The dissolved solids from the chemical weathering are removed in groundwater. On the upland surface water flow is diffuse, and linear flow (small streams) is limited in volume. Erosion is most active in trunk streams and least (very slow but possibly significant over time) in upland areas and interfluves.

The distribution of saprolite on the Piedmont landscape derived from field evidence appears to confirm such an erosional scenario. However, geochemical modelling suggests that two to four times the volume of saprolite may have been generated since Late Miocene time than currently is present on the uplands. Incision of the Susquehanna and Potomac Rivers from Late Miocene to Holocene into the Piedmont Upland of 120–150 m seems also to suggest that major removal of saprolite has occurred.

The major problem concerns identification of processes to remove saprolite from the upland surface of low relief. How is saprolite "stripped" from the uplands? Elements of a weathering model should include: incision into the upland surface by major rivers; local base levels and groundwater movement from upland areas to rivers such that saprolitization eventually proceeds to base level; lateral corrosion by streams plus mass movement on steep slopes and headward sapping by minor streams; pedogenic mass loss (?); and time.

Acknowledgements

The Triassic section benefited from the critical eye of Al Froelich (U.S. Geological Survey). The initial formulation of the manuscript was greatly assisted by conversations with Noel Potter (Dickinson College) and Bill Sevon (Pennsylvania Geological Survey).

The Permian landscape was recreated in part from pictures taken in the Himalayan Mountains by Harry Kent during a mountain climbing expedition in 1987.

References

Afifi, A.A. and Bricker, O.P., 1983. Weathering reactions, water chemistry and denudation rates in drainage basins of different bedrock types: I — sandstone and shale. In: Dissolved loads of river and surface water quantity quality relationships. Proc. Hamburg Symp., IAHS Publ. 141: 193–203.

Bowen, D.Q., 1973. Glaciations past and future. Geogr. Mag., 52: 62–67.

Cleaves, E.T., 1968. Piedmont and Coastal Plain geology along the Susquehanna Aqueduct: Baltimore to Aberdeen, Maryland. Md. Geol. Surv., Rep. Invest., 8, 45 pp.

Cleaves, E.T., 1983. Chemical weathering in a humid temperate environment: Sci. Geol. Mem., 72(2): 47–55.

Cleaves, E.T. and Costa, J.E., 1979. Equilibrium, cyclicity, and problems of scale — Maryland's Piedmont landscape. Maryland Geol. Surv., Inf. Circ., 29, 32 pp.

Cleaves, E.T., Godfrey, A.E. and Bricker, O.P., 1970. Geochemical balance of a small watershed and its geomorphic implications, Geol. Soc. Am. Bull., 81: 3015–3032.

Cleaves, E.T., Fisher, D.W. and Bricker, O.P., 1974. Chemical weathering of serpentinite in the eastern Piedmont of Maryland. Geol. Soc. Am., Bull., 85: 437–444.

Colman, S.M. and Mixon, R.M., 1988. The record of major Quaternary sea-level changes in a large coastal plain estuary, Chesapeake Bay, eastern United States, Paleogr., Paleoclimatol., Paleoecol, 68: 99–116.

Conant, L.C., in prep. Coastal plain beds if Cecil County. Cecil County Rep. Md. Geol. Surv.

Crickmay, C.H., 1975. The hypothesis of unequal activity. In: W.N. Melhorn and R.C. Flemal (Editors), Theories of Landform Development. State University of New York, Publications in Geomorphology, Binghamton, N.Y., pp. 103–110.

Crowley, W.P., Reinhardt, J. and Cleaves, E.T., 1976. Geologic map of Baltimore County and City. Md. Geol. Surv. Map 1/62,500.

Delacourt, P.A, and Delacourt, H.R., 1981. Vegetation maps for Eastern North America: 40,000 yr. BP to the present. In: R.C. Romans (Editor), Geobotany II. Plenum, New York, N.Y., pp. 123–164.

Faill, R.T., 1973. Tectonic development of the Triassic Newark-Gettysburg Basin in Pennsylvania. Geol. Soc. Am. Bull., 84: 725–740.

Fenneman, N.M., 1938. Physiography of Eastern United States. McGraw Hill, New York, N.Y., 689 pp.

Glaeser, J.D., 1966. Provenance, dispersal, and depositional environments of Triassic sediments in the Newark-Gettysburg basin. Pa. Geol. Surv., 4th Ser., General Geol. Rep., G-43, 168 pp.

Glaser, J.D., 1969. Petrology and origin of the Potomac and Magothy (Cretaceous) sediments, middle Atlantic Coastal Plain. Md. Geol. Surv., Rep. Invest., 11, 101 pp.

Godfrey, A.E., 1975. Chemical and physical erosion in the South Mountain anticlinorium, Maryland. Md. Geol. Surv., Inf. Circ., 19, 35 pp.

Hack, J.T., 1975. Dynamic equilibrium and landscape evolution. In: W.N. Melhorn and R.C. Flemal (Editors),

Theories of Landform Development. State University of New York, Publications in Geomorphology, Binghamton, N.Y., pp. 87-102.

Hack, J.T., 1980. Rock control and tectonism — their importance in shaping the Appalachian Highlands. U.S. Geol. Surv. Prof. Pap., 1126-B, 17 pp.

Hack, J.T., 1982. Physiographic divisions and differential uplift in the Piedmont and Blue Ridge. U.S. Geol. Surv., Prof. Pap., 1265, 49 pp.

Hansen, H.J., 1969. Depositional environments of subsurface Potomac Group in southern Maryland. Am. Assoc. Pet. Geol., Bull., 53: 1923-1937.

Hansen, H.J., 1984. Hydrogeologic characteristics of the Waste Gate Formation, a new subsurface of the Potomac Group underlying the eastern Delmarva peninsula. Md. Geol. Surv., Inf. Circ., 39, 22 pp.

Hay, W.H., Barron, E.J.., Sloan II, J.L. and Southam, J.R., 1981. Continental drift and the global patterns of sedimentation. Geol. Rundsch., 70: 302-315.

Hentz, T.F., 1985. Early Jurassic sedimentation of a rift-valley lake: Culpeper basin, northern Virginia. Geol. Soc. Am. Bull., 96: 92-107.

Jacobeen, F.H., Jr., 1972. Seismic evidence for high angle reverse faulting in the Coastal Plain of Prince George's and Charles Counties, Maryland. Md. Geol. Surv., Inf. Circ., 13, 21 pp.

Klitgord, K.D., Hutchinson, D.R. and Schouten, H., 1988. U.S. Atlantic continental margin: Structural and tectonic framework. In: R.E. Sheridan and J.A. Grow (Editors), Geological Society of America, The Atlantic Continental Margin, UG. I-2: 19-55.

Kutzback, J.E. and Wright, Jr., H.E., 1985. Simulation of the climate of 18,000 years BP: results for the North America/North Atlantic/European sector and comparison with the geologic record of North America. Quat. Sci. Rev., 4: 147-187.

LeTourneau, P.M., 1985. Alluvial fan development in the lower Jurassic Portland formation, central Connecticut — implications for tectonics and climate. U.S. Geol. Surv., Circ., 946: 17-26.

Lindholm, R.C., 1979. Geologic history and stratigraphy of the Triassic-Jurassic Culpeper basin, Virginia. Geol. Soc. Am., Bull., 90: Part I: 933-995; Part II: 1644-1701.

Lindholm. R.C., Hazlett, J.M. and Fagin, S.W., 1979. Petrology of Triassic-Jurassic conglomerates in the Culpeper basin, Virginia. J. Sediment. Petrol., 49: 1245-1262.

Manspeizer, W. and Cousminer, H.L., 1988. Late Triassic-Early Jurassic synrift basins of the U.S. Atlantic margin. In: R.E. Sheridan and J.A. Grow (Editors), Geological Society of America, the Atlantic Continental Margin, U.S. I-2: 197-216.

Martin, H., 1981. The late Paleozoic Gondwana glaciations. Geol. Rundsch., 70: 480-496.

McDonald, N.G. and LeTourneau, P.M., 1988. Paleoenvironmental reconstruction of a fluvial-deltaic-lacustrine sequence, lower Triassic Portland Formation, Suffield, Connecticut. In: A.J. Froelich and G.R. Robinson, Jr. (Editors), Studies of the Early Mesozoic Basins of the Eastern United States. U.S. Geol. Surv., Bull., 1776: 24-30.

Mixon, R.B. and Newell, W.L., 1977. Stafford fault system: structures documenting Cretaceous and Tertiary deformation along the Fall Line in northeastern Virginia. Geology, 5: 437-440.

Muller, P.D., 1985. Hereford Quadrangle Geologic Map. Md. Geol. Surv. Map, 1: 24,000.

National Geographic Society, 1988. Mt. Everest. Natl. Geogr. Mag., 174(5), Map Suppl., 1: 50,000.

Newell, W.L., 1984. Architecture of the Rappahannock estuary — neotectonics in Virginia. In: M. Morisawa and J.T. Hack (Editors), Tectonic Geomorphology. Allen and Unwin, Boston, Mass., pp. 321-344.

Newell, W.L. and Rader, E.K., 1982. Tectonic control of cyclic sedimentation in the Chesapeake Group of Virginia and Maryland. In: P.T. Lyttle (Editor), Central Appalachian Geology. Geol. Soc. Am. Guideb. Northeast-Southeast Meet., pp. 1-27.

Newell, W.L., Wyckoff, J.S., Owens, J.P. and Farnsworth, J., 1988. Cenozoic geology and geomorphology of southern New Jersey coastal plain. Southeast. Friend Pleistocene. Second Annu. Field Conf., Field Trip Guideb.

Nutter, L.J. and Otton, E.G., 1969. Ground-water occurrence in the Maryland Piedmont. Md. Geol. Surv., Rep. Invest., 10, 56 pp.

Oberlander, T.M., 1984. Origin of drainage transverse to structures in orogens. In: M. Morisawa and J.T. Hack (Editors), Tectonic Geomorphology. Allen and Unwin, Boston, pp. 155-182.

Olsson, R.K., Gibson, T.G., Hansen, H.J. and Owens, J.P., 1988. Geology of the northern Atlantic coastal plain: Long Island to Virginia. In: R.E. Sheridan and J.A. Grow (Editors), Geological Society of America, The Atlantic Continental Margin, U.S. I-2: 87-106.

Overstreet, W.C., White, A.O., Whitlow, J.W., Theobald, Jr., P.K., Caldwell, D.W. and Cupples, N.P., 1968. Fluvial monazite deposits in the southeastern United States. U.S. Geol. Surv., Prof. Pap., 568, 85 pp.

Owens, J.P., 1970. Post-Triassic tectonic movements in the central and southern Appalachians as recorded by sediments of the Atlantic coastal plain. In: G.W. Fisher, F.J. Pettijohn, J.C. Reed, Jr. and K.N. Weaver (Editors), Studies of Appalachian Geology. Interscience, New York, N.Y., pp. 417-427.

Owens, J.P. and Gohn, G.S., 1985. Depositional history of the Cretaceous series in the U.S. Atlantic coastal plain: stratigraphy, paleoenvironments, and tectonic controls of sedimentation. In: C.W. Poag (Editor), Geologic Evoluation of the United States Atlantic Margin. Van Nostrand Reinhold, New York, N.Y., pp. 25-86.

Owens, J.P. and Minard, J.P., 1979. Upper Cenozoic sediments of the lower Delaware Valley and the northern

Delmarva Peninsula, New Jersey, Pennsylvania, Delaware, and Maryland. U.S. Geol. Surv. Prof. Pap., 1067D 47 pp.

Pavich, M.J., 1986. Processes and rates of saprolite production and erosion on a foliated granitic rock of the Virginia Piedmont. In: S.M. Coleman and D.P. Dethier (Editors), Rates of Chemical Weathering of Rocks and Minerals. Academic Press, New York, N.Y., pp. 552–590.

Pavich, M.J. and Newell, W.L., 1987. The surficial mantle and landscape evolution in the Piedmont and inner coastal plain. In: W.L. Graf (Editor), Geomorphic systems of North America. Geol. Soc. Am., Boulder, Cent. Spec. Vol., 2: 17–21.

Poag, C.W., 1979. Stratigraphic and depositional environments if Baltimore Canyon Trough. Am. Assoc. Pet. Geol., Bull., 63: 1452–1466.

Poag, C.W., 1985. Depositional history and stratigraphic reference section for central Baltimore canyon trough. In: C.W. Poag (Editor), Geologic Evolution of the United States Atlantic Margin. Van Nostrand Reinhold, New York, N.Y., pp. 217–264.

Poag, C.W. and Sevon, W.D., 1989. A record of Appalachian denudation in post-rift Mesozoic and Cenozoic sedimentary deposits of the U.S. middle Atlantic continental margin. In: T.W. Gardner and W.D. Sevon (Editors), Appalachian Geomorphology. Geomorphology, 2: 119–157 (this volume).

Potter, P.E. and Pryor, W.A., 1961. Dispersal centers of Paleozoic and later clastics of the upper Mississippi Valley and adjacent area. Geol. Soc. Am. Bull., 72: 1195–1250.

Schlee, J.S., 1957. Upland gravels of southern Maryland. Geol. Soc. Am., Bull., 68: 1371–1409.

Secor, Jr., D.T., Snoke, A.W. and Dallmeyer, R.D., 1986. Character of the Alleghanian orogeny in the southern Appalachians: Part III. Regional tectonic relations. Geol. Soc. Am., Bull., 97: 1345–1353.

Slingerland, R. and Furlong, K.P., 1989. Geodynamic and geomorphic evolution of the Permo-Triassic Appalachian Mountains. In: T.W. Gardner and W.D. Sevon (Editors), Appalachian Geomorphology. Geomorphology, 2: 23–37 (this volume).

Smoot, J.P. and Olsen, P.E., 1985. Massive mudstones in basin analysis and paleoclimatic interpretation of the Newark Supergroup. In: G.R. Robinson, Jr. and A.J. Froelich (Editors), Proceedings of the Second U.S. Geological Survey Workshop in the Early Mesozoic Basins of the Eastern United States. U.S. Geol. Surv. Circ., 946: 29–33.

Sutter, J.F., 1985. Progress on geochronology of Mesozoic diabases and basalts. In: G.R. Robinson, Jr. and A.J. Froelich (Editors), Proceedings of the Second U.S. Geological Survey Workshop on the Early Mesozoic Basins of the Eastern United States. U.S. Geol. Surv., Circ., 946: 110–114.

Sutter, J.F., 1988. Innovative approaches to the dating of igneous events in the early Mesozoic basins of the eastern United States. In: A.J. Froelich and G.R. Robinson, Jr. (Editors), Studies of the Early Mesozoic Basins of the Eastern United States. U.S. Geol. Surv. Bull., 1776: 194–200.

Tiffney, B.H., 1985. Geological factors and the evolution of plants. In: B.H. Tiffney (Editor), Geological Factors and the Evolution of Plants. Yale University Press, New Haven, Conn., pp. 1–10.

Velbel, M.A., 1985. Geochemical mass balances and weathering rates in forested watersheds of the southern Blue Ridge. Am. J. Sci., 285: 904–930.

Whitehead, D.R. and Barghorn, E.S., 1962. Pollen analytical investigations of Pleistocene deposits from western North Carolina and South Carolina. Ecol. Monogr., 32: 347–369.

Geomorphology, 2 (1989) 181-196
Elsevier Science Publishers B.V., Amsterdam — Printed in The Netherlands

Regolith Residence Time and the Concept of Surface Age of the Piedmont "Peneplain"

M.J. PAVICH

U.S. Geological Survey, Reston, VA 22092 (U.S.A.)

(Received December 20, 1988; accepted after revision May 26, 1989)

Abstract

Pavich, M.J., 1989. Regolith residence time and the concept of surface age of the Piedmont "peneplain". In: T.W. Gardner and W.D. Sevon (Editors), Appalachian Geomorphology. Geomorphology, 2: 181-196.

The common upland weathering profile on foliated metamorphic rocks on the Appalachian Piedmont between Alabama and Maryland includes a 50 cm-1 m thick clay-rich soil B-horizon over 10-20 m of saprolite. The saprolite retains the volume, fabric and structure of the underlying parent rock, but has lost about one-third of the mass of the parent rock. Where the saprolite is thickest beneath upland drainage divides, those divides are of low slope and relief. The flatness of the Piedmont drainage divides provides the appearance of a planation surface of accordant summits, or peneplain. I interpret this flatness to be the result of three active processes: volume reduction during compaction of saprolite into soil, loss of mass in dissolved solids draining the soil B-horizon, and loss of mass from the soil surface by erosion of clay-size particles. The cumulative effects of these upland processes are to convert saprolite into soil, to limit soil thickness and residence time, and to limit the age of the oldest residual soils on the surface. This results in a continual lowering of the Piedmont "peneplain" surface. This interpretation is corroborated by evidence of topographic inversion of fluvial gravels since the Miocene near the Fall Line.

Study of baseflow dissolved solids draining Piedmont basins indicates a minimum rate of saprolite production of about 4 m Ma^{-1}. This minimum rate of production near the Fall Line is less than the estimated rate of surface lowering and saprolite consumption during the late Tertiary and Quaternary periods of topographic inversion. Measurement of cosmogenic ^{10}Be inventories in two 15-20 m cores of Piedmont upland regolith indicate a minimum residence time of residual soil and saprolite of about 1 Ma. This indicates a possible maximum rate of saprolite production and erosion of about 20 m Ma^{-1}.

Using the best available evidence, therefore, the typical Piedmont upland regolith has a residence time of between 1 and 5 Ma. Residence times may not be the same everywhere and ages may vary systematically across the surface. The geomorphic evidence of topographic inversion, and similar regolith profiles in other humid temperate and tropical regions, strongly suggests that production and removal of Piedmont saprolite has been an ongoing Quaternary process and that most of the Piedmont regolith is significantly younger than 5 Ma.

Independent geophysical evidence suggests that the rate of uplift, probably involving tilting, of parts of the Piedmont, Blue Ridge, and Valley and Ridge has averaged between 5 and 20 m Ma^{-1} through the Cenozoic.

Introduction

The purpose in this paper is to present a model for the continuous production and removal of Appalachian Piedmont (Fig. 1) upland regolith, and to compare the mass flux (mass/area-time) from the Piedmont, determined from independent geochemical and geomorphic measurements, with recent determinations of post-Triassic uplift of the Piedmont and other areas of the Appalachian orogen. Testing of this model will provide information necessary to defining the age range of the geomorphic surface of the Appalachian Piedmont

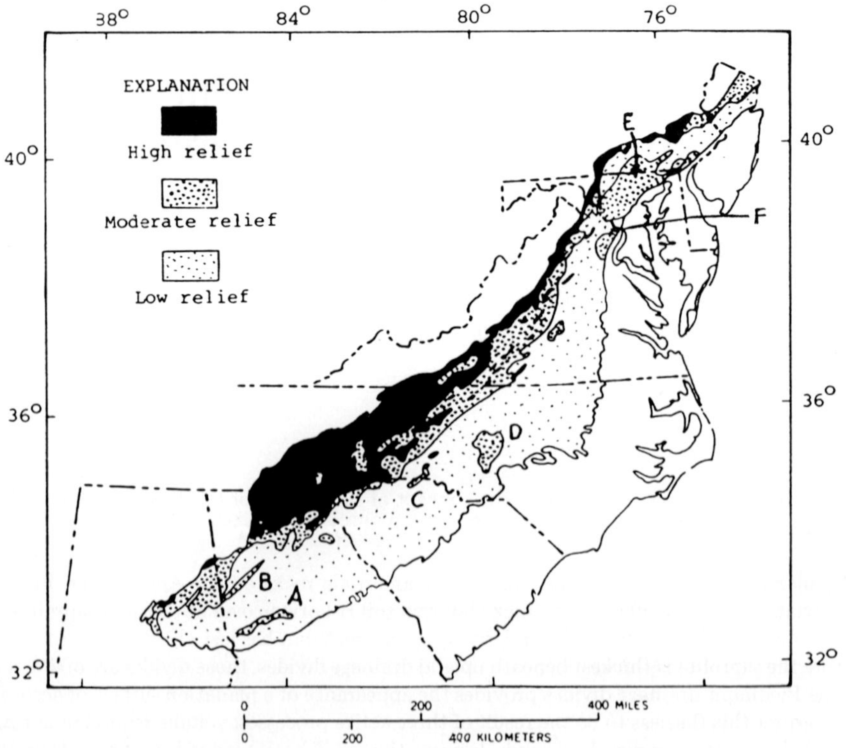

Fig. 1. Generalized relief map of the Piedmont and Blue Ridge provinces from Hack (1980). The relief classes are based on the number of contours within square areas 10 km on a side, as indicated on the 1:250,000-scale series of topographic maps by the U.S. Geological Survey. Less than 90–120 m in low-relief areas and 120–240 m in moderate-relief areas. High-relief areas are generally more rugged and have relief of more than 240 m and as much as 1200 m. *A*=Pine Mountain, Ga.; *B*=Brevard Zone; *C*=Kings Mountain Belt; *D*=Uwharrie Mountains; *E*=Westminster Highland; and *F*=Fairfax County, Va.

based on the concept of soil residence time. This model stresses that the definition of maximum age of an eroding surface should be based on the best available techniques for establishing the age, or residence time, of residual soil and saprolite (or weathered rock) at specific places, or points, on the least rapidly eroding positions of that surface. This contrasts with the geomorphic concept of establishing the minimum age of a depositional surface based on soil characteristics.

The concept of age, or age-range, of a depositional or erosional geomorphic surface is a fundamental geomorphic problem. The peneplain concept suggests that a synchronous erosional surface, or its remnants, exists over areas exceeding 10^4 km^2 and that such a surface, or its uneroded remnants, can persist unaltered over time periods exceeding 10^7 years. While useful as a concept, proving the present or former existence of peneplains, particularly from scattered erosional remnants, is difficult. I propose that our definition of age of an erosional surface of low relief (or planation surface) be based on techniques of dating regolith (i.e., soil and subsoil) profiles as independent points on the geomorphic surface, and comparison of soil characteristics on eroding surfaces with soil characteristics on "stable" fluvial terrace deposits.

The Appalachian Piedmont (Fig. 1) provides an important area to test such methods.

One of the basic elements of Davis' (1899) model of landscape cycles was that the Piedmont represented an "old" peneplain surface over deeply weathered rocks (Davis, 1899). The stage of deep weathering was thought to have occurred on the mature landscape under condition of little or no upland erosion well after the last major tectonic uplift, but prior to an uplift that caused stream (Davis, 1899). The relative age assignment given by Johnson (1931) to the Piedmont (Somerville) peneplain made it younger than the higher peneplain remnants of the Valley and Ridge (i.e. Harrisburg and Schooley), but suggested a middle Tertiary age at minimum (see Poag and Sevon, this volume, fig. 9). These concepts were unsupported by any direct dating, and were based on an arbitrary tectonic model of episodic uplifts. The argument that weathering of rock to saprolite exceeding 10 m thickness pre-dates the Quaternary, and, that surfaces above deeply weathered rocks are, at a minimum, of Late Miocene age has been made for the Maryland Piedmont by Costa and Cleaves (1984).

Hack (1960) presented a contrasting model for Piedmont relief and geomorphology based on the concept of dynamic equilibrium. In essence, a landscape in dynamic equilibrium maintains a state of balance between relief and erosional processes, thus landscape form is maintained while mass loss due to chemical weathering and mechanical erosion is continuous. This is analogous to the concept of a graded stream, one that maintains a gradient sufficient to transport its load. Inherent in models of graded erosional systems operating over long periods of geologic time is the implication that uplift must also proceed at a finite, though not necessarily constant, rate sufficient to maintain the landscape above base-level.

The approach in this paper is to go from the processes of regolith (saprolite and soil) production to the process of removal of that regolith, and the rate of removal. Those discussions will be followed by consideration of the crustal response to continued denudation.

Piedmont regolith

Regolith production

For the purpose of discussion of age of a particular part of an eroding surface, it is best to begin with a measurable process; for example the weathering of rock to saprolite beneath the upland, "peneplain", surface of the Piedmont. Figure 2 shows a schematic picture of saprolite distribution. Saprolite forms in place due to chemical weathering of rock, and its distribution beneath upland and side slopes may indicate that chemical weathering is active in all topographic positions. Studies of the upland weathering profile provide detailed information about chemistry and mineralogy of weathering (Pavich, 1986).

Chemical weathering of rock to saprolite

Samples for mineralogical and chemical analyses have been taken from several cores in Fairfax County, Virginia (Fig. 1), obtained by hollow-stem auger drilling. The entire regolith was sampled from the upland drainage divide (Fig. 2) using shelby tubes and drilling methods described in Pavich et al. (1989a). The analyses were performed on samples from a core taken from the upland near the head of drainage. The results of analyses from one of the cores site on granite gneiss are shown in Figs. 3 and 4. Depth of weathering in the granite is as much as 30 m (98 ft) but within the small drainage basin studied by Pavich (1986) the regolith is generally less than 15 m (49 ft) thick, which is close to the mean regolith thickness on the Fairfax County Piedmont (Froelich and Heironimus, 1977) and the mean regolith thickness reported by Hack (1976) for the entire Piedmont.

This core is considered to be representative of the regolith developed on the upland on slopes of less than 8°. The macroscopic properties of the core, analytical data, and photomicrographs all indicate that the A and B horizons of

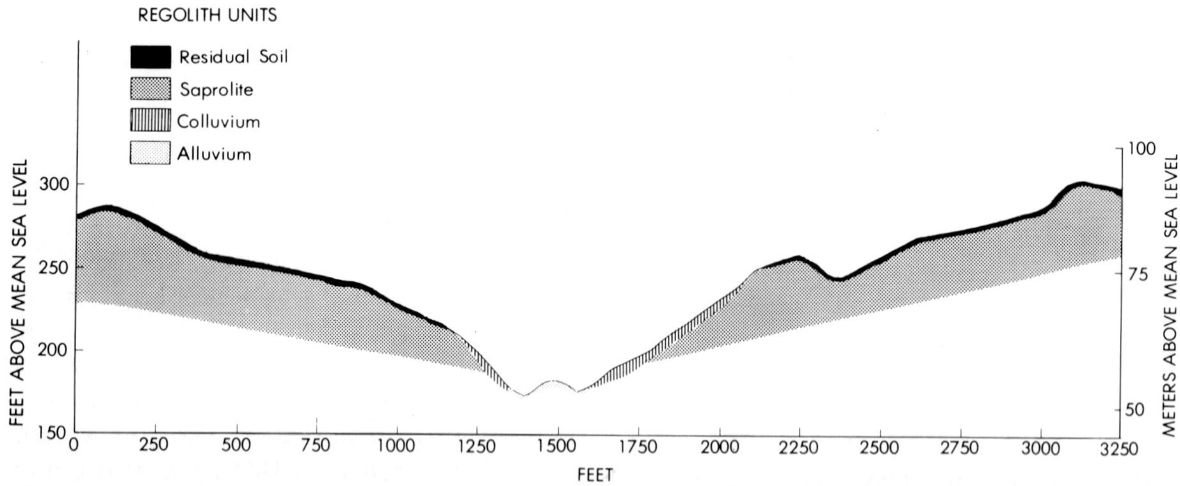

Fig. 2. Cross section from Pavich (1986) showing that the interface between saprolite and rock is flatter than the geomorphic surface. Regolith thins markedly beneath the steeper slopes of the valley sides. The regolith profiles are of similar thickness on either side of the stream. At about 76 m (250 ft) and below, the slopes are steep relative to the uplands. Portions of the steep valley side slopes are covered by colluvium. In the valley, the stream has cut down through a significant thickness of bedrock, not just saprolite, to attain its present elevation. Along the tributaries entering the main channel from the east, the lowest transition from ephemeral to perennial drainage is at 58 m (190 ft); the highest is at about 69 m (220 ft). The steepening of the valley side slopes occurs in this elevation interval. Average thickness of regolith is about 12 m (40 ft). There is a significant thickness of saprolite in areas of moderate slope.

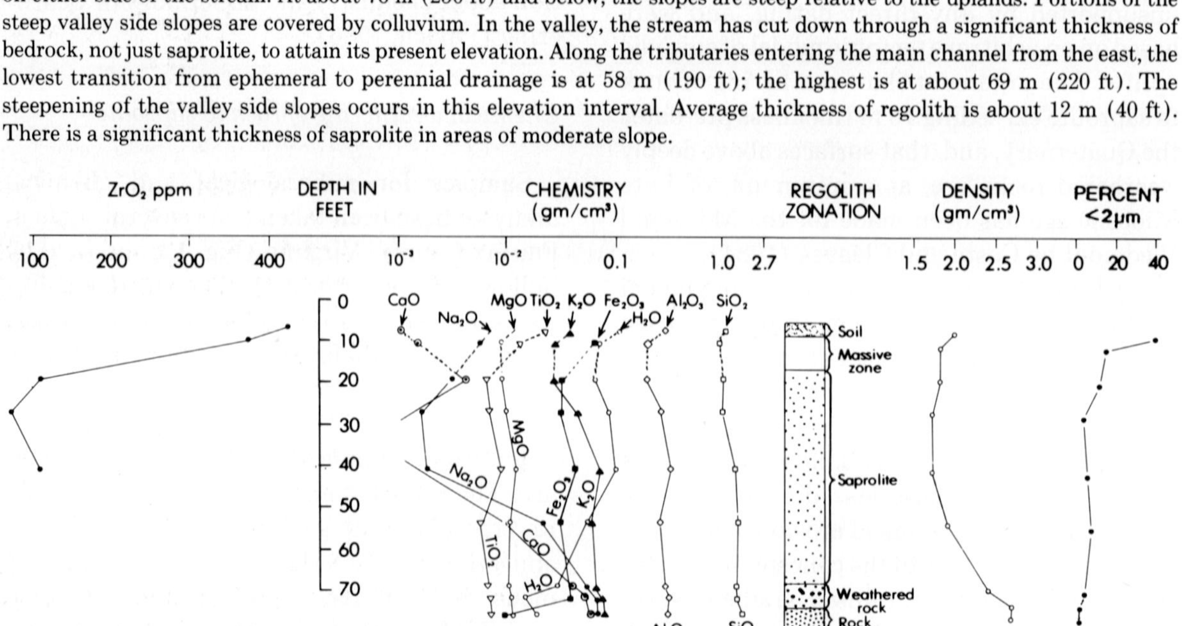

Fig. 3. Data from laboratory analyses of texture, bulk density, and chemical composition of upland regolith core from Pavich (1986).

the residual soil comprises the upper 1 m (3.3 ft) of the core, that saprolite, the isovolumetrically weathered rock, comprises most of the core between 1 and 18 m (59 ft), and that there is a thin weathered rock zone between 18 and 19 m (63 ft).

The weathered rock zone exhibits a major loss of FeO and increase in Fe_2O_3 correlative with visible staining of rock surfaces by oxidized iron phases and petrographically observed alteration of biotite and iron sulfides. These observations demonstrate that in situ alteration of ferrous iron silicate and sulfide phases is an important process initiating rock weathering. This

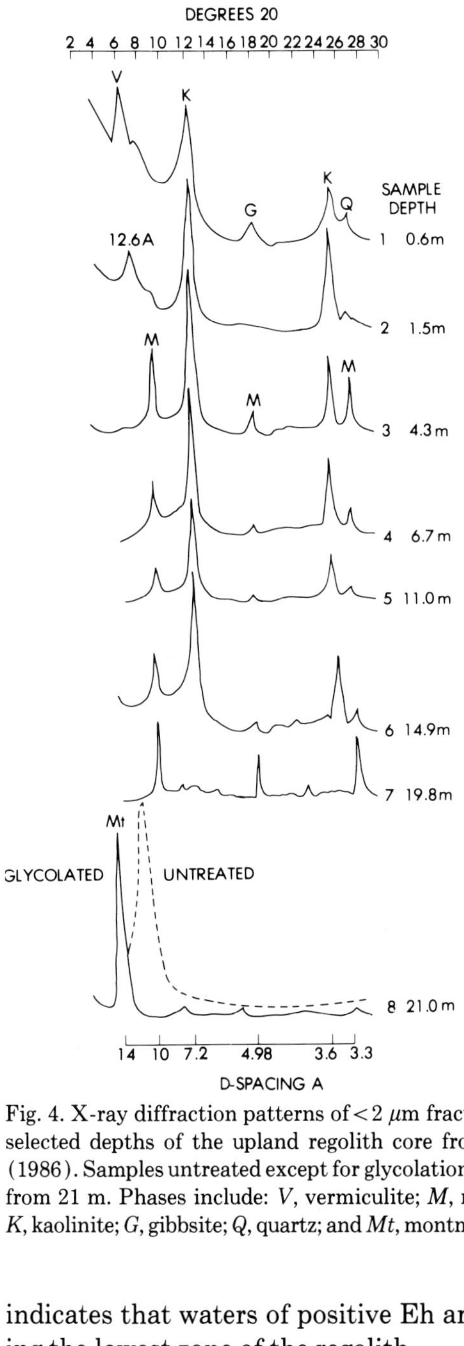

Fig. 4. X-ray diffraction patterns of < 2 μm fractions from selected depths of the upland regolith core from Pavich (1986). Samples untreated except for glycolation of sample from 21 m. Phases include: *V*, vermiculite; *M*, muscovite; *K*, kaolinite; *G*, gibbsite; *Q*, quartz; and *Mt*, montmorillonite.

indicates that waters of positive Eh are reaching the lowest zone of the regolith.

The saprolite is zoned mineralogically and chemically. The upper 10 m of the saprolite between 5 and 15 m is depleted of CaO and Na_2O. In the lower 6 m (20 ft) of the saprolite, which comprises about 30% of the regolith profile, the steep gradients in CaO, Na_2O, and H_2O indicate that this is the zone of significant chemical weathering reactions. This reactive zone is typical of lower saprolite from other Piedmont cores (Pavich et al., 1989a).

Chemically, most of the mass lost from the saprolite is CaO, Na_2O, and SiO_2. This correlates with petrographic observation of alteration and loss of plagioclase feldspar. Other phases, such as biotite and pyrite, are altered in the weathered rock zone, but they are less abundant phases and, therefore, comprise less of the mass loss than does feldspar dissolution. Base flow chemistry and discharge indicate a minimum rate of saprolite production of about 4 m Ma^{-1} (Pavich, 1986). This rate is similar to the rate calculated in the Maryland Piedmont by Cleaves et al. (1970).

Clay mineralogy (Fig. 4) correlates well with the chemical zonation in the weathered rock and saprolite. In the weathered rock zone, biotite is altered to a smectite, but this phase does not occur higher in the saprolite. Kaolinite is the dominant clay phase in the saprolite. The kaolinite is derived from the alteration of plagioclase feldspar. There is no petrographic or XRD evidence for alteration of muscovite to kaolinite in the saprolite.

Soil formation

Rebertus and Buol (1985a) presented evidence that the common residual soils of the Blue Ridge and Piedmont in North Carolina are formed from saprolite. This is also the case in the Georgia Piedmont (Markewich and Pavich, unpubl. data). Mineralogical and chemical data (Pavich, 1986) indicate that the solum (the A and B horizons of the soil profile) in this core and elsewhere on the Piedmont of Fairfax County, Va. (Pavich et al., 1989a) is derived from the underlying saprolite.

Alteration of saprolite to soil involves changes in volume and mass relative to the parent saprolite. Regolith bulk density reaches a minimum in the saprolite (Fig. 3). Density increases upward slightly with the slight rise in percent clay, and then rises markedly with the increase in percent clay in the B horizon. The volume reduction and density increase also correspond to changes in Al_2O_3, Fe_2O_3, and ZrO_2 (Fig. 3). The SiO_2 does not increase in proportion to the increases in density, percent of clay, or ZrO_2. This indicates an absolute loss of silica with relative increases in Al_2O_3, Fe_2O_3, and ZrO_2. The total mass lost during chemical alteration of saprolite to soil is about 75% as indicated by the ZrO_2 and clay fraction increases. The residual 25% of the mass derived from the saprolite is mainly clay-sized material.

Within the solum, a significant change in clay mineralogy takes place relative to the saprolite. Nonexpandable 14-A basal spacing clay identified as hydroxy interlayered vermiculite (HIV) is the dominant clay phase in the soil. The dominance of this hydroxy-interlayered clay in this residual solum is apparently due to in situ pedogenic reactions initiated by the alteration of 10-A muscovite and replacement of potassium in its interlayers by an aluminium–hydroxyl complex, as demonstrated by Rich (1958) for a soil on the Virginia Piedmont. The percentage of clay-size quartz also increases in the soil. Petrographic examination of the soil B-horizon shows sand-size quartz altering to silt and clay size particles (Pavich et al., 1989a).

Alteration of relatively stable quartz and muscovite in the soil and not in the saprolite could be a function of the longer period of weathering of the soil, and the longer residence time of solution in the relatively less permeable soil zone. HIV is not produced in the saprolite, indicating the lower intensity of chemical alteration in the saprolite relative to the soil environment.

The period of soil formation, or soil residence time, on the Piedmont upland is one of the key elements in defining the age of the Piedmont surface. On the upland erosional surface, the soil residence time is limited by erosion of fine particles from the surface. Figure 5 shows a schematic model of production of saprolite from rock and soil from saprolite (Pavich, 1986). Due to the downward movement of the weathering front (a) and the saprolite/soil boundary (b), there is, in effect, a net upward mass flux relative to those planes. If there were no erosion from the soil surface at the geomorphic surface, represented as x, then the soil (c) could thicken indefinitely. However, as stated by Ahnert (1970, p. 247): "...it is safe to assume that for long periods of denudational surface lowering the soil thickness at any locality remains approximately constant, since the processes of weathering and of removal tend toward a state of dynamic equilibrium...The lowering of the soil surface thus tends to be equal to the lowering of the bedrock surface underneath by weathering and by partial removal of mineral matter in solution."

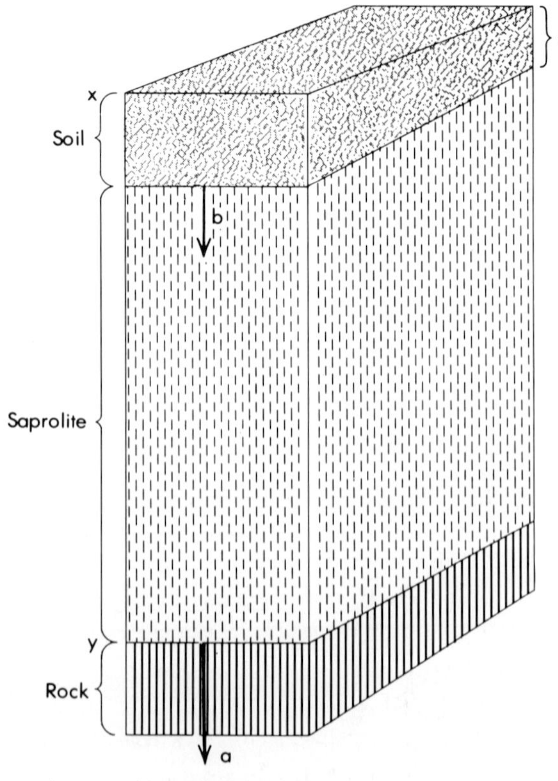

Fig. 5. Schematic model of upland saprolite production and erosion, from Pavich (1986).

There are two stratigraphic arguments that erosion does maintain a balanced thickness of soil. This concept was formalized by Budel (1977) in his model of "double planation". The thickness of the upland soil developed from saprolite tends to be quite uniform; generally less than 1 m thick. This is probably due to a balance of erosion and alteration of saprolite to soil, rather than some age-dependent limit on the thickness of soil that can be produced from saprolite. Comparison of Piedmont soil characteristics with Coastal Plain terrace soil characteristics (Fig. 6) supports this argument. The plot on Fig. 6 from Pavich et al. (1989b) shows characteristics of soil developed on fluvial and marine terraces of Delmarva and the Rappahannock River Basin determined by Markewich et al. (1987). The plots of soil thickness, thickness of argillic horizons, clay mass cm^{-2}, and Munsell color all follow linear trends plotted against a logarithmic time scale over 10^6 years. It is clear from these plots that the Piedmont soils are less well-developed than the 10^6 year old coastal plain terrace soils. This is interpreted as evidence that the Piedmont soil development is less than the oldest of the Coastal Plain soils because the soil surface erodes as the soil/saprolite plane (b on Fig. 5) descends, and not because of a time-limit on soil development on the Piedmont upland. If Piedmont soils were Miocene or older, they would be thicker than nearby coastal plain terrace soils.

In summary, the production of saprolite from rock and conversion of saprolite to soil are processes involving the movement of solutions into and out of these zones of regolith. They appear to be active processes as long as solution flux is maintained. The saprolite retains the volume of the parent rock, but the soil does not. Thus, the conversion of saprolite to soil involves a volume reduction, as well as mass loss in solution, both of which result in thinning the regolith profile and lowering of the geomorphic surface. As the surface is being lowered, the weathered rock zone is also descending. If sur-

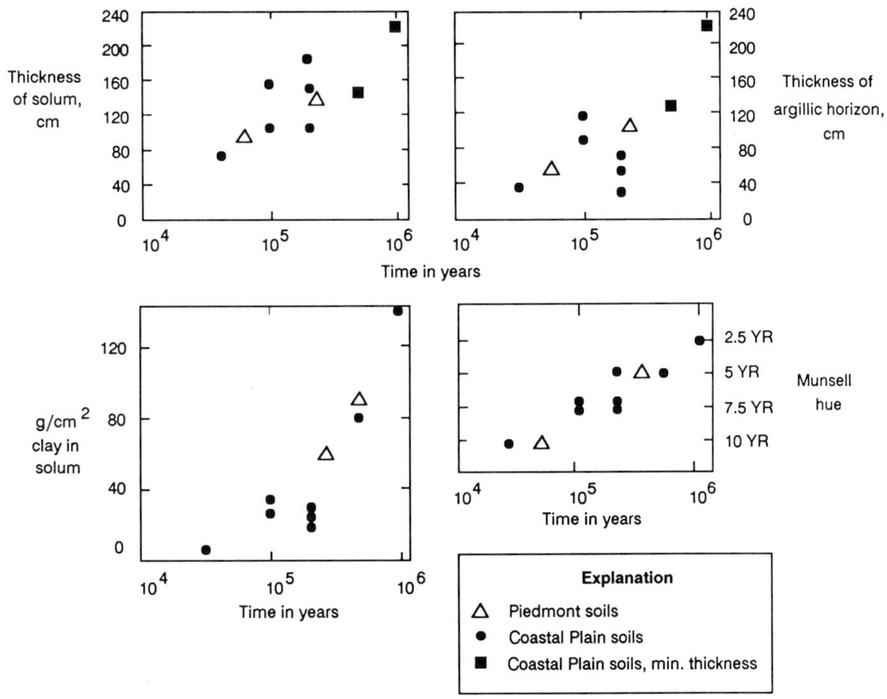

Fig. 6. Graphs showing comparison of characteristics of Piedmont and Coastal Plain soils from Pavich et al. (1989b).

face lowering occurred due to soil formation without erosion and without continued production of saprolite, eventually the saprolite would be consumed and a very thick, clayey B-horizon would result. The soil surface would continue to age as long as no physical erosion occurred. The additional constraint on soil surface age, therefore, is physical erosion from the upland soil surface. This conclusion can be evaluated by evidence about short and long-term rates of upland erosion.

Upland erosion

Short-term and long-term surface erosion

Most Piedmont slopes are convex in cross section. Figure 7 contrasts the ideal convex slope form used by Gilbert (1909) (Fig. 7A) to explain convex graded slopes undergoing erosion by creep with a flattened Piedmont upland form (Fig. 7B). The contrast is significant because flat and broad Piedmont upland slopes can be used as evidence of a planation process and were used by Davis (1899) as evidence for a Piedmont peneplain.

Flattening of Piedmont uplands can be explained by the processes of alteration of saprolite to soil and of erosion at the surface of the soil. In order for the process of regolith (saprolite and soil) production and removal to be balanced, soil must be eroded at roughly the rate it is produced, otherwise soil and saprolite thickness would increase without limit. An obvious requirement of this model is that upland soil erosion takes place beneath a forest cover. Typically, humid-temperature forested areas are considered to undergo little or no erosion. A theoretical model presented by Burt (1986) comparing concave, straight, and convex slopes shows, however, that the convex slope is the form that can be maintained as mass is continually removed. Upland erosion measurements are very difficult to make and uncertainty exists because of the unknown, but assumed, variation of erosion rate caused by variation of magnitude of rainfall events. Furthermore, extrapolation of short-term measurements over geologic time introduces untestable uncertainties. Despite these caveats, it is useful to try to determine the "natural" (e.g. non-anthropogenic) long-term rate of soil erosion. An argument for upland erosion of the uncultivated Piedmont is supported by three independent lines of evidence: (a) observations in contemporary temperate forests; (b) topographic inversion near the Fall line; and (c) ^{10}Be investigations.

Extrapolation of short-term rates of soil erosion in temperate forests

As pointed out by Trimble (1974) and Meade (1982) short-term erosion rates on the Piedmont and their extrapolation are highly uncertain. Recent investigations of soil erosion processes and rates include studies in the U.S. by Fowler and Heady (1981) and in Luxembourg by Imeson and Vis (1984) and Duijsings (1986). Most important to the question of long-term erosion of gently sloping uplands and upland hollows are the detailed measurements of rainsplash detachment and litter-transport of fine soil particles. Imeson and Vis (1984, p. 589) state that: "The dispersed clay [in stream runoff] is thought to originate from the A12 and upper B horizons of the soil and to be transported to the drainage channels by throughflow in the many macrobiopores and shrinkage cracks which typify the upper horizons of the Keuper soils."

Duijsings (1986) also stated that suspended clay sediment in storm peak runoff probably originated from the upper B-horizon of the soils. He observed that overland flow in upland hollows was a significant contributor to the storm hydrograph in the low-order basin studied, and that leaf litter, to which fine particles adhered, was transported downslope by overland flow.

Data presented by Fowler and Heady (1981) yield a soil erosion rate from forested land of the eastern U.S. of between about 0.001 and

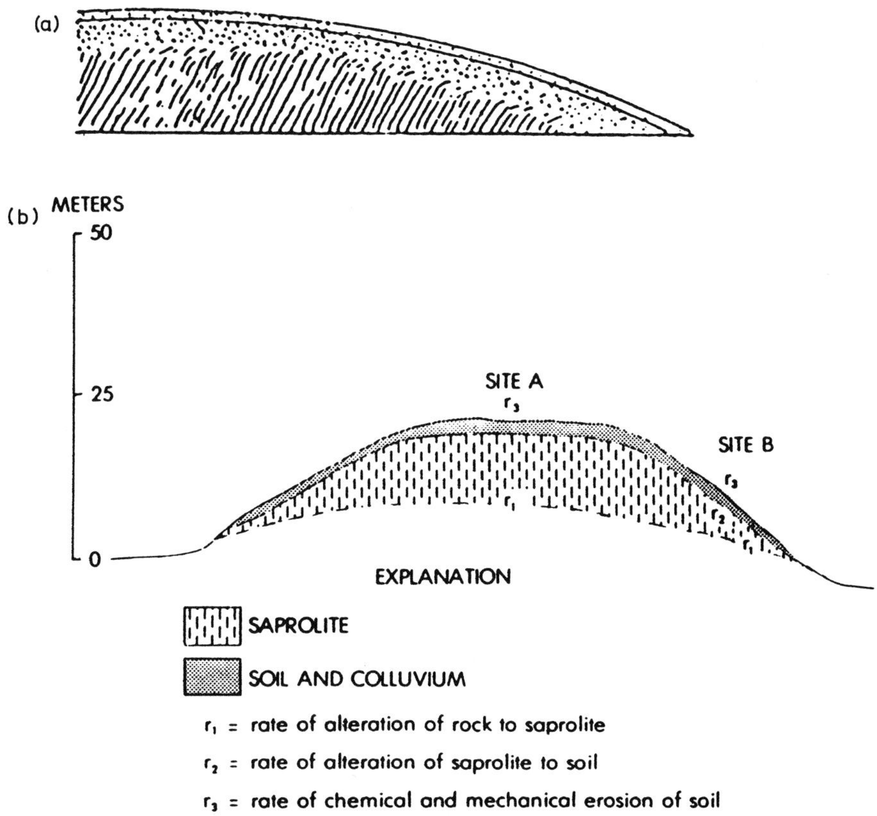

Fig. 7. Comparison of (a) idealized convex slope undergoing erosion by creep from Gilbert (1909), and (b) Piedmont drainage divide from Pavich (1986). On (b) the upland (site A) and colluvial slope (site B) are distinguished by their erosion rates (r_3) and rates of conversion of saprolite to soil and mass loss from soil in solution (r_2). The rate of rock weathering to saprolite is represented by r_1.

0.002 g cm^{-2} yr^{-1}. At those rates removal of a 100 cm thick soil at 1.7 g cm^{-3} density would require 85–170 ka; an erosion rate of 6 to 12 m Ma^{-1}. Similarly, the soil fluxes reported from Luxembourg translate to soil residence time of 500–700 ka. (Residence time is defined as mass/volume ÷ mass/area·time = time/thickness.) These are obviously uncertain extrapolations, but do seem to provide a reasonable estimate for the rate of soil erosion under forested land; or conversely an estimate for the soil residence time.

The mechanisms of Piedmont upland hollow soil erosion under forest have not been studied in detail, but the overland flow and litter transport processes described by Imeson and Vis (1984), and Duijsings (1986) have been inferred from observations following intense storms (Costa, 1974; Pavich, unpubl. data) and are likely to erode Piedmont upland soils. Piedmont soil characteristics, slopes and climate are similar to the area studied in Luxembourg. Soil detachment by splash, animal burrowing, and tree throw is probably an active process in all forests (Mitchell, 1988). Assuming that the sediment fluxes reported by Fowler and Heady (1981) and Duijsings (1986) are applicable to the Appalachian Piedmont, upland soils would erode at a rate that removes a minimum of about 1 m of soil in less than 1 Ma. This rate of soil erosion translates to a minimum surface lowering of 4 m Ma^{-1}. The rates are not precisely constrained, and may at best represent long-term averages of a highly variable rate. Despite the paucity of data and statistical uncertainty of extrapolating such measurements, the results presented fit the process of planation by surface wash presented by Budel (1977) from his observations on deeply weathered landscapes in under tropical, seasonal climates. The rates cited here can also be compared with independent estimates of long-term surface lowering by erosion.

Geomorphic and geochemical evidence for long-term erosion and surface lowering

Independent evidence of long-term Piedmont surface lowering by erosion includes topographic inversion near the Fall Line indicating that the Piedmont land surface has been lowered (Pavich et al., 1985). Topographic inversion occurs on the Piedmont relative to Miocene-age fluvial gravels. The minimum rate of late Cenozoic surface lowering deduced from this topographic relation is about 8 m Ma^{-1} (Pavich et al., 1985) over 10 Ma. This rate is perhaps fortuitously the same as the extrapolated rate of soil erosion from forested land taken from Fowler and Heady (1981). This total 80 m lowering cannot be due to chemical weathering or alteration of saprolite to soil alone; because the lowering far exceeds the minimum saprolite thickness ($\geq$40 m) that could be produced in 10 Ma. Therefore, erosion must have removed soil over the past 10 Ma.

Evidence of long-term erosion from cosmogenic ^{10}Be in Piedmont regolith

Another method of determining long-term denudation of the Piedmont is based on measurements of cosmogenic ^{10}Be in soil and saprolite profiles (Pavich et al., 1985; Brown et al., 1988). ^{10}Be produced by cosmic ray spallation reactions in the atmosphere is delivered to the terrestrial surface by rainfall. Due to its adsorption by soil and slow radioactive decay rate, ^{10}Be accumulates over long time periods (10^4–10^6 yr) in soils.

Inventories of ^{10}Be on the Piedmont indicate removal by erosion from upland soil of cosmogenic ^{10}Be attached to soil clay. In an infinitely old regolith constantly receiving ^{10}Be from the atmosphere, the inventory would build to an equilibrium value of:

$$N_0 = q/\lambda \tag{1}$$

where q=delivery rate in atom cm^{-2} time; λ=decay rate; and N_0=atom cm^{-2}. At the observed rates of q and λ, N_0 would be reached in about 5 Ma. The ^{10}Be inventories (N_0=atom cm^{-2}) in two cores yield minimum residence times of about 800 ka to 1.6 Ma for Piedmont regolith; uncertain by a factor of two (Pavich et al., 1985; Brown et al., 1988). Assuming a steady-state model, this indicates a maximum rate of production and removal of saprolite near the Fall Line of about 20 m Ma^{-1}; about five times the minimum rate of saprolite production (Cleaves et al., 1970; Pavich, 1986) based on base flow chemistry.

Solution concentrations of ^{10}Be (Brown et al., 1988) also indicate that the ^{10}Be inventories on the Piedmont must be depleted by soil erosion, and not by solution transport. A calculation of ^{10}Be removal by erosion (assuming a constant concentration of ^{10}Be on eroding soil) yields a long-term average soil erosion rate of about 5 m Ma^{-1} (Pavich et al., 1985) which is within two times of the estimate extrapolated from the data presented by Fowler and Heady (1981), and five times greater than the minimum 1 m Ma^{-1} extrapolated from the Luxembourg data.

The observed erosional efflux of ^{10}Be from the Piedmont, and the deficit of ^{10}Be inventory relative to the inventory expected if the upland regolith profiles were Tertiary remnants, indicates that the observed ^{10}Be inventory provides a minimum residence time of the regolith. If erosion from the soil surface is the main path of ^{10}Be efflux, then the observed inventory is also an accurate measure of the mean residence time of the regolith. This is a reasonable hypothesis based on the limited data presently available. It is important to note that the volumetric reduction that occurs in forming soil from saprolite, the solutional mass loss from soil not accounted for in base flow chemistry, and the thickness and erosion rate of upland Piedmont soil all appear to be elements of a dynamic equilibrium system. The soil erosion rate calculated from the ^{10}Be model of Brown et al. (1988) and the saprolite formation rate determined from various lines of evidence are consistent with the evidence for volumetric conversion of four volumes of saprolite to one volume of soil and erosion of the surface of that soil. Thus, ^{10}Be is a useful check on the model of continous production of saprolite, conversion of saprolite to soil, and of soil erosion.

Summary of evidence for upland erosion

In summary, the upland regolith near the Fall Line is produced at a rate of somewhere between 4 and 20 m Ma^{-1} and, thus, a typical 15 m thick Piedmont regolith profile in Fairfax County, Va. (Froelich and Heironimus, 1977) has a residence time between about 1 and 5 Ma. Both the ^{10}Be data and evidence for topographic inversion support a residence time closer to 1 Ma than to 5 Ma.

Assuming that the ^{10}Be model for determining regolith residence time is correct for the Piedmont in general, then saprolite production can keep pace with an erosion rate of up to 20 m Ma^{-1}, and perhaps even a higher erosion rate in a hydrologic environment conducive to faster rock weathering. This suggests that the average Cenozoic sedimentation on the Atlantic shelf at a rate of 5 m Ma^{-1} (Sevon, this volume) to 27 m Ma^{-1} (Mathews, 1975) may be largely the result of continued long-term stripping of a saprolitized Piedmont, in addition to the contributions from the Valley and Ridge and New England highlands for the Salisbury Embayment and the Blue Ridge for the southeast Georgia Embayment.

This hypothesized correspondence of weathering and erosion rates requires further investigation on a local scale, particularly along SE–NW transects across the Piedmont. One can predict, based on Hack's (1980) analyses of tilt, that the rate of uplift and erosion should increase from the Fall Line westward across the Piedmont; supported by Zimmerman's (1979) fission track determination of Piedmont uplift rates.

Tilt and uplift model for the Piedmont

Assuming a 20 m Ma^{-1} rate of soil erosion and surface lowering across the Piedmont, about 1400 m of rock has been removed in 70 Ma, roughly since the end of the Cretaceous. Even more crystalline metamorphic rock was removed by the saprolite/soil system if it also operated during the Cretaceous. Evidence that weathering and erosion of a similar regolith operated during the Cretaceous is supported by the observations that saprolite existed on crystalline rocks prior to burial of the basement during Cretaceous sedimentation (Pavich, 1974; Sigleo and Reinhardt, 1988), and that Piedmont rocks were a significant source area for nonmarine Cretaceous sediments in the Atlantic basin (Groot, 1955; Glaser, 1969).

Hack (1980) presented evidence that tilt of the basement is a monoclinal flexure to the southeast and that tilting has also been an ongoing process since the middle Cretaceous. This is consistent with cross sections presented by Johnson (1931) and Flint (1963). Hack also pointed out that an inland projection of the basement slope of the Fall Line Peneplain would put that surface about 700 m above the present surface only 50 km inland of the Fall Line. This would require a minimum average rate of surface lowering of 7 m Ma^{-1} since the deposition of the middle Cretacous sediments and the initiation of down-tilting of the buried basement. Maintenance of Piedmont relief and its elevation above base-level may be due largely to the isostatic response to erosion.

Isostatic uplift in response to erosion was suggested by Pavich (1985) to explain an uplift curve (Fig. 8) based on K/Ar thermal chronol-

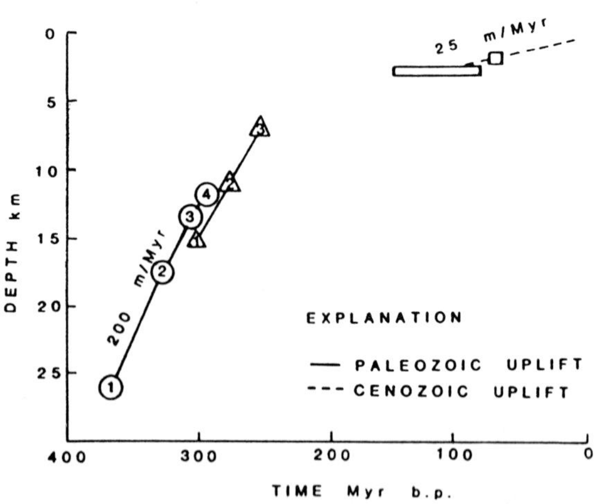

Fig. 8. Uplift curve for the Piedmont in the late Paleozoic and Cenozoic from Pavich (1985). *1* = regional metamorphic climax; *2* = argon retention by hornblende; *3* = argon retention by biotite; *4* = intrusion of Stone Mountain Granite in Georgia. Sources of data: ○ Dallmeyer (1978); ④ Whitney and others (1976); △ Durrant (1978); ▭ Zimmerman (1979); □ Mathews (1975).

ogy (Durrant, 1978) and fission track data (Zimmerman, 1979). Such an hypothesis was also proposed by Wellman (1988) in recent modelling of crustal flexure in Southeastern Australia.

Faulting versus tilting on the Piedmont

There is no evidence that the rate of faulting along the Fall Line (Mixon and Newell, 1977; Prowell, 1988) is sufficient to explain the present altitude or uplift rate of the Piedmont, or of other areas of the Appalachian orogen. Average long-term faulting rates are about an order of magnitude lower than the saprolite production and soil erosion rates discussed above. Even the major Blue Ridge escarpment is not clearly related to faulting (White, 1950; Hack, 1980). The simplest alternative hypothesis to explain Piedmont and Blue Ridge uplift is that: "tilting must have taken place to the southeast side of the Blue Ridge" (Hack, 1980, p. 12).

The significance of tilting can only be discussed qualitatively at this time. Long-term tilt rates are impossible to measure directly, and modern rates due to post-glacial isostatic rebound appear to exceed long-term rates by at least an order of magnitude. Hack (1980) presented an argument, based on a 40 m Ma^{-1} erosion rate, for the long-term average tilt rate of the Potomac River Basin of 2×10^{-4} mm/km yr. This seems a reasonable average for the down-to-the-coast tilting of the exposed basement and basement beneath the coastal plain in northern Virginia.

Figure 9 taken from Pavich and Obermeier (1985) presents cross sections of the present basement configuration near Washington, D.C. that are similar to one from New England presented by Flint (1963). Progressive downwarping of the southeastern Coastal Plain and buried basement as shown on Fig. 9a is demonstrated by biostratigraphy presented by Cronin (1981). At a constant tilt rate, the dip on the subsiding buried basement could increase while the exposed basement is weathered, eroded, uplifted, and maintained at a near horizontal dip. The significance of this is that the plane projected from the buried basement to the northwest goes well above the present topographic surface. The missing thickness of rock beneath that projected line has been removed by weathering and erosion. The evidence of surface lowering from topographic inversion near the Fall Line during the late Cenozoic is a recent manifestation of that ongoing surface lowering and uplift, as shown on Fig. 9b.

Such a tilt model is probably not applicable to every NW–SE cross section of the Appalachians. For example, the Piedmont in the area of the Carolina Slate Belt and the Cape Fear Arch appears to be a relatively stable area (Soller, 1989; Owens, 1989) compared with the area tilting into the Salisbury Embayment to the North and the S.E. Georgian Embayment to the South. Hack (1980, p. 15) pointed out that: "At the North Carolina–Virginia boundary, the gradient of the basement is only about 0.005. If projected inland as far as the Blue Ridge, the top of the basement would reach an

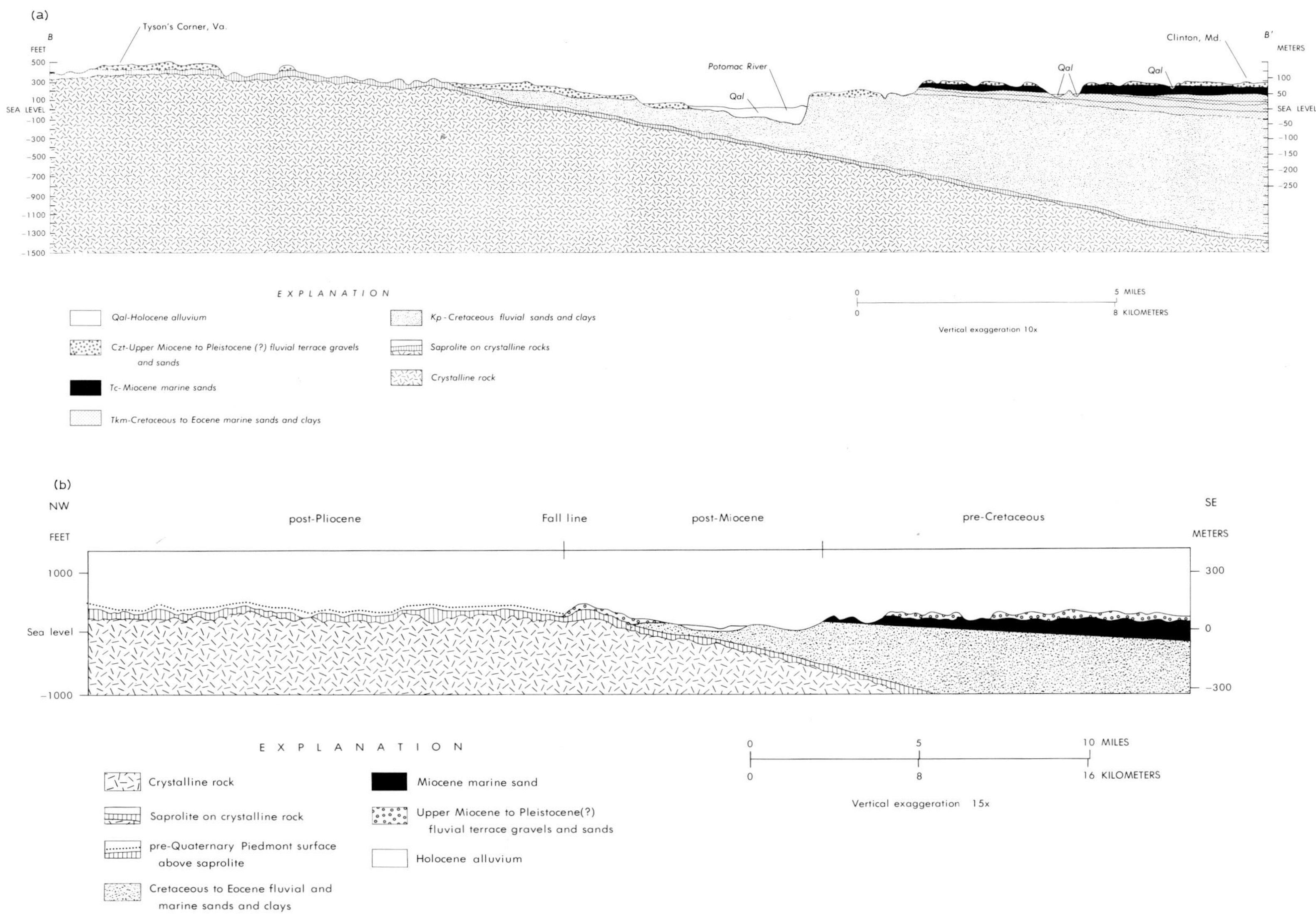

Fig. 9. Cross sections (from Pavich and Obermeier, 1985) of (a) Tyson's Corner, Va. to Clinton, Md., showing exposed Piedmont and buried crystalline surfaces, and (b) the relations of regolith age (labelled as post-Pliocene, post-Miocene, and pre-Cretaceous) on the basement, and of a pre-Quaternary Piedmont surface, with the present geomorphic surface.

altitude of only about 1200 m, almost equal to the crest of the present mountain range." Even there, however, tilting may significantly affect Blue Ridge and Piedmont physiography.

North and south of the Cape Fear Arch the rates of down-to-the-coast tilting may vary considerably and must be studied independently. Evidence of slight post-Triassic movement in the Fall Zone of North Carolina is presented by Atekwana et al. (1988). Although late Mesozoic burial and subsequent exhumation of Mesozoic sediments could have been as much as 4 km, it is also reasonable to assume little post-Mesozoic vertical movement. The Fall Zone near the Cape Fear Arch may be a relatively stable area in the Cenozoic uplift/erosion history of the Appalachians.

The weight of geomorphic and sedimentologic evidence favors continued uplift of the Appalachians, including the Piedmont, during the Cenozoic (Gardner, this volume). The most defensible model for the Piedmont includes saprolite production, soil erosion, and tilt to explain the regolith, relief, slope and altitude of the Piedmont "peneplain". Perhaps the most important general concept is a reiteration of the dynamic equilibrium concept presented by Hack (1965) in which the "equilibrium surface" has a variable age reflecting the balance of uplift, regolith production and denudation. Far from being the inactive "mature" landscape envisioned by Davis, the Piedmont appears to be an integral part of the geologically active and evolving Appalachian orogen.

Summary

The assumption of a "peneplain" is an incorrect starting point for investigation of a Piedmont or other surface of low relief. Until we date enough points on an erosional surface, we cannot assume a synchronous planation surface.

Future investigation should be directed at testing the idea of age-trends across "tilted" surfaces to determine the dynamics of uplift and denudation, and the range of rates that can obtain on what appears to be a geomorphic planation surface. Development of a "point-dating" method, such as with ^{10}Be inventories and detailed soil characterization, provides a method of determining the relative "activity" of different parts of an otherwise indistinguishable surface.

Acknowledgements

I extend thanks to H.W. Markewich for her constant willingness to discuss and critique my speculations and hypotheses.

References

Ahnert, F., 1970. Functional relationships between denudation, relief, and uplift in large mid-latitude drainage basins. Am. J. Sci., 268: 243–263.

Atekwana, E., Sutter, J. and Schwartzman, D., 1989. Thermochronology of the Pennsylvanian Liberty Hill, Pageland and Lilesville Plutons and the Wadesboro basin, North and South Carolina. Geol. Soc. Am., Abstr. Programs, 21(3), p. 2.

Brown, L., Pavich, M.J., Hickman, R.E., Klein, J. and Middleton, R., 1988. Erosion of the eastern United States observed with ^{10}Be. Earth Surf. Process. Landforms, 13: 441–457.

Budel, J., 1977. Climatic Geomorphology. Princeton University Press, Princeton, N.J., 443 pp.

Burt, T.P., 1986. Runoff processes and solutional denudation rates on humid temperate hillslopes. In: S.T. Trudgill (Editor), Solute Processes. Wiley, pp. 193–251.

Cleaves, E.T., Godfrey, A.E. and Bricker, O.P., 1970. Geochemical balance of a small watershed and its geomorphic implications. Geol. Soc. Am., Bull., 81: 3015–3032.

Costa, J.E., 1974. Response and recovery of a Piedmont watershed from tropical storm Agnes, June 1972. Water Resour. Res., 10(1): 106–112.

Costa, J.E. and Cleaves, E.T., 1984. The Piedmont landscape of Maryland: A new look at an old problem. Earth Surf. Process. Landforms, 9: 59–74.

Cronin, T.M., 1981. Rates and possible causes of neotectonic vertical crustal movements of the submerged southeastern U.S. Atlantic Coastal Plain. Geol. Soc. Am., Bull., 92: 812–833.

Dallmeyer, R.D., 1978. ^{40}Ar/^{39}Ar incremental release ages of hornblende and biotite across the Georgia Inner Piedmont; their bearing on late Paleozoic-early Mesozoic tectonothermal history. Am. J. Sci., 278: 124–149.

Davis, W.M., 1899. The peneplain. In: D.W. Johnson (Editor), Geographical Essays. Dover Publications, New York, N.Y., 1954, p. 777.

Duijsings, J.J.H.M., 1986. Streambank contribution to the sediment budget of a forest stream. Publ. Fys. Geogr. Bodemk. Lab., Univ. Amsterdam, No. 40, 185 pp.

Durrant, J.M., 1978. Structural and metamorphic history of the Eastern Virginia Piedmont province near Richmond, Va. Masters Thesis, The Ohio State University, 116 pp.

Flint, R.F., 1963. Altitude, lithology, and the Fall Zone in Connecticut. J. Geol., 71(6): 693-697.

Fowler, J.M. and Heady, E.O., 1981. Suspended sediment production potential on undisturbed forest land. J. Soil Water Conserv., 36: 47-53.

Froelich, A.J. and Heironimus, T.L., 1977. Thickness overburden map of Fairfax County, Va. U.S. Geol. Surv., Open File Rep., 77-797.

Gardner, T.W., 1989. Neotectonism along the Atlantic Passive Continental Margin - A review. In: T.W. Gardner and W.D. Sevon (Editors), Appalachian Geomorphology. Geomorphology, 2: 71-97.

Gilbert, G.K., 1909. The convexity of hilltops. J. Geol., 17: 344-350.

Glaser, J.D., 1969. Petrology and origin of Potomac and Magothy (Cretaceous) sediments, middle Atlantic Coastal Plain. Md. Geol. Surv., Rep. Invest., 11, p. 102.

Groot, J.J., 1955. Sedimentary petrology of the Cretaceous sediments of northern Delaware in relation to paleogeographic problems. Del. Geol. Surv., Bull., 5, p. 155.

Hack, J.T., 1960. Interpretation of erosional topography in humid temperate regions. Am. J. Sci., 258-A: 80-97.

Hack, J.T., 1965. Geomorphology of the Shenandoah Valley, Virginia and West Virginia, and origin of the residual ore deposits. U.S. Geol. Surv., Prof. Pap., 484, p. 84.

Hack, J.T., 1976. Geomorphic framework of the Piedmont and Blue Ridge. Beol. Soc. Am., Abstr. Programs, 8(6): 897-898.

Hack, J.T., 1980. Rock control and tectonism - their importance in shaping the Appalachian Highlands. U.S. Geol. Surv., Prof. Pap., 1126-B, p. 17.

Imeson, A.C. and Vis, M., 1984. The output of sediments and solutes from forested and cultivated clayey drainage basins in Luxembourg. Earth Surf. Process. Landforms, 9: 585-594.

Johnson, D.W., 1931. Stream Sculpture on the Atlantic Slope. Columbia University Press, New York, N.Y., 142 pp.

Markewich, H.W., Pavich, M.J., Mausbach, M.J., Hall, R.L., Johnson, R.G. and Hearn, P.P., 1987. Age relations between soils and geology in the coastal Plain of Maryland and Virginia, U.S. Geol. Surv., Bulletin 1589A, p. 34.

Mathews, W.H., 1975. Cenozoic erosion and erosion surfaces of eastern North America. Am. J. Sci., 275(7): 818-824.

Meade, R.H., 1982. Sources, sinks and storage of river sediments in the Atlantic drainage of the United States. J. Geol., 90: 235-252.

Mitchell, P., 1988. The influences of vegetation, animals, and microorganisms on soil processes. In: H.A. Viles (Editor), Biogeomorphology. Blackwell, Oxford, pp. 43-82.

Mixon, R.B. and Newell, W.L., 1977. Stafford fault system: structures documenting Cretaceous and Tertiary deformation along the Fall Line in northeastern Virginia. Geology, 5: 437-440.

Owens, J.P., 1989. Geology of the Cape Fear arch region, Florence 2° sheet and northern half of the Georgetown 2° sheet, North and South Carolina. U.S. Geol. Surv., Map I-1948-A.

Pavich, M.J., 1974. A study of saprolite buried beneath the Atlantic Coastal Plain in South Carolina. Ph.D. Diss., Johns Hopkins University, 132 pp.

Pavich, M.J., 1985. Appalachian Piedmont morphogenesis: weathering, erosion, and Cenozoic uplift. In: M. Morisawa and J.T. Hack (Editors), Tectonic Geomorphology, Proc. 15th Annu. Geomorphol. Symp. Ser., State University of New York, Binghamton, pp. 299-319.

Pavich, M.J., 1986. Processes and rates of saprolite production and erosion on a foliated granite rock of the Virginia Piedmont. In: S.M. Colman and D.P. Dethier (Editors), Rates of Chemical Weathering of Rocks and Minerals. Academic Press, Orlando, Fla., pp. 551-590.

Pavich, M.J. and Obermeier, S.F., 1985. Saprolite formation beneath Coastal Plain sediments near Washington, D.C. Geol. Soc. Am., Bull., 96: 886-900.

Pavich, M.J., Brown, L., Valette-Silver, J.N., Klein, J. and Middleton, R., 1985. ^{10}Be analysis of a Quaternary weathering profile in the Virginia Piedmont. Geology, 13: 39-41.

Pavich, M.J., Leo, G.W., Obermeier, S.F. and Estabrook, J.R., 1989a. Investigations of the characteristics, origin, and residence time of the upland residual mantle of the Piedmont of Fairfax County, VA. U.S. Geol. Surv., Prof. Pap., 1352, p. 114.

Pavich, M.J., Jacobson, R.B. and Newell, W.L., 1989b. Geomorphology, neotectonics, and process studies in the Rappahannock River Basin, Virginia. Int. Geol. Congr. Field Trip Guideb. T218, Am. Geophys. Union, Washington, D.C., p. 22.

Poag, C.W. and Sevon, W.D., 1989. A record of Appalachian denudation in postrift sediments of the U.S. Middle Atlantic continental margin. In: T.W. Gardner and W.D. Sevon (Editors), Appalachian Geomorphology. Geomorphology, 2: 119-157 (this volume).

Prowell, D.C., 1988. Cretaceous and Cenozoic tectonism on the Atlantic coastal margin. In: R.E. Sheridan and J.A. Grow (Editors), The Geology of North America, Vol. I-2, The Atlantic Continental Margin, U.S. Geol. Soc. Am., pp. 557-564.

Rebertus, R.A. and Buol, S.W., 1985a. Iron distribution in a developmental sequence of soils from mica gneiss and schist. Soil Sci. Soc. Am. J., 49: 713-720.

Rebertus, R.A. and Buol, S.W., 1985b. Intermittency of illuviation in dystrochrepts and hapludults from the Piedmont and Blue Ridge Provinces of North Carolina. Geoderma, 36: 277–291.

Rich, C.I., 1958. Muscovite weathering in a soil developed in the Virginia Piedmont. Clays Clay Miner., 5: 203–212.

Sevon, W.D., 1989. Erosion in the Junita River Drainage Basin, Pennsylvania. In: T.W. Gardner and W.D. Sevon (Editors), Appalachian Geomorphology. Geomorphology, 2: 303–318 (this volume).

Sigleo, W. and Reinhardt, J., 1988. Paleosols from some Cretaceous environments in the southeastern United States, Geol. Soc. Am., Spec. Pap., 216: 123–142.

Soller, D.R., 1988. Geology and tectonic history of the lower Cape Fear River valley, southeastern North Carolina. U.S. Geol. Surv., Prof. Pap., 1466-A.

Trimble, S.W., 1974. Man-induced soil erosion in the southern Piedmont. Soil Conserv. Soc. Am., Ankeny, Iowa, 180 pp.

Wellman, P., 1988. Tectonic and denudational uplift of Australian and Antarctic highlands. Z. Geomophol. N.F., 32(1): 17–29.

White, W.A., 1950. Blue Ridge Front - a fault scarp. Geol. Soc. Am., Bull., 61(12): 1309–1346.

Whitney, J.A., Jones, L.M. and Walker, R.L., 1976. Age and origin of the Stone Mountain Granite, Lithonia district, Georgia. Geol. Soc. Am., Bull., 87: 1067–1077.

Zimmerman, R.A., 1979. Apatite fission track age evidence of post-Triassic uplift in the central and southern Appalachians. Geol. Soc. Am., Abstr. Programs, 11(4), p. 219.

Geomorphology, 2 (1989) 197-207
Elsevier Science Publishers B.V., Amsterdam — Printed in The Netherlands

Computing Appalachian Geomorphology

RICHARD G. CRAIG

Department of Geology, Kent State University, Kent, OH 44242 (U.S.A.)

(Received March 16, 1989; accepted after revision June 12, 1989)

Abstract

Craig, R.G., 1989. Computing Appalachian geomorphology. In: T.W. Gardner and W.D. Sevon (Editors), Appalachian Geomorphology, Geomorphology, 2: 197-207.

A computer model is used to represent the stochastic effects of geomorphic processes acting on Appalachian structures, and to produce the various stages of Cenozoic Appalachian geomorphic history. That model incorporates the uncertainty in our current representation of elevation, lithology, dependence of slope on lithology, structural dip, percent of surface lithology uneroded, and stratigraphic thickness. Four theories of Appalachian history - defined by initial surface relief and adjustment to underlying lithology - are analyzed. Initial surface relief like that proposed by J.T. Hack yields the best fit to the current surface (mean elevation 410 m) when erosion proceeds from a surface initialized to the modern mean elevation of the ridgetops (760 m).

Introduction

No geomorphic problem is simplex than understanding the history of the Appalachian Mountains. Only a complete knowledge of their structure and the geomorphic processes operating at each stage of their development is required. In time, this ambitious goal may be achieved. Meanwhile, we are interested in representing our present knowledge about the problem as succinctly and usefully as possible. I describe here a computer model of landscape evolution which incorporates some of our current understanding of structure, process and stage. This model represents a mathematical kernel from which an evolutionary history of the Appalachians may be generated.

As described elsewhere in this volume, "structure" incorporates details of: topography, lithology, stratigraphy, discontinuities (faults, joints, lineaments), fold geometry, tectonics, soils, vegetation, climate, and groundwater. Each can be specified at some level of detail for the Appalachians. In modern parlance these could be termed boundary conditions.

The interaction of these various "structures" at a given point determine the processes at work at that point. No comprehensive map of the processes at work over a significant portion of the Appalachians currently exists. These processes must be inferred from examples meticulously evaluated at a few locations. The rules for such inferences are only vaguely understood. We can be sure that they require great expertise in their application and a systematic use of those rules throughout the Appalachians is yet to be achieved. Geomorphic processes can be thought of as the mathematical functions whose solution for a given set of boundary and initial conditions describes the evolution of the system.

The model used here represents an attempt to calculate the long-term (i.e. approximately

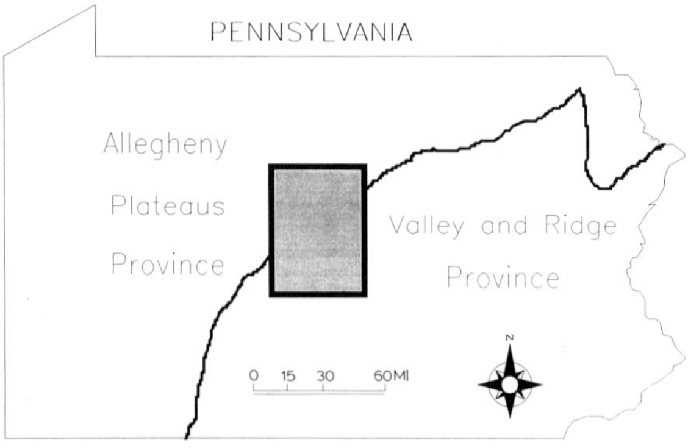

Fig. 1. Area chosen for analysis and physiographic provinces included.

Neogene) development of the Appalachians. It employes computer code to perform the calculations required, to consistently apply rules and, to present the results in a manner amenable to examination. The representation of process, structure, and stage is undoubtedly crude compared to what could be desired. At best, it may represent a "first attempt" at such computations. By computer reconstruction of hypothetical paleosurfaces and computation of the evolution of the resulting landscape until erosion reaches the modern level, several theories of landform evolution (such as Davis, 1889) will be compared. This comparison will be based upon their ability to fit the modern surface after evolving from a predefined paleosurface.

The area chosen for this study (36 quadrangles, covering over 5000 km^2) includes parts of the Valley and Ridge province and the Allegheny Plateaus province of central Pennsylvania (Fig. 1). It is a classical locale containing several sites specifically troublesome to Davis (1889; Morisawa, this volume) during the formulation of his monumental synthesis of the history of Appalachian landforms (e.g. Lewistown Narrows and the water gaps at Water Street).

How the model works

The model applied is an extension of the ideas presented by Craig (1982), where the slope to any point 1 km away depends upon the lithologies at the two points. The modern slope is a random sample from a distribution of slopes typical of that lithology pair. Any future or past slope will also be a random sample from the same distribution. Thus, our best estimate of the evolution of the landform can be achieved by random draws from that distribution.

In a normal erosional environment, these draws will be constrained. Evolution of the landform to form new slopes will not result in the addition of material (Fig. 2a). Neither will slopes form in which a point is lower than all of its neighbors (Fig. 2b). Within those constraints, slope changes and landform evolution

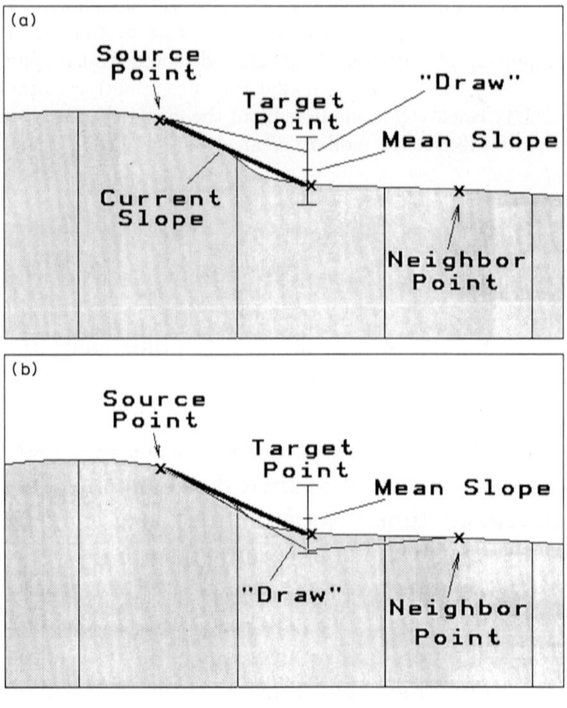

Fig. 2. Slopes are modeled as constrained "draws" from a frequency distribution of slopes. The mean slope for this lithology pair is labeled and the upper and lower 2σ ranges are shown by the end ticks on the vertical bar. Vertical lines show cell boundaries. Constraints eliminate certain possible slopes: (a) new slopes cannot result in addition of mass. This draw is within the expected range but lies above the current slope; (b) new slopes cannot result in elevations lower than all neighbors (closed depressions). This draw is within the expected range but lies below the elevation of its neighbor.

can be modeled. Thus, a first-order statement of landform evolution can be achieved knowing only the topography and the bedrock. Both of these are adequately mapped in the central Appalachians to allow this representation (Berg et al., 1980). Additional complexities are introduced in a dynamical model since erosion exposes new bedrock depending upon the geologic structure.

The relation of slope to underlying lithology is known to be strong in the central Appalachians (Craig, 1982). The analysis of variance reported by Craig suggests that 67% of the variance of slopes between elevations sampled at 1 km spacing is explained by knowing the lithologies at the ends of the slopes. Craig (1982) describes the rationale for using a 1 km spacing and the implications of this method for modeling changes in slopes. Craig (1980) gives a computer model that uses this information to describe landform evolution during "normal" erosion in the Appalachians.

To implement the model, erosional events called "cycles" are represented as follows. A point (x,y) on the 1 km^2 grid is selected at random. One of its four nearest neighbors (x', y') is selected at random as the hypothetical source point for the initial disturbance. The lithologies at those two points define a lithology pair $[L(x',y'), L(x,y)]$ abbreviated as $L'L$. A new slope is drawn at random from the distribution of slopes (in the modern landscape) having mean and variance typical of $L'L$. If the target point initially chosen actually changes elevation (the constraints discussed above allow the possibility of no change), then it is assumed that the neighbors of that point are now destabilized. Each of the three neighbors (the fourth was the hypothetical source) are considered in turn and new slopes drawn. This expansion of the erosion process continues as long as points change.

This cycle can be thought of as the diffusion of an initial disturbance and corresponds to a diffusion process described by Scheidegger (1961). Eventually no more points are added to the list of changed points and there are no more points to be considered. This completes one cycle of erosion. A summary of the changes is recorded (and is available for numerical or graphical analysis) and a new point is chosen at random to initiate another erosion cycle. These cycles continue until the stopping criterion (discussed below) is satisfied.

The constrained nature of the "draws" does not normally limit the variety of slopes that can be realized during a sequence of erosional events. This is because the constraints depend upon the elevation of adjacent points and these are constantly changing as they are eroded. It is only when the mean elevation of the landform approaches some base level, such as sea level, that the constraints become limiting. In that case slopes will more and more closely approximate the mean slope for that lithology (the terrain will become more predictable) and finally get lower and lower in unison. This has an interesting analogy to the predictions of the Davisian model (i.e. peneplanation). In this way, the predictions of the Davis and of the Hack models of landform development are reconciled.

Not all possible lithology pairs actually are observed in the modern landform. It is unlikely to observe a slope between a unit low in the stratigraphic section (e.g. the Gatesburg; see Table 1) and one high in the stratigraphic sec-

TABLE 1

Simplified stratigraphic sequence used in this analysis. For details of construction, see Craig (1979)

	Map unit		Map Unit
1	Conemaugh (youngest)	12	Marcellus-Keyser
2	Allegheny	13	Rose Hill
3	Pottsville	14	Tuscarora
4	Mauch Chunk and Loyalhanna	15	Juniata
5	Burgoon	16	Bald Eagle
6	"Tioga"	17	Reedsville
7	Catskill	18	Coburn-Loysville
8	Brallier	19	Axemann-Bellefonte
9	Lock Haven	20	Stonehenge-Nittany
10	Mahantango	21	Gatesburg
11	Onondaga	22	Warrior (oldest)

tion (e.g. the Pottsville). Indeed, most slopes occur between map units that are adjacent in the stratigraphic section.

For the purposes of this model the slope/lithology relation for lithology pairs not occurring today is based upon the slopes between lithologies having the most similar elevations today. For example, we wish to estimate the slope between lithologies A and B yet lithology A does not occur adjacent to B today. We use the observations of slopes between lithologies C and B if, of the lithologies that do occur adjacent to unit B, the mean elevation of lithology C is most similar to that of lithology A. Variances are taken to be the variances of slope of the substitute map unit pair. Most of this information is not required in the simulations since map units rarely are physically adjacent unless they are stratigraphically contiguous. Only in unusual circumstances, such as at a major fault, is the stratigraphic sequence disturbed in this area.

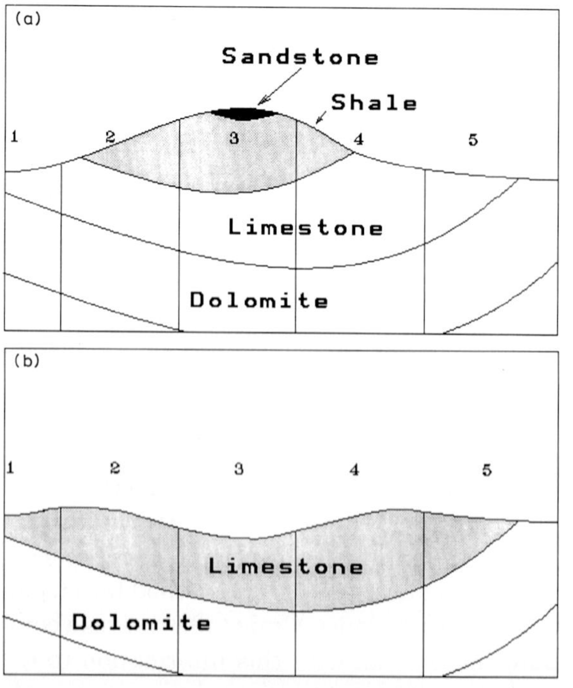

Fig. 3. Changes in the landform system as rock units are penetrated. (a) Thin caprock of sandstone (black) underlain by a weak shale (shaded) in cell 3; (b) penetration of caprock results in a high probability of removal of the underlying shale, exposing the subjacent limestone (shaded).

Boundary conditions

The stochastic process equations used in the model (Craig, 1982) require three important boundary conditions: topography, lithology, and structure. The first is defined by a digital elevation model (DEM) which represents the Appalachians in central Pennsylvania at a spacing of 1 km. Topography at times in the past can be reconstructed from the structural information as will be described below.

As erosion progresses, in any but the most homogeneous of areas it will pierce the original surface lithology. If the surface unit is a resistant sandstone and the subjacent unit a weak shale, the landform system will reorganize (Fig. 3). Fig. 3a shows the original landform. Only a thin caprock remains at point 3. At the next stage of erosion, the caprock is removed, exposing the shale beneath (Fig. 3b). This configuration is highly unstable; there is a low probability of such a slope. A major modification of the landform has occurred in response to a normal process.

To represent the potential for such changes, additional information is required. We must specify the actual thickness of the surface unit. Obviously a location where a unit is thin is likely to be penetrated quicker than a site where the same unit attains its full stratigraphic thickness. This full thickness may be considerably greater than the unit's stratigraphic thickness, depending upon the structural dip. Thus, to continue with our modeling exercise in an area both stratigraphically and structurally complex, as in the Appalachians, we must have detailed knowledge of: (1) stratigraphic thicknesses of each unit; (2) dip of strata at each point (at 1 km spacing); and (3) fraction of thickness already eroded from the surface units. In combination with the previously described information needs (elevation and lithology), over 25,000 data are needed for this relatively small area. Next we will see how to satisfy these massive data requirements.

Actual stratigraphic thickness varies slowly and systematically throughout the study area, so values for each map unit (Table 1) are assumed to be constant in each 7.5′ quadrangle. A detailed discussion of sources of such data and an analysis of variance of mean thickness is available in Craig (1979).

Rocks presently exposed at the surface have been eroded below the stratigraphic tops by some indeterminate amount. To represent the importance of this erosion, the percentage of this erosion is specified at each grid point of the modern surface. Since the amount is, in general, indeterminate the percentage was specified by drawing a random number in the range (0–100%) from a uniform distribution. Each time the model is run, these numbers will be different and slightly different numerical results will be obtained. This reflects some of the inherent uncertainty in our understanding.

Dips vary from 0° to moderately overturned throughout the area, so it is necessary to specify dip at each point of the grid. In most cases, dip was estimated from mapped fold axes (Berg et al., 1980). Uncertainty in dip is slight in the Plateau but larger in the Ridge and Valley Province. For this model, dips are chosen at random from a uniform distribution in the range 0° to 3° in the Plateau (Glass, 1974). In the southeast, dips are chosen from a normal distribution having a mean which is a function of the structural setting (proximity to fold axes) and standard deviation of 5°. Further details are given in Craig (1979). These values are fixed for a given model run. Other structural information, including the strike and outcrop pattern of these map units, is represented by a matrix of lithologies for the same points as in the DEM.

Special cases

As if these weren't complexities enough, other problems confront us. For example, certain points in the study area are underlain by the unit considered to be the lowest in the stratigraphic section in this area (the Cambrian Warrior Fm.). Erosion could penetrate these units. What will be exposed below? In this area, repetition of section is common (Gwinn, 1970) although the structure *above* the current landform is less well understood. For this model, the simplifying assumption is made that a fault occurs at the base of the lowermost unit. In the absence of additional information for the entire area (drill hole data, seismic records, etc.) the subjacent unit is modeled as a random selection from the stratigraphic section. This is not to imply that faults are really confined to such situations or that this method will correctly model even this situation. It is an expedient in the absence of a complete structural and seismic analysis. An example of application of this method is illustrated in Fig. 4.

Unless an entire continent is included in the model, there will always be a boundary. Since the model assumes there is a source neighbor for each point examined, we must specify how edge points are to be handled. It is possible that the source point is outside of the grid. This is especially likely if initiation of erosion cycles is

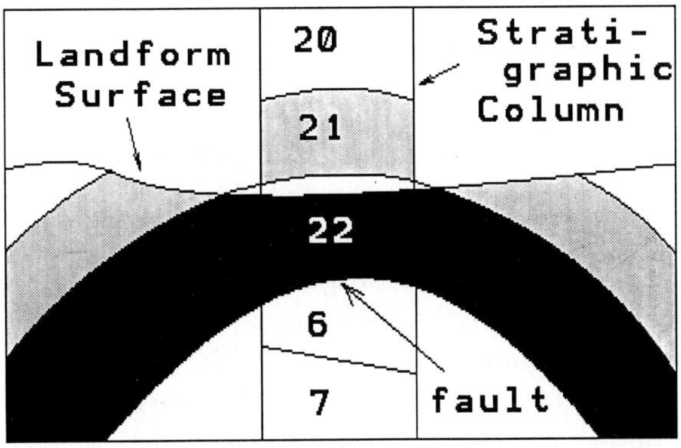

Fig. 4. Inferred fault and underlying stratigraphic section when the original stratigraphic section is exhausted. The stratigraphic section in one cell is illustrated. The lowest unit in the section is numbered 22 (Warrior Fm.). If erosion extends below that elevation, a fault is assumed and a new surface unit (unit 6, the "Tioga") and dip are chosen at random.

confined to the margin, an option that is available.

In the case of erosion at edges, the program assumes that the point outside of the grid has a lithology drawn at random from the adjacent stratigraphic section. The probability of each of the three choices (higher, lower, or the same unit) is defined by the observed transition probabilities in the lithology grid. The elevation at the point in question is defined as a random draw from the observed distribution of elevations for that lithology.

Stopping rule

The computer model (Craig, 1980) provides four stopping rules for simulations. In the present case, the intent is to start at a higher elevation and end the simulation when the landform is most like that of today. The stopping rule used terminates a simulation when the mean elevation is not significantly different than a specified ending value. Significance is judged with a one sample *t*-test. This method has the advantage of forcing the sums of deviations from the mean to be zero when the means of the two surfaces are equal. Use is made of this latter point in testing the various theories of landform evolution.

Data requirements

Five major matrices of information are specified at each gridpoint of the landscape: elevation, lithology, dip, percent eroded from the surface layer and, stratigraphic thickness. Reconstruction of the pointwise elevations of certain superjacent paleosurfaces (such as that inferred by Hack, 1960) – which form initial conditions for the model – requires the mean and standard deviation of elevation for each lithology. These data are also useful in estimating the expected slopes in cases where observational data are unavailable. lastly, the distributional parameters of the slope/lithology data are needed.

We are not yet in possession of geologic information systems that can provide such data on a systematic basis. There must be considerable uncertainty in their values and ignoring the impact of that uncertainty may produce deceptive results. The program represents the uncertainty in five boundary conditions by randomized selection of values prior to each simulation. The variables and the assumed distributions and (1σ) uncertainty are (Craig, 1979): elevation (normal, ± 3 m); surface lithology (normal, ± 1 unit); dips (uniform, $0°$–$3°$; normal, $\pm 5°$); thickness of each lithology (normal, σ computed for each unit from data values of the 36 quadrangles used); and initial thickness of surface unit (uniform, 0–100%).

Reconstructing previous surfaces

The computer program described above is used to test several theories of Appalachian landform development that have been propounded (see e.g., Melhorn and Flemal, 1975; Morisawa, this volume). This test is performed by using the model to compute changes over the vertical increment of erosion considered by the relevant theory. For example, Davis (1889) inferred that the Appalachian landform once appeared as a peneplain (the Schooley peneplain) at the level of the present ridgetops (about 760 m). If we initialize the surface at that level and give it the form of a peneplain, will erosion as represented in the model produce the current landform? To answer that question first requires reconstruction of a peneplain with a mean elevation of 760 m. Such reconstructions are available as the first element in the computer model.

The problem of reconstruction is more complicated than simply assuming a flat surface at the 760 m level. Even Davis' Schooley peneplain had relief; the resistant units formed monadnocks (Davis, 1889). Locations of monadnocks depend upon outcrop patterns on this surface, so the stratigraphic section must be restored at each point. This restoration depends

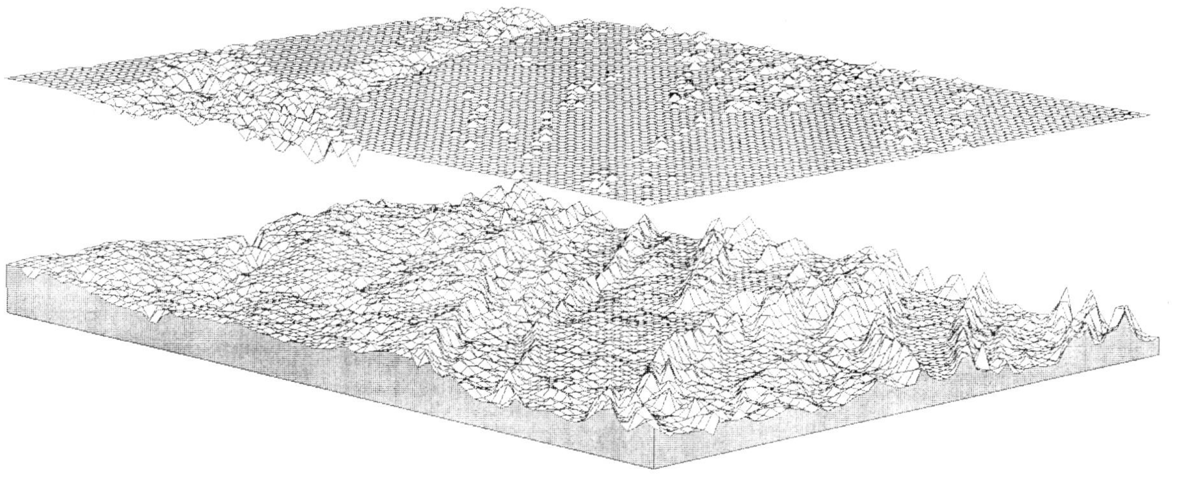

Fig. 5. Reconstruction of topography of a paleosurface (upper surface) according to the theory of Davis (1889). Monadnocks occur where the lithology is resistant (typically sandstones). Depressions occur where the stratigraphic column did not allow reconstruction to the level of the peneplain. Lower surface represents the modern landform.

upon the structure (in this case dip is used) and varying amount of uneroded material at the modern surface. The degree of relief at each point on the peneplain will be slight but depends upon the particular lithology present.

It is possible that the entire stratigraphic section is exhausted at some points before the restoration reaches the desired surface elevation (760 m). Fig. 5 illustrates the results of one such reconstruction (for Davis' theory). This shows the current surface (lower level) and an upper surface – at the level of the current ridgetops – representing a peneplain with monadnocks. Monadnocks occur where the lithology is resistant (sandstones). Depressions occur where the stratigraphic column did not allow reconstruction to the level of the peneplain. This type of reconstruction is the first step in computing a possible Appalachian geomorphic history. Four types of such reconstructions were made, one type for each of four theories studied. Each of these reconstructions is discussed next.

Initial conditions and landform theories

The numerous published theories of Appalachian landform development are grouped into three classes for the purposes of reconstruction of initial conditions. This classification is based upon: (1) the inferred relief of the land surface (low or "normal", i.e. as the modern relief) when that surface was at the level of the current ridgetops (about 760 m), and (2) the degree of adjustment of the terrain to underlying lithologies when the land surface was at that level (adjusted or not adjusted). The published theories (see Morisawa, this volume) are grouped as follows: (D) Davis (1889) and Barrell (1920); (J) Johnson (1931); and (H) Hack (1960), Thompson (1939; 1949), Meyerhoff and Olmsted (1936). This division (Table 2) also creates a fourth initial condition "Theory IV", in which the relief was "normal" but the streams were not adjusted (O). This theory

TABLE 2

Initial conditions for landform simulations

Relief	Adjustment	
	Adjusted	Not adjusted
Low	Davis, Barrell	Johnson
"normal"	Hack, Thompson, Meyerhoff and Olmsted	Theory IV

cannot be assigned to any previous author. It is retained in the interest of completeness.

Formulation of the four topographic surfaces corresponding to these initial conditions for the model is achieved as follows:

(1) Hack (1960), Thompson (1939), Meyerhoff and Olmsted (1936) – the relief on the initial surface is drawn from the distribution of elevations for the underlying lithologies. This is a departure in that the slope/lithology relation is not used, rather it is the elevation/lithology relation.

(2) Davis (1889) and Barrell (1920) – all terrain is flat except the most resistant lithologies stand up with elevations drawn from the elevation–lithology distribution with variance represented by the standard error of the mean (Fig. 5).

(3) Johnson (1931) – the terrain is a nearly totally flat surface of marine planation. It is assumed that the unconsolidated Cretaceous sedimentary cover has been stripped. The only departures from a flat surface occur where the available stratigraphic section does not allow reconstruction up to the desired level. Thus, the relief is low and not adjusted to lithology.

(4) Theory IV – Elevations are drawn at random from a distribution with mean equal to the surface level desired and standard deviation equal to the standard deviation of elevation today. No correction for underlying lithology is made. Thus, the landscape has "normal" relief and is not adjusted to lithology.

The four theories (Table 2) above have been examined by computing the landscape evolution as erosion progresses from a mean elevation ot 760 m to a mean elevation of 410 m (the present value). To determine whether the simulation results in a landscape that is "like" the modern landscape, it is reasonable to test the null hypothesis that the simulated landscape is a random sample drawn from a population of landscapes having the same mean and standard deviation of elevation as that observed today. A Chi-squared statistic is computed on a point-by-point basis. This allows an inference about which of the initial conditions is more likely to have preceded the current landform.

Results of landform simulations

Typical changes in the standard deviation of elevation at each cycle of the simulation are illustrated for one example for each theory in Fig. 6. This figure provides evidence that the simulations are working correctly. For both the Hack and Theory IV models, variance decreases at first and then remains stable. Stable behavior would be expected if the landform begins in an adjusted state and continues to evolve in a stable (stationary) process. The rapid adjustment in the first cycles can be attributed to the approximate nature of the initial state as described above. On the other hand, both the Davis and the Johnson model types begin with a low value of variance as would be expected if the landform is a peneplain or surface of marine planation. From that initial state, the variance of both model types continues to increase until it reaches values similar to those of the Hack and Theory IV models.

On average (for four runs of each case), the final standard deviation of elevation in the Hack (101 m) and Theory IV (109 m) runs is less than the current landform (125 m). Those of the Davis (141 m) and Johnson (155 m) runs end up larger than that of the current landform. It is not clear what meaning should be assigned

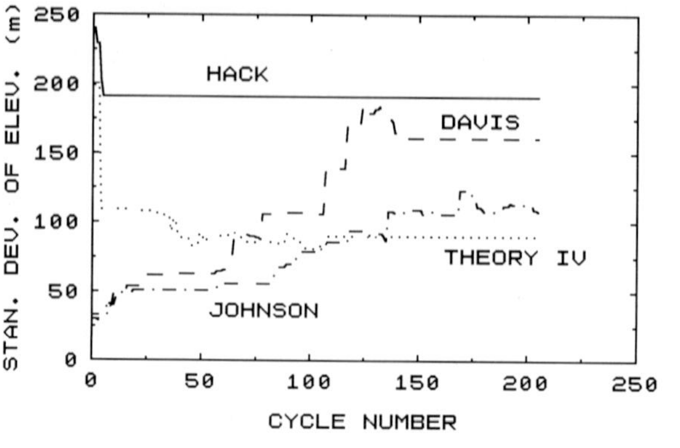

Fig. 6. Variability of elevation at each cycle of the landform erosion process.

to this; but, the differences in variance of the four models are statistically significant. It is interesting to note that it is the models that begin with "normal" relief (i.e. relief as at present) that end with the lowest relief (i.e. σ); those that begin with low relief end with the highest relief. Perhaps there is "overshooting" of the mark which would require more erosion before it returns to the expected value.

Fig. 7 shows example patterns (one run for each theory) of adjustment of the model landscape to the current landscape. Significantly different rates of adjustment are observed. The Hack and Theory IV models tend to begin in a nearly adjusted state and continue that way. Again, the Davis and Johnson models are far from adjustment in the initial cycles as would be expected on a surface of planation. They slowly assume a configuration more like the current landform as the surface continues to erode. On average (four runs for each), the Hack model takes only 37 cycles to achieve fit to the modern surface while Theory IV takes 124, Johnson 144 and Davis the most, 160 cycles. The final fit of the Hack model is much closer to the current landform. The average final z-value for each model is: Hack, 30; Theory IV, 50; Davis 76; and Johnson, 89.

These values represent the average goodness-of-fit of the computed surface. They are obtained by simulating adjustments of slope to underlying lithology as erosion progresses and

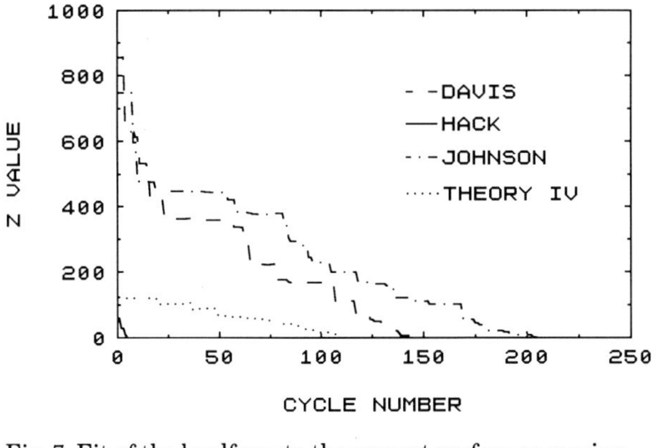

Fig. 7. Fit of the landform to the current surface as erosion progresses.

the landscape is lowered, exposing new lithologies. Adjustment is made for the thickness and dip of the units and for the uncertainty in these factors and the initial amount of erosion of the surface unit. Initial conditions (at a mean elevation of 760 m) are defined solely by the relief of the surface and the adjustment of that surface to underlying lithologies. Erosion is continued until the mean surface elevation matches that of the modern surface (410 m). A low value of goodness-of-fit implies that the computed surface matches the actual modern surface well. Some model runs produced matches that are nearly perfect ($z \approx 1$).

The results indicate that, on average, model runs that are initialized with the conditions referred to as the Hack model (i.e. relief as in the modern landscape and slopes adjusted to underlying lithology) tend to produce final landscapes that more closely match the modern landform than any of the other models tested. The differences between results for the four models are significant (at the $\alpha = 0.05$ level) and imply that three pairs of models can be distinguished (D–H, J–H, J–O) from one another; although it is less clear that three other pairs of models differ significantly (D–J, D–O, H–O). The hypothesis of equality can be rejected in favor of the alternative that the fit of the Hack model is significantly better than that of the other models. There is a clear degradation of fit when considering the next best model (Theory IV). This suggests that the Hack model provides a better description of the landscape when the mean elevation was at the level of the modern ridgetops (760 m).

In Figs. 6 and 7, the magnitude of relief of the initial surface seems to be most important in determining the path of adjustment as the surface erodes. It appears unlikely that the surface was flat when the mean elevation was 760 m. Apparently, it is even less likely for surfaces whose mean elevation is closer to that of the modern landform (410 m). These surfaces of low relief carry the memory of their initial state for a longer time than is represented by 350 m of erosion. If there ever was such a surface of low relief in the central Appalachians of Pennsylvania, more than 350 m of erosion separated it from the modern mean elevation. This does not necessarily imply that such a surface is older than Cretaceous. Mathews (1975) showed that sub-ridge-top erosion was probably completed in the Neogene. This could allow for the existence of a surface of planation of Cretaceous age with significant erosion occurring in the early Tertiary which had already produced a surface of modern relief and adjustment when the mean elevation was still 760 m.

The future of computational geomorphology

We have hardly begun to explore the opportunities that are opened by the ability to represent landscape evolution. Applications of this knowledge will surely include linkage to other models. For example, sedimentation models could use this to develop clastic sediment supply terms. Sediment supply is also a function of the basin size, shape and gradients, all of which can be provided dynamically by the model. Representing the amount of each rock type exposed at any given time allows more informed estimation of the geochemical supply term also. Surface form and lithology is one determinant of vegetation patterns, a key item in the geomorphic system. These and numerous other applications of geomorphic knowledge are important enough that improvements in this kind of modeling will be a high priority for the 1990's.

A plethora of improvements can be recommended. Studies of long-term evolution of landforms require precise knowledge of geologic structure. The reconstructions illustrated here make little use of the structural information that is available. All folds are assumed to maintain constant dip in vertical section without consideration of the "room problem". No faults are explicitly recognized. Variations in stratigraphic thickness are only poorly represented. Information about strike is not employed. Even more important, we will have to recognize that structure is not static; it develops at a separate rate but concurrent with the landform. Reconstructions are not the sole responsibility of the geomorphologist; they require input from the structural geologist. Such a collaborative program would be imperative for reconstructions intended to provide geomorphic information to others from the output of the erosion simulations. It may require "expert systems" to achieve this.

Another key variable of the reconstruction is very poorly known at this time – the remaining thickness of the surface rocks. Few data are available in the Ridge and Valley Province because drilling has been limited. Resolution of that information, even at the 1 km detail of the current model is far beyond the realm of any reasonable drilling program. Geophysical methods may be the solution; but, resolution of stratigraphic contacts is difficult without lithologic contrasts. Even this method will be expensive and time-consuming to apply. Without it a great deal of uncertainty remains in the reconstructions.

Perhaps even more important to the model is climate. Simulation of long time spans, of the order of the Cenozoic, must recognize that climate has also varied (Barron, this volume). Such climatic change affects the development of the landform. Carbonates that are today relatively easily eroded are resistant in more arid climates. The slope–lithology relations must be parameterized for the entire range of possible climates to allow representation of the response of the landform to climate change.

Parameterization of slope–lithology relations means sampling at other locations, where the climate represents the variations of interest. But, that eliminates the direct use of map units and requires simplification, perhaps to the basic lithologic classifications: sandstone, shale, carbonate. Finding areas where these lithologies all can be observed, where tectonic activity does not obscure the lithologic control and where climatic data are available is not a trivial task (Hoyt, 1981). It will be a major endeavor to provide such parameterizations.

Other climatic complications are not to be ignored. Quaternary glaciation covered portions of the Valley and Ridge Province, even within the area of this study. The impact of those glaciers upon the landform is poorly understood; however, it is clear that changes in processes and rates did accompany that glaciation (Braun, this volume). Representation of the impact of such extreme changes will be a challenge for the next decade. Indirect effects of global glaciation are also important. Variations in sea level in response to ice volume changes are well-documented. They impact base level and may be an important influence of the landform. We must learn how to represent such factors.

These challenges signal a resurgence of geomorphology as a subject of importance in the next century. We see the emergence of global studies of the earth system. Until now, geomorphology has eluded inclusion in such systems models. It is the awesome complexity of the field that has frustrated those who would embrace it in earth system science. Several decades of exponential growth in the capabilities of computers now allows programs capable of capturing this complexity. It is up to a new generation of geomorphologists to develop the logical, statistical and mathematical acumen and the enormous data bases that will become *Computational Geomorphology*.

Acknowledgements

I wish to thank Tom Berg of the Pennsylvania Geologic Survey for allowing access to the 1:24000 compilation maps of the State Geologic Map of Pennsylvania. Samuel Figuli helped adapt the FORTRAN code to the UNIX operating system.

References

Barrell, J., 1920. The Piedmont terraces of the northern Appalachians. Am. J. Sci., 4th ser., 49: 236–241.

Barron, E.J., 1989. Climate variations and the Appalachians from the late Paleozoic to the Present: Results from model simulations. In: T.W. Gardner and W.D. Sevon (Editors), Appalachian Geomorphology. Geomorphology, 2: 99–118 (this volume).

Berg, T.M., Edmunds, W.E. et al., 1980. Geologic map of Pennsylvania. Pa. Geol. Surv., 4th Ser., Map 1.

Braun, D.D., 1989. Glacial and periglacial erosion of the Appalachians. In: T.W. Gardner and W.D. Sevon (Editors), Appalachian Geomorphology. Geomorphology, 2: 233–256 (this volume).

Craig, R.G., 1979. A Simulation model of landform erosion. Ph.D. Thesis, The Pennsylvania State University, Univ. Park, Pa., 536 pp. (unpubl.).

Craig, R.G., 1980. A computer program for the simulation of landform erosion. Comput. Geosci., 6: 111–142.

Craig, R.G., 1982. The ergodic principle in erosional models. In: C.E. Thorn (Editor), Space and Time in Geomorphology. Allen and Unwin, London, pp. 81–115.

Davis, W.M., 1889. The rivers and valleys of Pennsylvania. Natl. Geogr. Mag., 1: 183-253.

Glass, G.B., 1974. Geology and mineral resources of the Philipsburg $7\frac{1}{2}'$ quadrangle, Centre and Clearfield counties, Pennsylvania. Pa. Geol. Surv. 4th ser., Atlas, 95a: 241 pp.

Gwinn, V.E., 1970. Kinematic patterns and estimates of lateral shortening, Valley and Ridge and Great Valley Provinces, central Appalachians, south central Pennsylvania. In: G.W. Fisher, F.J. Pettijohn, J.C. Reed, Jr. and K.N. Weaver (Editors), Studies of Appalachian Geology: Central and Southern. Wiley, New York, N.Y., pp. 127–146.

Hack, J.J., 1960. Interpretation of erosional topography in humid regions. Am. J. Sci., Bradley Vol., 258-A: 80–97.

Hoyt, B.R., 1981. The analysis of climatic effects on slopes on sedimentary rocks. M.S. Thesis, Kent State Univ. Kent, Ohio, 94 pp.

Johnson, D.W., 1931. Stream Sculpture on the Atlantic Slope. Columbia Univ. Press, Palisades, N.Y., 142 pp.

Matthews, W.H., 1975. Cenozoic erosion and erosion surfaces of eastern North America. Am. J. Sci., 275: 818–824.

Melhorn, W.N. and Flemal, R.C. (Editors), 1975. Theories of Landform Development. Publ. Geomorphol., State University of New York, Binghamton, N.Y., 306 pp.

Meyerhoff, H.A. and Olmsted, E.W., 1936. The origins of Appalachian drainage. Am. J. Sci., 232: 21–41.

Morisawa, M., 1989. Rivers and valleys of Pennsylvania, revisited. In: T.W. Gardner and W.D. Sevon (Editors), Appalachian Geomorphology. Geomorphology, 2: 1–22 (this volume).

Scheidegger, A.E., 1961. Mathematical models of slope development. Geol. Sci. Am., Bull., 72: 37–49.

Thompson, H.D., 1939. Drainage evolution in the southern Appalachians. Geol. Soc. Am. Bull., 50: 1323–1356.

Thompson, H.D., 1949. Drainage evolution in the Appalachians of Pennsylvania. N.Y. Acad. Sci., Ann., 52: 31–62.

Geomorphology, 2 (1989) 209-232
Elsevier Science Publishers B.V., Amsterdam — Printed in The Netherlands

Central and Southern Appalachian Water and Wind Gap Origins: Review and New Data

G. MICHAEL CLARK

Department of Geological Sciences, The University of Tennessee, Knoxville, TN 37996-1410 (U.S.A.)

(Received May 1989; accepted after revision June 1989)

Abstract

Clark, G.M., 1989. Central and southern Appalachian water and wind gap origins: Review and new data. In: T.W. Gardner and W.D. Sevon (Editors), Appalachian Geomorphology. Geomorphology, 2: 209-232.

The origin and evolution of transverse drainage were stumbling blocks to classical, historical geomorphologists, and many problems still remain. Décollements that upramp laterally-transported, at-depth, tectonic sheets in the central and southern Appalachians have produced structural geomorphic fronts that are prominent features, often of regional or subregional extent (cf. Blue Ridge, North Mountain, Allegheny). As these structural-geomorphic fronts likely were features that developed significant cover relief during the Alleghanian orogeny, they may hold clues to incipient landscape development during and soon after orogenesis, unless buried by thick sediment cover. Where these fronts are crossed by transverse drainage lines, spectacular water gaps that frequently expose complex structural features are produced. There are many other transverse water gaps not located across décollement structures in the region. These gaps display complex features that shatter bedrock, and must also be accounted for in any comprehensive hypothesis of transverse water gap origin. Lineaments, including both cross-strike structural discontinuities and disturbed zones are present in the Appalachians, and coincide with a number of transverse water gaps. The positioning of weakened bedrock zones high on structures where downcutting streams had access to them permits a localizing mechanism such as *structural ensnarement* to operate. Local sedimentary patterns may also have played roles in drainage localization. Major unsolved problems include: post-Alleghanian tectonic effects, rates and timings of uplift, erosion, and deposition, whether lithotectonic weaknesses are most pronounced in gap areas, and establishing a numerical chronology for drainage development. Whereas lithotectonic influence upon drainage development is stressed in this chapter, the importance of other factors (e.g., climate, catastrophic events), as documented by others in this volume is fully recognized.

Introduction

The origin of topography and drainage in fold-belt mountain systems is a central theme of regional structural geomorphology (cf. Oberlander, 1985). Following papers by McGee (1888a,b) and Davis (1888, 1889), the Appalachians became a theater for debate on the origin of regional landscapes and drainage. Transfixed by the Davisian nails of structure, process (which was neglected by the Davisian camp), and stage, the *Appalachian problem* became codified into its several subsets. Bryan et al. (1932/33, p. 313) stated that the Davisian viewpoint:

"...was based primarily on three lines of evidence: first, the existence of accordant ridge tops in the upland area together with discordant rock structure; second, a drainage pattern, that can be explained only on the assumption that the accordant levels are remnants of a once continuous surface; third, the existence, beneath the fine-grained Cretaceous beds of a deeply-weathered and smooth erosion surface that appears to have been once continuous from the Coastal Plain to the surface preserved on the accordant ridges of the upland. The second line of evidence was given much attention: it was emphasized that the courses of the present master streams are notably transverse to structure, and that such courses could have originated only on a land surface of low relief."

Hack (1975) pointed out the lack of accordance in ridge crests that are underlain by several different rock units in the area north of Harrisburg, Pennsylvania (Fig. 1). In this observation he was presaged by Monmonier (1967) who separated highlands developed on the Tuscarora Formation from those on other resistant rock units, and who distinguished between local broad upland (LBU) and ridge top (RT) samples. LBU's are high, level, extensive summit areas; RT's are the more common, narrower, and undulatory ridge crests that were described by Hack (1975). Monmonier (1967) used trend surface analysis to demonstrate that the LBU sample developed on the Tuscarora Formation is composed of summit areas that are highly accordant (percentage of variation explained: linear 81; quadratic, 89; cubic, 94). Clark (1989) described evidence favoring a cryoplanation origin for central Appalachian summit levels that bevel rocks of differing resistance to erosion, and noted the very different morphology of many summit levels in the Southern Blue Ridge area, where certain residual elevations – including many erosional domes – abound.

Aspects of Appalachian rivers that puzzled early workers were: (1) the reversal of the original northwestward-flowing drainage (recorded in many sedimentary rock packages of Paleozoic age); (2) the great length, relative straightness, and parallelism of today's major southeastward-flowing rivers in the Appalachians (see Fig. 1 for rivers most often cited by workers); and (3) drainage transverse to structural-lithologic trends where alternative courses were assumed to have been available. Judson (1975) dispatched with the problem of drainage reversal by resort to modern analogs of uplift during continental rifting, and the resulting regional tilting-away of highlands from borders of active rifts. River length, and parallelism were treated earlier (Thompson 1939; Strahler, 1945) and will not be discussed further here, nor will the Davisian model be critiqued (see Flemal, 1971; Morisawa, this volume). Sevon et al. (1983) reviewed the literature on the Appalachian peneplain construct, and noted three geologic consequences that heve been attributed to peneplanation. These are: (1) origin of present drainage; (2) origin of certain mineral resources; and (3) cementation of selected siliciclastic rock units.

Classic regional geomorphic subdivisions were made on the foundation of topography and drainage, and with later modifications are used in this paper (Fig. 1). Most of the provincial boundary lines within the Appalachian Highlands are well established, and with the exception of the Blue Ridge, sectional subdivisions and their boundaries seem to have more diversity and attract more interest. Hack (1982) subdivided the Piedmont province into objectively-based areas, and Outerbridge (1987) presented a new subdivision for part of the Appalachian Plateaus (Fig. 2). Much variety is evident in informal usages within the Ridge and Valley province as it was defined by Thornbury (1965). Later refinements generally have greater geologic significance (cf. Pike and Thelin, 1988). Another example, the Tallapoosa–Rappahannock line (Fig. 1) of Hack (1982), forms the southeastern border of Appalachian areas that have high relief and is expressed in the landscape by a slight steepening of the regional slope. In search of a more fundamental basis for overall apportionment, some researchers have turned to similarities and differences in lithotectonic style and history (cf. Lowry et al., 1971 table 1). Such features would include: lithologic variations, structural trends (cf. in Virginia: southern Appalachian "Valley and Ridge" ca. N60–65°E; central Appalachian "Ridge and Valley" ca. N30–35°E), nature and tectonic levels of thrust faults and their associated structures, presence or absence of "Fensters" and "Schurflingfensters", characteristics of "Klippen" and slices, and presence or absence of post-Cambrian magmatic activity and

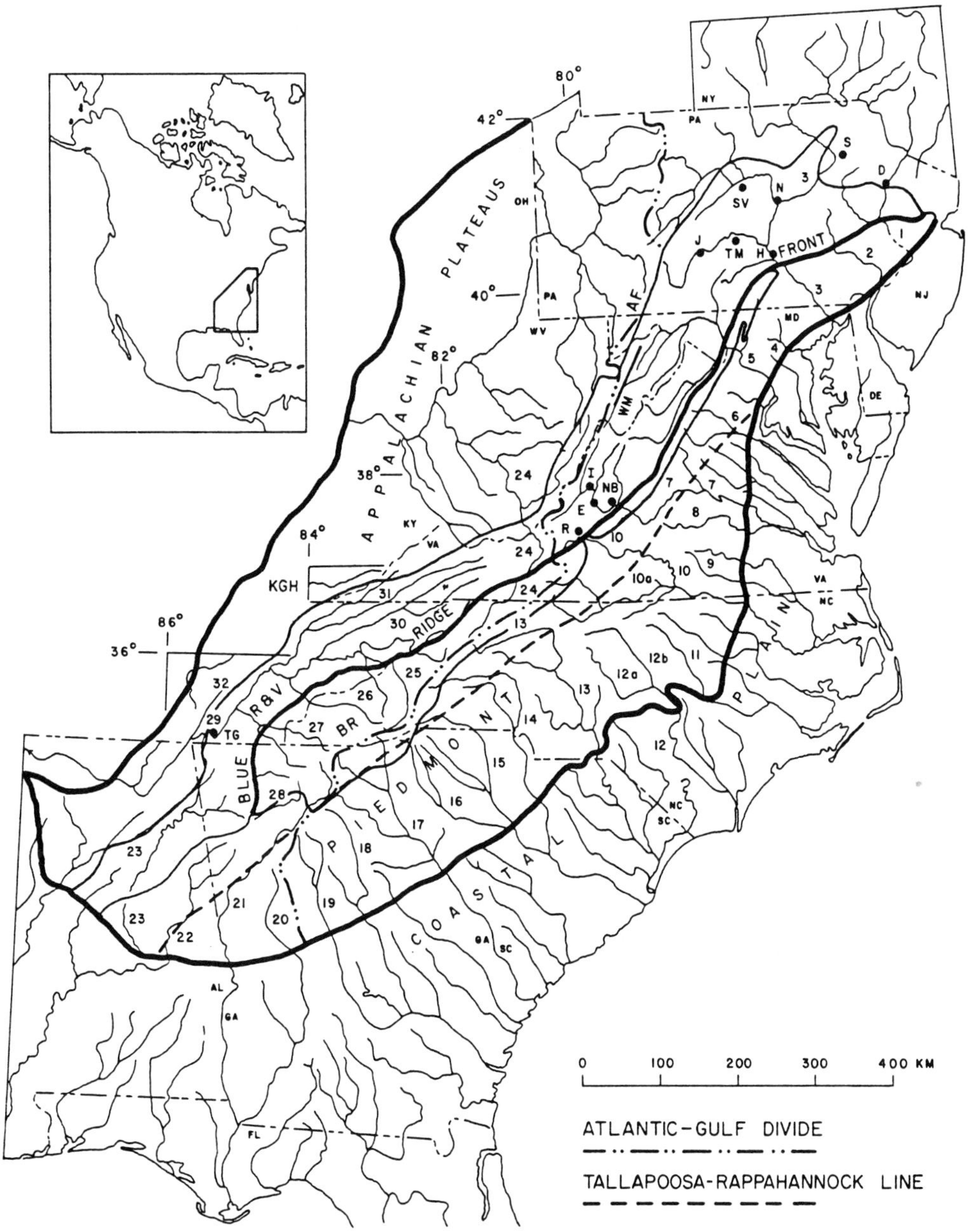

Fig. 1. Index map of study region. *AF* = Allegheny Front, *BR* = Blue Ridge, *R& V* = Ridge and Valley. Research areas identified by letters are: *D* = Delaware River Water Gap, *E* = Eagle Rock, *H* = Harrisburg, *I* = Iron Gate, *J* = Juniata River, *KGH* = Kentucky Gravity High (center), *N* = Northumberland, *NB* = Natural Bridge, *R* = Roanoke, *S* = Scranton, *SV* = Sugar Valley, *TG* = Tennessee Gorge, *TM* = Tuscarora Mountain, *WM* = portion of Wills Mountain anticlinorium discussed in text. Rivers identified by numbers are: *1* = Delaware, *2* = Schuylkill, *3* = Susquehanna, *4* = Patapsco, *5* = Potomac, *6* = Rappahannock, *7* = James, *8* = Appomattox, *9* = Meherrin, *10* = Roanoke, *10a* = Dan, *11* = Neuse, *12* = Cape Fear, *13* = Yadkin, *14* = Catawba, *15* = Broad, *16* = Saluda, *17* = Savannah, *18* = Oconee, *19* = Ocmulgee, *20* = Flint, *21* = Chattahoochee, *22* = Tallapoosa, *23* = Coosa, *24* = New, *25* = French Broad, *26* = Pigeon (to NE), Little (to S), *27* = Hiwassee, *28* = Etowah, *29* = Tennessee, *30* = Holston, *31* = Clinch-Powell, *32* = Sequatchie. (Modified from Hack, 1982).

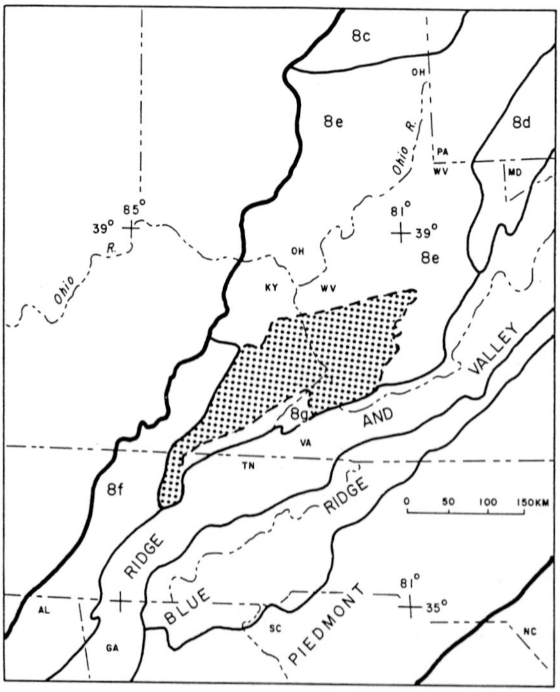

Fig. 2. The Logan Plateau (stippled), a newly-identified geomorphic section in the Allegheny Plateaus physiographic province. *8c* = Southern New York section, *8d* = Allegheny Mountain section, *8e* = Kanawha section, *8f* = Cumberland Plateau section, *8g* = Cumberland Mountain section. (Adapted from Outerbridge, 1987).

polymictic conglomerates. On these bases, the southern-central Appalachian transition would be placed farther northeast than New River, in the vicinity of Buchanan, Virginia, 37 km northeast of Roanoke. There, as noted by Thornbury (1965, p. 109), the Appalachian Valley or "Great Valley of Virginia", is only 2 mi (ca. 3 km) in width.

This paper deals with several aspects and examples of Appalachian transverse drainage near to and south of the glacial border, with emphasis on the Ridge and Valley province. Both the genesis and location of transverse water gaps are critical to the long-standing problem of the origin and evolution of Appalachian drainage. For the purpose of organizational clarity, and also because they are different problems, the topic of mechanisms for transcurrent drainage emplacement is separated from the subject of locational analysis; i.e. why are the gaps in their present positions, rather than elsewhere?

Brief lithotectonic history

Introduction

The portion of the central and southern Appalachians discussed in this paper is between the northwestern structural front of the Grenville province and the Atlantic–Gulf Coastal Plain. This continental crust records two Wilson cycles during Proterozoic and Phanerozoic time, comprised of Grenvillian orogeny, Late Proterozoic rifting, late Paleozoic collision, and Mesozoic rifting (Fig. 3). Any complete explanation of Appalachian topographic and drainage evolution will need to encompass rocks and structures produced during this time span, because this region has both varying erosional levels and tectonic throughprinting of structures from lower tiers, including Grenville basement. Many treatises on Appalachian geomorphic origin begin historical reconstruction with events related to the Alleghanian orogeny, and/or Mesozoic extensional tectonics, and this scenario is followed here for brevity, although it is recognized that earlier tectonic features, and/or the reactivation of them, have produced details that affect topography and drainage. This is especially true for those later orogenic episodes in the "Bewegungsbild" that created dynamic topographic relief, as they could have influenced earlier drainage development that may have recognizable inheritance in present major stream geometry. In this section I have not attempted a comprehensive review, but have chosen to highlight neglected or recent findings that bear on the drainage-development problem.

Central Appalachians

Sevon (1985) has outlined the complicated nature of geologic history as it bears on landscape and drainage evolution in Pennsylvania. Several additional studies highlight events that

	Ma.			
Jurassic	144		Rifting and Opening of Atlantic	
Triassic	208			
Permian	245 286	ALLECHANIAN	Collision with Africa	unconsolidated Permian cover(?)
Carb. Pa	320			Pottsville
Carb. Miss	360			Mauch Chunk
Devonian	408	ACADIAN	Terrane Accretion	Pocono
Silurian	438			Rose Hill Clinch - Tuscarora - Shawangunk
Ordovician	505	TACONIC PENOBSCOT	A-Subduction Obduction Arc Collision/ Accretion	
Cambrian	570			Hesse / Nebo - Erwin - Antietam Cochran - Unicoi - Weaverton
		AVALONIAN	Subduction Vulcanic Arc Generation	Catoctin
Precambrian	1000 - 1200	GRENVILLE	Rifting and Opening of Iapetos	Basement Complex

Fig. 3. Geologic time scale depicting the major orogenic episodes, tectonic events, and lithotypes thought to have affected master drainage origin and evolution in the Appalachians. Left column = geologic time units with ages, second column = orogeny names, third column = major formative tectonic events, right column = rock units. (Modified from Hatcher, 1987, and reproduced, with permission, from The Annual Review of Earth&Planetary Sciences, Vol. 15, 1987, by Annual Reviews, Inc.)

suggest drainage evolution may stem from an ancient tradition. Perry and de Witt (1977) and Perry (1978) implied that layers and lenses of Lower Mississippian polymictic conglomerate in Maryland and Pennsylvania signalled the beginning of an early phase of Alleghanian deformation that involved significant topographic relief generated by uplift or thrust faulting. From northeastern Pennsylvania to central West Virginia there is also a lacuna associated with the exotic-pebble beds that suggests early drainage development was already underway.

Balanced structural cross sections of the Ridge and Valley province that included the central Appalachians north to Pennsylvania (Woodward 1985) indicate that most major central Appalachian Ridge and Valley thrust faults are blind thrust systems with coeval cover deformation (Dunne and Ferrill, 1988). The form, relief, and structural weaknesses (e.g. see Billman, et al., 1989) developed in these roof sequences are of interest because they could have been exploited by early drainage development, if not buried by the major sediment cover postulated by Beaumont et al. (1988). Levine (1983, 1986) concluded that maximum deformation and burial in the anthracite region in Pennsylvania were confined to 285–270 Ma, and that much of the 6–9 km of overburden was emplaced tectonically. Levine inferred that extreme lithotectonic controls existed in the area and probably shaped early drainage trends. Hawman (1980), used the Scranton gravity high as evidence for a high-density mafic intrusion that may have loaded the crust to produce atypical flexural downwarp. (cf. Haworth et al., 1980).

Evidence that could be used to support a late-Alleghanian break-back sequence (northwest–southeast) was summarized briefly by Perry (1978) for the central Appalachians. The related back thrusts, when breached by erosion, could have been exploited by early drainage on the thrust sheets. Paleotopographic effects of the break-back sequence may also have been important in this late stage of deformation, but it is not certain what forms these may have taken. On the other hand, Beaumont et al. (1987, 1988) indicate Permian cover more than 4000 m thick throughout the central and southern Appalachians which may have masked much of the folding and allowed drainage initiation from an alluvial plain surface.

Faill (1973) presented sedimentological evidence for a subaerial deltaic fan of Mesozoic age in the Newark–Gettysburg Basin. The present course of the Schuylkill River trends across the mapped ENE portion of this fan (fig. 3, Faill, 1973). This pattern suggests that an overall southeast-flowing drainage system was already in existence at the time, although its full magnitude and degree of integration are unknown. Similar evidence indicates that the Susquehanna River opened at a somewhat later time, but still in the Late Triassic.

The above glimpses are suggestive of ancient drainage systems that may have influenced subsequent evolution, but the way in which it did so is unclear. Classical analysis might assume that later drainage evolved in place from former drainage lines, but this is not necessarily so. Major resistant channel fills when lithified become resistant rocks that later cap uplands; that is, one could envision a scenario of mega gully gravure in which former master river courses become lines of positive relief, diverting later drainage lines to one side or the other.

Southern Appalachians

Southern Appalachian bedrock geology differs markedly from that of the central Appalachians (Rodgers, 1970). Erosional levels are deeper into the geologic column, and normally expose Cambrian and Ordovician rocks in the Valley and Ridge, and rocks of Precambrian *X* and *Y* age in the Blue Ridge. The Southern Blue Ridge section is an allochthon containing numerous pervasive low-angle thrust faults that are often exposed in windows; the carbonate bedrock-floored coves of the Great Smoky Mountains constitute an epic nothwestern example. In the Valley and Ridge numerous imbricate sledrunner-like thrust faults rise from the subsurface décollements that have undergone long-distance transport from the present southeast (Roeder et al., 1978b; Woodward, 1985). These imbricate faults sometimes break surface to produce a topographic expression of structure and lithology in dramatic, long, subparallel, homoclinal ridges developed on relatively resistant rocks above the sledrunner-like thrusts. A few thrust faults reach land surface on the Appalachian Plateaus, but most do not. Hawman (1980) interpreted the Kentucky gravity high (KGH, Fig. 1) as a Precambrian (?) rift-filling mafic intrusion that has downloaded the crust. Levine (1986) hypothesized that downwarping by the KGH is an explanation for the high coal rank in southwestern Virginia coal fields that are east of the anomaly. In contrast, low sedimentation rates and shallow burial during Pennsylvanian time in Tennessee are indicated by many superposed channel sequences that suggest high sediment transport rates but small amounts of accumulation (Briggs and Minkin, 1977), at least when contrasted with the depocenters in Pennsylvania and Alabama. Thus the Tennessee portion of the Appalachians may represent a drainage development history that evolved closer to the topographic surface than that of the Anthracite region, for example.

Roeder et al. (1978b) constructed fourteen balanced structural cross sections between Roanoke, Virginia and Chattanooga, Tennessee that illustrate the thin-skinned, polyphase nature of Valley and Ridge deformation which preceded or accompanied development of the drainage. In the "Valley of East Tennessee" informal geomorphic district, Roeder et al. (1978a) noted the common and widespread presence of "Klippen" produced by late-stage (T4) thrusting of late Paleozoic age. All Klippen are positioned within 200 m of the same elevation, and seem to have been emplaced along shallow subhorizontal glide planes. Geomorphic implications of the Klippen drawn by Roeder et al. (p. 494) are:

"The widespread occurrence of these klippen in East Tennessee suggests that large parts of the present-day surface in the Valley and Ridge province of Tennessee must be positioned close to the average reliefated peneplain at the end of the Alleghanian orogeny. This conclusion is somewhat in contradiction to Cooper (1944) who demonstrated rapid erosion in the Valley and Ridge since at least the Plio-Pleistocene and to Rodgers (1970) who used Cooper's (1944) data to infer a mountainous morphology at the end of Alleghanian deformation. The present interpretation of the klippen would, in turn, argue for subdued relief through the latest Paleozoic and for a crustal reactivation since the Plio-Pleistocene."

The above apparent conflict between deep and shallow burial may be due to interregional differences in burial depths (see Hawman, 1980; Beaumont, et al., 1988).

Thomas (1989) presented evidence for a break-back sequence in the Valley and Ridge section of Alabama and Georgia, and suggested mechanisms for this late-stage event. These late structures are also of potential interest because structural weaknesses were produced high on culminations, and the paleoslopes may have had possible effects on drainage development.

Drainage development

Evidence

Early workers who followed Davis relied heavily on topographic and drainage "anomalies" apparently because they believed that certain landforms are relicts that hold clues to drainage history. Ver Steeg (1930, 1933) was a prominent proponent of using physiography to decipher fluvial development by the use of water and wind gap basal elevations noting that ridge crests descend gently toward the major steam courses (i.e., water gaps are notched into topographic sags, producing a "gap within a gap"). Meyerhoff and Olmstead (1934) also tabulated gap elevations and concluded (p. 416) that groupings of windgap elevations were indicators of former fluvial baselevels. Nickpoints, or sharp inflections in an interrupted stream profile (Thornbury, 1969, p. 109–110) were identified and ascribed to interruptions of a fluvial cycle of erosion by Hickok (1933). Drainage "anomalies" such as barbed tributaries, were used by a number of writers to infer prior stream captures. For example, Wright (1934, Fig. 1) illustrated barbed tributaries in Virginia between the New and Potomac Rivers. The hooked headwaters were then used as evidence, along with other criteria, to infer capture. Exemplary cases are the North and South Forks of Roanoke River and Cedar Creek above Natural Bridge (Fig. 1).

Strahler (1945) used the presence of two closely-spaced major water gaps through the same ridge as evidence for superpostition, citing the path of Schuylkill and West Branch Schuylkill Rivers across Second and Sharp Mountains and their confluence 3 km south of Second Mountain in which the gaps are only 1.6 km apart. Strahler also noted that the gaps of Juniata River and Frankstown Branch Juniata River (Fig. 1) through Tussey Mountain are separated only by the length of Short Mountain, and again asserted the need for regional superposition. At least in the case of the gaps through Tussey Mountain, however, there may be no need to call on an ancestral geomorphic scheme. The Tyrone-Mount Union cross-strike structural discontinuity (CSD), a major lateral ramp that Southworth (1986) suggested has origins both in basement and allochthon rocks, trends directly across these closely-spaced gaps.

Our knowledge of mechanisms of drainage development and evolution can benefit from studies that relate geologic and topographic controls to riverine topology (Abrahams and Flint, 1983; Phillips and Schumm, 1987) as well as plate tectonics.. Many of the problems of drainage initiation and propagation that seemed insuperable to earlier workers are approachable – or self evident – in these lights. Site specific problems can be attacked using tools of modern high-resolution remote sensing imagery, detailed structural and stratigraphic field studies, and statistical analysis. For example, mapping of cross-strike structural discontinuities (CSD's) and disturbed zones (DZ's) (cf. Southworth, 1986; Pohn and Purdy, 1988), and identification and facies interpretation of critical local stratigraphic variations (cf. Kreisa and Bambach, 1973; Gray et al., 1979; Theisen, 1983) often leads to recognition of structural-lithologic weaknesses that likely localized water and wind gaps. There are increasing numbers of findings of structural weaknesses and aberrant facies changes reported in the literature that are exposed in gap areas. Whether or not these anomalies are unique and causative phenomena is yet unknown, but the spatial coincidence lends credence to hypotheses that espouse geologic control for gap siting.

The rock and soil that give form and relief to the land comprise another class of evidence used in drainage studies. Variations in thickness, resistance, and dip angle of bedrock units were used by numerous workers to account for wind and water gap cross profiles, and to explain variations in intervening ridge crest topography (see discussion in Mills et al., 1987). Many of the earlier authors searched for thick zones of deeply weathered rock debris, remnants of alluvium, and other deposits at relatively high topographic levels. One successful investigator (Wright 1927b) reported high-level gravels (823–945 m) chiefly in the old U.S.G.S. Hillsville, VA–NC quadrangle (edition of 1896, scale 1:125,000), but also in the Yadkinville, NC quadrangle (edition of 1893, scale 1:125,000). Wright (1934) also located gravels in the Natural Bridge area and used these to support an origin by piracy. Another example is the limonite-cemented vein quartz and metaquartzite conglomerate in a wind gap (762 m) in the western Ridge and Valley province, Virginia (Gray et al. 1979). On the other hand, lowering of ridge crests since the development of a presumed former erosion surface would have eradicated easily-erodable materials, except in protected sites. For example, Denny (1956, p. 48–49) concluded that the upland surfaces in the areas of the Appalachian Plateaus he studied in Pennsylvania may have been lowered as much as 200 ft (61 m) since the West Branch Susquehanna River began to dissect its valley, and possibly 100 ft (30.5 m) since the Pliocene. If so, geomorphic materials and landforms now on many summits would either be in dynamic equilibrium with present conditions or would be palimpsests of the last major formative environments (Pleistocene?)

Transverse water gaps

Introduction

Sevon et al. (1983) listed the origin of present drainage as one of three geological phenomena assigned a dependence upon peneplanation, and the fixation of early Appalachian geomorphologists on this notion was noted in the Introduction. Spectacular water gaps in the Blue Ridge, Ridge and Valley, and Appalachian Plateaus provinces sparked early speculation on their origin. Major localities discussed are shown in Fig. 1.

Blue Ridge

Transverse water gap development across the Blue Ridge occurs only from Roanoke northeastward. Roanoke River was interpreted to be a relatively new, headward-eroding drainage basin (Wright, 1934) and may provide clues to the manner in which the older rivers to the northeast conquered the Blue Ridge structural front. James River exits the Valley of Virginia in a deep gorge through the Blue Ridge in Amherst, Bedford, and Rockbridge counties. Evidence for past and ongoing drainage changes in the Upper James River Basin include barbed tributaries, an alluvium-floored wind gap, and ongoing divide migration and intercision noted by Wright (1925).

South of Roanoke Valley the Blue Ridge is not bisected by rivers. Transverse drainage, such as New River, originates within it and flows northwest. Where detailed bedrock mapping has been completed by geologists interested in geomorphology (e.g., Hatcher, 1988) a close correspondence between bedrock structural weaknesses and drainage is reported. South of

New River drainage the main transverse drainages in Virginia and Tennessee are tributary to the Tennessee River system. Major cross-strike tributaries include the French Broad, Pigeon, Little Tennessee, and Hiwassee Rivers that exit the Southern Blue Ridge in deep gorges. Headwater tributaries of some rivers have lost drainage by piracy along the Blue Ridge Escarpment that forms part of the Southern Blue Ridge–Piedmont boundary (Acker and Hatcher, 1970; Hack, 1973, 1982), an area that is important because drainage is rapidly evolving there. A long-standing and detailed controversy surrounds the genesis and age of the Blue Ridge Escarpment. Hack (1980) presented several hypotheses of Blue Ridge Escarpment origin and gave an excellent discussion on evolution of this area (Hack, 1982).

Specific examples - Ridge and Valley province Most of the early fascination about Central and Southern Appalachian drainage centered on the Ridge and Valley Province. It is to this area that many modern specific case studies have been directed as well. Several areal inquiries that illustrate probable breaching mechanisms or localizing factors are given below, with emphasis on work that incorporates heretofore unpublished data and new or revamped hypotheses. Other examples are reported in recent guidebooks (e.g., Sevon, 1985; Clark et al., 1989).

Delaware Water Gap area Epstein (1966) listed structural weakness other than transverse faults that might influence resistance to erosion of resistant lithologies. In the Stroudsburg, Pennsylvania area he noted changes in bedding strike, dip, amount of jointing, other fracturing and shearing, thinning by faulting, and thinning or other alteration of resistance by sedimentation and facies changes as possible factors. Large-scale bedrock geologic mapping by Epstein (1966) on Blue and Kittatinny Mountains and on other ridges to their north established the presence of structural weaknesses in gap areas. Gap parallelism is related to the presence of prominent crossjoint sets that trend southeast across the mapped domain. Trend of the post-Taconic fold-thrust belt in the heart of the New York orocline changes abruptly in the Delaware Water Gap from ca. N65°E in Pennsylvania to ca. N35°E in New Jersey. Could this major regional structure also have assisted in the location of Delaware Water Gap?

Epstein (1966, 1969) also discussed stream capture related to Wind Gap and the Delaware River. Differences in structural level of the resistant Shawangunk Conglomerate exist between Wind Gap (higher) and present Delaware River (lower). Efficacious differential downcutting rates favored the Delaware River, establishing preferential conditions for capture.

Central and south-central Pennsylvania Ciciarelli (1971, 1984) studied factors that control the initial breaching location on the Sugar Valley and Tuscarora Mountain anticlines. Ciciarelli concluded that initial breaching zones on anticlines in this area are localized by fracture concentrations on or near the structural highs. After the resistant cap rock is breached, the weaker underlying rock units are exposed and drainage basin development can begin. Basins coalesce with growth, and Ciciarelli concluded that the drainage with the lowest local base level will be the most likely to capture the streamflow from the others.

Oberlander's (1985) explanation for the origin and age of the Susquehanna River emanates from fluvial development on the erodable Mauch Chunk Formation of Upper Mississippian age that is stratigraphically between the resistant rocks of the overlying Pottsville Group and underlying Pocono Formation. The thickness of the Mauch Chunk Formation locally exceeds fold amplitudes in many areas, but not in the anticlinorium west of the Susquehanna River, so there is no help there. What factors, then, located the overall course of Susquehanna River?

Theisen (1983) showed that the location of the Susquehanna River in Susquehanna Water Gap north of Harrisburg could be the result of both unusual weak facies in the normally resistant Tuscarora Formation as well as numerous faults. Hoskins (1987) described complex fracture systems, faults, folds, and overturned bedding in Susquehanna Water Gap and four others north of Harrisburg, and interpreted these structural weaknesses as localizing factors. Hoskins also pointed out the unsolved problems of: (1) whether or not equivalent weaknesses also occur between water gaps, and (2) whether a major regional feature controls the river's position. Sevon (1986) inferred that the southern edge of the Alleghanian Anthracite thrust sheet initially located this part of the Susquehanna River.

Liebling and Scherp (1984a) fit ridge crests at the 1500 ft (457 m) elevation and showed that the average position of a line between highlands west and east of the Susquehanna River spatially coincides with the present river course between Northumberland and Harrisburg. They concluded that low-relief broad open basins transecting rocks of differential resistance to erosion once existed in the area, and that gradual isostatic uplift and southeastward tilting funneled drainage into and across this localizing zone north of Harrisburg. If affirmation for such a former broad open basin could be devolved, the present general location of this part of Susquehanna drainage might be explained. The persistence through time, depth, and extent of the Baltimore Canyon Trough offshore, the Salisbury Embayment, and other features (Grow et al., 1988; Poag and Sevon, this volume) lends credence to the idea of a geographically related landward depression.

New Creek Mountain–North Fork Mountain, West Virginia The ridges of New Creek and North Fork Mountains in Mineral and Grant Counties, are crossed by North Fork South Branch Potomac River in North Fork Gap, and by three spectacular water gaps cut by smaller streams to the northeast. Northeast of these gaps is North Branch Potomac River. The three small-stream water gaps are separated from the major rivers by wind gaps, as noted by Davis (1889, p. 245). Bedrock geologic mapping of the Wills Mountain anticlinorium in these counties (Clark, 1967a, b) showed that the water gaps are associated with culminations and/or structural weaknesses high on the fold crests (Fig. 4–6). By contrast, faults – even transverse ones – that are not high on the anticlinorium do not support transverse water gaps. An excellent example of two of these faults is shown just north of North Fork Gap on the northwest limb of the fold (Cardwell et al., 1968). Crestal faults and tight folds are associated with highly-shattered Tuscarora Sandstone. Weakened bedrock would have been available to downcutting streams that were locally superposed from the Rose Hill Formation. This situation is encountered northeast down the plunge of the anticlinorium. Using the *ergodic hypothesis*, an entire scenario is present between the northern wind gap and North Branch Potomac River, where small transverse streams are in various stages of incision through the Rose Hill–Tuscarora contact. These observations suggest that, unless gaps are controlled by steeply dipping structures that pervade large stratigraphic intervals (as the Petersburg lineament of Sites, 1978), the evidence for former gap location is lost when folds become deeply dissected.

Also present on fold crests are remnants of the Rose Hill Formation (shown as Juniata Formation by Cardwell et al., 1968). This stratigraphy demonstrates that easily-erodable covermass was indeed present on the culminations, and could have been utilized by downcutting streams.

On the Wills Mountain anticlinorium in Pendleton County to the south, McKoy (1988, fig. 17) showed the topographic coincidence between small stream valleys and a transverse fault. Other faults north and south of Mouth of Seneca are spatially associated with indentations or offsets in the River Knobs, underlain by the Tuscarora Sandstone. Farther south on strike there are also faults in two small water

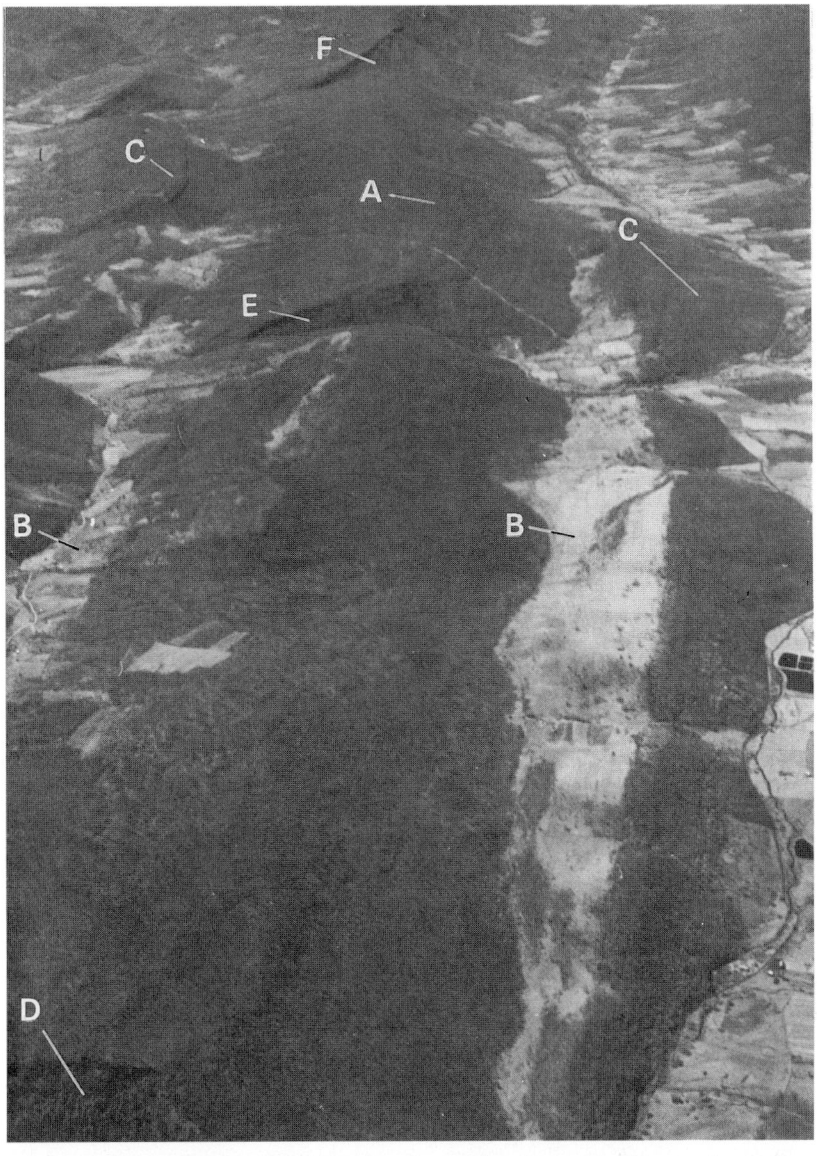

Fig. 4. Oblique infrared aerial photograph of Wills Mountain anticlinorium with view southwest from above Cosner Gap vicinity. A = New Creek Mountain, B = narrow valleys largely underlain by argillites and limestones of Tonoloway Limestone and Wills Creek Formation, C = homoclinal ridges underlain by rocks of Helderberg Group and capped by Oriskany Sandstone, D = Cosner Gap, E = Kline Gap, F = North Fork Gap. Cliff-forming rock unit in these three gaps is Tuscarora Sandstone; bedrock-shattering structural weaknesses are present near and/or at the crests of all three gaps. If structural ensnarement operated in the development of North Fork Gap, paleo-North Fork South Branch would have migrated down regional plunge of Wills Mountain anticlinorium toward the viewer, encountering Petersburg lineament at the present North Fork Gap.

gaps near Macksville, and also at Judy Gap as mapped by Perry (1971). The faults mapped by Perry are also large enough to be shown on the West Virginia State Geologic Map (Cardwell, et al., 1968).

The Iron Gate (Rainbow Gap)-Eagle Rock area, Virginia. Lesure (1957) mapped the geology of the Clifton Forge Iron District, and called attention (p. 78–79) to the considerable adjustment between structure, lithology, and drain-

Fig. 5. Highly sheared and shattered Tuscarora Sandstone in tight flexure on crest of southern end of New Creek Mountain, North Fork Gap. Rule in lower right-hand corner is 15 cm long.

age. The main drainages of Craig and Potts Creeks and James and Jackson Rivers flow primarily in the shale synclinoria underlain by relatively weak strata of Devonian age. The principal exception is the course of Jackson River across Rich Patch anticline in Rainbow Gap at Iron Gate. Lesure was perhaps prescient in his interpretation that Jackson River inherited its course from a higher topographic and stratigraphic level in Devonian shales that now wrap around the nose of Rich Patch anticline 3 km northeast of Rainbow Gap. Lesure cited the low

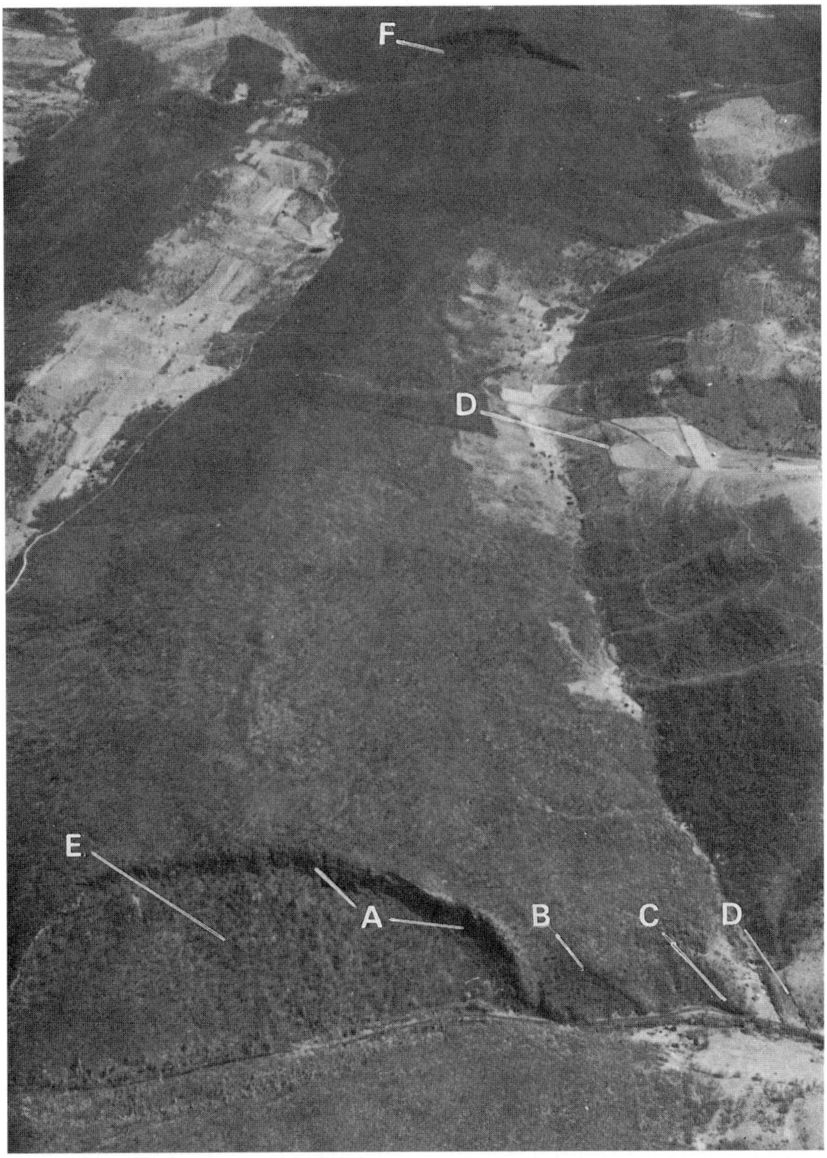

Fig. 6. Oblique infrared aerial photograph of Wills Mountain anticlinorium with view to northeast. Relatively resistant rock units with topographic expression are: *A* = Tuscarora Sandstone, *B* = Middle Member of Rose Hill Shale, *C* = Keefer Sandstone Member of Mifflintown Formation, *D* = Williamsport Sandstone Member of Wills Creek Formation; *E* = Cosner Gap, *F* = Greenland Gap. Cliff-forming rock unit in gaps is Tuscarora Sandstone, shattered in structural complexities between *A* and *B* and to right of *F*.

plunge of the anticline as the reason that Jackson River did not migrate downplunge from its present position.

The precise map location of Rainbow Gap is coincident with a northwest-dipping reverse fault (Lesure, 1957, plate 6), although he did not map the north side of Rainbow Gap. Lesure's map (plate 1) shows no fault along strike on the south side of Rainbow Gap. The local geometry suggests that Jackson River may have migrated down the 10° axial plunge on the Devonian Shale–Keefer Sandstone contact until the river intersected the fault, and incised down the highly-shattered rock.

Continued downcutting of Jackson River in Rainbow Gap has bisected the Rose Hill Formation and Tuscarora Sandstone and is now incising the Juniata Formation. In this case, the mechanism of down-plunge river migration at a weak-resistant lithologic contact, followed by siting on shattered rock, is a more reasonble explanation in an area where anticlinal axes are plunging. In this interpretation, the present course of Jackson River across the Tuscarora Sandstone in Rainbow Gap illustrates a localizing factor for transverse water gaps by a mechanism here termed *structural ensnarement*, as opposed to strictly vertical local superposition with chance encountering of rock weaknesses.

James River passes through Eagle Rock Gorge 17 km south of Rainbow Gap. Structural complexities exposed in Eagle Rock Gorge have long fascinated the tectonic community (see Rader and Gathright, 1986), and there are enough of them preserved here, even at this deep erosional level, to satisfy any proponent of structural control of gap locations (Fig. 7). Even thick-bedded sequences of Tuscarora Sandstone show pervasive shattering, a condition extremely favorable for disintegration during cold-climatic conditions (see Clark and Ciolkosz, 1988). Although no CSD's are shown by Southworth (1986) at this locality, there is a striking (7 km) straight-stream segment of its main tributary, Craig Creek is oriented S50°E aligned directly toward the north end of Crawford Mountain (Strom and Eagle Rock 7.5' quadrangles). Near the confluence with James River another 1.6 km straight kick in Craig Creek trends directly past the confluence into Eagle Rock Gorge, suggesting bedrock control.

From its confluence of formation James River flows to its entrance into the Great Valley of Virginia along the inflection line of the Roanoke orocline. In this only a spatial coincidence, or have tectonically-related structures along this sharp bend influenced river location?

Specific examples - Allegheny Front and Applachian Plateaus-Ridge and Valley transition

Introduction Another zone where transverse streams encounter major structural features is the boundary between the Ridge and Valley and Appalachian Plateaus provinces. This border is also a major structural front. It is produced in much of the central Appalachians by the effects of tectonic upramping along a décollement from the underlying Waynesboro sheet to the overlying Martinsburg sheet (McKoy, 1988). In the central Appalachians, Allegheny Front extends southwest from near Williamsport, Pennsylvania into the Appalachian Plateaus transition zone proximate to the West Virginia–Virginia state line in southwest Virginia and southeasternmost West Virginia. Allegheny Front attains its maximum elevation and local relief in Grant and Pendleton Counties (Fig. 8), before losing its identity in the transition zone to the southwest. In southeasternmost West Virginia and southwest Virginia, by contrast, the Valley and Ridge–Appalachian Plateaus transition is demarcated by surface thrusting of subthrust folded rocks above the St. Clair–Clinchport fault (Whisonant and Schultz, 1986). Streams exiting the Appalachian Plateaus across Allegheny Front in the central Appalachians include: the branches of Susquehanna River in Pennsylvania, and North Branch Potomac River at Keyser, and Seneca Creek at Mouth of Seneca, both in West Virginia. In the southern Appalachians, by contrast, both the New and Tennessee Rivers enter the Appalachian Plateaus province from the Valley and Ridge section.

New River Gorge A unique "embayment facies" of the Cloyd Conglomerate of Lower Mississippian age was described by Gray et al. (1979). Normally the Cloyd Conglomerate is a massive quartz-pebble conglomerate interpreted as having a southeastern provenance and deposition in a beach environment. At an exposure on

Fig. 7. View of south side of Eagle Rock Gorge (Eagle Rock, VA 7.5′ quadrangle), showing polydeformed Ordovician through Lower Devonian strata (see Rader and Gathright, 1986).

New River, however, this rock unit splits into a number of thin conglomerate beds separated by dark gray, clam-bearing siltstones. Could this locality have been a river-mouth bay of a forerunner of New River as suggested by Gray et al. (1979)?

During Early to Middle Pennsylvanian time in the central Appalachians of Virginia and West Virginia, southward and southwestward-flowing drainage systems and sediments from the continental interior became buried by a delta plain extending progressively west from the rising Appalachians (Rice and Schwietering, 1988). Deltaic deposition did not displace Early Pennsylvanian landforms and drainage in northeastern Kentucky until Middle Pennsylvanian time (Rice and Schwietering, 1988, p. B8). Were these channels also flowing southward and southwestward toward a low that also localized drainage ancestral to New River but

Fig. 8. Allegheny Front (right-center photograph edge to vanishing point, right-center skyline) underlain mainly by thick sequence of clastic sedimentary rocks of middle and upper Paleozoic age. In intermediate distance are the fold core and the northwest limb (double, faulted, "saw-toothed" symmetrical ridges which are underlain by Tuscarora Sandstone) of the Wills Mountain anticlinorium. View is from crest of dip slope on southeast limb of this fold at North Fork Mountain Horizontal control triangulation station (3553 ft = 1083 m), Grant County, (Hopeville 7.5′ quadrangle) looking southwest into Pendleton County, WV. Allegheny Front and Wills Mountain anticlinorium constitute a regional structural geomorphic complex that would have had major cover distortion and paleotopography which exerted strong effects on early drainage origin and evolution.

that flowed southwestward rather than to the present northwest? Sears (1957) reported lignite, dated as Late Cretaceous or early Tertiary in age, buried beneath colluvium near the present divide between the New and Tennessee Rivers on Brushy Mountain in Smyth County,

Virginia. The lignite, at elevation ca. 960 m, is well above the high-level gravels of the Blacksburg–Dublin Plains according to Gray et al. (1979).

Whisonant and Schultz (1986) described the Valley and Ridge–Appalachian Plateaus transition zone near and along New River gorge. The regional structural features are the St. Clair–Clinchport Thrust Fault and the Glen Lyn syncline, the axis of which divides the polydeformed structural style of the Valley and Ridge from the relatively flat lying rocks of the Appalachian plateaus province. Whisonant and Schultz (1986) also discussed sedimentological features in the Bluefield Formation (Mississippian–Middle Chester). Facies analysis indicates that two major regressions controlled sedimentation in this section of the Bluefield Formation. The exposures described by Whisonant and Schultz (1986) contain abundant information that will be useful in determining whether or not this part of New River Gorge has atypically weak rock units when additional sections northeast and southwest of the river are described.

Farther northwest along the course of New River Gorge, McColloch (1986) described the Mississippian–Pennsylvanian section exposed along Batoff Creek near its confluence with New River. McColloch (1986) noted that (p. 110): "Locally, the course of the New River, before rejuvenation, appears to have been inflenced by the presence of the Mann Mountain anticline. The river flows in a band of intrenched meanders that closely follow the west flank of the fold (Fig, 4) between Grandview (south of the site) and Gauley Bridge (shown near top of map in Fig. 4)." Indeed, McColloch's Fig. 4 (p. 110) shows a striking spatial alignment of New River in the area described above. If there is a genetic relationship, could this reach of New River be no older than syntectonic growth of the Mann Mountain anticline? Or, has New River simply encountered resistant beds during incision and then migrated laterally off the western flank of the anticline?

Tennessee River Gorge through Walden Ridge
In Tennessee, the departure of the Tennessee river from the Valley of East Tennessee to abruptly enter the gorge through Walden Ridge into Sequatchie Valley has puzzled geomorphologists since the time of Hull (1891), Hayes and Campbell (1894) and Johnson (1905). In its initial entry into the gorge, Tennessee River crosses over the complex fault trace of Missionary Ridge thrust fault. Reconstruction of the fault geometry prior to erosion will be difficult, however, because the greatest total relief in the gorge area is about 1500 ft (457 m) and the Missionary Ridge thrust fault is gently to moderately dipping (Tennessee Division of Geology, 1979). Milici (1968, p. 457) noted that: "Tennessee Valley Authority geologists have recorded numerous joints and faults in the Walden Ridge gorge (Kellberg, 1966, personal communication) which perhaps controlled the location of the ancestral Tennessee River through the anticlinal mountains." Milici (1968, Fig. 3) presented maps postulating the Mesozoic development of the Tennessee River. To test this plausible hypothesis, however, additional research will be needed.

Discussion

At this stage there are many facts, several hypotheses, but no single overall transverse drainage scheme that can be advanced as absolute (Sevon et al., 1983; Sevon, 1985). A speculative chronology of drainage evolution might initiate with Carboniferous paleoriver trends and Alleghanian deformation. Next, Mesozoic thermally-induced arching (Judson, 1975) had greater effect in the central Appalachians of Pennsylvania and perhaps northern Virginia because of the closer proximity of these rifts to the Blue Ridge. By analogy with studies of Atlantic Coastal Plain tectonics (Gardner, this volume), alternating basin and arch epeirogenic activity could also have been active in the highlands throughout the late Mesozoic and later. Mills (1986) has presented evidence that can be interpreted in favor of late Cenozoic warping of some New River Terraces, for example. I conjecture that emplacement mechanisms for transverse drainage lines were dominated by tectonics, headward erosion, general (*not* progressive) stream piracy, and by local (both vertical *and* down-plunge or down-dip migration to produce structural ensnarement) superposition from non-resistant covermasses (Oberlander, 1985) onto structural weaknesses. General piracy and headward erosion provide mechanisms for areal basin growth through time. Local superposition allows for vertical excavation, the incision of valley networks, and water and wind gap development. Specific gap siting has been influenced by structural-lithologic control.

Evidence for headward erosion is preserved in heavy minerals of nonmarine Cretaceous sediments of northern Delaware (Groot, 1955). The lower formations studied were all derived from the nearby Piedmont, whereas the overlying sedimentary units contain increasing detritus from the "folded Appalachians" to the northwest (p. 71). The continuing efficacy of upstream basin growth may be seen in the Logan Plateau, if it is viewed as a rapidly-disappearing vestige that is being attacked on all sides by modern river headwaters. Outerbridge (1987) explained the Logan Plateau as a young, highly dissected plateau with very active headward erosion, and described a deeply-weathered, mesa-like feature (Flatwoods) as a remnant of an older surface of the plateau. In fact, the Logan Plateau may well serve us as a meterstick for headward erosion rates when numerical age dating techniques permit quantification of landform and drainage ages.

Proof of general stream piracy can be found in many papers. For example, Davis (1889, p. 245) inferred the efficacy of piracy in the New Creek Mountain area, West Virginia. Liebling and Scherp (1984b) concluded that proficient piracy also operated at the base of Allegheny Front in Pennsylvania. These studies indicate that piracy is not only effective across the full sweep of the Ridge and Valley province, but is also quite vigorous near Allegheny Front as would be expected if piracy is more energetic in headward (newer) drainage basins. In like manner the Logan Plateau section of the Applachian Plateaus province has convincing examples of recent piracy as well as likely candidates for future capture (Outerbridge, 1987).

Confirmation of local vertical superposition is present in and bordering the gaps through New Creek Mountain, West Virginia (Clark, 1967a, b), where the shale cover is still present adjacent to the water gaps cut through the underlying resistant sandstone. If facies changes that are also lithologic weaknesses can be shown to have a significant relation to gap location, then lithology may become both an aid to superposition and a localizing factor.

Localizing structures for precise gap siting can be divided into vertically-pervasive structures that allow rivers to remain through time, strictly local (crestal) weaknesses that encourage structural ensnarement, and dynamic influences such as warping, tilting, and faulting. Deep-seated fundamental weaknesses include orocline inflexions, CSD's, DZ's, and other vertically-pervasive structures. These features have been documented extensively in the literature (e.g., Rodgers, 1975; Southworth, 1986; Pohn and Purdy, 1988). Structural control for some of the longer features has already been implied. Sites (1978) related the Petersburg lineament to a number of geologic features including drainage development in eastern West Virginia. Both Appalachian oroclines are utilized by rivers. The Delaware River in Delaware Water gap and the James River from the Great Valley of Virginia northwest to its confluence of origin flow along the inflection lines of the New York and Roanoke oroclines, respectively. Kent (1988) concluded that about half of the present curvature around the New York orocline is due to bending during the Alleghanian orogeny, and so abundant fracturing could have developed at that time.

A relatively large number of references deal

with lithologic (facies) changes and structural "anomalies" in gap areas. Are these reports of unusual bedrock features in the gaps just the result of the exposure bias, or are these characteristics indeed concentrated in the gap areas? If the latter is true, then there are valid rock weaknesses that evolving Appalachian drainage could have exploited, and the how and when of stream interaction with these lithologic defects will be interesting to study.

Suggestions, summary, and conclusion

Because of the focus of this paper, I have not touched on a much-neglected problem the localization of longitudinal drainage. Trellising streams appear to "prefer" weak rock trends, but why are these "subsequents" in certain strike belts and not others? A number of excellent maps show relations between cross-strike features (cf. Southworth, 1986), but longitudinal features, such as the New York–Alabama lineament (King and Zietz, 1978), seem to remain the intellectual property of tectonic geologists.

Following the idea of Keith (1894) and Barrell (1920), the old questions of wind and water gap levels and ridge crest topography (cf. Ver Steeg, 1930, 1933; Hickok, 1933; Meyerhoff and Olmstead, 1934) could be resolved, especially using detailed bedrock mapping where adequate exposures exist (cf. Epstein, 1966; Clark, 1967a, b). Quantitative analysis of precise gap elevations, three-dimensional digitized profiles and analysis of high-resolution remote sensing imagery could establish statistically valid associations in certain areas, so that verifiable tests could be run.

The importance of gap localization by structural ensnarement could be appraised in areas with adequate strata preservation and exposure. Gaps both produced and sited by structural and/or lithologic trapping are candidates for subsequent piracy by streams flowing on weaker rocks closer to local or regional base levels, unless the weaknesses are vertically pervasive. By examining examples of recent abstraction in areas with sufficient exposure, this hypothesis could be tested, and either modified or rejected.

The importance of stream capture in successive readjustment of drainage to structural-lithologic control encountered in both downcutting and lateral migration appears to be underplayed in the literature. Notable exceptions include Liebling and Scherp (1984b), Outerbridge (1987), and the early perceptive work of Wright (cf. 1925, 1927a,b, 1928, 1929, 1931, 1934, 1936) who, despite his proclivity for peneplains, recognized both the evidence for past captures and the present dynamic nature of competing streams. The brief sampling provided in this paper suggests that many present-day streams are engaged in the abstraction process in certain areas of the Appalachians. Unfortunately we neither know how the areas of intense piracy have shifted through time, nor if the cadence of divide migration has changed. When absolute rates and locations of neotectonic activity become available it will be interesting to compare modern epeirorogeny with stream activity, especially in areas where effects of lithologic resistance can be assessed.

Differing scenarios of structural geomorphic development of the central versus the southern Appalachians would explain some of the disparate topographic and drainage features present. These include, but are not limited to: (1) prominence of transverse drainage lines with spectacular water gaps in the Middle section of the Ridge and Valley province in contrast to dominance of longitudinal drainage within the Southern section; (2) a differing topography, perhaps also related to drainage, produced by erosion of surface folds in the central Appalachians versus dissection of thrust sheets in the southern Appalachians; (3) differing erosional levels with respect to both total amount of section removed, and lithology and age of rocks now exposed; and (4) differential timing, magnitudes, and rates of uplift and erosion. Jamieson and Beaumont (1988) concluded that post-orogenic unroofing of the central Appalachians likely reflects Mesozoic extension, and that post-orogenic unroofing of the southern Appalachians is compatible with predictions derived from flexural models.

Rapid advances in highspeed computer simulations allow geomorphological modelling on an unprecedented scale. Although there is research in this area, Craig (1979, and this volume) indicated a need for further work. Theoretical and empirical aspects of geomorphological models are well enough known (Ahnert, 1987; Thorn, 1988) that long-term drainage and evolutionary models can be reconstructed and run (see Slingerland and Furlong, this volume). Such studies could test our assumption and sharpen our perceptions of "what to look for" in both field and laboratory settings.

Many specific drainage basins can be recommended as ideal future study areas. Results from new absolute geophysical methods for quantifying recent crustal deformation will soon identify areas of interest to the student of neotectonics. Drainage basins in and near the Logan Plateau (Outerbridge, 1987) commended themselves for study of stream piracy. A systematic drainage and landscape study could be conducted in the thirty-eighth parallel fracture zone (Dennison and Johnson, 1971; Yeilding and Dennison, 1986) and the nearby major topographic high of Appalachian Highland summit areas mapped and discussed long ago by Wright (1925, Plate 1) as warping of the "Upland Peneplane" or "Schooley Peneplane" (Wright, 1934) to determine if drainage and landform evolution has followed the westward migration of igneous activity during Mesozoic time. Historical study could also center on the New River, and on adjoining basins as the Roanoke River drainage, to determine the long-term history of entrenchment, piracy, and other drainage adjustments. The quotation earlier in this paper from Roeder et al. (1978a) suggests that Valley and Ridge drainage in Tennessee may be nearer the paleosurface of deformation than elsewhere in the Appalachians. If so, is this longitudinal drainage pattern less evolved and a model for geologically much earlier drainage elsewhere? Anticlinal breaching without concomitant transverse drainage development across both limbs has received some study (cf. Ciciarelli, 1971, 1984), but there are many unstudied sites left. In Pennsylvania, Maryland and West Virginia, for example, there are localities on the northwest limbs of the Nittany and Wills Mountain anticlinoria where breaching of the Tuscarora Sandstone could be studied. If an evolutionary chain can be established, the unroofing of domes such as Burkes Garden (cf. Cooper, 1944) and elongate anticlinal valleys could be better understood. Can we, in fact, identify and date features, and then assemble a chronosequence of drainage basins in true evolutionary order? Only time, and the deployment of rapidly-developing numerical age dating techniques will tell.

Finally, some drainage problems that, to earlier workers, required base-level standstill, rapid uplift, and subsequent peneplanation may solve themselves. For example maturation of exquisite longitudinally-draining fluvial surfaces such as the Coosa strath, that bevels both lithology and structure (Bearce, 1978) could be expected given the tenets of uplift, tilting and erosion discussed herein. Phillips and Schumm (1987) studied the effect of regional slope on change in drainage network types. They found that, with steepening slope, drainage pattern changes from dendritic (1% slope), to subdendritic (2% slope), to subparallel (3% slope), to parallel (5% and > slope). Hack (1982, p. 20–22) suggested that significant uplift of the Coosa River is implied by the extreme convexity of its longitudinal profile, thus providing an epeirogenic source for both the energy and the internal fluvial feedback needed to develop the Coosa strath. Substantial uplift (or base-level lowering) is also implied by the Tennessee River Gorge through Walden Ridge as well as its lower course around the Nashville dome (Milici, 1968). Has the Nashville dome also been epeirogenically active in Cenozoic time? Why the Tennessee River does not drain directly to the Gulf of Mexico, however, is still unknown. Hack (1980) noted that the Appalachian Highlands, in a general way, owe their presence to uplift as the highlands pass under the Coastal Plain at their southern end.

Whereas this paper emphasizes lithotectonic influences on drainage development, I am cognizant of the importance of other factors that have affected Appalachian riverine evolution. These include, but certainly are not limited to, controls by climate (including glaciation), rate changes in erosion and deposition influenced by internal feedback mechanisms, and the role of catastrophic processes. Further, more complicated interactions among these different influences no doubt occured, and perhaps caused perturbations in the driving mechanisms of drainage evolution.

Still, lack of information on critical geochemical, geological, and geophysical datasets hampers further hypothesis erection and testing. Many of the same problems that dog the peneplain question (Sevon et al., 1983, p. 161) also plague the unravelling of Appalachian drainage history. Is it any wonder that Craig (1979) could find no commonality of origin? What was written in 1932–33, can still be quoted today: "The Appalachian problem, like the poor, we shall have with us always." (Bryan et al., 1932/33, p. 318).

Acknowledgements

Field mapping in West Virginia was supported by the West Virginia Geological and Economic Survey, and other research bearing on drainage development in the Great Smoky Mountains National Park and adjacent areas was aided by a grant from the Great Smoky Mountains Conservation Association. Permission to produce Fig. 3 was granted by Annual Reviews, Inc. R.D. Hatcher and W.A. Dunne made helpful suggestions on tectonic and structural topics.

References

Abrahams, A.D. and Flint, J.-J., 1983. Geological controls on the topological properties of some trellis channel networks. Geol. Soc. Am. Bull., 94: 80–91.

Acker, L.L. and Hatcher, R.D., Jr., 1970. Relationships between structure and topography in northwest South Carolina. S. C. Div. Geol. Geol. Notes, 14: 35–48.

Ahnert, F., 1987. Process-response models of denudation at different spatial scales. Catena, Suppl., 10: 31–50.

Barrell, J., 1920. The piedmont terraces of the northern Appalachians. Am. J. Sc., 4th Ser. 49: 227–258, 327–362, 407–428.

Bearce, D.N., 1978. Structure of the eastern Coosa Valley, Alabama. Am. J. Sc., 278: 461–476.

Beaumont, C., Quinlan, G.M. and Hamilton, J., 1987. The Alleghanian orogeny and its relationship to the evolution of the Eastern Interior, North America. Mem. Can. Soc. Pet. Geol., 12: 425–445.

Beaumont, C., Quinlan, G. and Hamilton, J., 1988. Orogeny and stratigraphy: numerical models of the Paleozoic in the Eastern Interior of North America. Tectonics, 7: 389–416.

Billman, D., Dunne, W., and Johnston, M., 1989. Penetrative strain contribution to cover deformation in the central Appalachians. Appalachian Basin Ind. Assoc., 14: 147–165.

Briggs, G. and Minkin, S.C., 1977. Depositional history and sedimentation rates of Pennsylvanian sediments of the Cumberland Plateau in northern Tennessee. Geological Society of America, Abstracts with Programs, 9: 122 (abstr.).

Bryan, K., Cleaves, A.B. and Smith, H.T.U., 1932/33. The present status of the Appalachian problem. Z. Geomorphol., 7: 312–320.

Cardwell, D.H., Erwin, R.B. and Woodward, H.P. (Compilers) 1968. Geologic map of West Virginia, scale 1: 250,000 W. Va. Geol. Econ. Surv..

Ciciarelli, J.A., 1971. Geomorphic evolution of anticlinal valleys in Central Pennsylvania. Ph.D. Thesis, The Pennsylvania State University, University Park, 98 pp.

Ciciarelli, J.A., 1984. Structural control of breaching of an Appalachian anticline in central Pennsylvania. Northeast. Geol. 6: 102–104.

Clark, G.M., 1967a. Structural geomorphology of a portion of the Wills Mountain Anticlinorium, Mineral and Grant Counties, West Virginia. Ph.D. Thesis, The Pennsylvania State University, University Park, 167 pp.

Clark, G.M., 1967b. Structural geomorphology of a portion of the Wills Mountain Anticlinorium, Mineral and Grant Counties, West Virginia. Geol. Soc. Am., Annu. Meet., New Orleans, p. 36 (abstr.).

Clark, G.M., 1989. Central and Southern Appalachian accordant ridge crest elevations south of glacial border: Regional erosional remnants or incipient periglacial surfaces? 28th Int. Geol. Congr., Abstr. Vol., pp. 1-299–1-300.

Clark, G.M. and Ciolkosz, E.J., 1988. Periglacial geomorphology of the Appalachian Highlands and Interior Highlands south of the glacial border - a review. Geomorphology, 1: 191-220.

Clark, G.M., Ciolkosz, E.J., Kite, J.S. and Lietzke, D.A., 1989. IGC Field Trip T150: Central and Southern Appalachian geomorphology - Tennessee, Virginia, and West Virginia. Am. Geophys. Union, Washington, D.C., 105 pp.

Cooper, B.N., 1944. Geology and mineral resources of the Burkes Garden quadrangle, Virginia. Va. Geol. Surv. Bull., 60, 299 pp.

Craig, R.G., 1979. A simulation model of landform erosion. Ph.D. Thesis, The Pennsylvania State University, University Park, 536 pp.

Craig, R.G., 1989. Computing landscape geomorphology. In: T.W. Gardner and W.D. Sevon (Editors), Appalachian Geomorphology. Geomorphology, 2: 197-207 (this volume).

Davis, W.M., 1888. Geographic methods in geologic investigations. Natl. Geogr. Mag., 1: 11-26.

Davis, W.M., 1889. The rivers and valleys of Pennsylvania. Natl. Geogr. Mag., 1: 183-253.

Dennison, J.M. and Johnson, R.W., Jr., 1971. Tertiary intrusions and associated phenomena near the thirty-eighth parallel fracture zone in Virginia and West Virginia. Geol. Soc. Am. Bull., 82: 501-508.

Denny, C.S., 1956. Surficial geology and geomorphology of Potter County, Pennsylvania. U.S. Geol. Surv., Prof. Pap., 288, 72 pp.

Dunne, W.M., and Ferrill, D.A., 1988. Blind thrust systems. Geology, 16: 33-36.

Epstein, J.B., 1966. Structural control of wind gaps and water gaps and of stream capture in the Stroudsburg area, Pennsylvania and New Jersey: U.S. Geol. Surv., Prof. Pap., 550B : B80-B86.

Epstein, J.B., 1969. Structural control of wind gaps and water gaps and of stream capture in the Stroudsburg area, Pennsylvania and New Jersey. In: Subitzky, S. (Editor), Geology of selected areas in New Jersey and eastern Pennsylvania. Rutgers University Press, New Brunswick, N.J., pp. 206-213.

Faill, R.T., 1973. Tectonic development of the Triassic Newark-Gettysburg Basin in Pennsylvania. Geol. Soc. Am., Bull., 84: 725-740.

Flemal, R.C., 1971. The attack on the Davisian system of geomorphology, a symposium. J. Geol. Educ., 19: 3-13.

Gardner, T.W., 1989. Neotectonism along the Atlantic passive continental margin. In: T.W. Gardner and W.D. Sevon (Editors), Appalachian Geomorphology. Geomorphology, 2: 71-97 (this volume).

Gray, D.R., Lowry, W.D., Read, J.F. and Schultz, A.P., 1979. Nature of thrusting along the Allegheny Front near Pearisburg and of overthrusting in the Blacksburg-Radford area of Virginia. Va. Div. Miner. Resour., Va. Geol. Field Conf., Guide., 11, 66 pp.

Groot, J.J., 1955. Sedimentary petrology of the Cretaceous sediments of northern Delaware in relation to paleogeographic problems. Del. Geol. Surv., Bull., 5, 157 pp.

Grow, J.A., Klitgord, K.D. and Schlee, J.S., 1988. Structure and evolution of Baltimore Canyon Trough. In: Sheridan, R.E. and Grow, J.A. (Editors), The Atlantic Continental Margin. U.S. Geol. Soc. Am., The Geology of North America, I-2: 269-290.

Hack, J.T., 1973. Drainage adjustment in the Appalachians. In: M. Morisawa (Editor), Fluvial Geomorphology. Publications in Geomorphology, State University of New York, Binghamton, pp. 51-69.

Hack, J.T., 1975. Dynamic equilibrium and landscape evolution. In: W.N. Melhorn and R.C. Flemal (Editors), Theories of Landform Development. Allen and Unwin, London, pp. 87-102.

Hack, J.T., 1980. Rock control and tectonism - their importance in shaping the Appalachian Highlands, U.S. Geol. Surv. Prof. Pap., 1126-B: B1-B17.

Hack, J.T., 1982.Physiographic divisions and differential uplift in the Piedmont and Blue Ridge. U.S. Geol. Surv. Prof. Pap., 1265, 49 pp.

Hatcher, R.D., Jr., 1987. Tectonics of the southern and central Appalachian Internides. Annu. Rev. Earth Planet. Sci., 15: 337-362.

Hatcher, R.D., Jr., 1988. Bedrock geology and regional geologic setting of Coweeta Hydrologic Laboratory in the eastern Blue Ridge. In: W.T. Swank and D.A. Crossley, Jr. (Editors), Ecological Studies, Vol. 66: Forest Hydrology and Ecology at Coweeta. pp. 81-92.

Hawman, R.B., 1980. Crustal models for the Scranton and Kentucky gravity highs: Regional bending of the crust in response to emplacement of failed rift structures. M.S. Thesis, The Pennsylvania State University, University Park, 107 pp.

Haworth, R.T., Daniels, D.L., Williams, H. and Zietz, I., 1980. Bouguer gravity anomaly map of the Appalachian orogen. Memorial University of Newfoundland, Map 3.

Hayes, C.W. and Campbell, M.R., 1894. Geomorphology of the Southern Appalachians. Natl. Geogr. Mag., 6: 63-126.

Hickok, W.O., 1933. Erosion surfaces in south-central Pennsylvania. Am. J. Sc., 5th Ser., 25: 101-122.

Hoskins, D.M., 1987. The Susquehanna River water gaps near Harrisburg, Pennsylvania. Geol. Soc. Am., Cent. Field Guide, 5: 47-50.

Hull, E., 1891. On the physical geology of Tennessee and adjoining districts. Geol. Soc. London, Q. J. 47: 69-77.

Hull, E., 1891. On the physical geology of Tennessee and adjoining districts. Am. Geol., 7: 345-348.

Jamieson, R.A. and Beaumont, C., 1988. Orogeny and metamorphism: A model for deformation and pressure-temperature-time paths with applications to the central and southern Appalachians. Tectonics, 7: 417-445.

Johnson, D.W., 1905. The Tertiary history of the Tennessee River. J. Geol., 13: 194-231.

Judson, S., 1975. Evolution of Appalachian topography. In: W.N. Melhorn and R.C. Flemal (Editors), Theories of Landform Development. Allen and Unwin, Boston, Mass, pp. 29-44.

Keith, A., 1894. Geology of the Catoctin belt. U.S. Geol. Surv., 14th Ann. Rep., 1892-93, part 2: 285-395.

Kent, D.V., 1988. Further paleomagnetic evidence for orocline rotation in the central folded Appalachians from the Bloomsburg and the Mauch Chunk Formations. Tectonics, 7: 749-759.

King, E.R. and Zietz, I., 1978. The New York-Alabama lineament; geophysical evidence for a major crustal break in the basement beneath the Appalachian basin. Geology, 6: 312-318.

Kreisa, R.D. and Bambach, R.K., 1973. Environments of deposition of the Price Formation (Lower Mississippian) in its type area, southwestern Virginia. Am. J. Sci., 273-A: 326-342.

Lesure, F.J., 1957. Geology of the Clifton Forge iron district, Virginia. Va. Polytech. Ins., Bull., 50, 130 pp.

Levine, J.R., 1983. Tectonic history of coal-bearing sediments in eastern Pennsylvania using coal reflectance anisotropy. Ph.D. Thesis, The Pennsylvania State University, University Park, 314 pp.

Levine, J.R., 1986. Deep burial of coal-bearing strata, Anthracite region, Pennsylvania: Sedimentation or tectonics? Geology, 14: 577-580.

Liebling, R.S. and Scherp, H.S., 1984a. Development of the Susquehanna River drainage across the folded Appalachians. Northeast. Geol., 6: 1-3.

Liebling, R.S. and Scherp, H.S., 1984b. Stream piracy at the base of the Allegheny Front, central Pennsylvania. Northeast. Geol. 6: 131-134.

Lowry, W.D., Tillman, C.G., Fara, M. and Kettren Jr., L.P., 1971. Guidebook to contrast in style of deformation of the Southern and Central Appalachians. Virginia Polytechnic Institute and State University, Dep. Geol. Sci, Blacksburg, Va., 185 pp.

McColloch, G.M., Jr., 1986. Batoff Creek section of Pennsylvanian-Mississippian strata, Raleigh County, West Virginia. Geol. Soc. Am., Cent. Field Guide, 6: 109-112.

McGee, W.J., 1888a. The geology of the head of Chesapeake Bay. U.S. Geol. Surv., 7th Annu. rep. pp. 537-646.

McGee, W.J., 1888b. Three formations of the middle Atlantic slope. Am. J. Sci., 3rd Ser., 35: 120-143, 328-330, 367-388, 448-466.

McKoy, M.L., 1988. Geology of the Seneca Rocks Recreation Area. Mountain State Geol., 1988: 18-28.

Meyerhoff, H.A. and Olmstead, E.W., 1934. Windgaps and watergaps in Pennsylvania. Am. J. Sci., 5th Ser., 27: 410-416.

Milici, R.C., 1968. Mesozoic and Cenozoic physiographic development of the Lower Tennessee River: in terms of the dynamic equilibrium concept. J. Geol., 76: 472-479.

Mills, H.H., 1986. Possible differential uplift of New River terraces in southwest Virginia. Neotectonics, 1: 75-86.

Mills, H.H., Brakenridge, G.R., Jacobson, R.B., Newell, W.L., Pavich, M.J. and Pomery, J.S., 1987. Appalachian mountains and plateaus. Geol. Soc. Am., Cent. Vol., 2: 5-50.

Monmonier, M.S., 1967. Upland accordance in the Ridge and Valley section of Pennsylvania. M.S. Thesis, The Pennsylvania State University, University Park, 58 pp.

Morisawa, M., 1989. Rivers and valleys of Pennsylvania, revisited. In: T.W. Gardner and W.D. Sevon (Editors), Appalachian Geomorphology, Geomorphology, 2: 1-22 (this volume).

Oberlander, T.M., 1985. Origin of drainage transverse to structures in orogens. In: M. Morisawa, and J.T. Hack (Editors), Tectonic Geomorphology. Allen and Unwin, Boston, Mass., 155-182.

Outerbridge, W.F., 1987. The Logan Plateau, a young physiographic region in West Virginia, Kentucky, Virginia, and Tennessee. U.S. Geol. Surv., Bull., 1620, 19 pp.

Perry, W.J., Jr., 1971. Structural development of the Nittany Anticlinorium, Pendleton County, West Virginia. Ph.D. Thesis, Yale University, New Haven, Conn., 227 pp.

Perry, W.J., Jr., 1978. Sequential deformation in the Central Appalachians. Am. J. Sc. 278: 518-542.

Perry, W.J., Jr. and De Witt, W., Jr., 1977. A field guide to thin-skinned tectonics in the central Appalachians. Am. Assoc. Pet. Geol./Soc. Econ. Paleontol. Mineral. Annu. Conv., Washington, D.C., Guideb., Post-Meet. Field Trip 4, 54 pp.

Phillips, L.F. and Schumm, S.A., 1987. Effect of regional slope on drainage networks. Geology, 15: 813-816.

Pike, R.J., and Thelin, G.P., 1988. New shaded relief map of the United States from digital elevations, with summary statistics on landform geometry for the physiographic provinces. Geol. Soc. Am. Abstr. Programs, 20: A284.

Poag, C.W. and Sevon, W.D., 1989. A record of Appalachian denudation in post-rift Mesozoic and Cenozoic sedimentary deposits of the U.S. middle Atlantic continental margin. In: T.W. Gardner and W.D. Sevon (Editors), Appalachian Geomorphology, 2: 119-157 (this volume).

Pohn, H.A. and Purdy, T.L., 1988. Disturbed zones: Indicators of subsurface faults and possible hydrocarbon traps in the Valley and Ridge and Appalachian Plateau provinces of Pennsylvania, Maryland, Virginia and West Virginia. U.S. Geol. Surv., Bull., 1839-C: C1-C11.

Rader, E.K. and Gathright, T.M., II, 1986. Stratigraphic and structural features of Fincastle Valley and Eagle Rock Gorge, Botetourt County, Virginia. Geol. Soc. Am. Cent. Field Guide, 6: 105-108.

Rice, C.L. and Schwietering, J.F., 1988. Fluvial deposition in the Central Appalachians during the Early Pennsylvanian. U.S. Geol. Surv. Bull., 1839-B: B1-B10.

Rodgers, J., 1970. The tectonics of the Appalachians. Wiley New York, N.Y., 271 pp.

Rodgers, J., 1975. Appalachian salients and recesses. Geol. Soc. Am., Abstr. Programs, 7: 111-112. (abstr.).

Roeder, D., Yust, W.W. and Little, R.L., 1978a. Folding in the Valley and Ridge province of Tennessee. Am. J. Sci., 278: 477-496.

Roeder, D., Gilbert, Jr., H.E. and Witherspoon, W.D., 1978b. Evolution and macroscopic structure of Valley and Ridge thrust belt, Tennessee and Virginia. University of Tennessee, Dep. Geol. Sci., Stud. Geol., 2,25 pp.

Sears, C.E., 1957. Late Cretaceous erosion surface in southwest Virginia. Geol. Soc. Am., 68: 1883 (abstr.).

Sevon, W.D., 1985. Pennsylvania's polygenetic landscapes. Harrisburg Area Geol. Soc., Guide. 4th Annu. Field. Trip., 55 pp.

Sevon, W.D., 1986. Susquehanna River water gaps; Many years of speculation. Pa. Geo. 17: 4-7.

Sevon, W.D., Potter, N., Jr. and Crowl, G.H., 1983. Appalachian peneplains: An historical review. Earth Sci. Hist., 2: 156-164.

Sites, R.S., 1978. Structural analysis of the Petersburg lineament, Central Appalachians. Ph.D. Thesis, West Virginia University, Morgantown, 171 pp.

Slingerland, R., and Furlong, K.P., 1989. Geodynamic and geomorphic evolution of the Permo-Triassic Appalachians. In: T.W. Gardner and W.D. Sevon (Editors), Appalachian Geomorphology. Geomorphology, 2: 23-37 (this volume).

Southworth, C.S., 1986. Side-looking airborne radar image map showing cross-strike structural discontiniuties and lineaments of the Central Appalachians. U.S. Geol. Surv., Misc. Field Stud., Map, MF-1891, scale 1: 500,000.

Strahler, A.N., 1945. Hypotheses of stream development in the folded Appalachians of Pennsylvania. Geol. Soc. Am. Bull. 56: 45-88.

Tennessee Division of Geology, 1979. Geology of Hamilton County, Tennessee. Tenn. Div. Geol., Bull., 79, 128 pp.

Theisen, J.P., 1983. Is there a fault in our gap? Pa. Geol., 14: 5-11.

Thomas, W.A., 1989. Break-back thrusting in the Southern Appalachians. Appalachian Basin Ind. Assoc., 14: 189-204.

Thompson, H.D., 1939. Drainage evolution in the Southern Appalachians. Geol. Soc. Am., Bull., 50: 1323-1356.

Thorn, C.E., 1988. An introduction to theoretical geomorphology. Unwin Hyman, Boston, Mass., 247 pp.

Thornbury, W.D., 1965. Regional geomorpholgy of the United States. Wiley, New York, N.Y., 609 pp.

Thornbury, W.D., 1969. Principles of Geomorphology. Wiley, New York, N.Y., 594 pp.

Ver Steeg, K., 1930. Wind gaps and water gaps of the northern Appalachians, their characteristics and significance. N.Y. Acad. Sci., Ann., 32: 87-220.

Ver Steeg, K., 1933. Windgaps and erosion surfaces. Am. J. Sci., 5th Ser., 26: 507-511.

Whisonant, R.C. and Schultz, A.P., 1986. Appalachian Valley and Ridge to Appalachian Plateau transition zone in southwestern Virginia and eastern West Virginia: Structure and sedimentology. Geol. Soc. Am., Cent. Field Guide, 6: 113-118.

Woodward, N.B., 1985. Valley and Ridge thrust belt; balanced structural sections, Pennsylvania to Alabama. University of Tennessee, Dep. Geol. Sci., Stud. Geol., 12, 64 pp.

Wright, F.J., 1925. The physiography of the upper James River Basin in Virginia. Va. Geol. Surv., Bull., 11, 67 pp.

Wright, F.J., 1927a. The Blue Ridge of Southern Virginia and Western North Carolina. Denison Univ. Bull., J. Sci. Lab., 22: 116-132.

Wright, F.J., 1927b. Gravels on the Blue Ridge. Dension Univ. Bull., J. Sci. Lab., 22: 133-135.

Wright, F.J., 1928. The erosional history of the Blue Ridge. Denison University Bulletin. Journal of the Scientific Laboratories, 23: 321-344, 10 plates.

Wright, F.J., 1929. Stream piracy near Asheville, North Carolina. Dension Univ. Bull., J. Sci. Lab., 24: 401-406.

Wright, F.J., 1931. The older Appalachians of the South. Dension Univ. Bull. J. Sci. Lab., 26: 143-250.

Wright, F.J., 1934. The newer Appalachians of the South (Part I). Between the Potomac and New Rivers. Denison Univ. Bull., J. Sci. Lab., 29: 1-105.

Wright, F.J., 1936. The newer Appalachians of the South (Part II). South of the New River, Denison Univ. Bull., J. Sci. 31: 93-142.

Yeilding, C.A. and Dennison, J.M., 1986. Sedimentary response to Mississippian tectonic activity at the east end of the 38th Parallel fracture zone. Geology, 14: 621-624.

Geomorphology, 2 (1989) 233-256
Elsevier Science Publishers B.V., Amsterdam — Printed in The Netherlands

Glacial and Periglacial Erosion of the Appalachians

DUANE D. BRAUN

Geography and Earth Science, Bloomsburg University, Bloomsburg, PA 17815 (U.S.A.)

(Received February 14, 1989; accepted after revision June 7, 1989)

Abstract

Braun, D.D., 1989. Glacial and periglacial erosion of the Appalachians. In: T.W. Gardner and W.D. Sevon (Editors), Appalachian Geomorphology. Geomorphology, 2: 233-256.

The marine ^{18}O record indicates that the first ice advance into the Appalachians equivalent to that of the late Wisconsin occurred in the Pliocene at about 2.4 Ma. Only periodic localized glacial and periglacial activity occurred in the northern Appalachians from 2.33 to 0.85 Ma. After 0.85 Ma (middle to late Pleistocene) eight out of ten glaciations covered the northern Appalachians and brought periglacial conditions to the southern Appalachians. During the last 850,000 years, glaciation extended to the late Wisconsin limit within the Appalachians for 15% of the time and extended to the early Wisconsin limit just north of the Appalachians for 25% of the time. Pleistocene glacial erosion of the Appalachians of 120-200 m is necessary to account for the estimated volume of Pleistocene ocean sediment derived from the Appalachians and the estimated volume of Pliocene-Pleistocene glaciogenic ocean sediment derived from the entire Laurentide ice sheet. This amount of glacial erosion, occurring over the maximum expectable duration of Pliocene-Pleistocene ice cover of 247,000 years, yields glacial erosion rates from 0.5-0.8 m ka^{-1}. Such depths of erosion by thick, warm-based ice streams were localized in the lowlands and estuaries of the northern Appalachians. In the Maritimes the presence of thin, cold-based ice caused minimal erosion of uplands. Periglacial landforms and tundra-boreal forest were distributed throughout the Appalachian Highlands and the northern Coastal Plain during the glacial maxima. Periglacial erosion rates of 0.15-0.30 m ka^{-1} near the ice in Pennsylvania decline to 0.1-0.2 m ka^{-1} in the southern Piedmont. Each cold period almost completely renewed the regolith by eroding slopes and filling valleys. The cumulative effect of the eight to ten major cold events could have produced as much as a few ten's of meters of periglacial ridge-top lowering during the Pleistocene. Both estimated glacial and periglacial erosion rates greatly exceed estimated current fluvial erosion rates. Therefore glacial and periglacial processes should be considered the dominant processes in shaping the present Appalachian landscape. The northern Appalachians may be evolving towards a form where the landscape provides just the shape necessary to permit the least work conveyance of the ice across the region during glacial maxima. The southern Appalachians may be evolving towards a form where the landscape provides just the hillslope form necessary to transport the debris provided by periglacial conditions.

Introduction

The purpose of this paper is to summarize current knowledge on the glacial and periglacial modification of the Appalachian landscape, to propose hypotheses, even outrageous ones, to stimulate further research, and to suggest areas for future investigation. The basic question addressed is: how much has glacial and periglacial activity shaped the Appalachian landscape? The Wisconsin ice border is adapted as the dividing line between the glacially modified northern Appalachians and the periglacially modified southern Appalachians (Fig. 1). The northern Appalachians are examined from Newfoundland to Pennsylvania. The southern Appalachians and Coastal Plain are examined from Pennsylvania to Alabama except for the

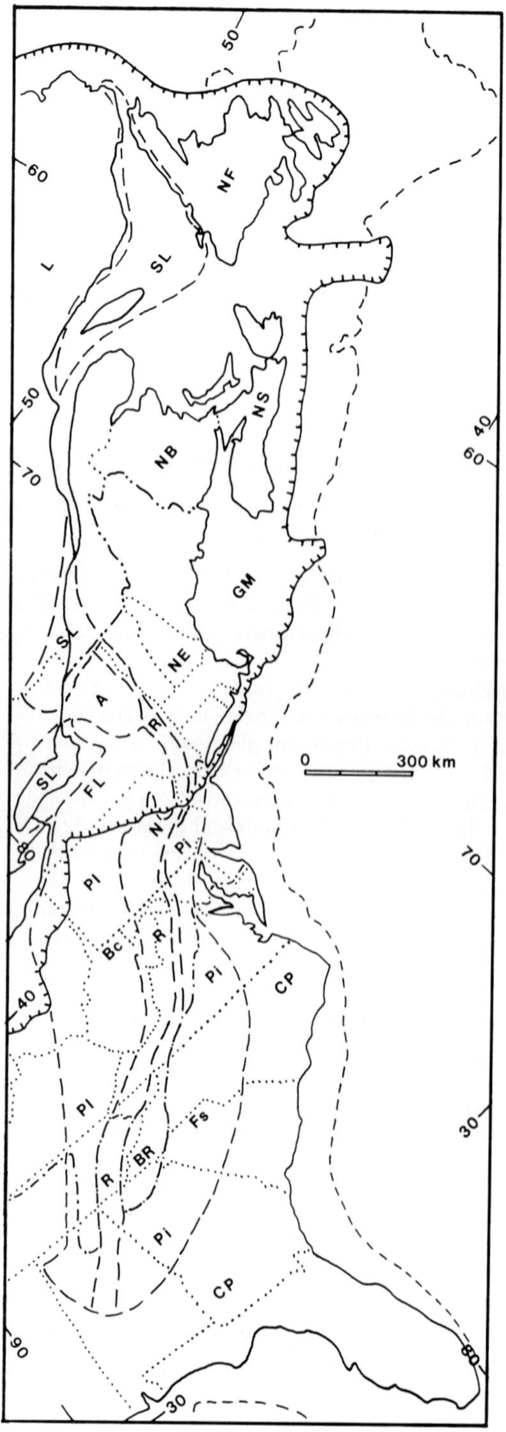

Coastal Plain south of Maryland. Much of the discussion relates the deep ocean sediment record to the terrestrial erosion record and the number, duration, and overall erosional effectiveness of Pliocene–Pleistocene events.

First, the marine ^{18}O oxygen isotope record in the North Atlantic is examined to determine the number and duration of Pliocene–Pleistocene cold climate events. Second, arguments are reviewed concerning the depth of glacial erosion of the Canadian shield and the Appalachians. Emphasis is placed on the amount of erosion necessary to account for the volume of glaciogenic marine sediment and on the duration of the glacial events that produced the sediment. A hypothesis is proposed that much of the northern Appalachians is far more deeply eroded by glaciation than presently recognized. Third, evidence will be summarized for extent and intensity of periglacial activity in the southern Appalachians. Three sites are used to exemplify the depth of periglacial erosion and the effectiveness of periglacial activity in modifying diverse parts of the southern Appalachians. It is proposed that periglacial activity during the Pleistocene has been considerably more effective than fluvial activity in modifying the southern Appalachians.

The marine ^{18}O record, a time–space diagram of glaciation

The onset of glacial activity in the Appalachians was the culmination of longer term Ter-

Fig. 1. Location map of Appalachian region. Appalachian physiographic provinces: *A*, Adirondack Mtns.; *BR*, Blue Ridge; *CP*, Coastal Plain; *L*, Laurentian Highlands; *NE*, New England; *Pi*, Piedmont; *Pl*, Plateau; *R*, Ridge and Valley; *SL*, St. Lawrence Lowland. Other features: *FL*, Finger Lakes; *GM*, Gulf of Maine; *NB*, New Brunswick; *NF*, Newfoundland; *NS*, Nova Scotia; *N*, Nesquehoning periglacial site; *Bc*, Buffalo Creek periglacial site; *Fs*, Flatwood School periglacial site. ⊥⊥⊥⊥ Late Wisconsin ice border (Hughes, 1987); - - physiographic province boundary; —·—·— mesic-thermic soil temperature boundary; —··— United States-Canada boundary; ...... state boundaries; ---- edge of continental shelf, 200 m contour.

tiary climatic cooling, especially in North America (Goudie, 1977; Frakes, 1979; Blackwelder, 1981; Donnelly, 1982; Prentice and Matthews, 1988). Glaciation began in North America in the Pliocene and intensified in the middle and late Pleistocene (Richmond and Fillerton, 1986). The most complete characterization of Pliocene–Pleistocene climatic change is the deep sea sediment foraminifera oxygen isotope (^{18}O) record. That record has been used to estimate ice volumes and climatic zonation for the Pleistocene (CLIMAP 1976, 1981, 1984; Weertman, 1976; Shackleton 1977, 1987; Imbrie and Imbrie, 1980; Birchfield et al., 1981; Mix, 1987).

While there is a consensus that the foraminifera ^{18}O variation is mostly from ice volume changes, some are uncertain about the amount of non-ice volume ^{18}O change, particularly water temperature effects (Mix, 1987). The use of tropical planktonic foraminifera minimizes possible water temperature effects and gives the most accurate ice volumes (Matthews and Poore, 1980; Crowley and Matthews, 1983; Prentice and Matthews, 1988), but those records have low sedimentation rates where bioturbation obscures details of the ^{18}O record (Shackleton, 1987). The most detailed ^{18}O record are from benthic foraminifera in areas of high sedimentation rate (Chappell and Shackleton, 1986; Shackleton, 1987).

The benthic ^{18}O record is affected by an abrupt bottom water temperature change going from interglacial to glacial conditions (Chappell and shackleton, 1986), but this temperature effect can be corrected for (Shackleton, 1987). With correction, the estimates of ice volume induced sea level lowering from the benthic ^{18}O record closely match sea level estimates from New Guinea marine terraces for the entire late Illinoian to Holocene interval (Shackleton, 1987, fig. 5). During that interval bottom water changed abruptly to the nearly minimum possible glacial temperature value at 115 ka, the beginning of the early Wisconsin, and then remained constant until the Holocene at 11 ka (Chappell and Shackleton, 1986; Shackleton, 1987). The change in bottom water temperature would only induce an error in the ^{18}O ice volume estimate at the transition back and forth from interglacial to glacial conditions. If only differing degrees of glacial conditions from one glaciation to another are being examined, as is the case in this paper, bottom water temperature is constant, and benthic ^{18}O records give an accurate estimate of ice volume that needs no water temperature correction. Thus, the benthic ^{18}O record can be considered a first or even second order approximation of ice volume throughout the Pleistocene (Shackleton, 1987).

Another implicit assumption is that the ^{18}O record is comparable from one glaciation to the next even if the linkage between ^{18}O record and ice volume is nonlinear and isotopic equilibrium is never reached at glacial maximum. The source of error is that average ^{18}O composition of ice sheets vary depending on their size and whether they are growing, decaying, or in steady-state (Mix and Ruddiman, 1984). As much as a 30% error in ice volume from one glaciation to the next can be caused by large differences (50%) in mean snow ^{18}O at maximum volume and large differences in ablation rate (200%) during ice growth (Mix and Ruddiman, 1984). If the differences in snow ^{18}O and ablation are small, then the ice volume error is correspondingly small. The close correlation of temperature corrected benthic ^{18}O sea levels to marine terrace sea levels (Shackleton, 1987) for differing degrees of glaciation from late Illinoian to late Wisconsin show no significant error relating to the size of glaciers or whether they are retreating or advancing. Also the basic similarity of the geographic areas covered from one glaciation to the next suggests that major differences between different glaciations in mean snow ^{18}O and ablation rate are unlikely. It is expected that ice volume errors from the nonlinear relationship of ^{18}O to ice volume, while real, are not more than a few percent of the total ice volume estimate for each glaciation.

A second related source of error is whether or

not the maximum cooling is of sufficient duration to produce the full steady-state volume of ice for that degree of cooling. The rapidity of ice volume change implied by the ^{18}O records themselves (Ruddiman and McIntyre, 1982) and the rapidity of the late Wisconsin Laurentide advance and retreat as shown by stratigraphic studies (Dreimanis and Goldthwait, 1973; Fullerton, 1986) suggest that only 5000–10,000 years are needed to bring ice to its maximum extent. Individual ^{18}O data points represent at least 5000 years and often 20,000 years, so it is expected that even single data point maximum spikes have enough duration to bring ice to its maximum volume for that degree of cooling.

The final assumption is that the ^{18}O amplitude proxy for ice volume is directly proportional to ice extent. The basic correspondence of the amplitudes of the ^{18}O record nearest to the Laurentide glacier and the terrestrial record of Laurentide extent suggest that ^{18}O amplitude is proportional to ice extent. The late Wisconsin stage 2 maximum, while relatively large, is exceeded by several much older maxima, and this pattern is what is observed in the Laurentide stratigraphic record of the midwestern United States (Richmond and Fullerton, 1986). This correspondence will be further examined in the discussion in the following section.

If it is accepted that the amplitudes of the ^{18}O record give an accurate representation of true ice volume and extent, then the ^{18}O record is, in effect, a space–time diagram of glacial advance and retreat. The remainder of the paper will treat the ^{18}O record as a first- or even second-order proxy of the true glacial space–time relationship, with ^{18}O maxima representing glacial maxima.

Number and magnitude of glacial events

For the Laurentide ice sheet that directly affected the Appalachians, the best record of climate change is in the adjacent North Atlantic deep sea record. The longest and most detailed North Atlantic ^{18}O record for the Pliocene–Pleistocene interval (Figs. 2A and 2B) is the benthic foraminifera record from DSDP site 552A (Shackleton et al., 1984; Shackleton and Hall, 1984; Zimmerman et al., 1984). Site 552A is one of the nearest sites to the North American Laurentide ice sheet and has a marked variation (Fig. 2A) in the ^{18}O amplitude of Wisconsin (stages 2 and 4), Illinoian (stage 6), and pre-Illinoian (stage 12 and 16) glaciations that matches the relative extent of those glaciations in the North American Laurentide record (Hallberg, 1986; Richmond and Fullerton, 1986). Other less northerly Atlantic ^{18}O records show similar but less marked differences in the amplitude of glacial maxima (Ruddiman et al., 1986a, b). In contrast, a composite planktonic foraminifera ^{18}O record (Fig. 3) (Imbrie et al., 1984) does not correspond well with the Laurentide record in that the late Wisconsin stage 2 amplitude equals or exceeds that of all older glaciations.

The onset of major glaciation in the North Atlantic region at site 552 is marked by a sharp increase in ice rafted sediment and ^{18}O values at 2.5 Ma (Fig. 2B) (Shackleton et al., 1984). This date for the onset of major glaciation has been verified at widely separated places in the North Atlantic: sites 607–609 (Ruddiman et al., 1986a, b), site 610 (Jansen and Sejrup, 1985) and site 646 (Shipboard Scientific Party, 1987). That first major glaciation reached its maxima at about 2.37–2.4 Ma and again at about 2.33 Ma. The maxima are similar in amplitude to the largest middle to late Pleistocene maxima of the last 850,000 yr (Fig. 2A). The large maxima between 2.33 and 2.4 Ma probably match the earliest evidence of extensive Laurentide ice in Nebraska (Hallberg, 1986) where till is overlain by eolian and fluvial deposits with a well developed paleosol that are in turn overlain by the 2.01 Ma Pearlette B ash bed. A 2.4 Ma age for that advance has been suggested by Boellstorff (1980). Before 2.5 Ma small scale ^{18}O oscillations indicate fluctuations in the Antarctic

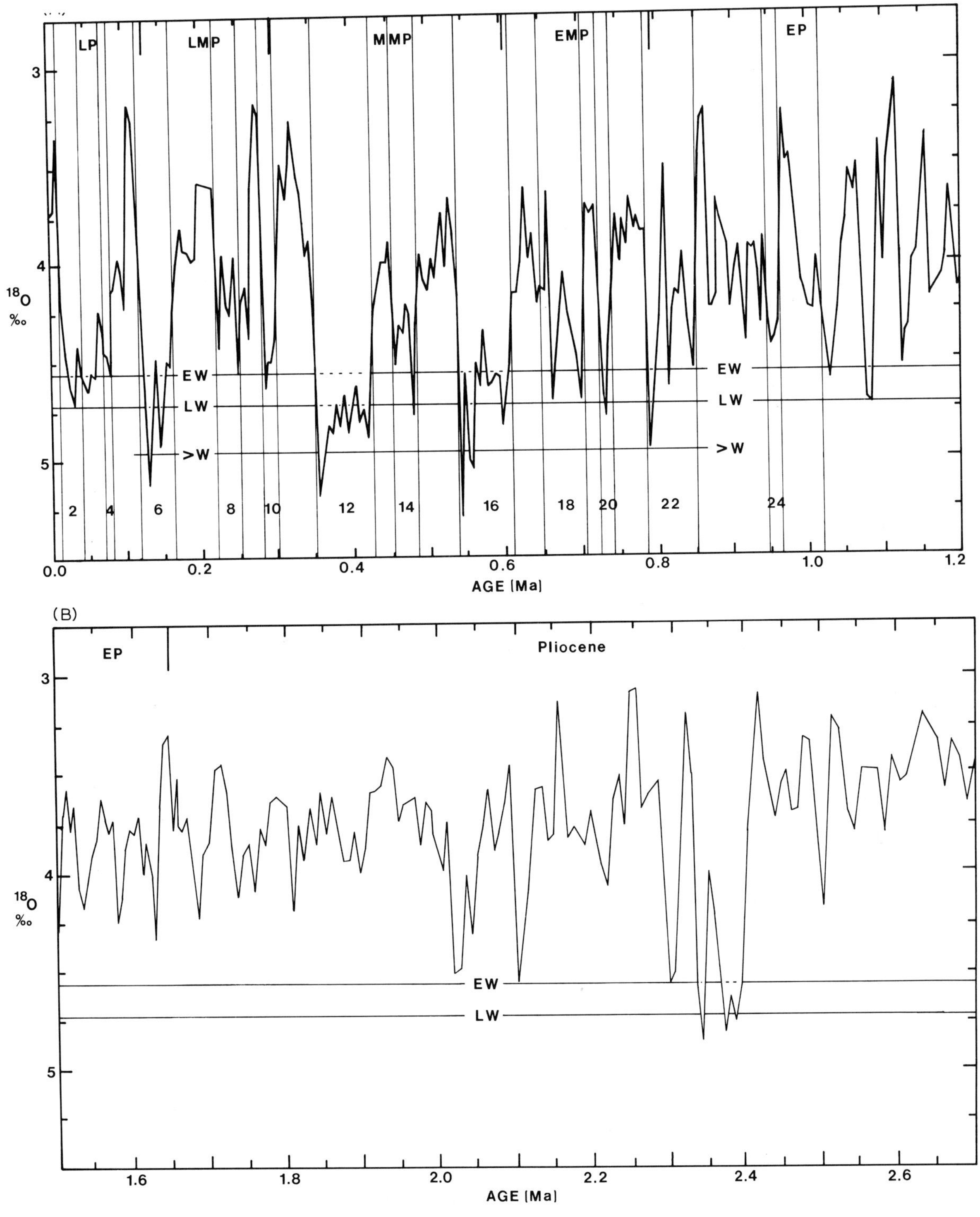

Fig. 2. (A) North Atlantic deep sea site 552A ^{18}O record, 0–1.2 Ma, showing large scale, variable amplitude ^{18}O maxima up to 0.85 Ma and smaller amplitude events before 0.85 Ma (modified from Shackleton and Hall, 1984; Zimmerman et al., 1984). *LP*, late Pleistocene; *LMP*, late middle Pleistocene; *MMP*, middle middle Pleistocene; *EMP*, early middle Pleistocene; *EP*, early Pleistocene; *EW*, early Wisconsin ^{18}O value; *LW*, late Wisconsin ^{18}O value; $> W$, 0.25% ^{18}O greater than late Wisconsin ^{18}O value. (B) North Atlantic deep sea site 522A benthic foraminifera ^{18}O record, 1.5–2.7 Ma, showing small amplitude ^{18}O events up to about 2.33 Ma, large amplitude ^{18}O events between 2.33 and 2.4 Ma, and yet smaller amplitude events before 2.4 Ma (modified from Shackleton and Hall, 1984; Zimmerman et al., 1984). *EP*, early Pleistocene; *EW*, early Wisconsin ^{18}O value; *LW*, late Wisconsin ^{18}O value.

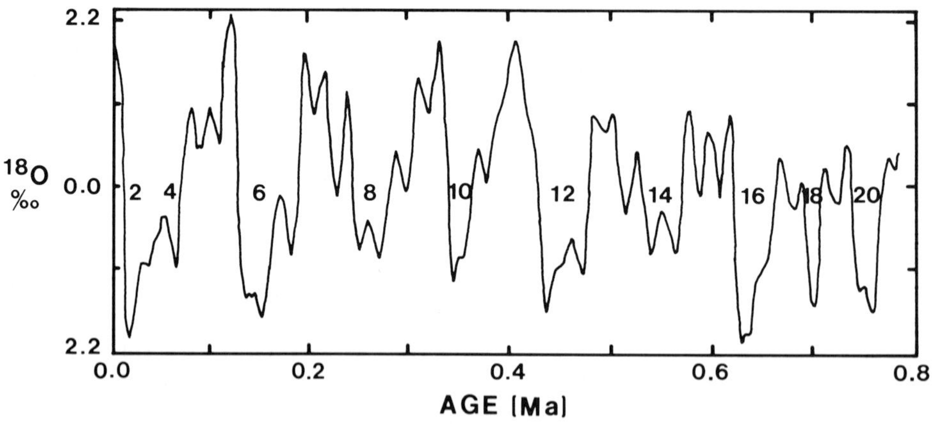

Fig. 3. Composite tropical planktonic ^{18}O record with a large-amplitude late Wisconsin stage 2 peak that equals or exceeds all other ^{18}O peaks of the last 0.8 Ma. The ^{18}O record is smoothed and stacked multiple record with ^{18}O values normalized to zero mean and plotted in standard deviation units (modified from Imbrie et al., 1984).

ice sheet, but the lack of ice rafted debris at site 552A suggests little North Atlantic ice before 2.5 Ma (Shackleton et al., 1984).

From 2.33 to 0.85 Ma at site 552A, the late Pliocene–early Pleistocene, there is a marked decline in the amount of glacially rafted sediment and the amplitude of the ^{18}O record is typically about one-half that of the onset of major glaciation at 2.4 Ma and glaciation after about 0.85 Ma (Fig. 2A and 2B) (Shackleton et al., 1984; Ruddiman et al., 1986b, c). This period of limited climate oscillation and restricted ice extent has been confirmed at North Atlantic DSDP sites 607 and 609 (Ruddiman et al., 1986a, b) and site 610 (Jansen and Sejrup, 1986). A lack of glacial till deposits between the Pearlette B ash at 2.01 Ma and the Pearlette S ash at 1.27 Ma beyond the late Wisconsin boundary in the Nebraska–Iowa region also support this period of limited ice extent (Hallberg, 1986; Fullerton and Richmond, 1986). The time of limited ice in the mid-continent probably extends to about 0.85 Ma because a particularly well-developed paleosol separates the Pearlette S ash from the next overlying till unit. Limited ice and warmer than present climate at about 2 Ma is indicated by the occurrence of spruce forest of that age in North Greenland, 2500 km north of its present position (Funder et al., 1985).

Within the 2.33–0.85 Ma interval of site 552A, four events reach early Wisconsin stage 4 amplitude and one, at about 1.1 Ma, reaches late Wisconsin stage 2 amplitude (Fig. 2A and 2B). The three events between 1.05 and 1.15 Ma are clearly present at site 607 and 609 but their amplitude is near that of the other oscillations in the 2.33–0.85 Ma interval. This suggests that the 1.05–1.15 Ma events may not be as significant as implied by site 552A. The earlier relatively high amplitude events at 2.0–2.1 Ma are correlated by Richmond and Fullerton (1986) with the oldest till in Nebraska on the basis of a polarity reversal event in the till that they thought indicated the 2.01 Ma transition from the Reunion normal subchron to the Matuyama reversed chron. This oldest "Nebraska" till is from one of the most extensive Laurentide advances and best matches the older large magnitude 2.33–2.4 Ma ^{18}O events.

The last 850,000 years of the ^{18}O record at site 552A is remarkably like other ^{18}O records in showing repeated large scale glacial–interglacial oscillations except that site 552A has more marked differences in relative ^{18}O maxima (Fig. 2A). North Atlantic sites 607 and 609, south of 552A, (Ruddiman et al., 1986a) show differences in maxima amplitude more similar to 552A than to the equatorial sites, particularly the relatively large amplitude of stage 12 and 16. Other ^{18}O records of tropical foraminifera (Shackleton and Opdyke, 1973, 1976) and the composite record of Imbrie et al. (1984) have differences that are generally in the same direction but less pronounced than those in 552A. During the last 850,000 years, back through ^{18}O stage 22, there have been 10 major ice advances of varying extent in the site 552A record (Fig. 2A), and there is a one to one match of these episodes with Laurentide ice advances in the mid-western United States (Fullerton and Richmond, 1986; Richmond and Fullerton, 1986, chart 1). Three ice advances, ^{18}O stages 6, 12, and 16, are clearly greater than the late Wisconsin stage 2 maximum (Fig. 2A). The same three stages also have the greatest maxima at sites 607 and 609 (Ruddiman et al., 1986a) and in a composite record from both the Atlantic and Pacific (Shackleton, 1987). These three stages have been tentatively correlated with glacial advances more extensive than the late Wisconsin in the mid-western United States (Hallberg, 1986; Johnson, 1986). Four ice advances, stages 18, 18a, 20, and 22, have amplitudes similar to or slightly larger than the late Wisconsin stage 2 (Fig. 2A). These four stages also have been tentatively correlated with glacial advances more extensive than the late Wisconsin in the mid-western United States (Hallberg, 1986; Johnson, 1986). Three ice advances, ^{18}O stages 8, 10, and 14, have distinctly smaller maxima than the late Wisconsin stage 2 (Fig. 2A). There is no stage 10 advance in the mid-western glacial record, suggesting that the advance was not as extensive as the late Wisconsin (Hallberg, 1986; Johnson, 1986; Fullerton, 1986). Stages 8 and 14 have been tentatively correlated with ice advances more extensive than the late Wisconsin (Hallberg, 1986, chart 1) and represent the main mismatch in the relative extent of ice between the Laurentide and site 552A records. The chronologic control of correlations of stage 8 through stage 14 Laurentide advances is uncertain and the correlations are open to alternate interpretations. Early Wisconsin ^{18}O stage 4 in most ^{18}O records and at site 552A is a relatively minor advance and this is supported by the Laurentide record (Eyles and Eyles, 1983; Andrews, 1987; Eyles and Westgate, 1987; Westgate et al., 1987; Winters et al., 1988; Curry, 1989). This basic correspondence of the marine and mid-western Laurentide record implies that the same number and extent of glacial advances should have also affected the Appalachians.

Duration of glacial events

The assumption that the ^{18}O record at site 552A represents a reasonably accurate space–time diagram of glaciation permits ready estimation of the duration of glacial events by the measurement of the width of each glacial maxima in the record. Stage 2 and 4 ^{18}O maxima values are used as baselines for measuring the width of ^{18}O maxima (Fig. 2 and Table 1) to determine the overall duration of glacial activity equivalent to or greater than that of the Wisconsin. As compared to other ^{18}O records, stages 12 and 16 in the site 552A ^{18}O record have exceptionally large amplitude, long duration ^{18}O maxima (Fig. 2A). The estimates of duration from the 552A record (Table 1) are regarded as maximum probable durations.

In contrast to the North Atlantic site 552A ^{18}O record, the composite tropical planktonic foraminifers ^{18}O record of Imbrie et al. (1984) minimizes the duration of ice advances equal to or greater than that of the late Wisconsin stage 2 (Fig. 3). This occurs because in the composite record the late Wisconsin stage 2 maximum is equalled by only one other glaciation, stage 16, and together the two glaciations are at the Wisconsin maximum extent for only a few percent of the last 0.85 Ma. The composite record is considered a minimum duration model of glacial advances.

Site 552A may overestimate the duration of glacial maxima if the sediment rate increases

TABLE 1

Time duration of glacial events in the Appalachians estimated from site 552A ^{18}O record

Glacial stage	Early Wisconsin extent (10^3 yr)		Late Wisconsin extent (10^3 yr)		Greater than late Wisconsin extent (10^3 yr)	
	A	B	A	B	A	B
2	24.4	19.6	2.0*	2.0*	0	0
6	26.7	42.8	16.7	26.7	4.4	7.1
8	0	0	0	0	0	0
10	2.2	1.6	0	0	0	0
12	73.3	52.8	72.2	52.0	10.0	7.2
14	4.4	7.1	2.2	3.6	0	0
16	46.7	37.4	31.1	24.9	10.0	8.0
18	8.9	10.0	0	0	0	0
20	7.8	10.3	3.3	4.4	0	0
22	13.3	13.3**	6.7	6.7**	0	0
Total	207.7	194.9	134.2	120.3	24.4	22.3
% of post 0.85 Ma	24.4	22.9	15.8	14.2	2.9	2.6
Late Pliocene	11.1		38.9			

*Assumed duration of ice at late Wisconsin limit.
**No astronomic date available.
A = duration from 552A time scale (Shackleton and others, 1984); B = duration from astronomical stage boundaries (Richmond and Fullerton, 1986; Johnson, 1982).

during glaciation due to ice rafting of debris. The time scale of the ^{18}O record shown in Fig. 2 has been constructed by assuming a constant sedimentation rate and linearly scaling the core depth time-wise between identified magnetic reversals. If there is a higher sedimentation rate during glaciation, the core thickness and the time scale derived from it would be too long. It is presently unknown how much different the glacial versus interglacial sedimentation rate is at site 552A. To evaluate the accuracy of the site 552A time scale, the astronomical ^{18}O stage ages (Johnson, 1982; Fullerton and Richmond, 1986) are also used in calculating the durations (Table 1). The duration of some individual maxima vary markedly depending on the time scale used, but the total duration differs by only a few percent between the two time scales (Table 1) and implies that the site 552A time scale is basically equivalent to other proposed ^{18}O time scales.

The calculated glacial maxima durations from site 552A indicate that during the last 850,000 years conditions have been essentially as cold and ice as extensive as the late Wisconsin stage 2 for about 134,000 years or 15% of the interval (Table 1). During the same 850,000 year interval, conditions have been essentially as cold and ice as extensive as the early Wisconsin stage 4 for about 208,000 years or 25% of the time. Conditions have been markedly colder, 0.25% ^{18}O, than the late Wisconsin for about 24,000 years or 3% of the last 850,000 years. From 2.4–2.3 Ma, ice reached to or just beyond the late Wisconsin limit for about 11,000 years.

Relating these events to the Appalachians, ice extensive as that of the late Wisconsin first covered the northern Appalachians in the Pliocene at about 2.4 Ma. From 2.33 to 0.85 Ma, less severe climate oscillation may have brought repeated episodes of periglacial and possibly localized glacial activity to the northern Appalachians but no major glaciations. During the last 850,000 years, effectively the middle to late Pleistocene (Richmond and Fullerton, 1986), ice was just north of the Appalachians for about

25% of the interval, near the late Wisconsin extent for about 15% of the interval, and beyond the late Wisconsin limit for 3% of the time. Essentially all of the observable glacial and periglacial erosion in the Appalachians is attributable to at most, only 208,000 years of the middle to late Pleistocene.

The durations of cold climate activity derived from site 552A gives a conservative estimate of the effectiveness of glacial and associated periglacial erosion. When the total change wrought by glacial and periglacial activity is divided by the long duration ^{18}O maxima of site 552A, a relatively small change per unit time or minimal erosional effectiveness results.

Glacial erosion

Entire Laurentide glacier

The effectiveness of glaciers in eroding the Appalachians must be considered in the context of the depth of glacial erosion of the Canadian Shield and of the total area covered by Laurentide ice. The depth of glacial erosion of the Shield has been estimated to vary from several hundreds of meters (White, 1972, 1988; Laine, 1980; Laine and Bell, 1982; Bell and Laine, 1980, 1985) to only a few meters or few ten's of meters (Ambrose, 1964; Flint, 1971; Gravenor, 1975; Sugden, 1976; Kaszycki and Shilts, 1979, 1980, 1987; Andrews, 1982).

Of the area covered by Laurentide ice 100–200 m of erosion is inferred by the total volume of late Pliocene–Pleistocene glaciogenic sediment in the oceans, conservatively estimated to be 1.62×10^6 km^3 (Laine, 1980; Laine and Bell, 1982; Bell and Laine, 1985). Several hundred meters of local to regional erosion is also inferred from deep, elongate lake basins and estuaries distributed around the periphery of the Shield (White, 1972, 1988; Hughes, 1981, 1987). Only 10–20 m of glacial erosion of the Shield is inferred from the little additional exhumation of the Precambrian that has occurred in the Pleistocene (Ambrose, 1964; Gravenor, 1975; Sugden, 1976; Higgs, 1978), the origin of Hudson Bay as a sedimentary basin rather than an erosional feature (Stockwell et al., 1970; Sugden, 1976; Kaszycki and Shilts, 1987), and from the limited volume of terrestrial glacial deposits, both regionally and from specific source areas (Flint, 1971; Kaszycki and Shilts, 1979, 1980, 1987).

Limited erosion of the Shield is also implied by various reconstructions of the Laurentide ice sheet (Sugden, 1978; Budd and Smith, 1981; Boulton et al., 1985; Fisher et al., 1985; Hughes, 1981, 1987; Hughes et al., 1985). These reconstructions differ considerably in detail but generally agree that the Shield is a zone of minimal ice velocity on or near ice divides. The models also tend to agree that the glacier is partly or completely frozen to its bed over much of the shield during its maximum extent and progressively less so during recession. This combination of slow ice and frozen bed would tend to minimize erosion of the Shield. The models disagree over the thermal conditions at the glacial bed in Hudson Bay. The models range from erosive ice sliding on a thawed bed (Sugden, 1978), to nonerosive ice frozen to the bed (Budd and Smith, 1981; Boulton et al., 1985; Hughes et al., 1985), including nonerosive ice on a highly deformable bed (Boulton et al., 1985), and thawed and frozen bed alternating time-wise, possibly maximizing erosion (Hughes, 1987). The reconstructions as a group do tend to agree that the southern margin of the Laurentide ice sheet had a thawed bed and was relatively erosive.

Bell and Laine's (1985) estimate of the volume of glaciogenic ocean sediment requires an average of 120–200 m of Pleistocene glacial erosion under the whole Laurentide ice sheet (Table 2). Kaszycki and Shilts (1987) argue for less than 20 m of Pleistocene glacial erosion of the crystalline rock of the shield and for essentially no erosion of the area north of the shield because the glacier was frozen to its bed. They do not dispute the total volume of glaciogenic ocean sediment, but argue that the ocean sediment is

TABLE 2

Erosion depths necessary to account for volume of glaciogenic ocean sediment

Geographic regions	Area (10^6 km^2)	Minimum erosion depth[a] (m)	Maximum erosion depth[b] (m)
Whole Laurentide area	13.4	120[c]	200[c]
Shield and area to north of shield	8.11	20[d]	20[d]
Hudson Bay and areas south and west of Shield[e]	5.28	300	500
Hudson Bay and areas south and west of Shield[f]	3.54	450	750

[a]Depth of erosion necessary to produce 1.6×10^6 km^3 of glaciogenic ocean sediment (Bell and Laine, 1985).
[b]Depth of erosion necessary to produce 2.7×10^6 km^3 of glaciogenic ocean sediment (Bell and Laine, 1985).
[c]Estimate of Bell and Laine (1985).
[d]Shield and sedimentary rock areas to the north where glacier was frozen to its bed. Both areas experienced only 20 m of Pleistocene glacial erosion according to Kaszycki and Shilts (1987).
[e]Areas of sedimentary rock within the shield and areas of all rock types around periphery of shield.
[f]One third reduction in area of the shield periphery region to account for areas of net glacial deposition and areas of crystalline rock.

too clay rich to be derived from crystalline Shield lithologies. It is their conception that the ocean sediment is almost exclusively from erodible fine-grained Paleozoic and younger sediments around the shield and beneath Hudson Bay and from a presumably thick preglacial regolith on the Shield. If there has been little Pleistocene glacial erosion of the Shield, then 300–750 m of erosion around the periphery of the Shield and under Hudson Bay are necessary to account for the volume of glaciogenic ocean sediment (Table 2). These depths of erosion are similar to those suggested by White (1972), for his arcs of exhumation containing "Zungenbecken" or tongue-shaped lakes. This amount of erosion around the periphery of the Shield also implies that ice is capable of carving much of the present relief of the Northern Appalachians.

Maritime provinces

Glacial erosion of the Canadian Maritime region of the Northern Appalachians (Fig. 1, GM, NB, NF, NS and SL) has been long thought to be insignificant (Flint et al., 1942; Flint, 1971). More recently, from evidence of ice flow directions, it has been proposed that Wisconsin Laurentide ice invaded only portions of New Brunswick and that thin local ice caps occupied Newfoundland and Nova Scotia (Prest and Grant, 1969; Grant, 1971, 1977; Boulton et al., 1985). The Gulf of St. Lawrence and the Gulf of Maine would be essentially ice free. Even the existence of local ice caps has been questioned due to the presence of weathered material and a lack of glacial scour in upland areas of Newfoundland and Nova Scotia (Grant, 1976, 1977; Brookes, 1982; Mayewski et al., 1981). If these views are correct, then minimal glacial erosion and modification of the Maritimes would be expected.

An alternative view holds that, at maximum extent, local ice domes as thick as 1.5 km existed over Newfoundland and Nova Scotia and helped feed major ice streams that occupied and scoured the Gulf of St. Lawrence and the Gulf of Maine (Borns, 1973; Borns and Hughes, 1977; Hughes, 1981; 1987; Hughes et al., 1985). Upon recession a local ice dome formed over the Gaspe Peninsula as ice was drawn-down by ice streams in the adjacent lowlands. The relative lack of glacial modification of the uplands is ascribed to relatively thin stagnant ice frozen to its bed, particularly near upland ice divides and uplands between ice streams (Sugden and John, 1976; Sugden and Watts, 1977; Sugden, 1978; Hughes, 1987).

Both minimum and maximum ice extent models imply that much of the upland interior of Newfoundland, Nova Scotia, and Gaspe Peninsula have undergone little glacial modification. The main disagreement is over whether or not the Gulf of Maine and the Gulf of St. Lawrence have been scoured by ice. The amount and distribution of glacio-isostatic uplift of coastal Maine (Stuiver and Borns, 1975) clearly indicates that Wisconsin ice occupied the Gulf of Maine rather than stopping at the present coast of Maine. The depth of channels and closed depressions in both gulfs indicates localized deep glacial scour (Oldale and Uchupi, 1970; White, 1988). The contrast in the depth of erosion between the estuaries and the uplands shows that glacial erosion can vary markedly in effectiveness due to the interplay of topography and ice dynamics.

New England

Several lines of evidence suggest that the depth of glacial erosion of crystalline rocks in New England (Fig. 1, NE) is only meters to tens of meters. The shallow depth is supported by the depth of plucking of the leeside of bedrock hills (Hahns, 1943), the incomplete removal of weathered bedrock regolith (Kaye, 1967; Flint, 1971) and the volume of glacial deposits mantling New England (Schafer and Hartshorn, 1965). Deeper erosion is possible if the 100 m depth of weathering observed at a few places (Kaye, 1967) was typical for the region (Flint, 1971) and that thickness was removed nearly everywhere to yield the present unweathered bedrock surfaces of New England (Goldthwait and Kruger, 1938).

The close correspondence of topography to rock type in New England has also been used as evidence that glaciation has produced only minor landform changes. Even though hills have been reshaped, valleys and lowlands deepened, and streams diverted by glaciation, the major topographic and drainage elements are primarily a response to the character and distribution of bedrock (Denny, 1982). The assumption is that preglacial, differential, fluvial erosion shaped the overall landscape and glaciation has only slightly modified that landscape. This reasoning implies that to be effective in modifying the landscape, glaciation must be nondifferential and erode different rock types equally. If anything, ice sliding on a thawed bed is a particularly effective agent of differential erosion and tends to accentuate rock type differences (Sugden and John, 1976).

Significant glacial modification of the submerged part of New England in the Gulf of Maine (Fig. 1, GM) is indicated by closed depressions with areas of thousands of km^2 and depths of more than 700 m (Oldale and Uchupi, 1970; Schlee and Pratt, 1970). While part of the basin's depth is from stripping of the Cretaceous and Tertiary cover from the crystalline basement, examination of available mapping (Klitgord et al., 1988) shows the lower 50–100 m of most of the basins are cut in Paleozoic and Mesozoic igneous and metamorphic rock. While the glaciers followed and did not destroy preglacial drainage patterns, they did significantly modify the region and scoured deep closed basins along valley axes (Oldale and Uchupi, 1970). This effective glacial modification may be related to particularly active ice streams in the Gulf of Maine (Borns, 1973; Hughes et al., 1985; Hughes, 1987), but while there is considerable ice flow convergence towards Northeast channel, the outlet for the Gulf of Maine, there is little convergence over much of the rest of the area of closed basins. It seems inescapable that glacial erosion has effectively sculptured the Gulf of Maine. The only question is whether or not the area was especially susceptible to such erosion.

A depth and effectiveness of glacial erosion similar to that of the Gulf of Maine for the terrestrial part of New England (Fig. 1, NE) is suggested by some peculiarities of the New England landscape. One such peculiarity is a 20 km wide belt along the north shore of Long Island Sound, where interfluves rise steeply to the north and at the same angle as the buried pre-Cretaceous basement (Denny, 1982). This indicates that, unlike areas to the south along the edge of the coastal plain, 200 m or so of Cretaceous material has been eroded to exhume the pre-Cretaceous erosion surface in geologically recent times (Denny, 1982). It has been suggested that this erosion was from Miocene and younger uplift and fluvial erosion (Denny, 1982). A likely alternate hypothesis is that Pleistocene glaciers did the eroding of the sediments since the erosion coincides with the only glaciated part of the Piedmont–Coastal Plain contact and is adjacent to the glacially scoured closed basin of Long Island Sound. A second landscape peculiarity is a series of notches in the drainage network elevation subenvelope (Denny, 1982, Fig. 11) that are transverse to the White and Boundary Mountain ranges. The notches represent narrow, lowlands projecting through and separating the highlands into individual mountain ranges. Their orientation subparallel to regional ice flow direction suggests the possibility that concentrated ice scour could have deepened shallow preexisting notches to their present "through valley" form.

Effective localized glacial erosion of the relatively resistant crystalline rocks of New England can also be postulated from the basal, thermal regime and overall glacial regimen of the New England sector of the Laurentide glacier. Greatest glacial erosion is associated with a thawed or refreezing bed, rapid ice motion, and great ice thickness (Boultin, 1972, 1974; Sugden and John, 1976). The maritime environment of New England and parts of the Canadian Maritime provinces would tend to generate the relatively warm snow temperature and high snow accumulation rates. The considerable local relief of much of New England would tend to localize the glacial erosion processes in the valleys leading to further differential erosion. The greatest effectiveness of ice erosion would be in increasing local relief through minimal upland erosion and maximal erosion in adjacent valleys or lowlands. The degree of glacial modification in the Gulf of Maine may be expressed in the terrestrial part of New England by more numerous and smaller closed basins, selective deepening of features oriented parallel to ice flow, and the erosion of Coastal Plain sediment from the edge of Piedmont.

Appalachian Plateau

The only part of the Appalachians where there is a consensus that glaciers have transformed the landscape is in the Finger Lakes region of the Appalachian Plateau (Fig. 1, FL). Early arguments did revolve around whether the erosion of the region was dominantly fluvial or dominantly glacial (Muller and Garrabrant, 1965; Coates, 1968). With recognition of the glacial origin of the Finger Lakes troughs, streamlined uplands, and "through valleys" (Tarr, 1906; Holmes, 1937; Fenneman, 1938; Von Engeln, 1961), recent arguments have centered on whether ice created an entirely new landscape (Clayton, 1965) or primarily enlarged and integrated preglacial, interglacial, and proglacial stream valleys (Coates, 1966, 1968, 1974; Clayton, 1966). Deepening of the Finger Lakes troughs is thought to be in excess of 300 m while intervening upland lowering is thought to be considerably less (Tarr, 1894, 1906; Clayton, 1965). Glacial modification of the landscape decreases southward across the Plateau (Clayton, 1965; Muller, 1965; Coates, 1974). Comparison of Plateau upland elevations to either side of the Wisconsin terminus suggest that glacial lowering of the plateau surface is less than 60 m (Denny, 1956; Muller, 1963, 1965). Consideration of both the lowering of uplands and the preferential erosion of valleys suggests an average erosion depth of more than 60 m (Coates and Kirkland, 1974).

The Finger Lakes region and the Appalachian plateau to the south of it exemplifies the interplay between topography, rock resistance, and ice thickness in determining the degree of glacial modification of the landscape. Where all factors work together, as in the Finger Lakes area, complete or near complete transformation of the landscape by glaciation occurs. Farther southward, more transverse topography, increasing rock resistance, and thinning ice progressively reduce obvious glacial shaping of the landscape (Clayton, 1965). The primary changes near the glacial limit are localized scouring of ice flow parallel valley segments and drainage diversions.

Erosion rates

Glacial erosion rates are best estimated from the volume of glaciogenic ocean sediment and the duration of glaciation as interpreted from the marine ^{18}O record. The ^{18}O record constrains the duration of ice cover available to do the erosion in the Northern Appalachians to between 134,000 and 208,000 years of the middle to late Pleistocene and 11,000–39,000 years of the late Pliocene (Table 1). If most of the erosion has been concentrated along the periphery of the Laurentide glacier, the 300–750 m of erosion necessary to account for the volume of glaciogenic ocean sediment occurred at most over 247,000 years, the early Wisconsin stage 4 duration baseline (Table 1), at a rate of 1.21–3.04 m ka^{-1}. These erosion rates are more than an order of magnitude higher than estimates of current fluvial erosion rates in the Appalachians of 0.02–0.03 m ka^{-1} (Sevon, this volume, table 1). The glacial erosion rates are certainly somewhat overestimated because some sediment was being produced by ice in Canada and interglacial times during the last 3 Ma covered by Bell and Laine's (1985) estimates. But if only the 120–200 m average erosion depth estimate for the entire ice sheet is used for the Appalachians, an average erosion rate of 0.49–0.81 m ka^{-1} is calculated, still more than an order of magnitude greater than the estimates of Appalachian fluvial erosion rates (Sevon, this volume, table 1). These calculations imply that Pleistocene glacial erosion is an order of magnitude more effective in modifying the Appalachian landscape than Pleistocene fluvial erosion.

The dominance of Pleistocene glaciation in eroding the Appalachians is supported by the total volume of Pleistocene marine sediment derived from the Maine to North Carolina region (Poag and Sevon, this volume). The Pleistocene sediment has been estimated to have accumulated at a rate of 38,000 km^3 Ma^{-1} (Poag and Sevon, this volume, fig. 3) and that requires about 170 m of erosion in the region during the 1.65 Ma of the Pleistocene. Current fluvial erosion rates, if representative of the remainder of the non-glacial portion of the Pleistocene, 1.44 ma, would provide only 29–43 m of erosion. This leaves about 125–140 m of erosion to be done by Appalachian glacial activity, essentially the amount estimated by Bell and Laine (1985) for the entire Laurentide glacier. Either Pleistocene glaciation has been an order of magnitude more effective in eroding the landscape of the Appalachians than previously thought or the interpretation of the marine sediment volume and ^{18}O record is in error by an order of magnitude.

An outrageous hypothesis?

A working hypothesis to consider is that the entire northern Appalachian landscape has been significantly eroded by ice and is evolving towards an equilibrium glacial landscape. Such a landscape is shaped to provide minimal resistance to maximal ice flow during full glacial conditions (Sugden, 1976; 1978). Whole landscapes are being streamlined and individual glacial troughs are being enlarged to steady-state form where they will be sufficient to handle all the glacial flow during glacial maximum conditions (Sugden, 1976; 1978). This process takes place most rapidly where ice is sliding across a thawed bed, the conditions inferred for much of the northern Appalachians (Sugden, 1978; Hughes, 1981, 1987; Boulton et al., 1985; Hughes et al., 1985). This process also takes place most rapidly where non-resistant rock forms topography parallel to ice flow and most slowly on more resistant rock and where topography is transverse to ice flow.

The most critical area for future research is

refining the Pleistocene marine sediment record. Further analysis of Site 552A to test the accuracy of the assumed constant sedimentation rate and linear time scaling is necessary to obtain better duration estimates. Additional North Atlantic sites with a complete Pliocene–Pleistocene record are needed to verify the accuracy of the Site 552A record. An updated glaciogenic sediment volume, keyed more closely to the period of glacial activity shown in the ^{18}O record, is needed to verify the present glaciogenic sediment volume estimates. The terrestrial glacial erosion record is much more difficult to interpret, but additional comparisons of areas to either side of the glacial limit having similar relief and geology may permit better characterization of differences between glacial and nonglacial landscape modification in the Appalachians.

Periglacial erosion

The extent of periglacial activity

The periglacial environment is characterized by intense freeze–thaw and mass wasting processes with or without permafrost and with or without tundra vegetation (Washburn, 1973, 1980). The Appalachians during glacial maximum would be expected to display a wide range of periglacial environments and a particularly wide range of latitudinal zones due to steepened southward climate gradients in front of the Laurentide ice (Pewe, 1983; Clark and Ciolkosz, 1988). Compilations of early work showed the widespread presence of periglacial features in the Appalachians (Brunnschweiler, 1962; Smith, 1962) but vegetation reconstructions questioned the presence of a periglacial environment, particularly the presence of tundra (Braun, 1950). A comparison of hillslope forms on the Appalachian Plateau near the glacial border and at a distance from the border showed distinct periglacial hillslope modification near the glacial border, but the depth of erosion was not addressed (Denny, 1951).

At present, periglacial features have been identified throughout the Appalachians south of the glacial border except for the Coastal Plain south of Delaware (Fig. 1, CP, Pi, B, R, and Pl) (Washburn, 1973, 1980; French, 1976; Pewe, 1983; Kaktins, 1986; Clark and Ciolkosz, 1988; Newell et al., 1988). The features that have been observed are patterned ground (sorted and nonsorted), rock and soil wedges, grezes litees and associated colluvial deposits, block fields and streams, cryoplanation terraces, and nivation hollows (Clark and Ciolkosz, 1988). Features indicative of permafrost, such as pingos and fossil ice wedge casts, have been described from New Jersey and Pennsylvania (Walters, 1978; Cronce, 1988; Marsh, 1987; Newell et al., 1988). The present mesic–thermic soil temperature boundary (Fig. 1) may mark the overall extent of the periglacial environment (Ciolkosz et al., 1986). That boundary encloses nearly all known periglacial features and appears to divide less well-developed soils on the north from thick, well developed soils on the south (Ciolkosz et al., 1986).

Vegetation reconstructions for the Wisconsin glacial maximum indicate a 50–150 km wide tundra existed in front of the ice in Pennsylvania and extended southward into West Virginia and Virginia at elevations above 1000 m (Watts, 1983). Spruce-pine boreal forest with local prairies extended to the southern edge of the Appalachians in Georgia, on the coastal plain to South Carolina, and interspersed with deciduous forest, may have reached the Gulf Coast (Watts, 1983). A somewhat more extensive tundra pattern is shown by using the 10% sedge and less than 20% spruce zones as an approximate open tundra marker and applying that to the 18 ka sedge and spruce pattern of Jacobson et al. (1987, plate 1). Using those criteria, the open tundra zone was about 150 km wide in New Jersey and widened westward to about 300 km in the Appalachian highlands where shallow permafrost indicators are also present. By 12 ka, when ice has receded to the north side of the Great Lakes and St. Lawrence lowland, tundra vegetation was still an important part of the Appalachian Highland flora, particularly on the Appalachian Plateau of Pennsylvania and West Virginia (Jacobson et al., 1987, plate 1). Also at 12 ka, New England and the Canadian Maritime Provinces were dominated by tundra sedge. At the same time, a large component of boreal forest was present in much of New England and the mid Atlantic States (Jacobson et al., 1987, plate 1).

While the periglacial landform and pollen records mutually support the presence of a periglacial environment in the Appalachian highlands at glacial maximum, this in itself says little about the effectiveness of that environment in modifying the landscape. The effectiveness question can only be answered by determining the amount of material moved, how long it took to move it, and how long the resulting landforms will survive as relics under a temperate fluvial erosion environment. There is a growing consensus (Pewe, 1983; Ciolkosz et al., 1986; Clark and Ciolkosz, 1988; Jacobson et al., this volume) that most if not all of the ubiquitous colluvial deposits in the Appalachians are the result of periglacial activity. What follows are three case studies from diverse parts of the Appalachians that have the colluvium volume data necessary to calculate periglacial erosion rates.

Erosion rates

The three sites are at different distances from the Pleistocene glacial limit and should have experienced different degrees of severity of periglacial activity. The most severe periglacial conditions would be expected at the Nesquehoning site (Fig. 1, N), a 450 m ridge-top within the Ridge and Valley of northeastern Pennsylvania, due to the proximity of the Pleistocene glacial limit. The area is within the pre-Illinoian border, 20 km south of the late Illinoian border, and 30 km south of the late Wisconsin border (Braun, 1988b). The ridge wraps around a valley where strip mines permit accurate determination of colluvium thickness (Fig. 4).

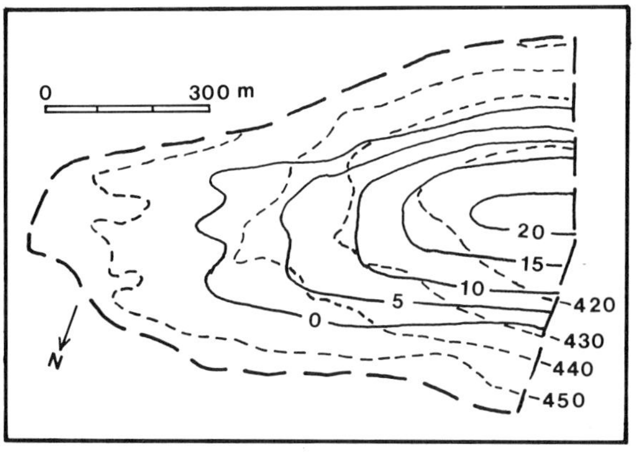

Fig. 4. Nesquehoning, Pennsylvania site used to calculate colluvium volume in order to estimate a periglacial erosion rate. The map shows the thickness, distribution, and upslope source area of the colluvium (Braun, unpubl. data). - - Drainage divide and basin boundary; ----- colluvium thickness, contour interval is 5 m; — elevation, contour interval is 10 m.

Somewhat less severe periglacial conditions would be expected at the 450 m Buffalo Creek basin site on the Appalachian Plateau in northern West Virginia (Fig. 1, Bc), a site about 150 km southwest of the late Wisconsin and the late Illinoian ice limits. Jacobson (1985, table 3.5) estimated colluvium thickness in 22 zero to first order coves in the basin. Minimum severity of periglacial conditions would be expected at the 290 m Flatwood School site (Eargle, 1977) on the Piedmont 30 km east of the Blue Ridge and a few km outside of the present mesic–thermic soil temperature boundary (Fig. 1, Fs). Data from Eargle (1977) was used to construct a colluvium thickness map for the site (Fig. 5).

The calculated periglacial erosion depth (Table 3) is greatest at the Nesquehoning site because of its proximity to the glacial limit, even though it has the most resistant rock of the three sites. The erosion depths at the Buffalo Creek and Flatwood School sites have similar ranges of values even though the sites are at markedly different elevations and distances from the glacial limit. The erosion depths at the Flatwood School site with its low elevation, gentle slopes, and distance from the glacial border may be due to the relative weakness of the unconsolidated

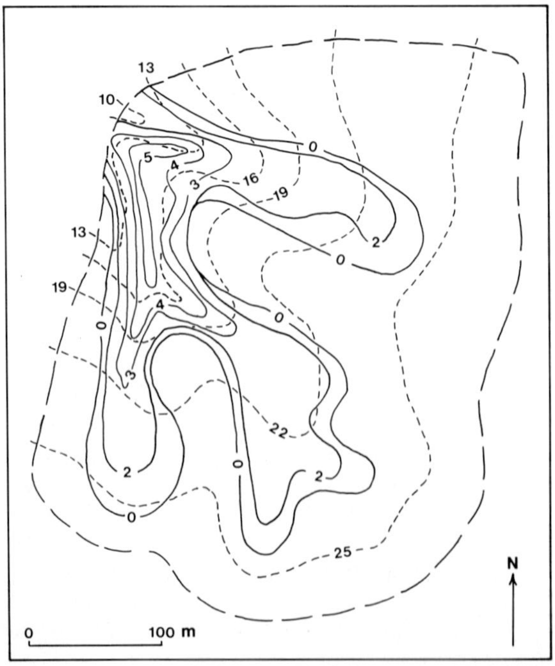

Fig. 5. Flatwood School, South Carolina site used to calculate colluvium volume in order to estimate a periglacial erosion rate. The map shows the thickness, distribution, and up-slope source area of the colluvium and was produced by combining Eargle's figs. 7 and 8 (1977) and the drainage divide from the Greer Quadrangle. - - Drainage divide and basin boundary; — colluvium thickness, contour interval is 1 m; ----- elevation, contour interval is 3 m.

saprolite under the site. All the calculated erosion depths are minimum estimates because only the volume of colluvium that presently remains at each site was used in the estimate. An unknown but probably significant amount of colluvial material left the sites during and after colluvium deposition.

The periglacial activity duration values (Table 3) were derived from the earlier discussion of the ^{18}O record and/or from regional glacial stratigraphic and age data. The deposition of colluvium at the Nesquehoning site started after the most extensive glaciation in Pennsylvania retreated from the site, probably ^{18}O stage 12 at about 440 ka (Braun, 1988b). The duration given for the Nesquehoning site was for cold climate activity from the end of ^{18}O stage 12 to present and was derived from the ^{18}O record (Table 1). At the Buffalo Creek site colluvium accumulation was thought to have occurred during the early Wisconsin (Altonian) over a period of 25,000 years (Jacobson, 1985). Recent work in Wisconsin and Illinois (Kempton et al., 1985; Winters et al., 1988; Curry, 1989), Ohio (Totten, 1987), New York (Braun, 1988a), Pennsylvania (Braun, 1988b), and Ontario, Canada (Eyles and Eyles, 1983; Eyles and Westgate, 1987; Westgate et al., 1987), indicates that early and middle Wisconsin ice was less extensive than late Wisconsin ice and did not extend beyond the north side of the Great Lakes. Likewise, the marine ^{18}O records consistently show the early and middle Wisconsin to be warmer than the late Wisconsin (Shackleton et al., 1983; Pisias et al., 1984; Martinson et al., 1987). Colluvium accumulation at the Buffalo Creek site would only be expected from when the ice approached the Appalachian Plateau at about 25 ka to when ice retreated from it at about 15 ka (Fullerton, 1986). At the Flatwood School site ^{14}C dating of peat under the colluvium (Whitehead and Barghoorn, 1962) suggests that the colluvium is only of late Wisconsin age. Periglacial activity would be expected only during the coldest part of the late Wisconsin when the ice was near its maximum extent.

The calculated periglacial erosion rates (Table 3, excluding Jacobson's highly conservative values) are of the order of magnitude necessary to account for the volume of Pleistocene aged ocean sediment derived from the Appalachians (Poag and Sevon, this volume, fig. 3) and are considerably higher than most of the 34 eastern United States fluvial erosion rates compiled by Sevon (this volume, table 1). The lowest periglacial erosion rate (3.4 cm ka^{-1}) exceeds all but the highest six of the 34 fluvial erosion rates. The periglacial erosion rates are from 3 to 13 times greater than the median value (2.2 cm ka^{-1}) of the 34 fluvial erosion rates. These high periglacial erosion rates suggest that periglacial activity is more effective in shaping the Appalachian landscape than temperate fluvial pro-

TABLE 3

Periglacial erosion rate data from three sites in the Appalachians

Site	Material being eroded	Up-slope source area (10^5 m^2)	Colluvium volume (10^5 m^3)	Average erosion depth (m)	Periglacial duration (10^3 yr)	Periglacial erosion rate (m 10^{-3} yr)
Nesquehoning (Fig. 4)	cong. and sandst.	2.17	19.2	8.84	30–60	0.293–0.146
Buffalo Creek mean of 22 coves	sandst. and shale	1.67[a]	2.13[b]	1.20[d]	25	0.018–0.048
			0.80[c]	0.45[d]		
(Jacobson, 1985)		1.13	2.13[b]	1.97[d]	10	0.075–0.197
			0.80[c]	0.75[d]		
Flatwood School (Fig. 5)	sandy-clay saprolite	0.82	1.00	1.22	5–10	0.244–0.122
Eastern U.S. erosion rates from 34 diverse sources (Sevon, this volume)			Median of all values			0.022
			mean excluding 2 highest			0.020
			Sevon value			0.024

[a]Total cove area (Jacobson, 1985, table 3.5).
[b]Assumed mean colluvium thickness = 4.0 m (Jacobson, 1985, table 3.5).
[c]Assumed mean colluvium thickness = 1.5 m (Jacobson, 1985, table 3.5).
[d]This mean erosion depth will differ slightly from the mean volume/mean area since the 22 individual data points were averaged as a final calculation.

cesses and are capable of providing the estimated Pleistocene volume of ocean sediment derived from the Appalachians.

Another outrageous hypothesis?

The effectiveness of periglacial erosion is borne out by the almost complete erosion and redeposition of regolith by periglacial activity during the last glacial maxima in the Buffalo Creek basin (Jacobson, 1985) and the Flatwood School basin (Eargle, 1977). The cumulative effect of three severe and two less severe periglacial events on the relatively resistant rock at the Nesquehoning site has been nearly 10 m of ridge-top erosion and 20 m of valley floor deposition. The eight to ten glacial and associated periglacial episodes of the last 850,000 yr are capable of repeatedly renewing the regolith and, collectively, are capable of having done ten's of meters of ridge-top reduction. A considerable amount of colluvium was produced by the fairly severe late Wisconsin episode; the much longer duration and more severe ^{18}O stage 12 and 16 events should have done proportionately more geomorphic work. Periglacial activity should be considered the dominant Pleistocene erosion process throughout the Appalachian Highlands and northern Coastal Plain.

The relative ineffectiveness of temperate fluvial processes in the Appalachians is borne out by the widespread preservation of periglacial landforms ranging from microforms such as patterned ground to macroforms such as smooth convexo-concave colluvium mantled slopes (Ciolkosz et al., 1986; Clark and Ciolkosz, 1988). Periglacial landforms and deposits persist as the most prominent features in the Central Appalachians despite the modifying effects of Holocene catastrophic fluvial and mass wasting events (Jacobson et al., this volume). The presence of argillic horizons and fragipans in colluvial soils is indicative of slope stability under Holocene temperate conditions. (Ciolkosz et al., 1986). The presence of older colluvial deposits in toe-slope positions buried by Wisconsin colluvium (Hoover, 1983; Waltman,

1985) is evidence for progressive reshaping of hillslopes by toeslope deposition and headslope erosion from repeated periglacial episodes that intervening fluvial processes have little altered. The ubiquity of relict periglacial forms in the Appalachians, particularly at high elevations and within 200 km of the glacial limit, suggests that the present landscape may be shaped more by periglacial activity than fluvial activity. A hypothesis that should be tested by further work is that much of the Appalachians south of the glacial limit are evolving towards an equilibrium periglacial form where the landscape is being shaped to provide just the slope form necessary to transport the debris provided by periglacial conditions.

The critical future research need is to obtain the three-dimensional geometry, stratigraphy, and age of the ubiquitous colluvium deposits in the Appalachians. A new generation of dating techniques such as thermoluminescence and resonance spin will be necessary to date the colluvium. Another key research area is to better determine the mode of deposition of the colluvium. Measured short-term mass wasting rates in present day periglacial environments are consistently too slow to produce the amount of colluvium in the time available during the glacial maxima (Washburn, 1980). Detailed analysis of Pleistocene deep sea terrigenous sediment is needed to provide a framework of times of greater and lesser sediment production from the Southern Appalachians. Again, the more continuous marine record should be invaluable in providing an overall chronology of Appalachian erosion.

Conclusions

The North Atlantic site 552A ^{81}O record is interpreted to represent a reasonably accurate time–space diagram of Laurentide glaciation (Fig. 2A and 2B). In that record, sufficient ice volume to reach to or beyond the late Wisconsin limit first occurred in the Pliocene between 2.4 and 2.33 Ma for a period of about 11,000 years (Table 1). Smaller scale ice volume changes from 2.33 to 0.85 Ma probably brought repeated periglacial and high elevation glacial activity to the Northern Appalachians. During the last 850,000 years, effectively the middle to late Pleistocene, eight out of ten glaciations covered the Northern Appalachians near to or somewhat beyond the late Wisconsin extent. For as much as 134,000 years (15%) of the last 850,000 years (Table 1), glacial and periglacial conditions as extensive as those of late Wisconsin ^{18}O stage 2 have affected the Appalachians. Ice as extensive as that of early Wisconsin ^{18}O stage 4 existed just north of the Appalachians for 208,000 years (25%) of the last 850,000 years (Table 1).

Pleistocene glacial erosion of the Appalachians of about 120–200 m is necessary to account for the estimated volume of Pleistocene ocean sediment derived from the Appalachians (Poag and Sevon, this volume) and the estimated volume of Pliocene–Pleistocene glaciogenic ocean sediment derived from the entire Laurentide ice sheet (Bell and Laine, 1985). This amount of glacial erosion of the Appalachians, occurring over the maximum expectable duration of Pliocene–Pleistocene ice cover of 247,000 years (Table 1), yields glacial erosion rates on the order of 0.5–0.8 m ka^{-1}. If glacial erosion of the Canadian Shield has been only 20 m, 300–750 m of erosion of the peripheral areas of the Laurentide glacier, including the Northern Appalachians, is necessary to produce the estimated volume of glaciogenic ocean sediment (Table 2). If this depth of erosion occurred during the estimated 247,000 year maximum duration of ice cover, calculated glacial erosion rates of 1–3 m ka^{-1} result. All these glacial erosion rates are more than an order of magnitude higher than estimates of current fluvial erosion rates (Sevon, this volume). This implies that glacial erosion during the Pleistocene has been far more important than fluvial erosion in shaping the Appalachians north of the glacial limit.

The Canadian maritime uplands have been only slightly eroded by thin, slowly flowing cold-based ice while adjacent lowlands and estuaries have been deeply scoured by thick, rapidly flowing warm-based ice. Closed basins in the Gulf of Maine indicate relatively deep erosion. A similar depth of erosion may be expressed in New England proper by more numerous, smaller closed basins, localized deepening of valleys parallel to ice flow, and stripping of Coastal Plain sediment. In the Finger Lakes region of the Appalachian Plateau, weak rock, ice parallel valleys, and thick warm-based ice provide optimum conditions for glacial sculpturing of the landscape.

The periglacial landform and boreal forest distributions support the conclusion that a periglacial environment existed throughout the Appalachian Highlands and the Coastal Plain into Delaware when ice was near the late Wisconsin limit. At a Ridge and Valley site near the ice front that experienced maximum climate severity, periglacial erosion rates of 0.15–0.30 m ka^{-1} (Table 3) show that periglacial activity could be an order of magnitude more effective in modifying the landscape than current fluvial erosion. On the Plateau and farther from the ice front but within the tundra zone, periglacial activity from a single glaciation has been effective in completely renewing the regolith and eroding the hillslopes at rates of 0.075–0.20 m ka^{-1} (Table 3), up to eight times present fluvial rates. Even where periglacial climate severity was least, as on the southern Piedmont, periglacial activity from a single glaciation has renewed much of the regolith by eroding slopes and filling valleys at rates exceeding present fluvial activity. A single cold period, the last one, has done 1–2 m of erosive work and implies that the cumulative effect of the eight to ten major cold events should produce as much as a few ten's of meters of ridge-top lowering during the Pleistocene. Periglacial activity should be considered the dominant Pleistocene erosion process throughout the Appalachian Highlands and northern Coastal Plain south of the glacial limit.

As the Pleistocene climatic oscillations continue, the dominance of periglacial and glacial activity in shaping the Appalachians should increase. The northern Appalachians may conceivably be evolving towards an equilibrium glacial form where the landscape is being sculptured to provide minimal resistance to maximal ice flow during full glacial conditions. Individual glacial troughs are being enlarged to maximum steady-state form where they will be sufficient to handle all the glacial flow provided by a glacial maximum, a concept suggested earlier by Sugden (1976, 1978). The Appalachians south of the glacial limit may be evolving towards an equilibrium periglacial form where the landscape is being shaped to provide just the slope form necessary to transport the debris provided by periglacial conditions. Where periglacial activity has been most effective, as in the Ridge and Valley in Pennsylvania, the present slopes may be so well adjusted to the more severe periglacial conditions that they are "super-stable" under present temperate fluvial conditions, a concept first suggested by Ciolkosz et al. (1986). Maximum disequilibrium and potential for rapid erosion would be expected during transitions from glacial to interglacial conditions but the duration of the transitions may not be sufficient to significantly affect the dominance of periglacial or glacial activity.

References

Ambrose, J.W., 1964. Exhumed paleoplains of the Precambrian Shield of North America. Am. J. Sci., 262: 817–857.

Andrews, J.T., 1982. Comment on "New evidence from beneath the Western North Atlantic for the depth of glacial erosion in Greenland and North America" by E.P. Laine. Quat. Res., 17: 123–124.

Andrews, J.T., 1987. The Late Wisconsin glaciation and deglaciation of the Laurentide Ice Sheet. In: W.F. Ruddiman and H.E. Wright, Jr. (Editors), North America and Adjacent Oceans During the Last Deglaciation. Geol. Soc. Am., Cent. Vol., K-3: 13–38.

Bell, M. and Laine, E.P., 1980. More evidence from the deep sea on the depth of glacial erosion in North America. EOS, 61 (17), p. 276.

Bell, M. and Laine, E.P., 1985. Erosion of the Laurentide Region of North America by glacial and glacio-fluvial processes. Quat. Res., 23: 154–174.

Birchfield, G.E., Weertman, J. and Lunde, A.T., 1981. A paleoclimate model of the northern hemisphere ice sheets. Quat. Res., 83: 462-472.

Blackwelder, B.W., 1981. Stratigraphy of Upper Pliocene and Lower Pleistocene marine and estuarine deposits of northeastern North Carolina and southeastern Virginia. U.S. Geol. Surv., Bull. 1502-B, 260: 427-438.

Boellstorff, John, 1980. North American pre-Wisconsinan stages in transition. Geol. Soc. Am., Abstr. Programs, 12(5) 220 (abstr.).

Borns, Jr., H.W., 1973. Late Wisconsin fluctuations of the Laurentide Ice Sheet in southern and eastern New England. In: R.F. Black, R.P. Goldthwait, and H.B. Willman (Editors), The Wisconsinan Stage. Geol. Soc. Am., Mem., 136: 37-45.

Borns, Jr., H.W. and Hughes, T.J., 1977. The implications of the Pineo Ridge readvance in Maine. Geogr. Phys. Quat., 31 (3-4): 203-206.

Boulton, G.S., 1972. The role of thermal regime in glacial sedimentation. Inst. Br. Geogr., Spec. Publ., 4: 1-19.

Boulton, G.S., 1974. Processes and Patterns of Glacial Erosion. In: D.R. Coates (Editor), Glacial Geomorphology. Publ. Geomorph., S.U.N.Y., Binghamton, N.Y., pp. 205-243.

Boulton, G.S., Smith, G.D., Jones, A.S. and Newsome, J., 1985. Glacial geology and glaciology of the last mid-latitude ice sheets. Geol. Soc. London, J., 142: 447-474.

Braun, D.D., 1988a. Wisconsinan deglacial history of Genesee Valley from the Terminal Moraine to the Valley Heads Moraine. In: W.J. Brennan (Editor), Late Wisconsinan Deglaciation of the Genese Valley. Guideb., 51st Ann. Meet., Friends Pleist., State Univ. Coll., Geneseo, N.Y., pp. 29-56.

Braun, D.D., 1988b. Glacial geology of the Anthracite and North Branch lowland regions. In: J.D. Inners (Editor), Bedrock and Glacial Geology of the North Branch Susquehana Lowland and the Eastern Middle Anthracite Field. Guideb., 53rd. Ann. Field Conf., Pa. Geol., pp. 3-25.

Braun, E.L., 1950. Deciduous forests of eastern North America. Blackston, Philadelphia, Pa., 596 pp.

Brookes, I.A., 1982. Ice marks in Newfoundland; A history of ideas. Geogr. Phys. Quat., 36 (1-2): 139-163.

Brunnschweiler, D., 1962. The periglacial realm in North America during the Wisconsin glaciation. Biul. Peryglacjalny, 11: 15-27.

Budd, W.F. and Smith, I.H., 1981. The growth and retreat of ice sheets in response to orbital radiation changes. Int. Assoc. Hydrol. Sci., Publ., 131: 369-409.

Ciolkosz, E.J., Cronce, R.C. and Sevon, W.D., 1986. Periglacial features in Pennsylvania. Pa. State Univ., Agron. Ser., 92, 15 pp.

Chappell, J. and Shackleton, N.J., 1966. Oxygen isotopes and sea level. Nature, 324: 137-140.

Chorley, R.J., Schumm, S.A. and Sugden, D.E., 1984. Geomorphology. Methuen, London, 605 pp.

Clark, G.M. and Ciolkosz, E.J., 1988. Periglacial geomorphology of the Appalachian Highlands and Interior Highlands south of the glacial border - A review. Geomorphology, 1: 191-220.

Clayton, K.M., 1965. Glacial erosion in the Finger Lakes region. z. Geomorphol., 9: 50-62.

Clayton, K.M., 1966. "Reply to Professor Coates". Z. Geomorphol., 10: 475-477.

CLIMAP Project Members, 1976. The surface of the ice-age Earth. Science, 191: 1131-1138.

CLIMAP Project Members, 1981. Seasonal reconstructions of the Earth's surface at the last glacial maximum. Geol. Soc. Am., Map and chart Ser., MC-36.

CLIMAP, Project Members, 1984. The last interglacial ocean. Quat. Res., 21 (2): 123-224.

Coates, D.R., 1966. Discussion of K.M. Clayton "Glacial erosion in the Finger Lakes region (New York, U.S.A.)". Z. Geomorphol., 10: 469-474.

Coates, D.R., 1968. Finger Lakes. In: R.W. Fairbridge (Editor), The Encyclopedia of Geomorphology. Reinhold, New York, N.Y., pp. 351-357.

Coates, D.R., 1974. Reappraisal of the glaciated Appalachian Plateau. In: D.R. Coates (Editor), Glacial Geomorphology, Publ. Geomorphol., S.U.N.Y., Binghamton, N.Y., pp. 205-243.

Coates, D.R. and Kirkland, J.T., 1974. Applications of a glacial model for large-scale terrain derangements. Symp. Quat. Environ., Proc., York Univ., Toronto, Ont., pp. 108-124.

Cronce, R.C., 1988. The genesis of soils overlying dolomite in the Nittany Valley of central Pennsylvania. Ph.D. Thesis, Pa. State Univ., Univ. Park, Pa., 389 pp.

Crowley, T.J. and Matthews, R.K., 1983. Isotope-plankton comparisons in a Late Quaternary core with a stable temperature history. Geology, 11: 275-278.

Curry, B.B., 1989. Absence of Altonian glaciation in Illinois, Quat. Res., 31(1): 1-13.

Denny, C.S., 1951. Pleistocene frost action near the border of the Wisconsin drift in Pennsylvania. Ohio J. Sci., 51: 116-125.

Denny, C.S., 1956. Surficial geology and geomorphology of Potter County, Pennsylvania. U.S. Geol. Surv., Prof. Pap., 288, 72 pp.

Denny, C.S., 1982. Geomorphology of New England. U.S. Geol. Surv., Prof. Pap., 1208, 16 pp.

Donnelly, T.W., 1982. Worldwide continental denudation and climatic deterioration during the late Tertiary: Evidence from deep-sea sediments. Geology, 10: 451-454.

Dreimanis, A. and Goldthwait, R.P., 1973. Wisconsin glaciation in the Huron, Erie, and Ontario lobes. In: R.F. Black, R.P. Goldthwait, and H.B. Willman (Editors), The Wisconsinan Stage. Geol. Soc. Am., Mem., 136: 71-106.

Eargle, D.H., 1977. Piedmont Pleistocene soils of the Spartaburg area, South Carolina. S.C. Div. Geol., Geol. Notes, 21: 57-74.

Eyles, C.H. and Eyles, N., 1983. Sedimentation in a large lake: A reinterpretation of the late Pleistocene stratigraphy at Scarborough Bluffs, Ontario, Canada. Geology, 11: 146-152.

Eyles, N. and Westgate, J.A., 1987. Restricted regional extent of the Laurentide Ice Sheet in the Great Lakes basins during early Wisconsin glaciation. Geology, 15: 537-540.

Fenneman, N.M., 1938. Physiography of Eastern United States. McGraw-Hill, N.Y., 691 pp.

Fisher, D.A., Reeh, N. and Langley, K., 1985. Objective reconstruction of the Late Wisconsinan Laurentide ice sheet and the significance of deformable beds. Geogr. Phys. Quat., 39(3): 229-238.

Flint, R.F., Demorest, M. and Washburn, A.L., 1942. Glaciation of the Shickshock Mountains, Gaspe Peninsula. Geol. Soc. Am., Bull., 53: 1211-1230.

Flint, R.F., 1971. Glacial and Quaternary Geology. Wiley, New York, N.Y., 892 pp.

Frakes, L.A., 1979. Climates throughout Geologic Time. Elsevier, Amsterdam, 453 pp.

French, H.M., 1976. The Periglacial Environment. Longman, London, 309 pp.

Fullerton, D.S., 1986. Stratigraphy and correlation of glacial deposits from Indiana to New York and New Jersey. In: V. Sibrava, D.Q. Bowen and G.M. Richmond (editors), Quaternary Glaciations in the Northern Hemisphere. Quat. Sci. Rev., 5: 23-38.

Fullerton, D.S. and Richmond, G.M., 1986. Comparison of the marine oxygen isotope record, the eustatic sea level record, and the chronology of glaciation in the United States of America. In: V. Sibrava, D.Q. Bowen and G.M. Richmond (Editors), Quaternary Glaciations in the Northern Hemisphere. Quat. Sci. Rev., 5: 183-196.

Funder, S., Abrahamsen, N., Bennike, O. and Feyling-Hanssen, R.W., 1985. Forested Arctic: Evidence from North Greenland. Geology, 13: 542-546.

Goldthwait, J.W. and Kruger, F.C., 1938. Weathered rock in and under the drift in New Hampshire. Geol. Soc. Am., Bull., 49: 1183-1198.

Goudie, A.S., 1977. Environmental Change. Oxford Univ. Press, Oxford, 234 pp.

Grant, D.R., 1971. Glaciation of Cape Breton Island, Nova Scotia. Geol. Surv. Can., Pap. 76-1B: 100-102.

Grant, D.R., 1976. Reconnaissance of early and middle Wisconsinan deposits along the Yarmouth-Digby coast of Nova Scotia. Geol. Surv. Can., Pap., 76-1B: 363-369.

Grant, D.R., 1977. Altitudinal weathering zones and glacial limits in western Newfoundland with particular reference to Gros Morne National Park. Geol. Surv. Can., Pap., 77-1A: 455-463.

Gravenor, C.P., 1975. Erosion by continental ice sheets. Am. J. Sci., 275: 594-604.

Hallberg, G.R., 1986. Pre-Wisconsin glacial stratigraphy of the Central Plains region in Iowa, Nebraska. In: V. Sibrava, D.Q. Bowen and G.M. Richmond (Editors), Quaternary Glaciations in the Northern Hemisphere, Quat. Sci. Rev., 5: 183-196.

Higgs, R., 1978. Provenance of Mesozoic and Cenozoic sediments from the Labrador and west Greenland continental margins. Can. J. Earth Sci., 1850-1860.

Holmes, C., 1937. Glacial erosion in a dissected plateau. Am. J. Sci., 33: 217-232.

Hoover, M.T., 1983. Soil development in colluvium in footslope positions in the Ridge and Valley Area of Pennsylvania. M.S. Thesis, Pa. State Univ., Univ. Park, Pa., 271 pp.

Hughes, T., 1981. Numerical reconstruction of paleo ice sheets. In G.H. Denton and T. Hughes (Editors), The Last Great Ice Sheets. Wiley, New York, N.Y., pp. 221-261.

Hughes, T., 1987. Ice dynamics and deglaciation models when ice sheets collapsed. In: W.F. Ruddiman and H.E. Wright Jr. (Editors), North America and Adjacent Oceans During the Last Deglaciation. Geol. Soc. Am., Cent. Vol., K-3: 183-219.

Hughes, T., Borns, H.W., Jr., Fastook, J.L., Hyland, M.R., Kite, J.S. and Lowell, T.V., 1985. Models of glacial reconstruction and deglaciation applied to Maritime Canada and New England. Geol. Soc. Am., Spec. Pap., 197: 139-150.

Imbrie, J. and Imbrie, J.Z., 1980. Modeling the climatic response to orbital variations. Science, 207: 943-953.

Imbrie, J., Hays, J.D., Martinson, D.G., McIntyre, A., Mix, A.C., Morley, J.J., Pisias, N.G., Prell, W.L. and Shackleton, N.J., 1984. The Orbital theory of Pleistocene climate; Support from a revised chronology of the marine ^{18}O record. In: A.L. Berger, J. Imbrie, J. Hays, G. Kukla and B. Salzman (Editors), Milankovitch and Climate. Boston, Massachusetts, Part 1. Reidel, Dordrecht, pp. 269-305.

Jacobson, G.L., Jr., Thompson, W., III. and Grimm, E.C., 1987. Patterns and rates of vegetation change during the deglaciation of eastern North America. In: W.F. Ruddiman and H.E. Wright Jr. (Editors), North America and Adjacent Oceans During the Last Deglaciation. Geol. Soc. Am., Cent. Vol. K-3, pp. 277-288.

Jacobson, R.B., 1985. Spatial and temporal distributions of slope processes in the upper Buffalo Creek drainage basin, Marion County, West Virginia. Ph.D. Thesis. The Johns Hopkins Univ., Baltimore, Md., 485 pp.

Jacobson, R.B., Miller, A.J. and Smith, J.A., 1989. The role of catastrophic geomorphic events in central Appalachian landscape evolution. In: T.W. Gardner and W.D. Sevon (Editors), Appalachian Geomorphology. Geomorphology, 2: 257-284 (this volume).

Jahn, R.H., 1943. Sheet structure in granites. Its origin and use as a measure of glacial erosion in New England. J. Geol., 51: 71-98.

Jansen, E. and Sejrup, H.P., 1986. Stable isotope stratigraphy and amino-acid epimerization for the last 2.4 m.y. at Site 610, Holes 610 and 610A. In: W.F. Ruddiman,

R.B. Kidd et al. (Editors), Initial Report of the Deep Sea Drilling Project. U.S. Gov. Print. Off., Washington, D.C., 94: 879-887.

Johnson, R.G., 1982. Brunhes-Matuyama magnetic reversal dated at 790,000 yr B.P. by marine-astronomical correlations. Quat. Res., 17(2): 135-147.

Johnson, W.H., 1986. Stratigraphy and correlation of the glacial deposits of the Lake Michigan lobe prior to 14 ka BP. In: V. Sibrava, D.Q. Bowen and G.M. Richmond (Editors), Quaternary Glaciations in the Northern Hemisphere. Quat. Sci. Rev., 5: 183-196.

Kaktins, T.L., 1986. Fluvial teraces of the Juniata River in central Pennsylvania. M.S. Thesis, Pa. State Univ., Univ. Park, Pa., 283 pp.

Kaszycki, C.A. and Shilts, W.W., 1979. Average depth of glacial erosion, Canadian Shield. Geol. Surv. Can., Pap., 79-1B: 395-396.

Kaszycki, C.A. and Shilts, W.W., 1980. Glacial erosion of the Canadian Shield; Calculation of average depths. At. Energ. Can. Ltd., Tech. Rep., TR-106, 37 pp.

Kaszycki, C.A. and Shilts, W.W., 1987. Glacial erosion of the Canadian Shield. In: Shilts, W.W., Aylsworth, J.M., Kaszycki, C.A. and Klassen, R.A., Canadian shield. In: W.L. Graf (Editor), Geomorphic Systems of North America. Geol. Soc. Am., Cent. Vol., 2: 155-161.

Kaye, C.A., 1967, Kaolinization of bedrock of the Boston, Massachusetts, area. U.S. Surv., Prof. Pap., 575: C165-172.

Kempton, J.P., Berg, R.C. and Follmer, L.R., 1985. Revision of the stratigraphy and nomenclature of glacial deposits in central northern Illinois. In: Illinoian and Wisconsinan stratigraphy and environments in northern Illinois, the Altonian revised. Ill. Dep. Energ. Nat. Resour., Guideb., 19: 1-20.

Klitgord, K.D., Hutchinson, D.R. and Schouten, H., 1988. Tectonic features and geophysical lineaments. In: R.E. Sheridan and J.A. Grow (Editors), The Atlantic Continental Margin. Geol. Soc. Am., Cent. Vol. I-2, Plate 2C.

Laine, E.P., 1980. New evidence from beneath the western North Atlantic for the depth of glacial erosion in Greenland and North America. Quat. Res., 14: 188-198.

Laine, E.P. and Bell, M., 1982. Reply to Andrew's Comment. Quat. Res., 17: 125-127.

Marsh, B., 1987. Pleistocene pingo scars in Pennsylvania. Geology, 15: 945-947.

Martinson, D.G., Pisias, N.G., Hay, J.D., Imbrie, J., Moore, T.C. and Shackleton, N.J., 1987. Age dating and the orbital theory of the ice age: Development of a high resolution 0 to 300,000-year chronostratigraphy. Quat. Res., 27: 1-27.

Matthews, R.K. and Poore, R.Z., 1980. Tertiary ^{18}O record and glacio-eustatic sea-level fluctuations. Geology, 8: 501-504.

Mayewski, P.A., Denton, G.H. and Hughes, T., 1981. Late Wisconsin ice sheets in North America. In: G.H. Denton and T. Hughes (Editors), The Last Great Ice-Sheets. Wiley, New York, N.Y., pp. 67-178.

Mix, A.C.., 1987. The oxygen-isotope record of glaciation. In: W.F. Ruddiman and H.E. Wrigth Jr. (Editors), North America and Adjacent Oceans During the Last Deglaciation. Geol. Soc. Am., Cent. Vol., K-3: 111-135.

Mix, A.C. and Ruddiman, W.F., 1984. Oxygen-isotope analyses and Pleistocene ice volumes. Quat. Res., 21: 1-20.

Muller, E.H., 1963. Reduction of preglacial erosion surfaces as a measure of the effectiveness of glacial erosion. Rep. VIth Int. Congr. Quat., Vol. III: Geomorphol. Sect., Warsaw, 1961, ICQ Publ., pp. 233-243.

Muller, E.H., 1965. Quaternary geology of New York. In: H.E. Wright Jr. and D.G. Frey (Editors), The Quaternary of the United States. Princeton Univ. Press, Princeton, N.J., pp. 99-112.

Muller, E.H. and Garrabrant, W.A., 1965. Historical note on studies of New York Quaternary geology. In: E.H. Muller (Editor) Bibliography of New York Quaternary Geology. N.Y. State Museum Sci. Serv., Bull., 398: 1-10.

Newell, W.L., Wyckoff, J.S., Owens, J.P. and Farnsworth, J.F., 1988. Cenozoic geology and geomorphology of the southern New Jersey Coastal Plain. 2nd Annu. Field Conf., SE. Friends Pleist., Guideb., 49 pp.

Oldale, R.N. and Uchupi, E., 1970. The Glaciated Shelf off the northeastern United States. U.S. Geol. Surv., Prof. Pap., 700-B: B167-B173.

Pewe, T.L., 1983. The periglacial environment in North America during Wisconsin time. In: S.C. Porter (Editor), The Late Pleistocene; Late-Quaternary Environments of the United States. Univ. of Minn. Press, Minneapolis, Minn., pp. 157-189.

Pisias, N.G., Martinson, D.G., Moore, T.C., Jr., Shackleton, N.J., Prell, W., Hays, J. and Boden, G., 1984. High resolution stratigraphic correlation of benthic oxygen isotopic records spanning the last 300,000 years. Mar. Geol., 56: 119-136.

Poag, C.W. and Sevon, W.D., 1989. A record of Appalachian denudation in post-rift Mesozoic and Cenozoic sedimentary deposits of the U.S. middle Atlantic continental margin. In: T.W. Gardner and W.D. Sevon (Editors), Appalachian Geomorphology. Geomorphology, 2: 119-157 (this volume).

Prentice, M.L. and Matthews, R.K., 1988. Cenozoic ice-volume history: Development of a composite oxygen isotope record. Geology, 16: 936-966.

Prest, V.K. and Grant, D.R., 1969. Retreat of the last ice sheet from the maritime provinces - Gulf of St. Lawrence region. Geol. Surv. Can., Pap., 69-33, 15 pp.

Richmond, G.M. and Fullerton, D.S., 1966. Summation of Quaternary glaciation in the United States of America. In: V. Sibrava, D.Q. Bowen and G.M. Richmond (Editors), Quaternary Glaciations in the Northern Hemisphere. Quat. Sci. Rev., 5: 183-196.

Ruddiman, W.F. and McIntyre, A., 1982. Severity and speed of Northern Hemisphere glaciation pulses. The limiting case? Geol. Soc. Am., Bull., 93: 1273-1279.

Ruddiman, W.F., Backman, J., Baldauf, J., Hopper, P., Keigwin, L., Miller, K., Raymo, M. and Thomas, E., 1986a. Leg 94 Paleoenvironmental Synthesis. In: W.F. Ruddiman, R.B. Kidd et al. (Editors), Initial Report Deep Sea Drilling Project. U.S. Gov. Print. Off., Washington, D.C., 94: 1207-1215.

Ruddiman, W.F., McIntyre, A. and Raymo, M., 1986b. Paleoenvironmental results from North Atlantic sites 607 and 609. In: W.F. Ruddiman, R.B. Kidd et al. (Editors), Initial Report Deep Sea Drilling Project, U.S. Govt. Print. Off., Washington, D.C., 94: 855-878.

Ruddiman, W.F., Raymo, M. and McIntyre, A., 1986c. Matuyama 41,000-year cycles: North Atlantic Ocean and northern hemisphere ice sheets. Earth Planet. Sci. Lett., 80: 117-129.

Schafer, J.P. and Hartshorn, J.H., 1965. The Quaternary of New England. In: H.E. Wright Jr. and D.G. Frey (Editors), The Quaternary of the United States. Princeton Univ. Press, Princeton, N.J., pp. 113-128.

Schlee, J. and Pratt, R.M., 1970. Atlantic continental shelf and slope of the United States - Gravels of the northeastern part. U.S. Geol. Surv., Prof. Pap., 529: H53-67.

Sevon, W.D., 1989. Erosion in the Juniata River drainage basin. In: T.W. Gardner and W.D. Sevon (Editors), Appalachian Geomorphology. Geomorphology, 2: 303-318 (this volume).

Shackleton, N.J., 1977. The oxygen isotope stratigraphic record of the late Pleistocene. Phil. Trans. R. Soc., 280: 169-179.

Shackleton, N.J., 1987. Oxygen isotopes, ice volume and sea level. Quat. Sci. Rev., 5: 183-190.

Shackleton, N.J. and Hall, M.A., 1984. Oxygen and carbon isotope stratigraphy of Deep Sea Drilling Project Hole 552A: Plio-Pleistocene glacial history. In: D.G. Roberts, D. Schnitker et al. (Editors), Initial Report of the Deep Sea Drilling Project. U.S. Gov. Print. Off., Washington, D.C. 81: 599-609.

Shackleton, N.J. and Opdyke, N.D., 1973. Oxygen isotope and paleomagnetic stratigraphy of Equatorial Pacific core V28-238: Oxygen isotope and paleomagnetic temperature and ice volumes on a 10,000 year and 100,000 year time scale. Quat. Res., 3: 39-55.

Shackleton, N.J. and Opdyke, N.D., 1976. Oxygen-isotope and paleomagnetic stratigraphy of Pacific core V28-238: late Pliocene to latest Pleistocene. Geol. Soc. Am., Mem., 145: 449-453.

Shackleton, N.J., Imbrie, J. and Hall, M.A., 1983. Oxygen isotope record of East Pacific core VI 9-30: Implications for the formation of deep water in the late Pleistocene North Atlantic. Earth Planet. Sci. Lett., 65: 233-234.

Shackleton, N.J., Backman, J., Zimmerman, H., Kent, D.V., Hall, M.A., Roberts, D.G., Schnitker, D., Baldauf, J.G., Desprairies, A., Homrighausen, R., Huddlestun, P., Keene, J.B., Kaltenback, A.J., Krumsiek K.A.O., Morton, A.C., Murray, J.W. and Westberg-Smith, J., 1984. Oxygen isotope calibration of the onset of ice-rafting and history of glaciation in the North Atlantic region. Nature, 307: 620-623.

Shipboard Scientific Party, 1987. Site 646. In: S.P. Srivastava, M. Arthur et al. (Editors), Proc., Init. Rep. (Pt. A), O.D.P., 104: 419-476.

Smith, H.T.U., 1962. Periglacial frost features and related phenomena in the United States. Biul. Peryglacjalny, 11: 325-342.

Stockwell, C.H., McGlynn, J.C., Emslie, R.F., Sanford, B.V., Norris, A.W., Donaldson, J.A., Fahrig, W.F. and Currie, K.L., 1970. Geology of the Canadian Shield. In: R.J.W. Douglas (Editor), Geology and Economic Minerals of Canada. Geol. Surv. Can., Econ. Geol. Rep., 1: 45-150.

Stuvier, M. and Borns, Jr., H.W., 1975. Late Quaternary marine invasion in Maine: Its chronology and associated crustal movement. Geol. Soc. Am., Bull., 86: 99-104.

Sugden, D.E., 1976. A case against deep erosion of shields by ice sheets. Geology, 4: 580-582.

Sugden, D.E., 1978. Glacial erosion by the Laurentide Ice Sheet. J. Glaciol., 20: 367-391.

Sugden, D.E. and John, B.S., 1976. Glaciers and Landscape, A Geomorphological Approach. Arnold, London, 376 pp.

Sugden, D.E. and Watts, S.H., 1977. Tors, felsenmeer and glaciation in northern Cumberland Peninsula, Baffin Island. Can. J. Earth Sci., 14(4): 514-534.

Tarr, R.S., 1894. Lake Cayuga, a rock basin. Geol. Soc. Am., Bull., 5: 339-356.

Tarr, R.S., 1906. Glacial erosion in the Finger Lake region of central New York. J. Geol., 14: 18-21.

Totten, S.M., 1987. Stratigraphy of tills in northern Ohio. In: Pre-Woodfordian stratigraphy of north-central Ohio. Midwest Friends Pleist., 34th Field Conf., Guideb., pp. 1-25.

Von Engeln, O.D., 1961. The Finger Lakes region: Its Origin and Nature. Cornell University Press, Ithaca, N.Y., 156 pp.

Walters, J.C., 1978. Polygonal patterned ground in central New Jersey. Quat. Res., 10: 42-54.

Waltman, W.J., 1985. The stratigraphy and genesis of pre-Wisconsinan soils in the Allegheny Plateau. Ph.D. Thesis, Pa. State Univ., Univ. Park, Pa., 353 pp.

Watts, W.A., 1983. Vegetational history of the eastern United States 25,000 to 10,000 years ago, In: S.C. Porter (Editor), The Late Pleistocene; Late-Quaternary Environments of the United States. Univ. of Minn. Press, Minneapolis, Minn., pp. 294-310.

Washburn, A.L., 1973. Periglacial Processes and Environments. Arnold, London, 320 pp.

Washburn, A.L., 1980. Geocryology. A Survey of Periglacial Processes and Environments. Wiley, New York, N.Y., 406 pp.

Weertman, J., 1976. Milankovitch solar radiation varia-

tions and ice age ice sheet sizes. Nature, 261: 17-20.
Westgate, J.A., Chen, F.J. and Delorme, L.D., 1987. Lacustrine ostracodes in the late Pleistocene Sunnybrook diamict of southern Ontario, Canada. Can. J. Earth Sci., 24: 1244-1253.
White, W.A., 1972. Deep erosion by continental ice sheets. Geol. Soc. Am., Bull., 83: 1037-1056.
White, W.A., 1988. More on deep glacial erosion by continental ice sheets and their tongues of distributary ice. Quat. Res., 30: 137-150.
Whitehead, D.R. and Barghoorn, E.S., 1962. Pollen analytical investigations of Pleistocene deposits from western North Carolina and South Carolina. Ecol. Monogr., 32: 347-369.
Winters, H.A., Alford, J.J. and Rieck, R.L., 1968. The anomalous Roxana Silt and mid-Wisconsinan events in and near southern Michigan. Quat. Res., 29: 25-35.
Zimmerman, H.B., Shackleton, N.J., Backman, J., Baldauf, J.G., Kaltenbach, A.J. and Morton, A.C., 1984. History of Plio-Pleistocene climate in the northeastern Atlantic, Deep Sea Drilling Project Hole 552A. In: D.G. Roberts, D. Schnitker et al. (Editors), Initial Report of the Deep Sea Drilling Project. U.S. Gov. Print. Off., Washington, D.C., 81: 861-875.

Geomorphology, 2 (1989) 257-284
Elsevier Science Publishers B.V., Amsterdam — Printed in The Netherlands

The Role of Catastrophic Geomorphic Events in Central Appalachian Landscape Evolution

ROBERT B. JACOBSON[1], ANDREW J. MILLER[2] and JAMES A. SMITH[3]

[1]*U.S. Geological Survey, Mailstop 926, Reston, VA 22092 (U.S.A.)*
[2]*Department of Geography, University of Maryland, Baltimore County, Baltimore, MD 21228 (U.S.A.)*
[3]*Hydrologic Research Lab, National Weather Service, Silver Spring, MD 20910 (U.S.A.)*

(Received February 20, 1989; accepted after revision June 7, 1989)

Abstract

Jacobson, R.B., Miller, A.J. and Smith, J.A., 1989. The role of catastrophic geomorphic events in central Appalachian landscape evolution. In: T.W. Gardner and W.D. Sevon (Editors), Appalachian Geomorphology. Geomorphology, 2: 257-284.

Catastrophic geomorphic events are taken as those that are large, sudden, and rare on human timescales. In the nonglaciated, low-seismicity central Appalachians, these are dominantly floods and landslides. Evaluation of the role of catastrophic events in landscape evolution includes assessment of their contributions to denudation and formation of prominent landscape features, and how they vary through space and time.

Tropical storm paths and topographic barriers at the Blue Ridge and Allegheny Front create significant climatic variability across the Appalachians. For moderate floods, the influence of basin geology is apparent in modifying severity of flooding, but for the most extreme events, flood discharges relate mainly to rainfall characteristics such as intensity, duration, storm size, and location. Landslide susceptibility relates more directly to geologic controls that determine what intensity and duration of rainfall will trigger slope instability.

Large floods and landslides are not necessarily effective in producing prominent geomorphic features. Large historic floods in the Piedmont have been minimally effective in producing prominent and persistent geomorphic features. In contrast, smaller floods in the Valley and Ridge produced erosional and depositional features that probably will require thousands of years to efface. Scars and deposits of debris slide-avalanches triggered on sandstone ridges recover slowly and persist much longer than scars and deposits of smaller landslides triggered on finer-grained regolith, even though the smaller landslides may have eroded greater aggregate volume.

The surficial stratigraphic record can be used to extend the spatial and temporal limits of our knowledge of catastrophic events. Many prominent alluvial and colluvial landforms in the central Appalachians are composed of sediments that were deposited by processes similar to those observed in historic catastrophic events. Available stratigraphic evidence shows two scales of temporal variation: one related to Quaternary climate changes and a more-recent, higher-frequency variation due to rare events during the Holocene. In much of the central Appalachians, landforms related to Quaternary climate changes persist as the most prominent features, despite the modifying effects of late-Holocene catastrophic events.

Introduction

Hack and Goodlett (1960) hypothesized that rare, large, geomorphic events may be the dominant sculptors of the central Appalachian landscape. At nearly the same time, Wolman and Miller (1960) developed the concept that moderate-sized events occurring with moderate frequency were responsible for transporting most of the sediment load on some landscapes. Both of these concepts, and Wolman and Gerson's (1978) elaboration of the magnitude/fre-

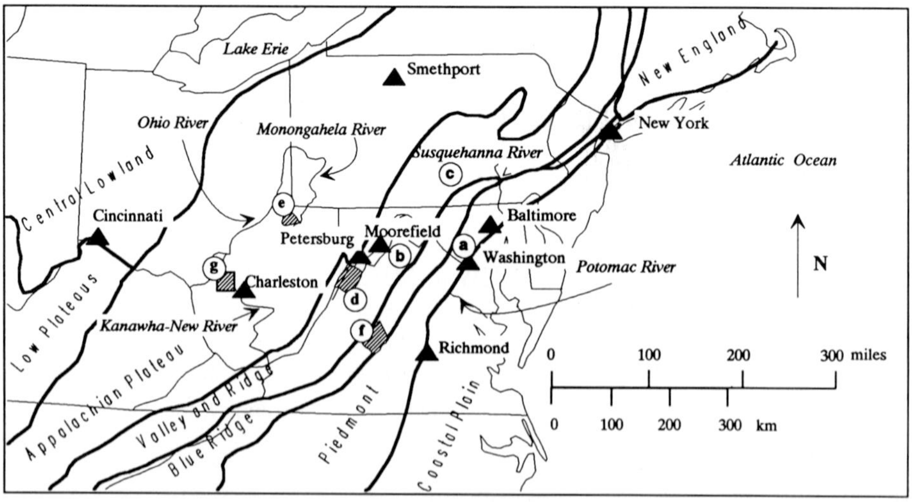

Fig. 1. Physiographic provinces and locations in the central Appalachians. a=Seneca Creek, b=Passage Creek, c=Little Lehigh Creek, d=Pendleton County, W. Va., e=Buffalo Creek basin, f=Nelson County, Va., g=St. Albans archeological site.

quency argument, were substantiated with data and examples from the central Appalachians, an area with great physiographic variety. Our goal in this paper is to examine the current state of knowledge of the role of catastrophic events in evolution of the landscapes of the central Appalachians. In this paper, the central Appalachians are considered to include the nonglaciated portions of the Piedmont, Blue Ridge, Valley and Ridge, and Plateau physiographic provinces of Pennsylvania, Maryland, Virginia, and West Virginia (Fig. 1).

Almost any geomorphic event could be considered catastrophic, depending on one's reference frame in time and space. From social and economic perspectives, however, there is little debate over the catastrophic effects of a few dozen flood and landslide events that have affected parts of the central Appalachians during historic time. Therefore, to provide a reference frame for discussion, and to emphasize applications in hazards mitigation, human time and spatial scales are adopted here for definition of the term "catastrophic". This paper considers events to be catastrophic that are rare on human timescales (one or two per generation to several during the Holocene) and sufficiently intense and (or) extensive as to threaten lives and property. In the central Appalachians these are usually storm-induced landslides and floods. An understanding of the role of these events in shaping the Appalachian landscape will allow better prediction of where and how often catastrophic flood and landslide hazards occur.

Assessment of the role of catastrophic geomorphic events in shaping the landscape should be cast in terms of magnitude, frequency, and effectiveness of geomorphic processes (Wolman and Miller, 1960; Wolman and Gerson, 1978). In these terms, two separable questions are apparent: are catastrophic events important in total geomorphic work and denudation? and, are catastrophic events important in producing prominent landforms? The answers to these questions vary with the highly variable geology and climate of the central Appalachians. Lack of sediment transport data precludes quantitative consideration of the first question, but observations of the causes and effects of extreme geomorphic events across the region apply to the second question. In the first part of this paper we consider some examples of historic catastrophic flood and landslide events to generalize about the spatial and temporal controls on present-day catastrophic events. In the second part of the paper we consider selected examples of the surficial stratigraphic record to evaluate the long-term effect of catastrophic events on the Appalachian landscape.

Controls on the spatial and temporal distributions of catastrophic geomorphic events

Variations in structure, lithology, and regolith have produced four distinctly different physiographic provinces in the central Appalachians (Fig. 1). The geologic framework and variable meteorological settings determine a wide range of factors controlling hydrologic and hillslope processes across the region.

Geologic framework

The Piedmont physiographic province is dominated by gently rolling topography underlain by thick saprolite regolith formed from schists, gneisses, and other metamorphic rocks. Small areas underlain by carbonates and mafic rocks have thinner regolith and the Mesozoic sedimentary rocks have thin soils and low relief.

In the Blue Ridge province, regolith thickness and relief vary markedly with bedrock lithology and texture (Pavich, 1986). Generally, quartzites, mafic rocks, carbonates, and massive siliceous igneous rocks have thin residuum and underlie a steep landscape. Metasedimentary and metaigneous rocks have thick saprolite and underlie a gently rolling landscape similar to the Piedmont. Thick colluvium occurs on some hillslopes and alluvium and debris are common in fans along the Blue Ridge. Debris is defined as poorly sorted to diamictic hillslope sediment transported in part by debris flow with variable fluvial reworking (Jacobson, 1988a).

In the folded and faulted Valley and Ridge, thin regolith is dominant on shaley lithologies and on tops of steep, quartzite ridges. Thicker colluvium and debris have collected at the bases of many slopes and may be tens of meters thick where deposited over dissolving limestone.

Some of the steepest land in the central Appalachians occurs in the Plateau province where major rivers have incised nearly flat lying Paleozoic sandstones; relief is more subdued in areas underlain by shales and limestones. Thin residual weathering profiles are dominant in the Plateau although thicker colluvium exists at high elevations, in hillslope hollows, and at the bases of slopes.

Meteorological patterns and hydrologic responses

Mean annual precipitation varies markedly across the four physiographic provinces (Fig. 2), from a maximum of over 1420 mm (56 in) in the Allegheny Mountains at the eastern edge of the Appalachian Plateau and along the crest of the Blue Ridge, to a minimum of less than 914 mm (36 in) at low elevations in the western Valley and Ridge. Approximately 125 mm (5 in) of the mean annual precipitation in the Appalachian Plateau and 30 mm (1.2 in) in the Blue Ridge is from snow.

The role of topography in determining precipitation characteristics is complex. More important than total altitude are the effects of the two major topographic boundaries, the Blue Ridge and Allegheny Front. The mechanisms by which the Blue Ridge and Allegheny Front influence precipitation are quite different. The Blue Ridge produces an elongate area with high values of mean annual precipitation which varies with that of the two neighboring provinces, as shown by high correlations of annual precipitation among three representative rain gages in the Piedmont, Blue Ridge, and Valley and Ridge, Table 1. In contrast, the Allegheny Front separates much more distinctly the climate of the Appalachian Plateau from the eastern physiographic provinces. Mean annual precipitation increases more than 500 mm over a distance of less than 48 km and, more significantly, annual precipitation in the eastern Appala-

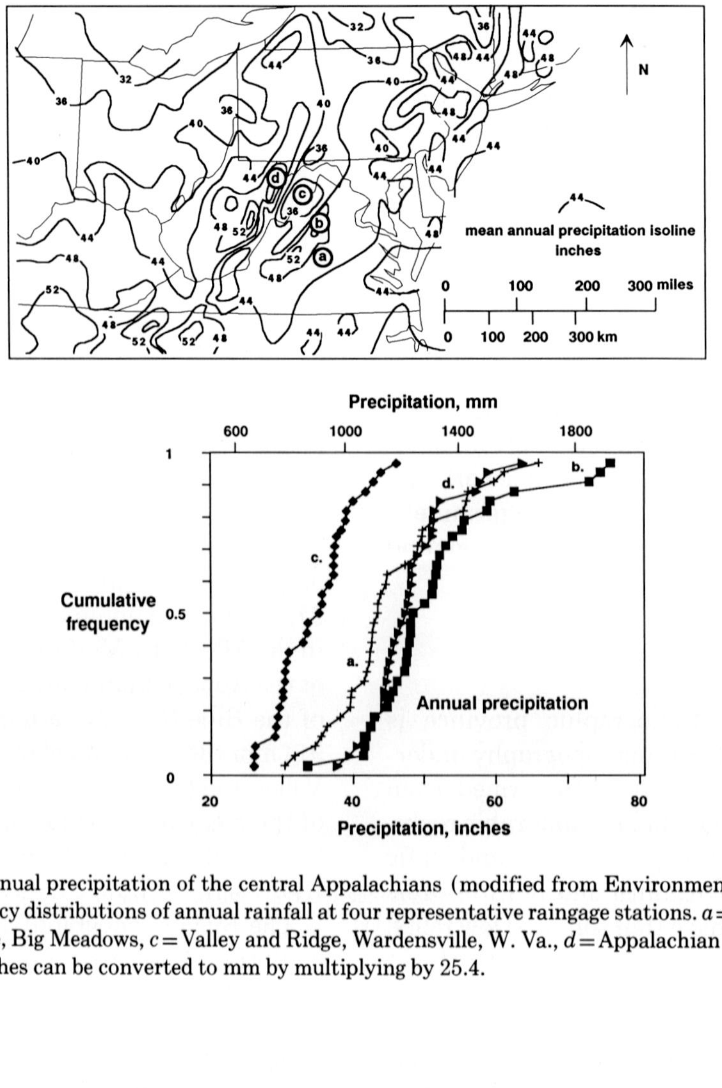

Fig. 2. Map of mean annual precipitation of the central Appalachians (modified from Environmental Data Service, 1968) and cumulative frequency distributions of annual rainfall at four representative raingage stations. a = Piedmont, Charlottesville, Va, b = Blue Ridge, Big Meadows, c = Valley and Ridge, Wardensville, W. Va., d = Appalachian Plateau Bayard, W. Va. Rainfall isohyets in inches can be converted to mm by multiplying by 25.4.

TABLE 1

Cross correlation of annual rainfall for four stations in the central Appalachians

	Big Meadows (Blue Ridge)	Charlottesville (Piedmont)	Wardensville (Valley and Ridge)	Bayard (Plateau)
Big Meadows	1			
Charlottesville	0.64	1		
Wardensville	0.75	0.70	1	
Bayard	0.27	0.29	0.31	1

chian Plateau is very poorly correlated with gages in the other provinces despite small distances between gages (Table 1, Fig. 2). The Appalachian Plateau experiences a significantly different precipitation regime on the annual time scale from the other provinces.

Values of extreme, short-duration rainfall generally decline in magnitude east to west across the central Appalachians, increasing locally along the Blue Ridge topographic boundary but not at the Allegheny Front (Fig. 3). Contrasts in extreme rainfall between the east-

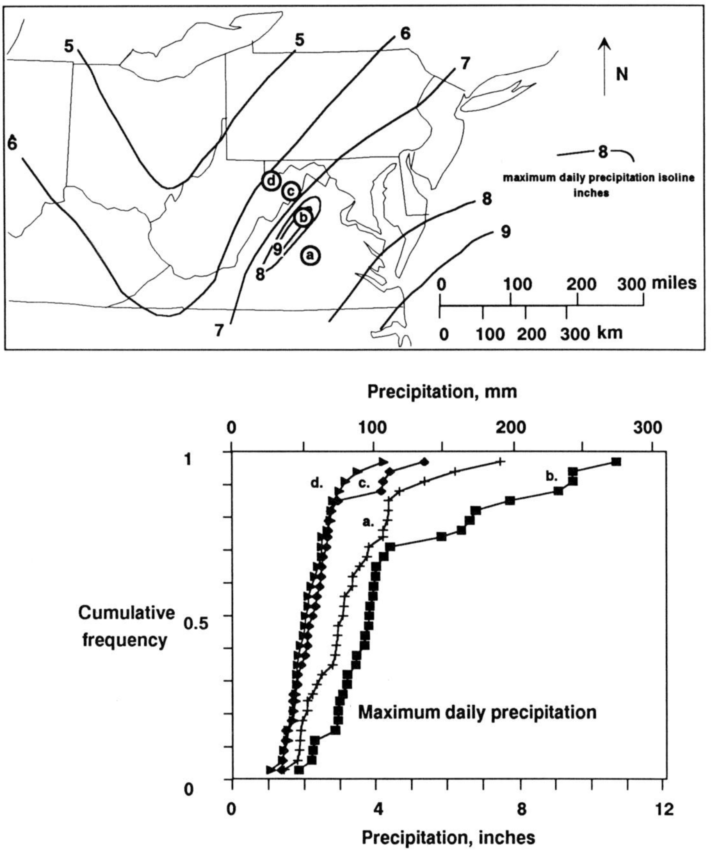

Fig. 3. Map of maximum daily rainfall with 100 yr recurrence interval (modified from Hershfield, 1961) and cumulative frequency distributions of annual maximum daily precipitation at four representative raingage stations: *a* = Piedmont, Charlottesville, Va, *b* = Blue Ridge, Big Meadows, *c* = Valley and Ridge, Wardensville, W. Va., *d* = Appalachian Plateau, Bayard, W. Va. Rainfall isohyets in inches can be converted to mm by multiplying by 25.4.

ern and western central Appalachians can be explained in large part by the typical paths of tropical storms along the coast (Neumann et al., 1981), that impose an east to west gradient in mean rainfall patterns.

To assess geomorphic effects of extreme storm rainfall at the basin scale, spatial variability of storms must also be considered. A statistical model of storm rainfall for the Valley and Ridge portion of the central Appalachians (based on storm data 1949–1981) depicts storms as consisting of intense cells representative of convective complexes (Smith and Karr, in press). Applying this model to Valley and Ridge storm data yields the following statistical description of storms: Storms with rainfall exceeding 125 mm over three days (measured at a point) occur with a frequency of 1 per 2.2 years over the 10,000 km^2 area; each storm consists of a mean density of 3 cells/10,000 km^2. The average rainfall at cell centers is 64 mm and the radial distance over which cell rainfall decreases by one half is 28 km. Virtually all of the rainfall of a single cell falls within a 50 km radius, corresponding to the meteorological small mesoscale (50–1000 km^2). Using the model, and assuming a stationary climate over the Holocene, we obtained the probability distribution of maximum 3-day totals at a point for the 10,000 km^2 area (Fig. 4). These results show that extrapolation of historic rainfall data over 1000–10,000 years predicts high probabilities

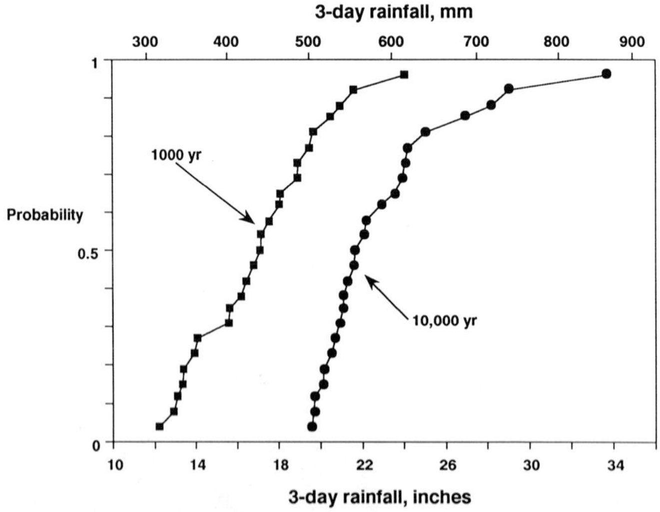

Fig. 4. Calculated probability distributions of maximum point rainfall for the central Appalachians over periods of 1000 and 10,000 yr.

that each point in the central Appalachians will be (and has been) subjected to extreme rainfall events on the order of 305-760 mm (12–30 in) over a 3-day period.

The tendency for extreme rainfall to occur in cells of 50–1000 km^2 area strongly determines the basin scales at which rainfall produces geomorphically significant floods. Decrease in maximum unit discharge with increasing basin size is controlled in part by organization of extreme rainfall at the small mesoscale (Fig. 5a). Note that maximum unit discharge and variability in maximum unit discharge decrease sharply beyond approximately 400 km^2; basin areas of maximum unit discharge are generally less than corresponding storm cell areas because storm cells usually distribute rainfall to two or more basins.

The role of spatial inhomogeneities of extreme rainfall in determining extreme flood characteristics is even more pronounced when contrasting only Appalachian Plateau and Piedmont basins (Fig. 5b). The contrast in values of extreme, short-duration rainfall between the Plateau and the Piedmont (Fig. 3) are clearly reflected in extreme floods. Unit discharges at small drainage areas are much larger in the Piedmont than in the Plateau. Except for the anomalous 1985 storm, which resulted from complex interactions of a tropical cyclone with two low-pressure systems (Jacobson et al., 1989), big floods in the Appalachian Plateau are much smaller than big floods in the Piedmont. For all the Piedmont basins included in Fig. 5b, the flood of record is associated with a tropical storm. In contrast, the floods of record for the Plateau basins are associated with a variety of meteorological events, including convective thunderstorms, frontal rainfall, and snowmelt.

In a discussion of the relative magnitudes and geomorphic effectiveness of several historic floods in the Potomac River Basin and adjacent watersheds, Miller (in press) describes basin-scale hydrologic response as a function of the intensity, duration, and area of precipation. Three discharge–drainage area envelope curves enclosing the maximum recorded values of discharge for each of three great floods on the Potomac River illustrate differences in extreme floods attributable to storm differences (Fig. 6, Table 2). Each flood has a steep rising limb; peak discharge increases downstream with increasing size of area contributing runoff, and the rising limbs of all three curves have similar slopes. On each curve there is a break in slope at a drainage area above which little or no in-

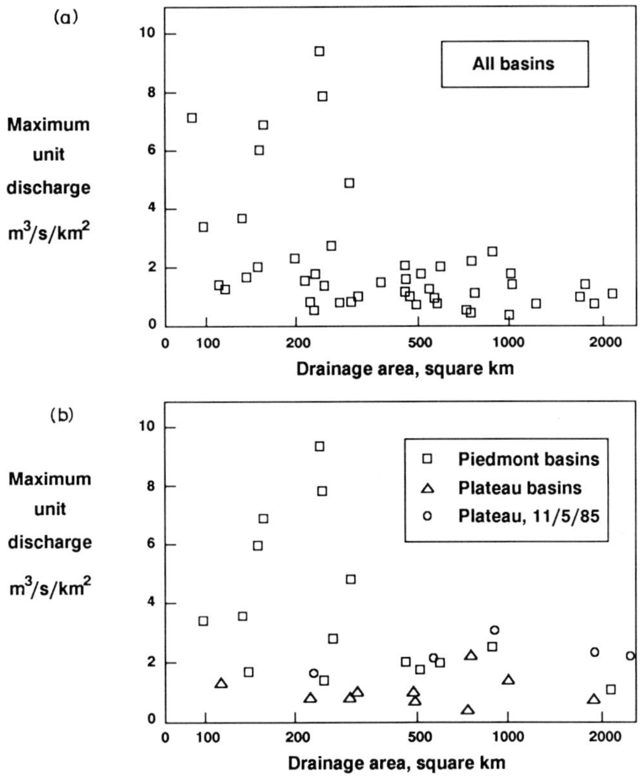

Fig. 5. (a) Plot of maximum unit discharge versus drainage area for 70 streamgage stations in the central Appalachians. (b) Plot of maximum unit discharge versus drainage area showing contrasts between Piedmont and Plateau floods. Modified from Smith (1988).

crease in peak discharge is observed. The flat limb of each curve represents translation and possibly some attenuation of the flood wave, with only modest contributions from tributary basins outside the areas of greatest precipitation depth.

The three floods were generated by three different kinds of precipitation events. The greatest intensity and smallest area of precipitation were observed in the 1949 storm, a conventional complex astride the Shenandoah and South Branch Potomac River basins (Hack and Goodlett, 1960), that affected basin areas less than 200 km^2. This storm triggered more than 100 debris avalanches and caused reworking of valley floors in small basins (Stringfield and Smith, 1956; Hack and Goodlett, 1960). The lowest intensity and largest area affected were observed in the 1936 flood produced by a series of winter–spring frontal storms, partly on snow and frozen ground (Grover, 1937). Despite widespread inundation of river valleys and extensive property damage, there are no reports indicating major geomorphic changes on slopes or on valley floors resulted from this event. The intermediate event, the 1985 storm, was produced by a tropical cyclone interacting with two extratropical low-pressure systems. Peak discharge of this storm occured at drainage areas between those of the 1936 and 1949 storms; valleys with drainage areas between 500 and 2000 km^2 incurred extreme channel widening and flood-plain erosion (Miller, 1987).

Analysis of other storms in the central Appalachians shows that flood response is not simply related to the meterological context of the storm, because storms of similar genesis can produce very different floods depending on in-

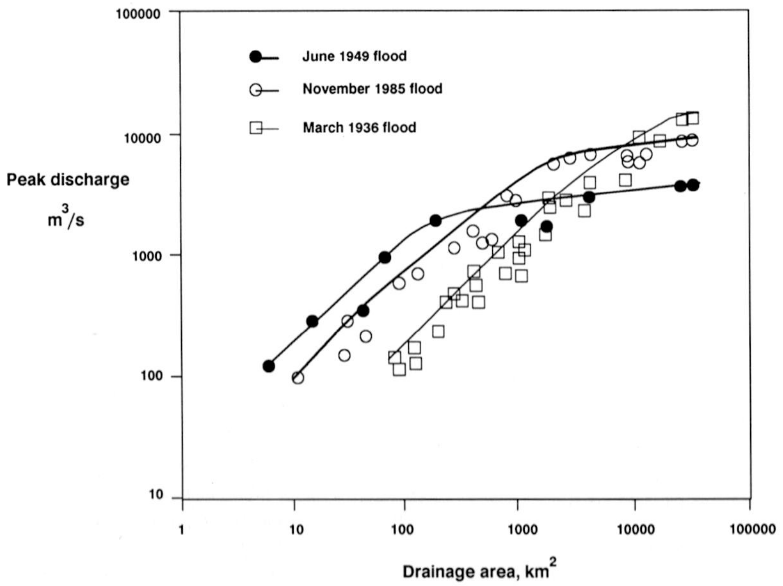

Fig. 6. Plot of peak discharge versus drainage area for three extreme floods on the Potomac River. Modified from Miller (in prep).

TABLE 2

Peak two-day precipitation totals for selected storms

Storm	Date	Official peak (mm)	Unofficial peak (mm)
Potomac, James, and upper Ohio floods (Grover, 1937)	March 1936	179	no data
Smethport, Pa. (Eisenlohr, 1952)	July 1942	209	909
Petersburg, W. Va. and North River basin, Va. (Stringfield and Smith, 1956)	June 1949	305	413
Johnstown, Pa. et al (Hoxit, 1982)	July 1977	305	no data
Hurricane Camille, Nelson County, Va. (Camp and Miller, 1970)	August 1969	278	711
Tropical storm Agnes, Md., Pa., and Va., (Bailey, 1975)	June 1972	442	406
November 1985 storm (Lescinsky, 1987)	November 1985	310	no data

tensity, duration, and area of the storm (Fig. 7). The Smethport storm (Fig. 1, July 1942), like the June 1949 storm, was a short-lived convectional system producing high precipitation intensity in a limited area. Camille and Agnes were both tropical cyclones, but Camille moved through the Appalachians very quickly and delivered all of its precipitation in a small area in less than 12 h, whereas Agnes, like the November 1985 storm, combined with an extratropical low-pressure system and, stalled by a high-pressure system off the east coast, delivered

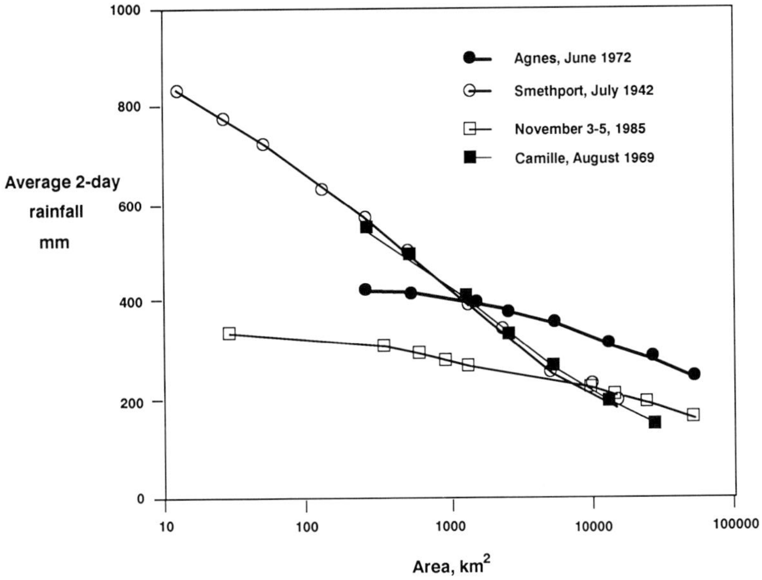

Fig. 7. Plot of rainfall depth (over two-day-duration) and area for four extreme storms in the central Appalachians. Modified from Miller (in press).

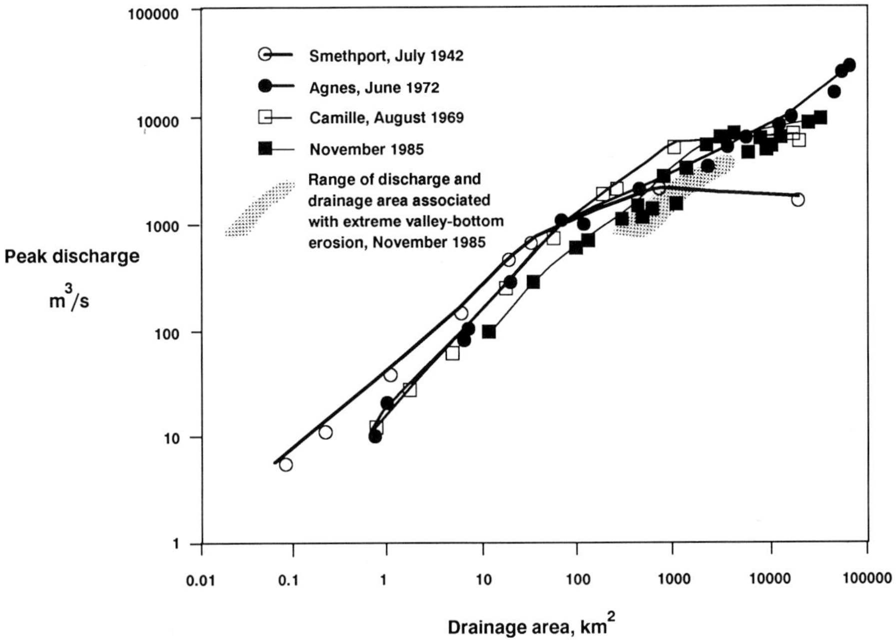

Fig. 8. Plot of peak discharge versus drainage area for four extreme floods in the Central Appalachians. Modified from Miller (in press).

precipitation over a broad region for several days.

Although the four storms form two distinct classes of rainfall depth–area curves, discharge–drainage area envelope curves for the resulting floods (Fig. 8) show greater diversity in flood responses, with peak discharges occurring over a range of 700–70,000 km^2. The differences in discharge–drainage area curves are presumably related to different combinations

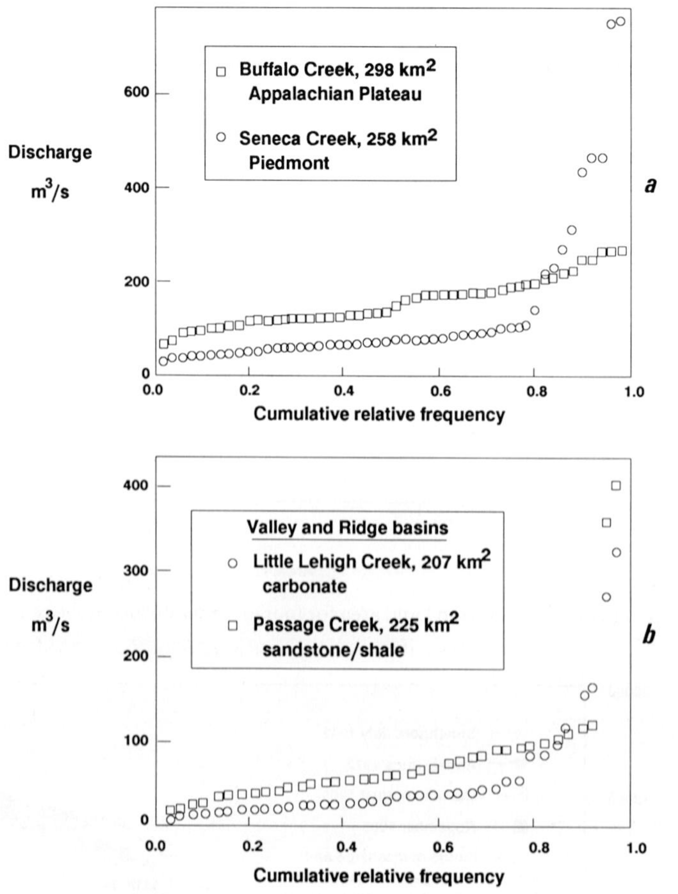

Fig. 9. (a) Plot of discharge versus probability that the discharge is equal to or less than the given value for two similar size drainage basins in the Piedmont and Plateau. (b) Plot of discharge versus probability that the discharge is equal to or less than the given value for two similar size drainage basins in the Valley and Ridge on different bedrock types. Modified from Smith (1988).

of rainfall intensity and duration, and differing landscape hydrologic responses.

Geologic controls are capable of modifying flood responses to moderate and high-magnitude storms but are ineffective in mediating the most extreme events. In a comparison of the annual peak flow distributions for two streamgage stations of similar drainage area, Seneca Creek in the Piedmont, and Buffalo Creek in the Appalachian Plateau, annual peak quantiles for Seneca Creek are lower than Buffalo Creek for discharges less than the 80th percentile (Fig. 9a). This reflects the higher storage capacity of thick saprolite in the Piedmont compared to the thin discontinuous, colluvial and residual soils of the Plateau. In the upper tail of the annual peak distribution, however, Seneca Creek discharges are much higher, reflecting the dominant effect of extreme rainfall magnitudes associated with tropical storms.

In a comparison between two Valley and Ridge basins of similar size, Little Lehigh Creek, draining an area dominated by carbonates, and Passage Creek, draining an area dominated by sandstone and shale, the effects of geologic control are apparent for annual discharges up to the 80th percentile but are less important for the extreme, infrequent storms (Fig. 9b). For the small and moderate-size events, the annual discharges from Passage Creek are larger than

Little Lehigh Creek, evidently reflecting the higher relief and lower storage available in the non-carbonate, Passage Creek basin. Similar dramatic effects of karst features on the central portion of annual flood peak distributions have been described by White (1976). For extreme floods, however, the discharge differences between the carbonate and sandstone/shale basins are negligible.

Landslide susceptibility and meteorological triggers

Landslides are often catastrophic geomorphic events. When landslides are triggered during a storm that also produces catastrophic flooding, the sediment supplied to the flood can increase its potential for geomorphic change.

As with hydrologic processes, landslide processes vary with physiography and climate across the central Appalachians. Factors influencing landslide susceptibility are interdependent with one another and rainfall characteristics. Intrinsic factors operating at a slope site include regolith shear strength, regolith thickness, regolith permeability, slope morphology, and shear strength provided by roots. The intrinsic factors interact with extrinsic factors, such as antecedent soil moisture levels, and intensity and duration of rainfall, to determine landslide susceptibility. At the slope scale, interaction of these factors make prediction of landslide susceptibility at a site extremely difficult during any particular storm (Mills et al., 1987; Jacobson et al., 1989). At a regional scale, where vegetation influences can be averaged out and rock types can be assumed to have characteristic slope morphologies, landslide susceptibility can be described in terms of characteristic regolith infiltration and drainage rates that determine the accumulation of soil pore-water pressures under given rainfall intensity and duration.

Landslides are uncommon on natural slopes of the Piedmont. Even the prolonged, heavy rainfall of Hurricane Agnes did not trigger noticeable slope instability in the Piedmont of Maryland (J.E. Costa, unpubl. data; Clark, 1987). In contrast to the Piedmont, the steeper slopes of the Blue Ridge have been subjected repeatedly to landsliding, especially in the southern Blue Ridge of North Carolina and Tennessee (Clark, 1987). One of the most dramatic cases of landsliding in the central Blue Ridge was the regionally extensive destabilization caused by intense rainfall from Hurricane Camille in Nelson County, Va., in 1969. Landslides were dominantly debris slide-avalanches that delivered large amounts of coarse sediment to streams and fans (Williams and Guy, 1973; Johnson, 1983; Kochel, 1987; Gryta and Bartholomew, 1987).

Landslide processes and spatial distributions in the Valley and Ridge have been related to rock and regolith, and variations in their hydrologic response to different types of storms (Jacobson, 1988b; Carr and Jacobson, 1988). For example, in November of 1985, 200–240 mm rainfall over 48 h in Pendleton County, W. Va. produced thousands of shallow landslides on thin colluvial and residual soils formed on shale, and eight debris slide-avalanches on quartzite ridges. In contrast, when the same landscape experienced much more intense, short-duration rainfall (approx. 356 mm in 24 h, Stringfield and Smith, 1956) from the 1949 convective thunderstorm, dozens of debris slide-avalanches were triggered on quartzite ridges. Presumably, long-duration/low-intensity rainfall is required to fill soil moisture reservoirs in fine-grained regolith, whereas, short-duration/high-intensity rainfall is required to fill soil-moisture reservoirs in quartzite regolith where infiltration and drainage rates are high to coarse materials and fractures in underlying bedrock.

At the slope scale, other factors appear to control landslide location, but these factors have variable influence depending on the storm characteristics. Locations of the 1949 debris avalanches were insensitive to bedrock dip (Clark, 1987) whereas most of the debris slide locations in 1985 were on dip slopes where bedding plane weaknesses and down-dip flow of groundwater may have been important contributors to slope instability (Jacobson et al.,1989). Smaller landslides triggered in 1985 were strongly biased toward cleared land but were fairly insensitive to hillslope topography (Jacobson et al., 1989).

The eastern edge of the Allegheny Plateau has steep slopes on Mississippian and Pennsylvanian sandstones along major river valleys. These slopes, like those documented by Pomeroy (1980), are subject to debris slide-avalanches triggered by intense rainfall. In contrast, shaley Pennsylvanian and Permian lithologies in the Plateau have gentler slopes characterized by shallow slips, slumps, and earth flows (Jacobson, 1985; Pomeroy, 1982). On interbedded sandstones and mudstones of the Permian Dunkard Group, ages of a population of landslides dated by tree ring growth correlated best with long, wet periods, mostly spring months with high values of runoff over 30-day durations, rather than with any specific storm event (Jacobson, 1985).

Geologic controls and meteorological variations determine a broad range of conditions that can trigger major landslide events. As indicated by Wieczorek (1987), slope instability is triggered by pore-water pressures that rise as a function of regolith thickness, infiltration rate, and drainage rate, plus convergence or divergence of shallow groundwater flow, as well as intensity and duration of rainfall. Because of the complex conditions that determine rainfall intensity and duration and how they trigger slope instability, flood-producing events and landslide-producing events are not necessarily the same. For example, Hurricane Agnes produced extreme amounts of rainfall over durations from 12 to 96 h, which resulted in extreme flooding (Fig. 8) but negligible landsliding (J.E. Costa, unpubl. data; Clark, 1987) because it mainly affected the Piedmont (Fig. 10, lines i, j). Much lower storm totals at comparable durations (Fig. 10, lines a–h) produced thousands of small landslides (Jacobson et al., 1989) and

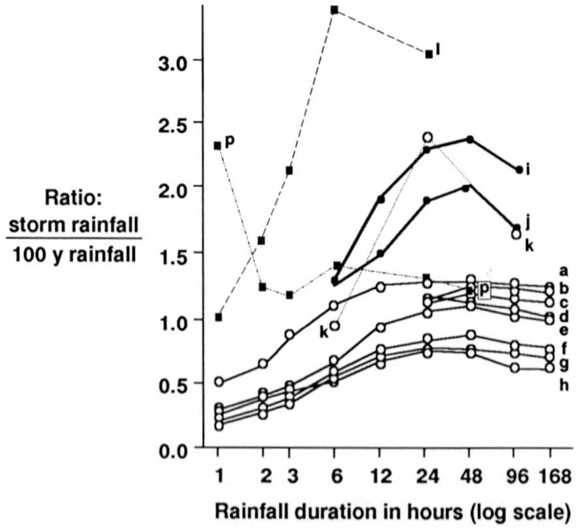

Fig. 10. Plot of ratio of storm rainfall to estimate of 100 y recurrence rainfall (from Hershfield, 1961) versus storm duration for extreme storms in the central Appalachians. *a–h*=Nov. 1985 storm; *i–j*=tropical Storm Agnes, June 1972; *k*=June 1949 storm; *l*=Hurricane Camille, August 1969; *p*=July 1942 storm (Smethport, Pa.).

extreme flooding (Figs, 6, 8) in the Valley and Ridge in November, 1985. In the same area, smaller storm totals at higher intensity produced dozens of large debris slide-avalanches (Stringfield and Smith, 1956) with less-extreme flooding (Fig. 6) in June, 1949 (Fig. 10, line k). The extremely intense rainfall recorded during the Smethport storm (Fig. 10, line p) evidently produced some small "blowout" landslides but was not noted for triggering extensive slope instability (Eisenlohr, 1952) probably because rainfall intensities were too great and duration too short to allow deep infiltration. Extremely high-intensity/moderate-duration rainfall from Hurricane Camille (Fig.10, line 1) was very effective in triggering landslides in Nelson County, Va. and in producing extreme flooding at small drainage areas (Figs. 7, 8).

History of landsliding in an area also influences the response to extreme rainfall. After a hillslope site has failed and regolith has been removed, some time interval is necessary to replenish regolith by bedrock weathering or accumulation of colluvium in the landslide scar to critical thickness for subsequent failure (c.f. Reneau and Dietrich, 1987). Lack of hillslope sites with critical regolith thicknesses may have prevented extensive slope instability in Nelson County, Va. in 1985 when 325 mm of rain fell over 48 h. Alternatively, rainfall intensity and duration may not have been adequate to trigger landslides during the 1985 storm.

Geomorphic work and effectiveness

Wolman and Miller (1960) concluded that most of the geomorphic work accomplished by rivers, as measured by the sediment loads they transport, is performed by flood discharges with frequencies of one or two per year in many areas. However, their data did not extend to consider catastrophic events with frequencies of 1 in 100 years and smaller. Data for assessing the work performed for these events are rare and estimated recurrence intervals for threshold of transport are tenuous. This is especially true for rare events in steep mountainous terrain where debris flows are common and flood transport is dominated by coarse bedload.

Existing data on sediment transport by catastrophic floods indicate that they transport a very large percentage of total annual suspended sediment. Floods from Hurricane Agnes transported 135–435% of the estimated annual sediment load from three small basins (<2 km^2) in Pennsylvania (Reed, 1980) and 300% of the annual sediment load of the Susquehanna River (Gross et al., 1978). No comparable data are available for steep basins in the central Appalachians, although rough estimates of debris-flow denudation rates in Central Virginia (Osterkamp and Costa, 1986) are close to long-term denudation rates, implying that infrequent debris flows are the dominant denudational process in that area. However, debris-flow denudation rates are also ten times greater than present-day erosion rates based on river sediment loads, indicating disequilibrium between slopes and rivers (Osterkamp and Costa, 1986).

While available data are insufficient to adequately characterize and compare geomorphic work performed by catastrophic events in the central Appalachians, we can characterize their geomorphic effectiveness. Geomorphic effectiveness was defined by Wolman and Gerson (1978) as the ability of events to form or shape the landscape, regardless of how those events rank in denudational importance. Effectiveness also is defined relative to the time required to restore the landscape to the form it had before the event, that is, a recovery time. A geomorphic event is effective if the resulting erosional and (or) depositional landforms are large, cover a broad area, and last a long time. For most of the historic, catastrophic events discussed here, absolute erosion attributable to the event, and event frequency, are poorly known, but thicknesses of flood deposits, channel widening, and depths of landslide scars attest that the events produced many times the mean annual erosion.

Effectiveness in the fluvial regime

Within any one basin or physiographic province, geomorphic effectiveness of a storm event on river channels and flood plains may vary with intensity, duration, and spatial extent of precipitation. For example, no major geomorphic changes were reported from the flood of largest contributing area and lowest rainfall intensity in the Potomac River Basin (1936 flood, Fig. 6). The smaller, more intense storm of June 1949 caused channel widening and valley-floor erosion at small drainage areas ($<$ 50 km^2) or at sites where debris-avalanche deposits blocked river channels in larger valleys (A.J. Miller, unpubl. data; Hack and Goodlett, 1960; Stringfield and Smith, 1956). In contrast, size and intensity of the November 1985 storm produced flood peaks capable of widespread valley-floor erosion at drainage areas of several hundred km^2. Flood features documented by Miller (1987), Clark et al. (1987), McKoy (1988), and Miller (in press) include cobble bars, cobble levees, sand splays, stripped flood-plain surfaces, extremely widened channels, and chutes cut across flood plains. Sites of extreme erosion and stripping of flood-plain surfaces are slowly recovering since the 1985 flood by incremental deposition of clay- to sand-size material from moderate floods. Depositional features, like cobble bars and levees, however, probably will persist much longer as prominent features.

The contrast of geomorphic effects of the 1949 and 1985 floods is typified by the sequence of events occuring at a location along the South Branch Potomac River in the Smoke Hole, a canyon reach near Petersburg, W.Va. (Fig. 1). This site was located in the area of most intense precipitation during the 1949 storm, and debris avalanches occurred in most of the local headwater valleys. Debris avalanches in the Redman Run basin delivered a large volume of coarse sediment directly into the channel of the South Branch Potomac River; discharge on the South Branch was incapable of removing this material, and the channel was diverted to the right, cutting into the alluvial bottomland on the opposite bank. Comparison of aerial photographs from 1945 (Fig. 11a) and 1952 (Fig. 11b) shows the resulting change in configuration of the channel and valley floor. During the subsequent 36-year period the only notable change was the growth of vegetation on the surface created in the 1949 flood (Fig. 11c shows the site in 1980). In November 1985, only a few debris avalanches occurred in the South Branch Potomac River basin and all of them came to rest well upstream of tributary confluences with the main trunk river. At the site illustrated in Fig. 11, some reworking of tributary valley deposits was observed but the volume of new sediment delivered to the South Branch was relatively small. Peak discharge on the South Branch was much larger than in 1949, however, and resulted in complete removal of the surface created in the 1949 storm and formation of a large gravel bar along the right side of the valley (Fig. 11d).

Although the comparison of storms and flood effectiveness cited above is valid for that part of the Valley and Ridge, storms of comparable intensity and size occurring in different geomorphic settings may not be equally effective. The range of peak discharges measured at stations in the Potomac River basin in areas that experienced severe erosion in 1985 is indicated by the shaded area in Fig. 8. The other envelope curves included in the figure indicate that Agnes, Camille, and the Smethport storm all produced discharge peaks falling within or above the shaded area. Geomorphic impacts of Camille and of the Smethport storm have been documented (Eisenlohr, 1952; Williams and Guy, 1973; Johnson, 1983) and are characterized chiefly by slope failures and reworking of valley floors in small drainage basins that are directly affected by slope failures. Fluvial effects in larger basins affected by Camille were concentrated in the vicinity of bridges and roads (Williams and Guy, 1973) and were not comparable in magnitude to the fluvial impacts observed in the Potomac and Cheat River basins at drainage areas of 250–2000 km^2 in 1985. Similarly, geomorphic effects of Tropical Storm Agnes were also meagre. Most rivers recovered fairly quickly from any channel widening occurring during the flood (Costa, 1974; Gupta and Fox, 1974; Moss and Kochel, 1978). Gravel and sand were deposited only at scattered sites on flood plains, and extensive erosion of the flood plain was not documented.

Differences in geomorphic effectiveness appear to result in part from the controlling influence of valley physiography on flood hydraulics in the Valley and Ridge province (Miller, in press). The three forks of the South Branch Potomac River are narrow and steep, with average gradients between 0.002 and 0.009 and average width of the valley floor varying between 200 and 400 m for drainage areas up to about 800 km^2. Structural and lithologic controls effectively prevent any systematic increase in valley width as drainage area increases. Furthermore, these valleys also have many local expansions and contractions of val-

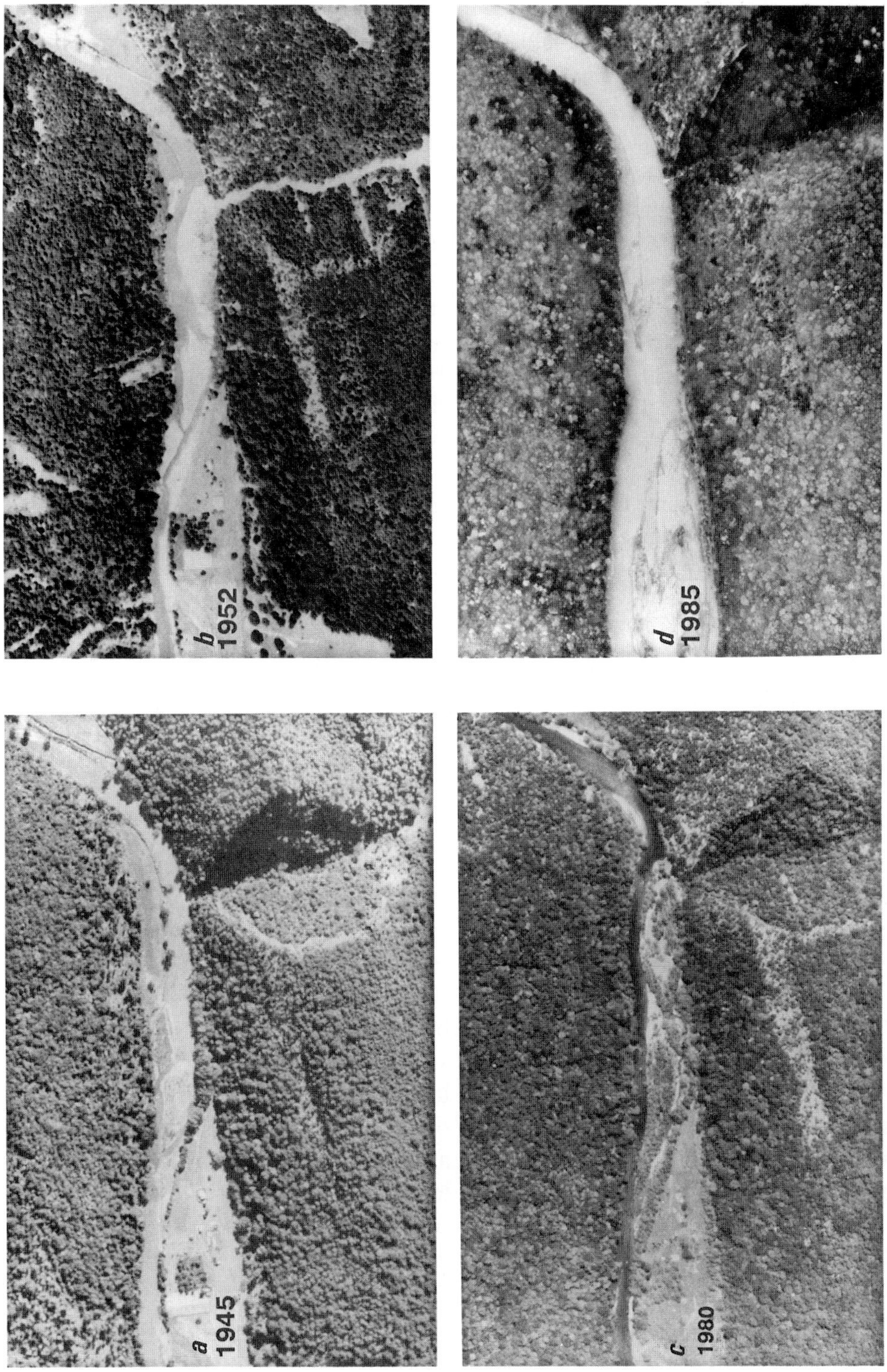

Fig. 11. Sequence of four aerial photographs of site along the South Branch River upstream of Petersburg, W. Va. described in text. The valley width is approximately 50 m; flow is to left.

ley width that create backwaters upstream from the contractions and severe erosion of the valley floor by jet-shaped flow patterns downstream from the valley expansions. Preliminary investigations suggest that unit stream power, which is a function of both slope and discharge and varies inversely with width, may be more useful than discharge alone as an indicator of geomorphic effectiveness.

Effectiveness in the hillslope regime

Geomorphic effectiveness of landsliding is at a minimum in the Piedmont where, apparently, thick, well-drained saprolite rarely accumulates pore-water pressures sufficient for failure. In other areas where a variety of types and sizes of landslides occur, relative effectiveness must be measured in terms of the size of the resulting landforms and their recovery time.

For landslide scars, recovery time is the time required to re-establish critical thicknesses of colluvium and (or) residuum on a hillslope site. Re-establishment of vegetation is not considered recovery of a landslide scar. Failure susceptibility generally increases with the time interval between triggering events as failure thresholds are steadily lowered.

The time required for recovery of landslide scars is variable depending on the rock type involved and the capability of high-frequency, low-magnitude processes to operate on filling the scar. For example, shallow landslides triggered in 1985 on steep pastured slopes underlain by thin, shaley residual and colluvial soils in Pendleton County, W. Va., have substantial natural revegetation after only three years. Ravel and wash from steep scarps around slip scars have filled portions of scars with 2–5 cm of fine colluvium. As scarp steepness declines, however, rates of filling are expected to decrease and total recovery to prefailure thickness may require many years. On shaley rocks in the Buffalo Creek basin (Fig. 1), diffusional colluvial transport by bioturbation and creep, and weathering of bedrock into failure-susceptible residuum were calculated to require several thousand years for a landslide scar to recover to mean failure conditions, although head and toe scarps disappear much more rapidly (Jacobson, 1985). Scars from large debris slide-avalanches in Nelson County, Va. (1969) and Grant County, W. Va. (1949) persist as prominent landscape features. In both cases, landslides stripped colluvium and weathered bedrock from hillslope sites; refilling with colluvium derived from side slopes has been minimal and weathering of bedrock negligible.

Recovery of landslide deposits is a fundamentally different process from recovery of landslide scars. Many landslides produce sediment that falls or flows and accumulates in debris cones, colluvial benches, or debris fans. Unlike scars, these deposits can have much longer residence times if they are on low-angle, inactive slopes and (or) if the deposit itself is resistant to erosion. Some clayey mudflows triggered in the November 1985 storm in Pendleton County, W. Va. have been in corporated into pre-existing colluvial cones and benches. After three years, only very subtle turf levees remain; however, the colluvial landforms may integrate many such events over very long time intervals, and thus establish aggregate effectiveness. In contrast, individual bouldery debris deposits from 1949 debris slide-avalanches still retain primary depositional features like lobes, levees, and fans. Older, prehistoric deposits also are evident. The most dramatic examples of Appalachian landslide features with long residence times are the extremely large rock-block slides and debris avalanche features identified by Schultz (1986) and Southworth (1987). The size (several million to 1 billion cubic meters) and resistant materials that comprise these features, assure that the landforms will remain prominent for thousands of years.

The surficial geologic record of catastrophic events

Surficial deposits contain a discontinuous, transient record of depositional geomorphic events. The history contained in these deposits is essential for understanding the spatial distribution and long-term recurrence of rare catastrophic events. An example of the utility of such data is the formal statistical procedures that have been developed recently to incorporate paleoflood information directly into frequency analyses of rare floods (e.g., Hupp, 1988). In other cases, however, progress would be made if we could simply count the number of large events over a time interval and assign heuristic recurrence intervals.

As the time interval of interest is extended backward into the middle Holocene, it is increasingly tenuous to assume stationary climatic conditions. One of the most important goals of surficial stratigraphic research is to determine what portions of the episodic, surficial stratigraphic record are attributable to extreme events under the present-day climate as opposed to major Quaternary climatic changes. This section reviews selected surficial stratigraphic studies in the central Appalachians to consider the role of catastrophic floods and landslides in landscape evolution during the Quaternary.

Valley bottoms and alluvial terraces

Interpretation of catastrophic flood histories from valley-bottom alluvium and alluvial terraces requires comparisons to effects of historic catastrophic flooding. In the Piedmont, the Hurricane Agnes flood in 1972 significantly widened channels but produced only minor flood-plain deposition in rare areas that were hydraulically favorable (Gupta and Fox, 1974; Costa, 1974; Moss and Kochel, 1978). Reasoning from the Agnes example, we would not expect isolated extreme flood events like the Agnes flood to be extensively recorded in the Piedmont valley-bottom stratigraphy.

Few studies exist on prehistoric alluvial stratigraphy of Piedmont valley bottoms. Wolman and Leopold (1957) used a conceptual model that flood plains form from floods of 1–2 year recurrence and radiocarbon dates from Pennsylvania and North Carolina to suggest that Piedmont flood plains are young (less than 2000 yr) and adjust rapidly to frequent flood events. In view of paleobotanical evidence that the eastern U.S. has been dominated by woodlands for most of the Holocene (Knox, 1983; Delcourt and Delcourt, 1986), and the historic insensitivity of Piedmont streams to extreme events, it is unlikely that Holocene climate changes or catastrophic floods have been recorded unambiguously in flood plain strata of stream basins located entirely within the Piedmont. In contrast, agriculturally induced upland erosion produced a distinct, broadly distributed layer of silty flood-plain sediment throughout the Piedmont of Maryland (Jacobson and Coleman, 1986), Georgia (Trimble, 1974), and presumably elsewhere in the Piedmont. In the Maryland Piedmont, agricultural sediments are much more extensive and distinct that any produced by major floods, including the Tropical Storm Agnes flood. Hence, it would appear that a widespread, intense, and long-lasting disturbance equivalent to agricultural disturbance is necessary to produce an unambiguous stratigraphic record in the Piedmont.

In contrast to the Piedmont, stream valleys of the Blue Ridge, Valley and Ridge, and Plateau, and streams draining from these areas across the Piedmont, exhibit more evidence of episodic deposition. Historical observations, again, help to calibrate the surficial record. Hurricane Camille in 1969 produced dramatic widening of stream channels with small drainage areas and extensive deposition of alluvium and debris-flow in channels and fans (Williams and Guy, 1973; Johnson, 1983). Many of these features are distinct today. In the Valley and Ridge, the 1949 storm around Harrisonburg,

Virginia produced channel widening, flood-plain incision, and deposition of levees and flood-plain bars (Hack and Goodlett, 1960). Similar features were created by the 1985 storm in the same area (Kochel et al., 1987); the effects of the 1985 storm on the South Branch Potomac River basin have been described above. All these storms occurred in mountainous areas with thin soils, steep stream gradients, and coarse sediment supply. The 1949 and 1969 events also were characterized by extensive delivery of coarse landslide debris to the floods. These conditions appear to be the most favorable for creation and preservation of a stratigraphic record of extreme events.

There have been only a few alluvial stratigraphic studies to document the prehistoric record of mountainous parts of the central Appalachians. Most of these studies have attributed flood-plain strata or terraces to climate change rather than individual catastrophic events but the number of observations are insufficient to generalize over the region.

Multiple fans and terraces have been studied by Kite et al. (1986) along the west flank of the Blue Ridge and in the Valley and Ridge of Virginia. They observed post settlement alluvium, an indistinct Holocene record, and extensive alluvial terraces and fans of undetermined age, some of which are extensively weathered and are probably pre-Holocene. Terraces along the Shenandoah River in the vicinity of the Thunderbird archeological site also have been attributed to major Pleistocene climate changes (Segovia and Franco, 1977; Foss, 1977).

Alluvium of the South Branch Potomac River valley near Moorefield and Petersburg, W. Va. appears typical of the Valley and Ridge. A complex series of cut and fill episodes is represented by strata in the valley bottom (Fig. 12). The oldest valley bottom unit is presumed to be early Holocene to Late Wisconsin based on relative weathering criteria. A younger aggradational sequence has yielded basal dates of 7060 +/− 230 yr BP (W-6077) and 2170 +/− 180 yr BP (W-6076) at two separate locations (R.B. Jacobson, unpubl. data). Without additional radiocarbon dates, episodic deposition within the valley bottom sediments cannot be attributed reliably to either climatically induced periods of increased flooding or individual catastrophic floods. Similarity of the cut and fill scour features in valley bottom alluvium to flood-plain chutes produced by the 1985 flood, however, suggest that they may have been incised by a catastrophic flood event and subsequently filled by moderate flood events. At least three higher, undated, highly weathered terraces flank the valley bottom. We interpret the major terraces to reflect aggradation under Pleistocene climates when the capacity of streams to transport sediment eroded from slopes was proportionally less compared to present conditions.

Valley-bottom alluvium in the Buffalo Creek drainage basin (Fig. 1) may be typical of small drainage basins in the Appalachian Plateau. Alluvium there is composed of uniform silty and sandy units, lacking buried weathering profiles, and dominated by late-Holocene radiocarbon ages of less than 4000 yr. (Jacobson, 1985). This was interpreted as evidence that the area had experienced no major climate changes over this time interval and that extreme storms, if they had affected the area, had not left distinct stratigraphic evidence.

In another alluvial stratigraphy study in the Plateau, an archeological excavation along the Kanawha River (26,700 km^2 drainage area, Kanawha County, W. Va.) exposed 5.5 m of stratified sediment with seventeen occupation zones spanning the years 9850–8160 yr BP (Broyles, 1971). If each occupation zone corresponds to a depositional hiatus between flood events, the average recurrence interval is approximately 100 years. Variations in the amount of sediment between occupation zones suggest that flooding conditions were not uniform during the time interval and the entire sequence is thought to reflect a climatically controlled period of aggradation unlike the late Holocene (Olafson, 1971; Barlow, 1971).

Older and higher alluvial terraces along ma-

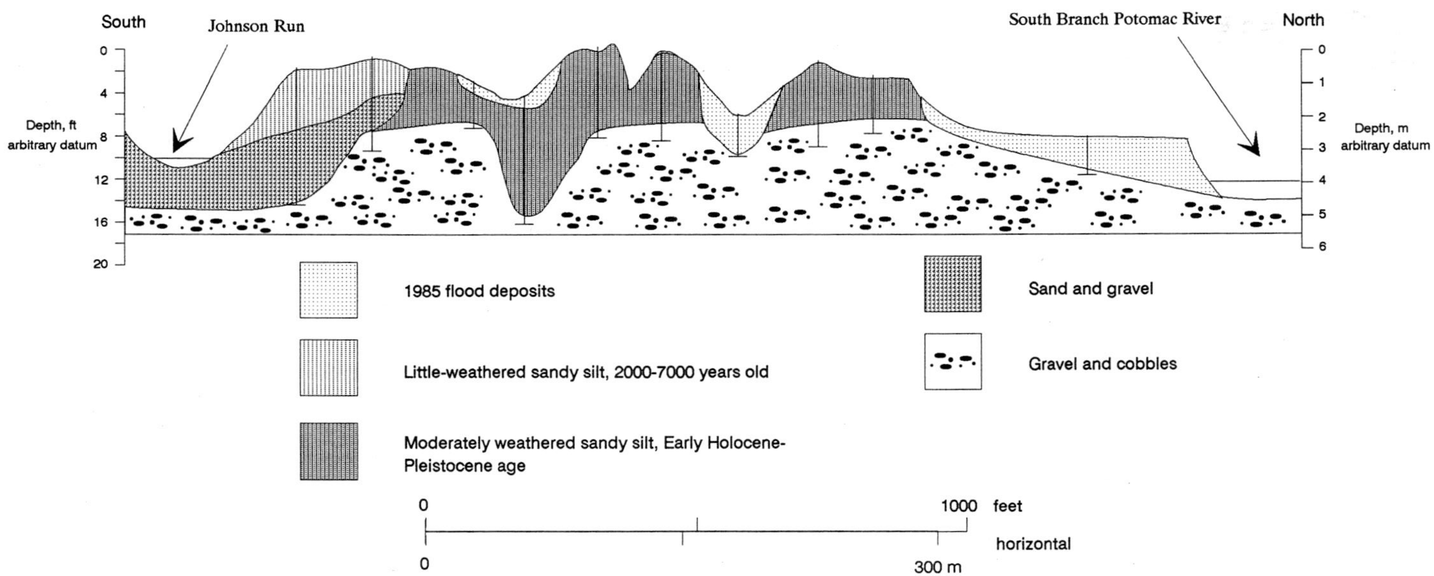

Fig. 12. Cross section of valley-bottom alluvium near Petersburg, West Virginia, showing 1985 flood deposits and erosional features compared to underlying deposits.

jor rivers in the central Appalachians are considered generally to be Pleistocene and pre-Pleistocene in age. These include terraces along the New River (Mills and Wagner, 1985), Monongahela River (Jacobson et al., 1988), Rappahannock River (Dunford-Jackson, 1978), Rappahannock Estuary (Newell, 1985), and Potomac River (Reed, 1981). Episodic terrace formation for these rivers over the late Cenozoic has been ascribed variably to climate change and tectonism.

In summary, the alluvial stratigraphic record in the central Appalachians has two contrasting components. The high-level terrace remnants that comprise some of the most dramatic elements of the landscape were deposited during the Pleistocene or earlier, as the result of climatically induced changes in sediment supply and (or) hydrology, or tectonism. Meagre studies of alluvium in the valley flats suggest that aggradational episodes have been associated with both climatically induced periods of more frequent flooding and with catastrophic floods. Based on historic flood effects and preserved flood-plain morphologic features and deposits, it is apparent that catastrophic flooding is more effective in sculpting flood plains in steep basins of the Valley and Ridge and Blue Ridge, than in the Piedmont and Plateau.

Hillslope sediments

The surficial stratigraphic record of colluvium, debris, and related slope deposits is the most direct available evidence for the sediment supply side of the geomorphic system. As such, hillslope sediments may comprise a more sensitive record of extreme storms and climatic events than alluvial strata. Additionally, colluvium and debris can have longer residence times, especially in cases where colluvium and debris are composed of coarse materials that resist erosion.

In the Piedmont, scattered, thick, colluvial deposits have been attributed to episodic depostion, but lack of datable materials or a reliable stratigraphic framework has precluded a definite paleoclimatic interpretation (Whittecar, 1985), although thick, colluvial deposits in North and South Carolina have been attributed to cool-wet paleoclimates (Whitehead and Barghoorn, 1962; Eargle, 1977).

Debris and alluvial fans along Blue Ridge of Virginia have been described by Kochel and Johnson (1984), Kochel and Simmons (1986), and Kochel (1987). Radiocarbon dates on debris flow deposits were used to distinguish three or four debris-flow events in the last 11,000 years, for an average recurrence interval of 3000–4000 yr for the area upslope from the fans. Kochel (1987) interpreted the fan stratigraphy to reflect increased debris-flow activity due to the influence of tropical air masses following the northward migration of the Polar Front at the end of the Holocene. In contrast to the relatively active debris fans east of the Blue Ridge, Kochel (1987) describes inactive, coalescent alluvial fans along the western flank that are interpreted to be pre-Holocene age based on weathering criteria.

In the Valley and Ridge of West Virginia, toeslope colluvial deposits with a source area in fine shale bedrock have revealed an episode of colluviation radiocarbon dated from 12,000–10,000 yr BP (Fig. 13). These dates may indicate a series of landsliding events at this time, or alternatively, they may mark the transition from a time of general hillslope instability during the late Wisconsin, to a period of less frequent landsliding during the Holocene. Overlying the thick, late Wisconsin colluvium are at least two debris flows separated by a distinct soil horizon: the top one is from 1985, the lower was radiocarbon dated as 315 yr BP. On adjacent slopes situated below quartzite ridges, extensive debris terraces composed of mixed matrix-supported diamicton and clast-supported gravel, ascend in steps to the tops of spur ridges at elevations up to 50 m above modern channels (Fig. 14). Although undated, elevations of these terraces above present-day stream channels and lack of weathering suggest the higher terraces

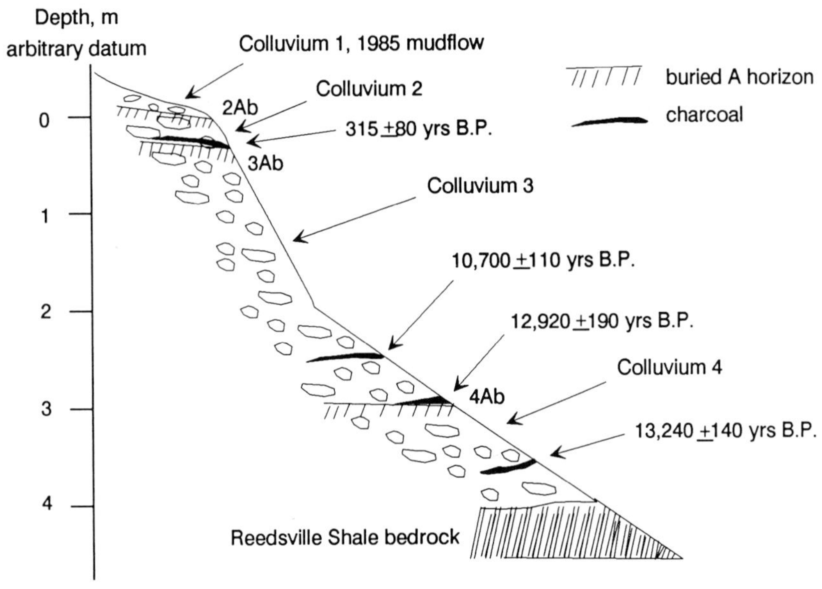

Fig. 13. Stratigraphic section of colluvium in Penleton County, W. Va. showing episodic debris-flow deposition. Radiocarbon identification numbers are: AA-4350 (315 yr BP), AA-4351 (10,700 yr BP), AA-4352 (12,900 yr BP), and AA-4353 (13,240 yr BP).

may be as old as middle Pleistocene or older. Lack of weathering of the lowest, youngest unit suggests it is probably Holocene age. Significantly, all but the youngest debris terrace are much more extensive in width than large debris flows triggered nearby in the same landscape setting in 1949 and 1985. Hence, the older remnants were either deposited by much larger single events, or they were produced by lateral aggradation of numerous smaller events. Stratigraphic exposures are not sufficient to chose between these alternatives. However, the large apparent time intervals between periods of debris terrace aggradation suggest episodicity with a frequency comparable to Pleistocene climate change.

Similar debris deposits have been noted in southwestern Virginia (Mills et al., 1987; Mills, 1989). Mills (1989) subdivided debris deposits into those produced by debris flow and those produced by gelifluction. Based on relative weathering characteristics, Mills (1989) also concluded that younger deposits tended to be emplaced by debris flows and older deposits showed more influence of periglacial processes.

Studies of colluvium in the Valley and Ridge of Pennsylvania also have documented episodic colluvial deposition (Ciolkosz et al., 1979; Sevon, 1985; Hoover and Ciolkosz, 1988). Although no absolute dates exist, regional climatic models and the existence of periglacial sedimentary features suggest that these units were deposited during Pleistocene cold periods.

Similarly, thick weathered colluvial deposits in the Buffalo Creek drainage basin have been attributed to accelerated colluviation under early Wisconsin paleoclimatic conditions (Jacobson, 1985). These thick diamicton colluvial deposits were then entrenched and eroded materials were incorporated in younger debris fans that grade to late Wisconsin (?) to late Holocene age alluvial units. The younger deposits have a higher percentage of clast-supported units compared to the early Wisconsin deposits, thus indicating a change in dominant transport processes from the early Wisconsin to the late Wisconsin. Holocene fans have multiple units but individual dates for events could not be obtained. Because the associated alluvial stratigraphic record did not disclose regional

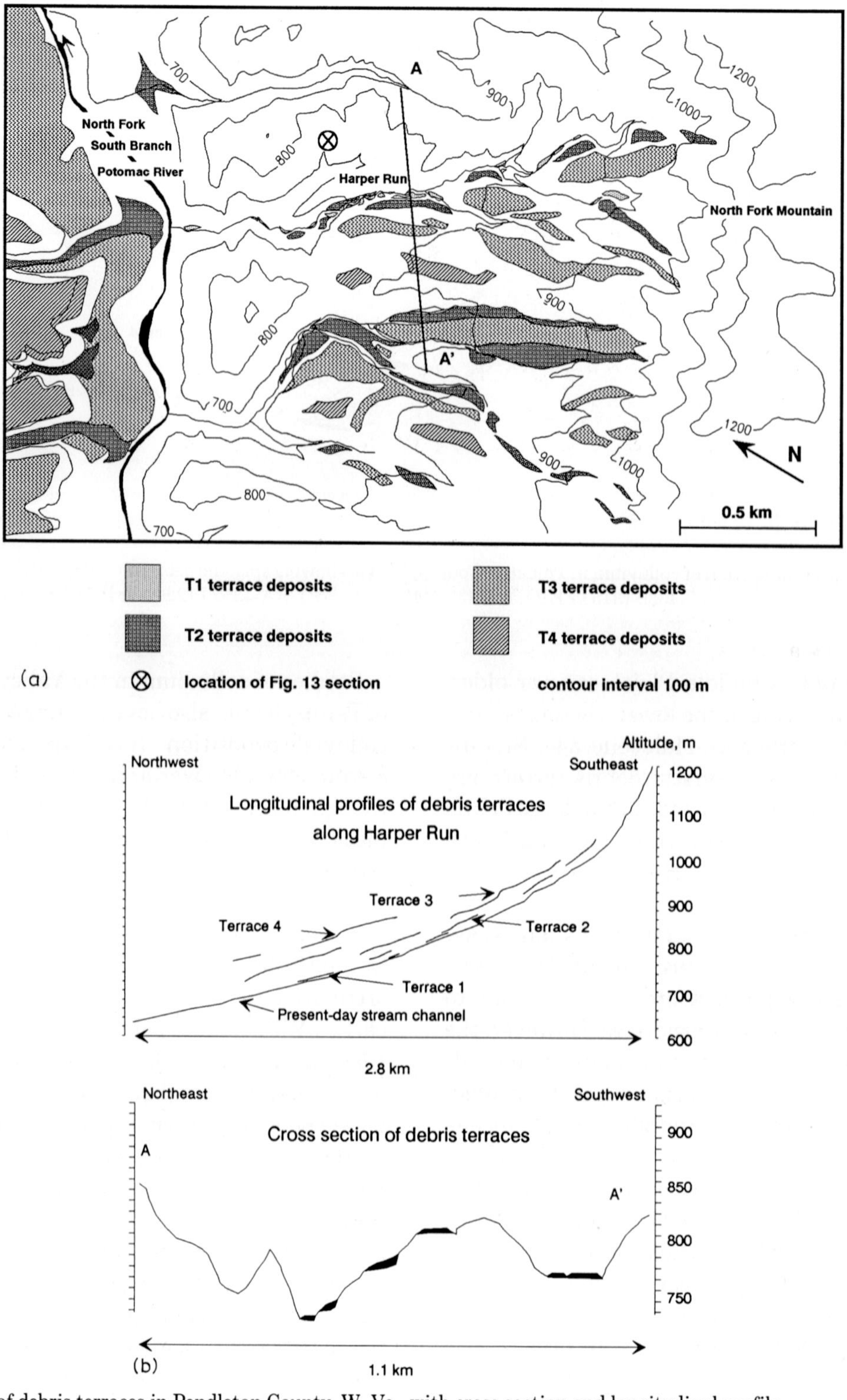

Fig. 14. Map of debris terraces in Pendleton County, W. Va., with cross section and longitudinal profile.

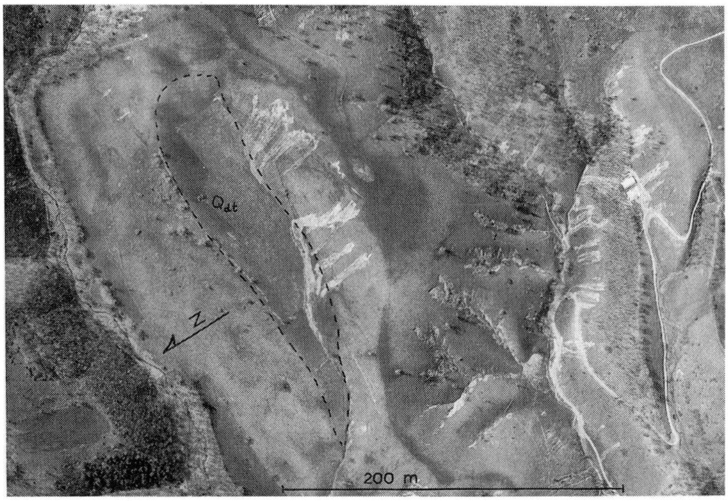

Fig. 15. Aerial photograph of muddy debris flows triggered in Pendleton County, W. Va. in November, 1985. Bulk of debris-flow volume in central part of photo was intercepted and stored on a Pleistocene debris terrace, Q_{dt}.

depositional events, Jacobson (1985) interpreted the fan stratigraphy to reflect asynchronous destabilization by small landslides scattered in time and space. Relict prehistoric debris flows were also recognized in southwestern Pennsylvania by Pomeroy (1983), who interpreted them as the product of the Late Pleistocene climate. Pleistocene climates are also invoked to explain thick colluvial deposits on the eastern edge of the Appalachian Plateau (Jordan et al., 1987).

Under present-day conditions Pleistocene deposits continue to influence moderate and extreme geomorphic events. The debris terraces described above (Fig. 14) act as long-term sediment storage sites that intercept sediment produced by landslides on adjacent slopes, thus mitigating the effects of extreme events (Fig. 15) Miller (1987) noted that valley constrictions formed by debris fans were important hydraulic controls that influenced erosional and depositional effects of 1985 flood waters on the South Branch Potomac River.

In summary, the surficial stratigraphic record of hillslope processes in the central Appalachians is dominated by deposits that owe their origin to climate change. In the Piedmont, where slopes presently are insensitive to catastrophic storms, the stratigraphic record shows old, episodically deposited sediments. In the Plateau, scattered occurrences of landslides under present-day climatic conditions have produced thin, scattered deposits and landforms while most of the colluvium present is Pleistocene age (see also Braun, this volume). In the Valley and Ridge and Blue Ridge provinces, where modern-day events have produced distinct and dramatic deposits, colluvial and debris deposits record large Holocene events, but even these are much less extensive than those produced during the Pleistocene.

Discussion and conclusions

Most historic catastrophic geomorphic events in the central Appalachians have been triggered by extreme rainfall events from either tropical storms or from small mesoscale features associated with frontal systems and extra-tropical cyclones. Our model of spatially varying rainfall indicates high probability that all points within the Valley and Ridge province have been subjected during the Holocene to rainfall amounts that historically have been sufficient to trigger catastrophic geomorphic events (Fig. 4). Although extreme rainfall amounts will vary across the region with climate and physiography, this conclusion suggests that all parts of the central Appalachians have been affected by one or more extreme rainfall events during the Holocene.

Tropical storm tracks and topographic barriers of the Blue Ridge and Allegheny Front act to influence the spatial distribution of the extreme rainfall. These topographic and climatic controls are superimposed on the highly variable physiography of the Central Appalachians to determine variable hydrologic responses and landslide susceptibilities. Geologic and physiographic controls strongly modify flood discharges and determine landslide locations when floods and landslides are triggered by moderate storm events, but intrinsic controls are less important for truly rare, catastrophic events when storm characteristics dominate. Size of storm, track and location of storm elements with respect to basin boundaries, and intensity and duration of rainfall combine to determine peak discharges occurring at different drainage areas. Locations and spatial densities of landslides are primarily determined by rainfall intensity and duration relative to the infiltration and drainage rates of regolith.

In contrast to the direct dependency of extreme flood discharges on storm characteristics, the geomorphic effectiveness of central Appalachian floods are significantly modified by basin geology. Extreme discharges in the Piedmont have left little lasting geomorphic record because the deposits can be reworked by subsequent moderate flows and sufficient fine sediment is available to fill in widened channels. In contrast, smaller discharges in steep areas of the Blue Ridge and Valley and Ridge have been effective in producing geomorphic change because of the direct sources of coarse sediments supplied by hillslopes and high stream powers produced in steep channels. Furthermore, geologic constraints like the water gaps in Valley and Ridge impose hydraulic controls that increase stream power relative to similar discharges elsewhere in the Appalachians.

Catastrophic landslides are effective geomorphic events when they are triggered in hard rocks on slopes, producing prominent scars and deposits with long residence times. Despite relatively common occurrence of long-duration, high-intensity rainfall from tropical storms, the Piedmont landscape is minimally susceptible to landsliding because the thick and highly permeable saprolite there allows infiltration and drainage of rainfall without allowing soil moisture to accumulate to critical levels. The same storms can trigger many landslides on steep slopes of the Blue Ridge. In the Valley and Ridge, shaley regolith is more likely to be destabilized by long-duration, low-intensity rainfall whereas the steeper, better-drained sites on sandstone ridges are more likely to destabilized by short-duration, high-intensity rainfall. However, scars and deposits from the small landslides on shale recover quickly whereas those from the sandstone ridges remain prominent for much longer. Thin, residual and colluvial soils on shaley lithologies of the Plateau have not been susceptible to catastrophic landsliding because extreme rainfall totals are relatively rare in the region; instead, most hillslopes are adjusted to more frequent movement triggered during long wet periods in the spring.

The ultimate geomorphic effect of a storm results from the combination of landslides and floods. The importance of sediment supplied directly from landslides to flood waters was obvious in Camille (1969) and the 1949 storm when sediment from debris slide-avalanches was entrained in the flood and was deposited downstream on fans and flood plains (Johnson, 1983; Schleiff and Kite, 1989). Less directly, correlations of the locations of thousands of landslides triggered by the November 1985 storm with areas of flood erosion, suggests that sediment introduced from many small landslides can contribute in aggregate fluvial erosion, probably by increasing flow density (Harper et al., 1987). Thus, the most catastrophic of central Appalachian geomorphic events will be those in which conditions simultaneously promote landslides and high flood discharges.

The answer to the question of whether catastrophic events or smaller, more-frequent events are the most important in modifying the central Appalachian landscape under present-day conditions cannot be answered reliably until more data are gathered on meteorology, hydrology, and sediment transport. Without these data, recurrence intervals and geomorphic work cannot be evaluated. However, the foregoing discussion of spatial variations of intrinsic and extrinsic geomorphic controls across the central Appalachians implies that the answer will vary with location. Our preliminary conclusion is that catastrophic events are currently less important in shaping the Piedmont and Plateau than the Blue Ridge and Valley and Ridge. This results primarily from the interplay of decreasing importance of tropical storms east to west and increasing sensitivity to catastrophic events in areas of hard rocks and high relief.

Because the historic record is too short for adequate assessment of the long-term importance of catastrophic events, the surficial stratigraphic record becomes an important source for additional data. Most alluvial stratigraphic studies attribute terrace formation to aggradation triggered by changing climates during the Holocene and Pleistocene. However, the detailed mechanisms of such aggradational events have not been explored. In particular, we do not know how changing meteorological patterns and seasonal climate changes have interacted to determine relative changes in sediment supply, hydrologic response, and dominant transport processes. Periods of increased flooding during the Holocene have been documented at other sites in the eastern U.S. (Knox, 1983, 1988; Brakenridge, 1980) and should be expected in the central Appalachians. Conditions for preservation of stratigraphic evidence vary widely across the Appalachians because of variable hydrology and hydraulic settings. Conditions for preservation of a long stratigraphic record are more favorable on low-energy rivers like the Kanawha (Barlow, 1971) than high-energy rivers like the South Branch Potomac River or smaller streams like Buffalo Creek.

Evidence for individual extreme floods has not been distinguished stratigraphically from climatically induced periods of aggradation in the central Appalachians. Detailed stratigraphic studies like those of Knox (1987) have been successful in documenting extreme and moderate floods in the midwest U.S. and could be applied to the central Appalachians. Dendrochronological paleoflood analyses have been used successfully to date individual extreme floods (Hupp, 1988) but slackwater methods appear less promising, because few hydraulically suitable sites are available (Kite and Linton, in prep).

The surficial stratigraphic record of hillslope sedimentation is dominated by discrete time intervals of aggradation. Weathering of the most extensive deposits suggests that they formed during the Pleistocene, probably due to an increase in the rate of regolith production on hillslopes relative to the transport capability of streams. The surficial record also shows that Holocene episodes have been smaller in aggregate than Pleistocene episodes. Furthermore, sedimentological data (Mills, 1989) and observations of remnant periglacial features (Clark and Ciolkosz, 1988) show that in some parts of the central Appalachians, hillslope transport was at least in part by periglacial processes, presumably during Pleistocene cold periods. Therefore, aggradational episodes probably are not solely due to increased magnitude and frequency of catastrophic events relative to the present.

In general, the surficial stratigraphic record of the central Appalachians indicates two timescales of variation. Most of the region exhibits multiple, prominent hillslope and alluvial terrace deposits indicative of climate changes with durations and recurrence frequencies of tens of thousands of years or more. Historic and late Holocene catastrophic events create, add to, or erode these features, but their effects are generally smaller. Thus, the events that have been the most effective in sculpting the Appalachian landscape are largely the result of events associated with what appears to have been a very different geomorphic climate in the Pleistocene. Present-day catastrophic events, although effective in modifying the landscape in some parts of the central Appalachians, have not yet erased much of the legacy of previous climates.

Acknowledgements

We are indebted to M. Rubin, G. Burr, and D. Donohue for providing radiocarbon dates.

References

Bauley, J.F., Patterson, J.S. and Paulhus, J.L.H., 1975. Hurricane Agnes rainfall and floods, June–July 1972. U.S. Geol. Surv. Prof. Pap., 924: 403 pp.

Barlow, J.A., 1971. Recent geological historical of the St. Albans site. In: B.J. Broyles, (Editor). The St. Albans Site, Kanawha County, West Virginia. W. Va. Geol. Econ. Surv., Rep. Archeol. Invest., 3: 80–92.

Brakenridge, G.R., 1980. Widespread episodes of stream erosion during the Holocene and their climatic cause. Nature, 283: 655–656.

Braun, D.D., 1989. Glacial and periglacial erosion of the Appalachians. In: T.W. Gardner and W.D. Sevon (Editors), Appalachian Geomorphology. Geomorphology, 2: 233–256 (this volume).

Broyles, B.J., 1971, The St. Albans site, Kanawha County, West Virginia. W. Va. Geol. Econ. Surv. Rep. Archeol. Invest., 3: 1–76.

Camp, J.D., and Miller, E.M., 1970, Flood of August 1969 in Virginia. U. S. Geol. Surv. Open File Rep., 120 pp.

Carr, C.E. and Jacobson, R.B., 1988. Physical properties of regoliths and the distribution of slope movements triggered by the November 3–5, 1985, storm, Pendleton County, West Virginia. EOS, Am. Geophys. Union, 69: p. 348 (abstr.).

Ciolkosz, E.J., Petersen, G.W., Cunningham, R.L. and Matelski, R.P., 1979. Soils developed from colluvium in the ridge and valley area of Pennsylvania. Soil Sci., 128: 153–162.

Clark, G.M., 1987. Debris slide and debris flow historical events in the Appalachians south of the glacial border. In: J.E. Costa and G.F. Wieczorek, (Editors), Debris flows/avalanches: process, recognition, and mitigation. Geol. Soc. Am., Rev. Eng. Geol., VII: 125–138.

Clark, G.M. and Ciolkosz, E.J., 1988. Periglacial geomorphology of the Appalachian highlands and interior highlands south of the glacial border – a review. Geomorphology, 1: 191–220.

Clark, G.M., Jacobson, R.B., Kite, J.S. and Linton, R.C., 1987. Storm-induced catastrophic flooding and related phenomena in Virginia and West Virginia, November, 1985. In: L. Mayer and D. Nash (Editors), Catastrophic Flooding. Allen and Uwin, Boston, pp. 355–379.

Costa, J.E., 1974. Response and recovery of a Piedmont watershed from tropical storm Agnes, June 1972. Water Resour. Res., 10: 106–112.

Delcourt, H.R. and Delcourt, P.A., 1986. Late Quaternary vegetation change in the central Atlantic states. In: J.N. McDonald and S.O. Bird (Editors), The Quaternary of Virginia. Va. Div. Miner. Resour., Publ., 75: 23–35.

Dunford-Jackson, C.S., 1978. The geomorphic evolution of the Rappahannock River basin. Unpubl. M.S. thesis, University of Virginia, Charlottesville, Va., 92 pp.

Eargle, D.H., 1977. Piedmont Pleistocene soils of the Spartanburg area, South Carolina. Geol. Notes, S.C. Div. Geol., 21: 57–74.

Eisenlohr, Jr., W.S., 1952. Floods of July 18, 1942 in north-central Pennsylvania. U.S. Geol. Surv., Water Supply Pap., 1134B: 98 pp.

Environmental Data Service, 1968. Climate atlas of the United States. U.S. Dep. Commer., Environ. Sci. Serv. Admin., 43 pp.

Foss, J.E., 1977. The pedological record at several paleoindian sites in the northeast. New York Acad. Sci., Ann., 228P: 234–244.

Gross, M.G., Karweit, M., Cronin, W.B. and Schubel, J.R., 1978. Suspended sediment discharge of the Susquehanna River to northern Chesapeake Bay, 1966 to 1976. Estuaries, 1: 106–110.

Grover, N.C., 1937. The floods of March 1936 - 3 Potomac, James, and upper Ohio Rivers. U. S. Geol. Surv., Water Supply Pap., 800: 351 pp.

Gryta, J.J. and Bartholomew, M.J., 1987. Frequency and susceptibility of debris avalanching induced by Hurricane Camille in central Virginia. In: A.P. Schultz and C.S. Southworth (Editors), Landslides of eastern North America. U. S. Geol. Surv., Circ., 1008: 16–18.

Gupta, A. and Fox, H., 1974. Effects of high-magnitude floods on channel form. A case study in the Maryland Piedmont. Water Resour. Res., 10: 499–509.

Hack, J.T. and Goodlett, J.C., 1960. Geomorphology and forest ecology of a mountain region in the central Appalachians. U. S. Geol. Surv., Prof. Pap., 347: 66 pp.

Harper, J.M., Jacobson, R.B., Howard, A.D. and McGeehin, J.P., 1987. Relationship between landslides, and tributary channel changes from an extreme flood, North Fork, South Branch Potomac River basin, West Virginia. Program Abstr., Catastrophic Flooding Symp., Oxford Ohio, 1987, 32 pp. (abstr.)

Hershfield, D.M., 1961. Rainfall frequency atlas of the United States. U. S. Weather Bur., Tech. Pap., 40: 114 pp.

Hoover, M.T. and Ciolkosz, E.J., 1988. Colluvial soil parent material relationship in te ridge and valley physiographic province of Pennsylvania. Soil Sci., 145: 163-172.

Hoxit, L.R., Maddox, R.A., Chappell, C.F. and Brua, S.A., 1982. Johnstown-western Pennsylvania storm and floods of July 19-20, 1977. U. S. Geol. Surv., Prof. Pap., 1211: 68 pp.

Hupp, C.R., 1988. Plant ecological aspects of flood geomorphology and paleoflood history. In: V.R. Baker, R.C. Kochel and P.C. Patton (Editors), Flood Geomorphology. Wiley, New York, N.Y., pp. 335-356.

Jacobson, R.B., 1985. Spatial and temporal distributions of slope processes in the upper Buffalo Creek drainage basin, Marion County, West Virginia. Unpubl. Ph. D. thesis, The Johns Hopkins University, Baltimore Md., 485 pp.

Jacobson, R.B., 1988a. Debris deposits and Quaternary drainage basin evolution in the central Appalachian Mountains of West Virginia. EOS, Am. Geophys. Union, 69: p. 1225 (abstr.).

Jacobson, R.B., 1988b. Rainfall and slope movements in the central Appalachian Mountains of West Virginia. EOS, Am. Geophys. Union, 69: p. 348 (abstr.).

Jacobson, R.B. and Coleman, D.J., 1986. Stratigraphy and recent evolution of Maryland Piedmont flood plains. Am. J. Sci., 286: 617-637.

Jacobson, R.B., Elston, D.P. and Heaton, J.W., 1988. Stratigraphy and magnetic polarity of the high terrace remnants in the upper Ohio and Monogahela Rivers in West Virginia, Pennsylvania, and Ohio. Quat. Res., 29: 216-232.

Jacobson, R.B., Cron, E.D. and McGeehin, J.P., 1989. Slope movements triggered by heavy rainfall, November 3-5, 1985, in Virginia and West Virginia, In: A.P. Schultz and R.W. Jibson (Editors), Landslides of Eastern North America. Geol. Soc. Am., Spec. Pap., 236: 1-13.

Johnson, R.A., 1983. Stream channel response to extreme rainfall events. the Hurricane Camille storm in central Nelson County, Virginia. Unpubl. Ma. Thesis, University of Virginia, Charlottesville, Va., 109 pp.

Jordan, M.K., Behling, R.E. and Kite, J.S., 1987. Characterization and late Quaternary history of the alluvial and colluvial deposits of the Pendleton Creek valley, Tucker County, WV. Geol. Soc. Am., Abstr. Programs, 19:p. 92 (abstr.).

Kite, J.S. and Linton, R.C., in prep. Depositional aspects of the November 1985 flood on Cheat River and Black Fork, West Virginia. U.S. Geol. Surv., Bull.

Kite, J.S., Bell, A. and Allamong, M., 1986. Preliminary observations on floodplain, terrace, and fan deposits in the upper Potomac River basin of Virginia and West Virginia, In: J.N. McDonald and S.O. Bird (Editors), The Quaternary of Virginia. Va. Div. Miner. Resour., Publ., 75: 121-122.

Kochel, R.C., 1987. Holocene debris flows in central Virginia. In: J.E. Costa and G.F. Wieczorek (Editors), Debris Flow/Avalanches: Process, Recognition, and Mitigation. Geol. Soc. Am., Rev. Eng. Geol., VII: 139-4155.

Kochel, R.C. and Johnson, R.A., 1984. Geomorphology and sedimentology of humid-termperate alluvials fans, central Virginia. In: E.H. Koster and R.J. Steele (Editors), Sedimentology of Gravels ans Conglomerates. Can. Soc. Pet. Geol., Mem., 10: 109-122.

Kochel, R.C. and Simmons, D.W., 1986. Quaternary alluvial fans in central Virginia. In: J.N. McDonald and S.O. Bird (Editors), The Quaternary of Virginia. Va. Div. Miner. Resour., Publ., 75: 123-125.

Kochel, R.C., Ritter, D.F. and Miller, J., 1987. Role of tree dams on the construction of pseudo-terraces and variable geomorphic response to floods in the Litter River valley, Virginia. Geology, 15: 718-721.

Knox, J.C., 1983. Responses of river systems to Holocene climates. In: H.E. Wright (Editor), Late-Quaternary Environments of the United States - Volume 2, The Holocene. University of Minnesota Press, Minneapolis, Minn., pp. 26-41.

Knox, J.C., 1987. Stratigraphic evidence of large floods in the upper Mississippi Valley. In: L. Mayer and D. Nash (Editors), Catastrophic Flooding. Allen and Unwin, Boston, Mass., pp. 155-180.

Knox, J.C., 1988. Climatic influence on upper Mississippi valley floods. In: V.R. Baker, R.C. Kochel and P.C. Patton (Editors), Flood Geomorphology. Wiley, New York, N.Y., pp. 279-300.

Lescinsky, J.B., 1987. Flood of November 1985 in West Virginia, Pennsylvania, Maryland, and Virginia. U.S. Geol. Surv., Open File Rep., 86-486: 33 pp.

McKoy, M.L., 1988. The November 1985 storm and flood and resulting geomorphic effects on South Branch Basin of the Potomac River, West Virginia. W. Va. Geol. Econ. Surv., Open File Rep. OF8803: 143 pp.

Miller, A.J., 1987. What does it take to make a geomorphically effective flood? Some lessons from the November 1985 flood in West Virginia. In: J.S. Kite (Editor), Research on the Late Cenozoic of the Potomac Highlands. W. Va. Open File Rep., OF8801: Geol. Econ. Surr., 3-30.

Miller, A.J., in press. Flood hydrology and geomorphic effectiveness in the central Appalachians. Earth Surf. Process. Landforms.

Mills, H.H., 1989. Surficial geology and geomorphology of the Mountain Lake Area, Giles County, Virginia, including sedimentological studies of colluvium and boulder streams. U.S. Geol. Surv., Prof. Pap., 1469: 57 pp.

Mills, H.H. and Wagner, J.R., 1985. Long-term change in the regime of New River indicated by vertical variation in extent and weathering intensity of alluvium. J. Geol., 93: 131-142.

Mills, H.H., Brakenridge, G.R., Jacobson, R.B., Newell, W.L., Pavich, M.J. and Pomeroy, J.S., 1987. Appalachian mountains and plateaus. In: W.L. Graf (Editor),

Geomorphic Systems of North America. Geol. Soc. Am., Cent. Spec. Vol., 2: 5-50.

Moss, J.H. and Kochel, R.C., 1978. Unexpected geomorphic effects of the Hurricane Agnes storm and flood, Conestoga drainage basin, south-eastern Pennsylvania. J. Geol., 86: 1-11.

Neumann, C.J., Cry, G.W., Caso, E.L. and Jarvinen, B.R., 1981. Tropical cyclones of the North Atlantic Ocean, 1871-1980. National Climatic Data Center, 152 pp.

Newell, W.L., 1985. Architecture of the Rappahannock estuary - neotectonics in Virginia. In: M. Morisawa and J.T. Hack (Editors), Tectonic Geomorphology. Allen and Uwin, Boston, Mass., pp. 321-342.

Olafson, S., 1971. Late Pleistocene climate and the St. Albans site. In: B.J. Broyles (Editor), The St. Albans site, Kanawha County, West Virginia. W. Va. Geol. Econ. Surv., Rep. Archeol. Invest., 3: 77-80.

Osterkamp, W.R. and Costa, J.E., 1986. Denudation rates in selected debris-flow basins. Fourth Fed. Interagency Sediment. Conf., Proc. 1: 4-91-4-99.

Pavich, M.J., 1986. Processes and rates of saprolite production and erosion on a foliated granitic rock of the Virginia Piedmont. In: S.M. Colman and D.P. Dethier (Editors), Rates of Chemical Weathering of Rocks and Minerals. Academic Press, Orlando, Fl., pp. 552-590.

Pomeroy, J.S., 1980. Storm-induced debris avalanching and related phenomena in the Johnstown area, Pennsylvania, with reference to other studies in the Appalachians. U.S. Geol. Surv., Prof. Pap., 1191: 23 pp.

Pomeroy, J.S., 1982. Mass movement in two selected areas of Washington County, Pennsylvania. U.S. Geol. Surv., Prof. Pap., 1170-B: 17 pp.

Pomeroy, J.S., 1983. Relict debris flows in northwestern Pennsylvania. Northeast. Geol., 5: 1-7.

Reed, Jr., J.C., 1981. Disequilibrium profile of the Potomac River near Washington, D.C. - A result of lowered base level or Quaternary tectonics along the Fall Line? Geology, 9: 445-450.

Reed, L.A., 1980. Suspended-sediment discharge in five streams near Harrisburg, Pennsylvania, before, during, and after highway construction. U.S. Geol. Surv., Water Supply Pap., 2072: 37 pp.

Reneau, S.L. and Dietrich, W.E., 1987. Size and location of colluvial landslides in a steep forested landscape. Erosion and Sedimentation in the Pacific Rim (Proceddings of the Corvallis Symposium, August, 1987). Int. Assoc. Hydrol. Sci., Publ., 165: 39-48.

Schleiff, K.A. and Kite, J.S., 1989. Geomorphology of debris slides, debris flows, and related phenomena developed during the June 1949 flood in eastern West Virginia. Geological Society of America, Abstracts with Programs, 21: p. 56 (abstr.).

Schultz, A.P., 1986. Ancient, giant rockslides, Sinking Creek Mountain, Southern Appalachians, Virginia. Geology, 14: 11-14.

Segovia, A.V. and Franco, L, 1977. Archaeologic implications of the geomorphologic evolution of the lower Shenandoah Valley, Virginia. New York Acad. Sci., Annal., 228: 189-193.

Sevon, W.D., 1985. Pennsylvania's polygenetic landscape. Harrisburg Geol. Soc., 4th Annual. Field Trips Guidebook., 55 pp.

Smith, J.A., 1988. Regional flood frequency analysis using upper order statistics of the annual peak record. Water Resour. Res., 25: 311-317.

Smith, J.A. and Karr, A.F., in press. A statistical model of extreme storm rainfall. J. Geophys. Res.

Southworth, C.S., 1987. Prehistoric giant rockslides of the Appalachian valley and ridge province. types and characteristics. In: A.P. Schultz and C.S. Southworth (Editors), Landslides of Eastern North America. U.S. Geol. Surv., Circ., 1008: 24-32.

Stringfield, V.T. and Smith, R.C., 1956. Relation of geology to drainage, floods, and landslides in the Petersburg area, West Virginia. W. Va. Geol. Econ. Surv., Rep. Invest., 13: 19 pp.

Trimble, S.W., 1974. Man-induced soil erosion of the southern Piedmont, 1700-1970. Soil Conserv. Soc. Am., Ankeny, Iowa, 180 pp.

Wieczorek, G.F., 1987. Effect of rainfall intensity and duration on debris flows in central Santa Cruz Mountains, California. In: J.E. Costa and G.F. Wieczorek (Editors), Debris flows/avalanches: process, recognition, and mitigation. Geol. Soc. Am., Rev. Eng. Geol., VII: 93-104.

Williams, G.P. and Guy, H.P., 1973. Erosional and depositional aspects of Hurricane Camille in Virginia, 1969. U.S. Geol. Surv., Prof. Pap., 804: 80 pp.

White, E., 1976. Role of carbonate rocks in modifying flood flow behavior. Water Resour. Bull., 12: 351-170.

Whitehead, D.R. and Barghoorn, E.S., 1962. Pollen analytical investigations of Pleistocene deposits from western North Carolina and South Carolina. Ecol. Monogr., 32: 347-369.

Whittecar, G.R., 1985. Stratigraphy and soil development in upland alluvium and colluvium north-central Virginia Piedmont. Southeast. Geol., 26: 117-129.

Wolman, M.G. and Gerson, Ran, 1978. Relative scales of time and effectiveness of climate in watershed geomorphology. Earth Surf. Process., 3: 189-208.

Wolman, M.G. and Leopold, L.B., 1957. River flood plains. some observations on their formation. U.S. Geol. Surv., Prof. Pap., 282-C: 87-109.

Wolman, M.G. and Miller, J.P., 1960, Magnitude and frequency of forces in geomorphic processes. J. Geol., 68: 54-74.

Geomorphology, 2 (1989) 285-302
Elsevier Science Publishers B.V., Amsterdam — Printed in The Netherlands

Distribution and Genesis of Soils of the Northeastern United States*

EDWARD J. CIOLKOSZ[1], WILLIAM J. WALTMAN[2], THOMAS W. SIMPSON[3] and ROBERT R. DOBOS[1]

[1]*Department of Agronomy, The Pennsylvania State University, University Park, PA 16802 (U.S.A.)*
[2]*Department of Agronomy, Cornell University, Ithaca, NY 14853 (U.S.A.)*
[3]*Department of Agronomy, Virginia Polytechnic Institute and State University, Blacksburg, VA 24061 (U.S.A.)*

(Received March 16, 1989; accepted after revision June 1, 1989)

Abstract

Ciolkosz, E.J., Waltman, W.J., Simpson, T.W. and Dobos, R.R., 1989. Distribution and genesis of soils of the northeastern United States. In: T.W. Gardner and W.D. Sevon (Editors), Appalachian Geomorphology. Geomorphology, 2: 285-302.

The soils of the northeastern United States (central and northern Appalachians) north of the glacial border are Spodosols, Inceptisols, Alfisols, and some Entisols and Histosols. These soils are relatively young (12-18 ka). Climate (frigid soil temperatures), parent material (acidic, sandy), and vegetation (conifers) are the major soil forming factors affecting the genesis of the Spodosols, while parent material (calcareous material) is a determining factor affecting the genesis of the Alfisols. The Inceptisols of southern New England are slightly different (better developed cambic horizon) than those of the Appalachian Plateau. The genesis of the Alfisols of the unglaciated Appalachian Plateau and Triassic Lowlands are also significantly influenced by carbonates and basic minerals in their parent material. Ultisols are found only south of the glacial border. Ultisols on the Appalachian Plateau and in the northern Ridge and Valley are young soils and should be identified as parent material Ultisols, while those of the southern Ridge and Valley, Piedmont and Coastal Plain show significant soil development and should be identified as genetic Ultisols. The soils south of the glacial border, on steep slopes, have been affected significantly by erosion during the cold cycles of the Pleistocene, while soils on gentle slopes have been affected to a lesser degree. Much of the material eroded during the Late Pleistocene has accumulated on lower sideslopes as colluvium. The erosional and depositional processes have resulted in a complex mosaic of moderate to well developed soils on the more stable landscape surfaces, and moderate to weakly developed soils on the erosional and depositional surfaces.

Introduction

The distribution and genesis of the soils of northeastern United States (central and northern Appalachians) is a complex subject. To facilitate the presentation of this material, a section on soil classification and soil genesis precedes the discussion of the distribution and genesis of the soils of the region. In addition, only the major soils of the region will be discussed.

Soil classification

The soil classification system used in the this presentation is Soil Taxonomy (Soil Survey Staff, 1975). This system has been the official soil classification system of the United States

*Authorized for publication as paper no. 8126 of the journal series of The Pennsylvania Agricultural Experiment Station, University Park, PA 16802, USA.

TABLE 1

Soil orders, formative elements (order, suborder, and great group), and a brief description of the main characteristics of the formative elements of the major soils of the Northeast region

Formative element (order)	Description
Order	
Ent (Entisol)	Weakly developed (young) soil without a B horizon, A/C profile or if very sandy a weakly developed color B horizon
Ept (Inceptisol)	Weak to moderately well developed (relatively young) soil with a color and/or structural B horizon (cambic) or a fragipan B horizon
Alf (Alfisol)	Moderately well developed soil with a B horizon of illuvial clay accumulation (argillic horizon) and relatively high subsoil base (Ca+Mg+Na+K) saturation status (>35% of the cation exchange capacity), can also have a fragipan B horizon
Ult (Ultisol)	Moderately well to very well developed soil with a B horizon of illuvial clay accumulation (argillic horizon) and low subsoil base saturation status (<35% of the cation exchange capacity), can also have a fragipan B horizon
Od (Spodosol)	Weak to moderately well developed, sandy soil with a spodic B horizon (subsurface horizon of humus, Fe and Al accumulation) can also have a fragipan B horizon
Ist (Histosol)	Soil composed of organic materials (peats and mucks)
Suborder and great group	
Aqu	Wet-aquic moisture regime (somewhat poorly and poorly drained)
Dystr	Low base saturation (acid) soil
Eutr	High base saturation soil
Fibr	Composed of mostly undecomposed plant material (peats)
Fluv	Composed of recent alluvium (on floodplain)
Fol	Freely drained organic soil
Frag	Fragipan-dense impermeable subsurface B horizon
Hem	Composed of partially decomposed plant material (peaty muck)
Hum	High organic matter content
Ochr	Thin, light colored, A horizon
Orth	Most common type of profile
Psamm	Sand or loamy sand texture
Sapr	Composed of mainly decomposed plant material (muck)
Ud	Humid climate
Umbr	Thick, acid, dark colored A horizon

since 1965. Soil Taxonomy has six hierarchal levels (order, suborder, great group, subgroup, family, and series) and a soil can be discussed or mapped at any of the levels. At the order level the soil name is made up of a specific formative element joined to the suffix-sol (Latin-solum, soil), e.g., Incepti-sol, Alfi-sol, and Ulti-sol (Table 1). At the suborder through subgroup levels, other formative elements are added to the order formative element to give these categories distinctive names, and each of the formative elements provides a significant amount of information about the soil. The criteria used for classification in the upper categories of Soil Taxonomy were chosen to indicate the genetic pathway of soil development (Smith, 1983). Thus, Soil Taxonomy is an appropriate tool to use in discussing the distribution and genesis of soils of a region. Additional information on the structure of Soil Taxonomy is presented by Soil Survey Staff (1975; 1987), Buol et al. (1989), Birkeland (1984), and Witty and Arnold (1987).

Soil genesis

Definition of soil

The soil is an unconsolidated three-dimensional natural body that mantles the landscape. The lateral boundaries of the soil are bedrock or water too deep to support the growth of rooted plants. Its upper boundary is the atmosphere and its lower boundary is the rooting depth of the natural vegetation of the area or the depth to which the soil forming reactions have significantly altered the underlying material. The characteristics and properties of soils are the result of a complex interaction of physical, chemical and biological reactions (soil forming reactions). The rate and extent to which these reactions proceed is controlled by the soil forming factors (climate, organisms, parent material, topography, and time).

Soil forming factors

Introduction

The effect of the various soil forming factors is not equal on the development of all soils. Table 2 gives an estimate of the relative influence of the soil forming factors on soil formation in the Northeast. This relative rating is only applicable for comparisons between soils within the Northeast and should not be applied to soil of other regions.

Time

As indicated in Table 2, time is one of the major factors influencing soil genesis in the Northeast. The northern half of the region was glaciated (Fig. 1), and the deposits in this area vary in age from about 18 ka in Pennsylvania to about 12 ka in northern New York and northern New England (Mickelson et al., 1983). Since deposition of the glacial materials, only minor changes in the landscape of this area have occurred (Schafer and Hartshorn, 1965; Sevon, 1985). Thus, the age of these deposits can be taken as the age of the soils of this area.

The age of the soils south of the glacial border is a more complex issue, due to a complicated landscape evolutionary history. Although no attempt will be made to discuss the long-term geomorphic history of this area, a brief discussion is needed of the most recent landscape history. Because of its location and the reported occurrence of periglacial features in the area (Ciolkosz et al., 1986b; Clark and Ciolkosz, 1988), the most appropriate model to describe these landscape relationships appears to be a combination of those presented by Budel (1982) and Osterkamp et al. (1987). Budel's model developed in Europe in an area analogous to the Northeast indicates that with the exception of slopes of greater than 50 to 60%, today's Northeast landscapes have changed very little in the last 10 ka (Holocene), Budel's model also indicates that Northeast landscapes are the result of periglacial erosion and deposition during the last glacial period and that 95% of current landforms existed prior to the Holocene. The multi-erosion-stability cycle model of Osterkamp et al. (1987) indicates many cycles in the Pleistocene with most soil formation occurring during stable interglacial periods. The combination of these two models into a general geomorphic model provides and overall framework for viewing the soils of the unglaciated area.

The application of this general model to soils of the region requires consideration of the processes than encompass periglacial erosion, in particular the degree of landscape truncation with steepness of slope. According to Budel (1982) and Goldthwait (1976), little down-slope mass movement occurs on slopes of less than about 3% in a periglacial environment. They also indicate that some down-slope movement occurs on slopes of 3–10% and significant movement occurs where slopes exceed 10%. This indicates that soils on gentle slopes would persist, although possibly cryoturbated during a glacial period (Goodman, 1953), while soils on steeper slopes would be truncated on upper slopes with a significant amount of the truncated material accumulating on the lower slopes as colluvium. The larger quantity of colluvium on the footslopes of the region attests to this process. For example, in the Ridge and Valley area of Pennsylvania a typical county has about 27% of its area covered with footslope colluvium (Ciolkosz et al., 1986b). The magnitude of this process is also illustrated by Pennsyl-

TABLE 2

Relative influence of the various soil forming factors on the development of soils in the Northeastern region

Soil order	Parent material	Time	Topography	Climate	Organisms
Entisols	***	*	**	*	*
Inceptisols	***	**	**	*	*
Alfisols	***	**	*	*	*
Ultisols (PM) +	***	**	*	*	*
Ultisols (GEN) + +	*	***	*	**	*
Spodosols	***	**	*	**	***
Histosols	***	*	***	**	*

* = weak influence ** = moderate influence; *** = strong influence; PM = parent material; GEN = genetic.

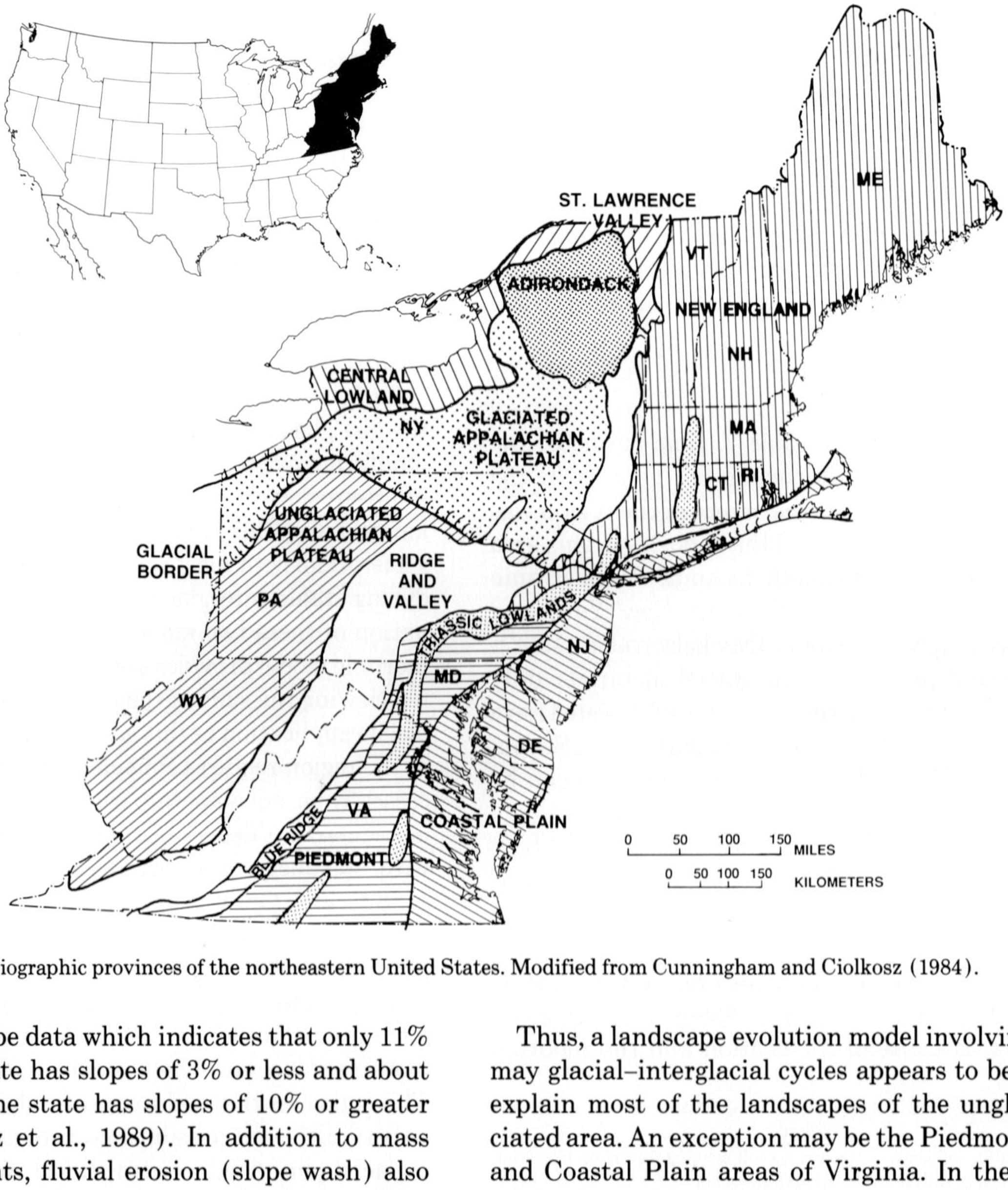

Fig. 1. Physiographic provinces of the northeastern United States. Modified from Cunningham and Ciolkosz (1984).

vania slope data which indicates that only 11% of the state has slopes of 3% or less and about 53% of the state has slopes of 10% or greater (Ciolkosz et al., 1989). In addition to mass movements, fluvial erosion (slope wash) also contributed to denudation during the cold periods (Lewkowicz, 1988), particularly on steep slopes (Budel, 1982). The results of this fluvial process have been observed in areas adjacent to footslopes in central Pennsylvania. In these areas, silty alluvium with a few pea sized sandstone gravels (from adjacent ridges) has filled drainage ways and low areas adjacent to the colluvial mantled footslopes with up to 3 m of material (Ciolkosz et al., 1986a).

Thus, a landscape evolution model involving may glacial–interglacial cycles appears to best explain most of the landscapes of the unglaciated area. An exception may be the Piedmont and Coastal Plain areas of Virginia. In these areas the cycle may still have operated (Whittecar, 1985; Costa and Cleaves, 1984); but, because of less relief and a warmer climate, the landscape may have undergone less severe landscape truncation than in other parts of the region. The Piedmont and Coastal Plain areas of Virginia today have a thermic soil temperature regime (Fig. 2), and the present-day mesic-thermic soil temperature boundary is thought to mark an approximate zone that separates an

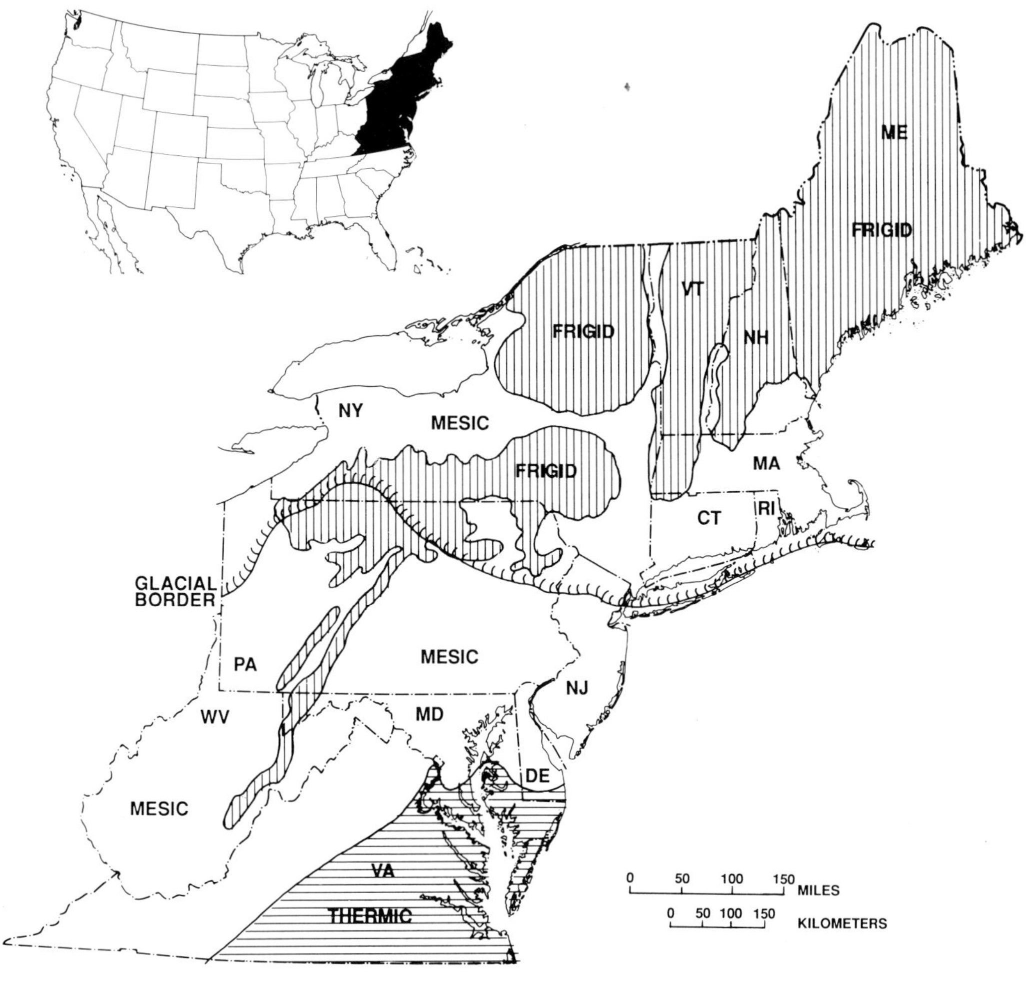

Fig. 2. Soil temperature regimes of the northeastern United States. Frigid = 0–8 °C mean annual soil temperature (MAST). Mesic = 8–15 °C MAST and Thermic = 15–22 °C MAST. Modified from the map of Smith (1984), and Carter and Ciolkosz (1980).

area to the north that has undergone a higher degree of soil truncation than the area to the south (Ciolkosz, 1986b). In the Ridge and Valley area, Clark et al. (1989) reported that a typical county in southern Virginia has about half as much footslope colluvium as a typical county in central Pennsylvania. These data also indicate less landscape truncation in the southern part of the area than in the northern part.

Thus, the age of the soils in the unglaciated area may vary from old on stable parts of the Coastal Plain and Piedmont (Pavich, this volume) to an age similar to that of the glacial soils on truncated bedrock surfaces, in footslopes colluvium and on low alluvial terraces. Although some of the soils may be old, many have had the upper part of their profiles partially truncated or altered by frost processes (Cronce, 1988). In addition, some of the partially truncated soils have had colluvium and/or aeolian deposits added to their surfaces creating a welded profile in which the present day soil forming processes have reached through the added material into the underlying truncated paleosol (Hoover, 1983; Waltman, 1985; Ciolkosz et al., 1988; and Darmody and Foss, 1982).

Parent material

Parent material is also a very important soil forming factor in the Northeast region. The glacial deposits in New England and northern New York are derived primarily from granites and schists and are acid materials with sandy loam textures. In central New York, the deposits are derived from sandstone, shales, and limestones. In southern New York and Pennsylvania the glacial deposits are also derived from sandstone and shale but little, if any, limestone. The texture of these materials are loam to silt loam, and their reaction is calcareous in central New York and northeastern Pennsylvania. The parent materials in the area south of the glacial border are more diverse than north of the border. The bedrock of the Unglaciated Plateau is sandstone and shale with exception of southwestern Pennsylvania and northwestern West Virginia, which has in addition to sandstone and shale, interbedded limestone and calcareous shale. The bedrock of the Ridge and Valley area is primarily sandstone in the ridges and shale and limestone in the valleys. The bedrock of the Blue Ridge and Piedmont is primarily granites and schists. Inset into the Piedmont are areas of Triassic lowlands with red shales and sandstones which have been intruded by diabase dikes and sills. The Coastal Plain is composed of multilevels of weakly consolidated to unconsolidated acid sands and clays.

Topography and climate

The topographic factor of soil formation encompasses mainly erosional and soil drainage relationships. With the exception of very steep slopes (greater then 50–60%), the present landscape is relatively stable and has not been significantly affected in the Holocene by surface erosion other than during some catastrophic events such as debris flows triggered by unusual precipitation events (Mills et al., 1987; Jacobson, this volume) or by accelerated erosion caused by man's activities. Thus, present day surface erosion is a minor factor while episodic erosion in the past is the major erosion factor that has affected regional soil formation.

The water relations within soils (well vs. poorly drained) is partially determined by the topographic factors of slope steepness and landscape position. Soils on gentle slopes as well as on lower or footslope areas which receive surface run-on as well as subsurface discharge, tend to be more poorly drained than other soils. The type and rate of physical and chemical reactions that occur in poorly drained conditions produces a much different soil morphologically and chemically than develops under well drained conditions. In particular, poorly drained soils tend to be less leached and less weathered than associated well drained soils (Ranney et al., 1974). They also tend to have a thinner solum and a secondary clay mineral suite with a greater abundance of expandable silicates than their well drained counterparts (Ciolkosz, 1967).

Present day climate (temperature and precipitation), particularly in relation to its impact on vegetation, is significant in soil formation in the Northeast. One of the major impacts is the effect of cool temperatures on the development of coniferous vegetation in the northern and higher elevation parts of the region. The coniferous vegetation in conjunction with the cool temperatures has a major effect on the developments of Spodosols. In addition, some slight climatic variations in the recent past in the Northeast have been important. It has been reported that the warmer and drier climate in the period 4–8 ka was responsible for the occurrence of tall grass prairies and their associated soils (Mollisols) in parts of Pennsylvania (Waltman, 1988). This climate trigger may also be responsible for the occurrence of prairie soils in other parts of the region (Table 3). Earlier climate changes and their impact on soil formation are more difficult to evaluate. Any soil that is mid or early Pleistocene in age has undergone many episodes of temperature and precipitation change, landscape stability change, as well as vegetative change (Delcourt

TABLE 3

Percentage of soil orders and suborders in the northeastern U.S. by region and by state (data from Ciolkosz and Dobos, 1989)

Soil order and suborder	Northeast region	States												
		ME	NH	MA	RI	CT	VT	NY	PA	NJ	DE	MD	WV	VA
Entisols	5.79	0.19	0.19	0.75	0.07	0.24	0.22	1.21	0.80	0.62	0.15	0.27	0.38	0.70
Aquents	2.00	0.11	0.02	0.04	**	0.02	0.05	0.40	0.46	0.16	0.01	0.15	0.12	0.43
Fluvents	0.32	-	**	0.03	-	0.01	0.03	0.05	0.05	**	-	0.02	0.04	0.10
Orthents	1.56	0.08	0.08	0.24	0.06	0.17	0.08	0.27	0.27	0.02	-	**	0.19	0.10
Psamments	1.92	**	0.09	0.44	0.01	0.04	0.06	0.49	0.02	0.44	0.14	0.10	0.03	0.06
Inceptisols	36.90	3.44	1.03	1.87	0.36	1.76	1.10	11.67	7.94	0.47	0.03	0.68	3.72	2.83
Aquepts	8.35	2.82	0.28	0.34	0.06	0.21	0.33	3.08	1.02	0.12	0.02	0.02	0.01	0.05
Ochrepts	28.51	0.62	0.75	1.53	0.30	1.55	0.77	8.58	6.92	0.35	0.01	0.66	3.68	2.78
Umbrepts	0.04	-	-	**	-	-	-	0.01	-	-	-	-	0.03	**
Alfisols	14.50	-	-	-	-	-	0.27	5.60	3.71	0.39	0.01	0.38	2.18	1.96
Aqualfs	3.28	-	-	-	-	-	0.10	1.90	1.00	0.05	**	0.02	0.04	0.17
Udalfs	11.22	-	-	-	-	-	0.17	3.70	2.71	0.34	0.01	0.36	2.14	1.79
Ultisols	25.73	-	-	-	-	-	-	0.01	6.28	1.37	0.62	2.59	3.79	11.07
Aquults	2.51	-	-	-	-	−1	-	-	0.61	0.12	0.24	0.58	0.05	0.91
Udults	23.22	-	-	-	-	-	-	0.01	5.67	1.25	0.38	2.01	3.74	10.16
Spodosols	15.28	8.92	2.41	0.49	-	**	2.23	0.97	0.07	0.16	**	0.02	0.01	**
Aquods	0.53	0.14	0.04	0.01	-	**	0.06	0.12	-	0.16	**	0.01	-	**
Orthods	14.74	8.78	2.37	0.48	-	-	2.17	0.85	0.07	**	-	0.01	0.01	**
Histosols	1.63	0.47	0.14	0.18	0.02	0.05	0.07	0.32	0.01	0.13	**	0.16	**	0.08
Fibrist	0.15	0.14	-	-	-	-	**	**	-	-	-	-	-	-
Folist	0.06	0.01	0.05	-	-	-	**	**	-	-	-	-	-	-
Hemist	0.51	0.16	0.09	0.04	**	0.01	**	0.05	-	**	-	0.13	-	0.02
Saprist	0.91	0.15	-	0.14	0.02	0.04	0.07	0.27	0.01	0.13	**	0.03	**	0.06
Mollisols	0.17	-	-	-	-	**	**	0.09	0.03	**	-	0.01	0.03	0.01
Aquolls	0.08	-	-	-	-	**	-	0.07	0.01	**	-	**	**	**
Udolls	0.09	-	-	-	-	-	**	0.02	0.02	-	-	0.01	0.03	0.01

** <0.005%; - = none.

and Delcourt, 1983, 1986). Precipitation changes may have in some cases altered the drainage condition of the soil while temperature changes have affected the rate of various soil forming reactions. In addition, some reactions may have been triggered which have temperature dependent thresholds. For example, the presence of red (rubified) soils in Pennsylvania has been attributed to a hotter climate during the Sangamonian interglacial (Waltman, 1985) than at present. The rubification process has been associated with a mean annual temperature of >15°C (Bullock, 1985). The present average temperature in State College, Pennsylvania (Central Pa), is 10°C. If Bullock (1985) is correct, the mean annual temperature was about 15°C in central Pennsylvania during the Sagamonian, and this region has experienced a temperature variation of from 15°C (Sagamonian) to −5 to −10°C (Woodfordian glacial) (Ciolkosz et al., 1986b) to 10°C today. These type of climatic variations have significantly affected the soil forming reactions, but their effect is too complex and too poorly understood to be discussed here, although the subject holds many interesting opportunities for future research.

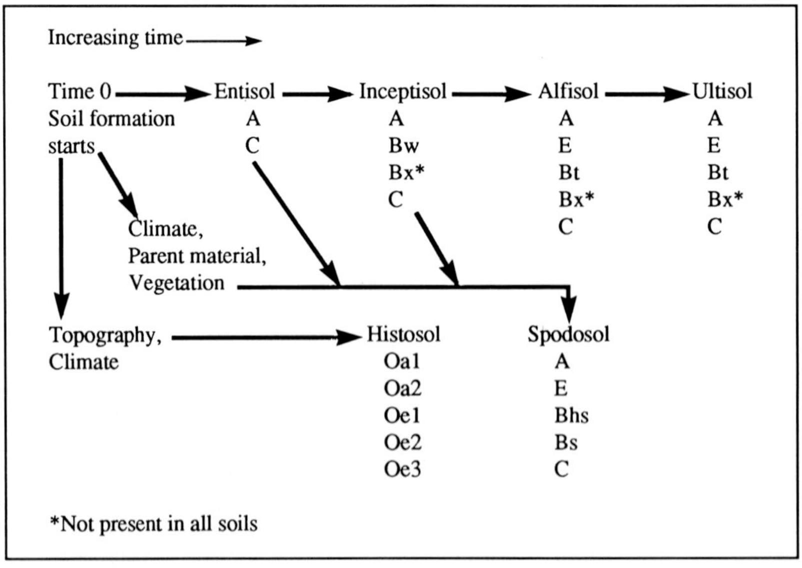

Fig. 3. Idealized soil development sequence for soils of the northeastern United States.

Developmental model

The effect of most soil forming reactions decreases from the surface downward. Thus soil characteristics vary from the surface downward from one zone or horizon to another with the major soil horizons being the A, E, B, and C. Figure 3 gives an idealized developmental sequence for Northeastern soils. This sequence will be used as a conceptual model of soil formation in the following discussion. For additional discussion of the principles of soil genesis see the texts of Birkeland (1984), Buol et al. (1989), Jenny (1980), Wilding et al., (1983a, 1983b), Fanning and Fanning (1989), Tedrow (1977), Rieger, (1983), and the Soil Survey Staff (1975).

Distribution and genesis

Entisols

Entisols are pedologically very young soils (Grossman, 1983) and are found mainly on the Coastal Plain (Fig. 4) in the glaciated area, and on floodplains throughout the region. They are developed in well drained, very sandy glacial or Coastal Plain deposits (Psamments: 60% are in the glacial area and 40% are in the Coastal Plain area), in poorly drained Coastal Plain or flood plain deposits (Aquents: 80% are on floodplains and 20% are on the Coastal Plain) and in well drained glacial fluvial or man-disturbed areas such as strip mines (Orthents). Only the Psamments occur as extensive soil areas. Although the Aquents and Orthents cover about the same total area as the Psamments (Table 3) they occupy smaller individual areas intermixed with other soils and can not be shown in Fig. 4. The Entisols of the region show very weak development because: (1) their parent material is of recent origin (glacial material or floodplain deposits); and/or (2) they are saturated with water most of the time; and/or (3) they are very resistant to change because of sandy texture and quartzose mineralogy. In Pennsylvania, the floodplain soils which are mainly silt loam texture have been dated by Bilzi and Ciolkosz (1977) and range between a few hundred to a few thousand years old. In this span of time, the well drained soils have developed into Dystrochrepts and Eutrochrepts while the poorly drained soils are classified as Fluvaquents. This relationship generally holds for the entire re-

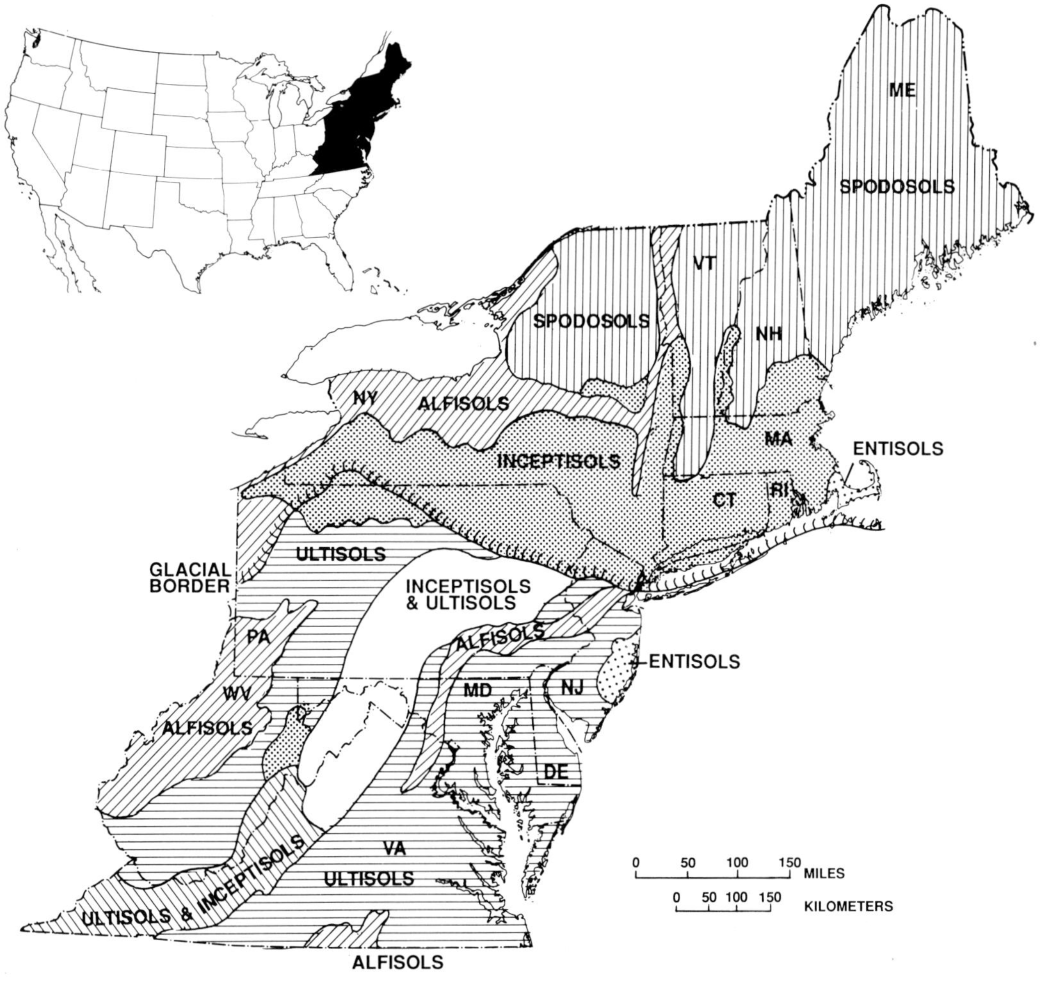

Fig. 4. Soils of the northeastern United States. Modified from the map of Smith et al. (1984).

gion. With additional time and landscape stability many of the Entisols of the region will develop into Inceptisols, Alfisols or possibly Ultisols (Fig. 3).

Inceptisols

Inceptisols are pedologically more developed than Entisols, and they have cambic and/or fragipan B horizons. They are found in glacial deposits in southern New England, southern New York, and northeastern Pennsylvania and intermixed with Alfisols and Ultisols in the unglaciated Ridge and Valley area from Pennsylvania through Virginia (Fig. 4).

The Inceptisols of southern New England, with the exception of those in the Connecticut river valley, have developed in glacial till derived primarily from igneous and metamorphic rocks; are acid; have fine sandy loam to sandy loam textures; and have dense subsoil horizons or layers (Soil Survey Staff, 1968). These Inceptisols (Dystrochrepts), which were previously classified as Brown Podzolics (Lyford, 1952), have a well-expressed color Bw cambic horizon. These color B's are the result of deposition in the B horizon of Fe and some humus which has been eluviated from upper horizons (Flach, 1960; Lyford, 1952; Baur and Lyford, 1957). Weathering of the crystalline rock minerals in the glacial materials of these young soils (about 18 Ka; Mickelson et al., 1983) under the climatic conditions of this area apparently has not caused the release if very much iron within the B horizon to form the more typical cambic color Bw horizon. The lack of weathering of the parent material is also indicated by the lack of illuvial clay in these soils. In order to accumulate clay in the B horizon, the clay, particularly the fine clay, must be present in the parent material or it must be produced by weathering. The material in the upper part of the soils in many areas of southern New England underwent sorting and redistribution by wind shortly after deposition (Schafer and Hartshorn, 1965; Colby et al., 1953). This sorting has given many of these soils very similar textures (fine sandy loam) particularly in the upper part of the soil profile. The dense subsoil horizons in the northern and southern New England soils and in the soils of Adirondacks of New York have in the past been identified as fragipans. Recent studies have questioned whether these dense horizons are true genetic fragipans or just dense basal till (Lindbo and Veneman, 1989), and the soil survey of the USDA-Soil Conservation Service has reclassified the majority of the fragipan soils in New England and the Adirondacks as non-fragipan soils. This reclassification may have been premature, particularly for the Inceptisols of southern New England (Lindbo and Veneman, 1989) and a more comprehensive study is needed to resolve this issue.

The Inceptisols (Fragiochrepts and Fragiaquepts) of the glaciated Appalachian Plateau have developed from glacial till derived from acid sandstones and shales. These materials have imparted a silt loam to loam texture to these soils as compared to the coarser sandy loam textures of the Inceptisols of New England. These soils were previously classified as Sols Bruns Acides (Baur and Lyford, 1957; Flach, 1960) and do not have as strong a degree of cambic color B horizon development as do the the New England Inceptisols. The reason for this difference is the Appalachian Plateau Inceptisols have little illuvial Fe and humus, and their color is obtained primarily from limited in situ mineral weathering in the B horizon. Although the cambic horizon is not well developed, the fragipan in these till-derived soils shows very strong development and is recognized as a true genetic fragipan (Lindbo and Veneman, 1989). Historically, these fragipans form the core of the fragipan concept developed by Dr. Marlin Cline and his students at Cornell University. In addition to having formed in glacial till, fragipans also have formed in other transported materials that are not too sandy, too clayey, or calcareous. These materials include loess, colluvium, and old alluvium, but not recent alluvium (Ciolkosz et al., 1979). Movement of the material apparently enhances packing of the soil particles giving the fragipan its characteristic high bulk density. Fragipans are not found in recent alluvium on floodplains but they are found on older fluvial and outwash terraces above the floodplains, and it has been estimated that it takes more than 2–5 ka for them to form (Hall et al., 1982). They also are not permanent fixtures in soils. Fragipans in well drained glacial till soils form and are then degraded by oxidation, leaching, and argillic horizon formation. This conclusion is based on their presence in Woodfordian age till soils in northeastern Pennsylvania and their absence in Pre-Wisconsinan Age tills soils just outside the Woodfordian border (Ciolkosz et al., 1985). Supporting evidence comes from Louisiana where fragipans formed in loess under more rigorous weathering conditions are now degrading (Bartelli, 1973; Miller, pers. commun., 1985). Fragipans persist in poorly drained soils of the Pre-Wisconsinan till catena (Waltman, 1981). Apparently leaching and argillic horizon formation have not proceeded to the point of destroying the fragipan in the more poorly drained soils, although it is found deeper in the profile and is less dense than in Woodfordian age soils

(Waltman, 1981). With adequate landscape stability these fragipans may degrade following the pathway noted by Steele et al. (1969) for fragipans on the North Carolina Coastal Plain. Smeck and Ciolkosz (1989) give an up-to-date presentation on recent work done on fragipans.

The Inceptisols (Dystrochrepts) of the unglaciated Ridge and Valley area are developed on the sandstone ridges and in the shale valleys. These soils have cambic B horizons without fragipans. Both ridge tops and shale valleys underwent significant erosion during the late Pleistocene. The erosion and the resistance of the parent material to pedological development has resulted in weakly developed soils on these landscapes (Ciolkosz et al., 1986a; Carter and Ciolkosz, 1986; Carter, 1983). In particular, soils that developed on acid brown and gray shales show very weak development as indicated by profiles which are shallow or moderately deep to bedrock and contain a high content of rock fragments. In contrast to the grayish brown shales, red shales on the same landscape weather more rapidly giving deeper soils with fewer rock fragments and weakly expressed argillic horizons (Clark et al., 1989). Soils formed from calcareous shales tend to be deeper and better developed than soils formed from acid shales. This apparently results from rapid leaching of carbonate cement which then releases a relative large mass of rock material to soil development. An example of this relationship occurs near Burnham, Pennsylvania (central Ridge and Valley area) where the vertical bedrock comprises and alternating sequence of thin (2–5 m thick) acid brown shale and weakly calcareous shale beds. The depth to the hard bedrock (R horizon) is about 1 m in the acid shale and 3–4 m in the calcareous shale.

Alfisols

In an idealized sequence of development, Alfisols are a step more developed than Inceptisols (Fig. 3). They have argillic horizons and a relatively high subsoil base saturation. They are found in the glaciated areas of central New York and northwestern Pennsylvania, as well as in the unglaciated areas of southwestern Pennsylvania and northwestern West Virginia. Alfisols are also found in the Triassic lowland area of southeastern Pennsylvania and the limestone valleys of the Ridge and Valley area in Pennsylvania, West Virginia, and Maryland.

The occurrence of Alfisols in central New York forms an apparent anomaly because they are younger in age (about 12 ka) than the Inceptisols found in southern New York and northeastern Pennsylvania (about 18 ka). However, these Alfisols have developed from calcareous glacial deposits, and the carbonates have significantly enhanced the development of these young soils. As the carbonate is dissolved and leached from the upper horizons during the development of these soils, the clay, which is an insoluble residue, is released and eluviated from the A and accumulates in the B horizon forming an argillic horizon. The depth of the clay accumulation represents the depth of movement of water carrying the clay or a zone of clay flocculation. The work of Cronce (1988) indicates that a base (Ca+Mg) saturation of 40% or greater forms a threshold for the flocculation of clays in argillic horizon formation. Thus calcareous parent materials, which are very susceptible to dissolution, foster argillic horizon formation, even in young glacial deposits (Rostad et al., 1976; Smeck and Wilding, 1980). The calcareous drifts of central New York vary in their carbonate contents from high (>25%) to medium (5–14%) to low (<5%). Argillic horizon development parallels this sequence with the most developed argillic horizons having the highest carbonate content (Merritt, 1958). Much of the clay that is dispersed in the upper horizons of an acid (low base saturation) soil may be moved completely out of the soil by the percolating water. The reason argillic horizons have not formed in the Inceptisols of southern New York is probably twofold. Firstly, the acid shale and sandstone till parent material is not releasing much clay which can be dispersed and moved; and secondly, some of the clay that is being moved may be completely removed from the soil by percolating water. The first reason is probably dominant because these soils have fragipan B horizons which do not allow movement of much material. Although the soil developed in the calcareous till are Alfisols they are not strongly developed soils, and have thin solums (A+B horizon) that are only 60–75 cm thick, with weakly expressed argillic horizons.

The Alfisols of northwestern Pennsylvania are also developed in glacial till. These Alfisols have marginal argillic horizons and are close to being Inceptisols. The till of northwestern Pennsylvania has a lower carbonate content (2–3%; White et al., 1969) than that in central New York, and this is probably the reason for the weakly developed argillic development in these soils. In addition to the weakly developed argillic horizons, these soils also have well expressed fragipans (Ciolkosz et al., 1989).

The Alfisols in southwestern Pennsylvania and northwestern West Virginia have developed from interbedded shale, limestone, and calcareous shale bedrock. These parent materials are one of the major factors controlling the development of argillic horizons in these soils as well as giving them a high base status. These soils show somewhat thicker argillic horizons that the calcareous till soils, but they are still only weak to moderately well developed, which is apparently a reflection of their development on recently truncated landscapes (Ciolkosz et al., 1989). The amount of truncation on these landscapes, as well as others in the unglaciated area varies directly with slope. Also, either a significant amount of rock disaggregation predated the last truncation or frost churning processes have contributed significantly to the disaggregation of the bedrock. The importance of these two processes needs further study to determine their relative importance on the development if these residual soils.

The Alfisols of the Triassic lowland area are developed from red shale, diabase, hornfels, and some limestone. The presence of deep, well drained soil on the diabase of southeastern Pennsylvania indicates that landscape truncation over time has been a complex process because these soils are too thick to have developed since the last glacial period. Most of the Alfisols in the northern Ridge and Valley area are formed on limestone. These Alfisols, which can be many meters thick, occur on gently rolling landscapes and may represent some of the oldest soils in the Northeast. An estimate based on the accumulation of insoluble residues for one meter of soil material indicates a 300 ka age (Ciolkosz et al., 1986a). Similar results were obtained by Cronce (1988) using additional indicators such as quartz and Fe_2O_3 content. Cronce also indicates that aeolian material added to these soils complicates the interpretation of their genesis, but helps to account for the very common silt loam texture of their surface horizons. Cronce's work also indicates that the upper part of the limestone soils has been significantly affected by frost processes giving them a very uniform upper profile horizon sequence.

Ultisols

All the Ultisols of the region are found in the unglaciated area (Fig. 4). They are found on the Appalachian Plateau, the Piedmont , the Coastal Plain, and intermixed with Inceptisols in the Ridge and Valley area. These Ultisols can be grouped into two categories: parent material Ultisols and genetic Ultisols (Clark et a., 1989). Parent material Ultisols show weak development in residual acid shale materials on the Appalachian Plateau and in the Ridge and Valley and weak to moderate development in colluvial deposits in the Ridge and Valley area (Ciolkosz et al., 1979). The low base status of these soils results from a very low base status inherited from the parent material and not from extreme leaching. These soils contrast with the well developed genetic Ultisols developed on the igneous and metamorphic rocks of the Piedmont, on the limestones as well as some older alluvium and colluvium in the southern Ridge and Valley, and on the sands and clays of the Coastal Plain. The argillic horizon of genetic Ultisols differs from that found in Alfisols and in parent material Ultisols. In uneroded genetic Ultisols the argillic horizon occurs deeper in the profile, has a more gradual transition from eluvial to illuvial zone, has a thicker zone of accumulation, and has more clay than can be accounted for by eluviation from overlying horizons than is found in Alfisols and parent material Ultisols (Ciolkosz et al., 1985). These differences are a result of a greater period of time for clay to move down the soil profile as well as adequate time for clay to be stripped from the upper part of the argillic and translocation downward. In addition, the longer period of time has allowed clay to form within the argillic horizon by weathering. The age of the soils and pedogenetic trends on the Coastal Plain have recently been studied by Markewich et al. (1987). These studies indicate progressive development from the youngest age material (about 30 ka) to the oldest age material (>1 Ma). The most highly developed soils in the Northeast region occur on the high Coastal Plain and on nearby high terrace deposits. Solum thicknesses in excess of 5 m and the presence of plinthite horizons have been observed in these soils. An interesting future comparison between the youngest soils of the Coastal Plain and the slightly younger glacial till soils should be made, particularly in relation to the effect of climate on soil development.

The Ultisols formed on the granites and schists of the Piedmont do not show the thick solum (A+B horizon) development that is observed in soils on the middle and upper Coastal Plain on similar landscapes. Although these soils do not have thick solums, they do have very thick C horizons that frequently are called saprolite. This material has been estimated to be about 15 m thick in northern Virginia (Froelich and Heironimus, 1977; Pavich, 1986). The Piedmont is unique in this respect, because no other residual soils of significant extent in the region have thick saprolite C horizons. An example of this relationship occurs in southeastern Pennsylvania where the Piedmont is in close proximity to a long belt of acid shale (Martinsburg) in the great valley north and south of Harrisburg. The shale area and the Piedmont have similar landscapes, but the shale soils lack the thick saprolite C horizons of the Piedmont soils. Thus, parent material appears to be a significant factor in the development of the thick saprolite zones. The thick saprolites and thin solums of the Piedmont soils has been discussed by Pavich (1986). He suggests that the soils on the Piedmont are in a dynamic equilibrium and will always have about the same development that we see today. If this is true, the Piedmont is a very unique pedological system that is not operating in the same manner as the Coastal Plain or other areas of the region. A more plausible explanation is given by Costa and Cleaves (1984), Cleaves (this volume), and Poag and Sevon (this volume) who evoke episodic erosion to explain these relationships.

The Ultisols in the Ridge and Valley of southern Virginia are developed dominantly from limestones. Although developed from limestone, they have undergone a greater degree of leaching of bases and weathering than the limestone soils found in the northern Ridge and Valley which are Alfisols.

Spodosols

Spodosols are found primarily in the northern part of the glaciated area associated with acidic, sandy textured parent materials and frigid soil temperature regimes. Some Spodosols occur in the unglaciated area and are associated with acidic, sandy (quartzose sandstone) parent materials and frigid soil temperatures (Fig. 2). This association also extends southward into the southern Appalachians at higher elevations (Lietzke and McGuire, 1987). The only Spodosols found outside of the frigid areas are poorly drained (Aquods). They occur primarily on the lower and middle Coastal

Plain from New Jersey through Virginia and in glacial outwash plains in Central New York. The temperature and landscape relationships noted for the Northeast also hold for the United States as a whole (Rourke et al., 1988). Some Aquods also occur in the frigid areas but they account for only about 2–3% of the total Spodosol area (Table 3). Although very minor in occurrence, Aquods tend to have the best expressed ortsteins (cemented spodic horizons) (McKeague et al., 1983). In addition to cold temperatures, and acidic sandy texture parent material, coniferous vegetation is also a major factor influencing Spodosol development (McKeague et al., 1983).

Spodosols, previously called Podzols, were originally defined in Russia on the basis of their E horizons, but later the illuvial B horizon (spodic) of humus plus aluminium with or without iron became the definitive horizon. This change is of major significance because some of the soils classified today as Spodosols have little or no E horizon development. Although the reactions involved in spodic horizon formation have been studied for years, a clear understanding of the exact mechanism of their development has not been established. The most feasible mechanism encompasses the complexation of iron and aluminium with organic acids to form soluble complexes. These complexes move downward in percolating water until they are precipitated by increases in the concentration of metal ions, by desiccation or by oxidation of the organic complex (Mokma and Buurman, 1982). This mechanism seem appropriate for Orthods but not for Aquods. Although less is known of spodic horizon genesis in Aquods, some progress is being made in this area of research (Lee et al., 1988).

Climate, particularly temperature, in concert with vegetation appears to be very important in the development of Spodosols in the Northeast. Stanley and Ciolkosz (1981) reported a trend of increasing spodic horizon development with decreasing soil temperature for Spodosols developed from sandstone on the unglaciated Appalachian Plateau. This trend continues into New England and helps explain the genesis of the very well developed spodic horizons which have been called "super spodics" in some Spodosols of northern New England and northern New York. A general trend of decreasing E horizon (Albic horizon) thickness in Spodosols from the unglaciated Appalachian Plateau to New England has also been observed. This trend is believed to be associated with a much higher quartz content in the sand size material of the sandstone of the plateau than in the till of the glaciated area, and a much higher content of weatherable minerals in the till. The sand size quartz provides the whitish mineral surfaces that give E horizons their light color and the weatherable minerals release bases which precipitates the soluble organic complexes near the soil surface (DeConinck, 1980). It also has been speculated that windthrow of trees, which tends to mix the upper horizons, also may contribute to this trend.

Most well drained Spodosols in the northeastern U.S. have a solum thickness of a meter or less, and according to Franzmeier and Whiteside (1963), this amount of soil formation could have taken place in 10,000 years. Thus, the Spodosols in the unglaciated area may be only about 10 ka or they may be older and in a state of equilibrium. If the mechanism of Spodosol genesis proposed by McKeague et al. (1983) is correct, the E horizon, and probably the spodic horizon also should become thicker with age. This mechanism seems to apply in Aquods genesis in tropical and subtropical regions (McKeague et al., 1983). Thus, it appears that the Spodosols of the unglaciated area are only 10–15 ka. This conclusion supports the contention that these areas were eroded during the last glacial period, otherwise thicker, better developed Spodosols should be found in the unglaciated area.

Histosols

Histosols are organic soils and are found throughout the region, but they are more abundant in the glaciated area and on the lower Coastal Plain. The organic matter of Histosols accumulated because its decomposition is retarded by water saturation and/or cold temperatures (Everett, 1983).

Histosols can be divided into two groups: topographic and climatic. Topographic Histosols are found in depressional areas that are saturated with water most of the year. The glaciated area, with its disrupted drainage and many undrained depressions, and the lower Coastal Plain, with its high water table and tidal inundation are major areas of topographic Histosol formation. Climatic Histosols (Folists) are freely drained organic soils. They have formed in climates that are cold and humid. In the northeast region they are of minor extent and are found on moderate to steep slopes at high elevations in northern New England and the Adirondacks. In this region they also tend to form directly on bedrock and are relatively thin (usually <0.5 m thick). Many of the salt water Histosols, when drained, become very acidic due to the oxidation of sulfides that have accumulated with the organic material. Fresh water Histosols do not tend to accumulate sulfides and many have been drained and are used for agricultural purposes.

With drainage, Histosols oxidize rapidly and are transformed from Fibrists to Hemists to Saprists. Under natural conditions this conversion is greatly reduced, although, there is a general trend of more rapid conversion with increasing temperature from northern New York to the southern part of the region.

Conclusions

The soils of the northeastern United States have followed a varied and complex pathway of development. The pathway would not be nearly as complex if the soil forming factors had been constant for all soils of the region. Of particular importance in the region has been the impact of glaciation and associated variations in climate. The impact of glaciation in the northern part of the region has been to create new parent material and set the pedological clock of soil formation at a known time. Since the deposition of the glacial materials, the variation of climate, vegetation, and other soil forming factors has been minor. Thus, the glaciated area provides a less complex area than the unglaciated region and probably explains, why more is known of the genesis of glacial soils than of nonglacial soils.

In the unglaciated area, the pathway of soil formation for the area as a whole is more complex because there are many different ages of surfaces. Consequently, the pedological clock of soil formation has started at different times in different places. One of the major factors causing the complexity in the age of the surfaces is the changes in climate associated with glacial–interglacial cycles and its impact on landscape stability. A particular complication associated with instability and erosion is the complete removal of previous soils on one landscape and partial to nonremoval on other landscapes or slopes. Consequently, soil formation can start in a new soil material, in a partially truncated soil or in a soil that has not been truncated.

In addition to the challenge of learning more about soil genesis in the glacial area is the even larger challenge of sorting out the genetic pathways of soils in the unglaciated area. In particular, what are the relationships of slope under present conditions to past conditions with respect to possible thresholds of erosion, and what impact has varying climatic conditions over time had on the rate, extent, and direction of soil formation reactions. In addition, what can the soil properties observed today tell us about the processes that have shaped the soils and landscapes in the northeastern United States.

References

Bartelli, L.J., 1973. Soil development in loess in the southern Mississippi river valley. Soil Sci., 115: 254–260.

Baur, A.J. and Lyford, W.H., 1957. Sols Bruns Acides of the Northeastern United States. Soil Sci. Soc. Am. Proc., 21: 533–536.

Bilzi, A.F. and Ciolkosz, E.J., 1977. Time as a factor in the genesis of four soils developed in recent alluvium in Pennsylvania. Soil Sci. Soc. Am. J., 41: 122-127.

Birkeland, P.W., 1984. Soils and Geomorphology. Oxford University Press, New York, N.Y., 372 pp.

Budel, J., 1982. Climatic Geomorphology. Princeton University Press, Princeton, N.J., 443 pp.

Bullock, P., 1985. The role of micromorphology in the study of soils. In: J. Boardman (Editor), Soils and Quaternary Landscape Evolution. Wiley New York, N.Y., pp. 45-68.

Buol, S.W., Hole, F.D. and McCracken, R.J., 1989. Soil genesis and classification. Iowa State University Press, Ames, Iowa, 456 pp.

Carter, B.J., 1983. The effect of slope gradient and aspect on the genesis of soils on a sandstone ridge in central Pennsylvania. Ph.D. Thesis, Pennsylvania State University, University Park, Pa., 233 pp.

Carter, B.J. and Ciolkosz, E.J., 1980. Soil temperature regimes of the central Appalachians. Soil Sci. Soc. Am. J., 44: 1052-1058.

Carter, B.J. and Ciolkosz, E.J., 1986. Sorting and thickness of waste mantle material in a sandstone spur in central Pennsylvania. Catena, 13: 241-256.

Ciolkosz, E.J., 1967. The mineralogy and genesis of the Dodge catena of southeastern Wisconsin. Ph.D. Thesis, University of Wisconsin, Madison, Wisc., 133 pp.

Ciolkosz, E.J. and Dobos, R.R., 1989. Distribution of soils of the northeastern United States. Pa. State Univ. Agron. Ser., 103, 20 pp.

Ciolkosz, E.J., Petersen, G.W., Cunningham, R.L. and Matelski, R.P., 1979. Soils developed from colluvium in the Ridge and Valley area of Pennsylvania. Soil Sci., 128: 153-162.

Ciolkosz, E.J., Cronce, R.C., Cunningham, R.L. and Petersen, G.W., 1985. Pre-Woodfordian soil development in northeastern Pennsylvania. Geolog. Soc. Am. Abstr. Programs, 17(1), p. 44.

Ciolkosz, E.J., Cronce, R.C., Cunningham, R.L. and Petersen, G.W., 1986a. Geology and Soils of Nittany Valley. Pa. State Univ., Agron. Ser., 88, 52 pp.

Ciolkosz, E.J., Cronce, R.C. and Sevon, W.D., 1986b. Periglacial features in Pennsylvania. Pa. State Univ., Agron. Ser., 92, 15 pp.

Ciolkosz, E.J., Gardner, T.W. and Dobos, R.R., 1988. Paleosols in Pennsylvania. Pa. State Univ., Agron. Ser., 100, 9 pp.

Ciolkosz, E.J., Cronce, R.C., Day, R.L. and Dobos, R.R., 1989. Soils. In: C,H. Shultz (Editor), The Geology of Pennsylvania. Pa Geol. Surv. (In press).

Clark, G.M. and Ciolkosz, E.J., 1988. Periglacial geomorphology of the Appalachian Highlands and Interior Highlands south of the Glacial Border — A review. Geomorphology, 1: 191-220.

Clark, G.M., Ciolkosz, E.J., Kite, J.S. and Lietzke, D.A., 1989. Central and southern Appalachian geomorphology — Tennessee, Virginia, and West Virginia. Am. Geophys. Union, Washington, D.C., 150 pp.

Cleaves, E., 1989. Appalachian Piedmont landscapes from the Permian to the Holocene. In: T.W. Gardner and W.D. Sevon (editors), Appalachian Geomorphology. Geomorphology, 2: 159-179 (this volume).

Colby, W.G., Light, M.A. and Bertinuson, T.A., 1953. The influence of windblown material on the soils of Massachusetts. Soil Sci. Soc. Am., Proc., 17: 395-399.

Costa, J.E. and Cleaves, E.T., 1984. The Piedmont landscape of Maryland: A new look at an old problem. Earth Surf. Processes Landforms, 9: 59-74.

Cronce, R.C., 1988. The genesis of soils overlying dolomite in the Nittany Valley of Central Pennsylvania. Ph.D. Thesis, Pennsylvania State University, University Park, Pa., 389 pp.

Cunningham, R.L. and Ciolkosz, E.J. (Editors), 1984. Soils of the Northeastern United States. Pa. Agric. Exp. Stn., Bull., 848, 47 pp.

Darmody, R.G. and Foss, J.E., 1982. Soil-landscape relations in the Piedmont of Maryland. Soil Sci. Soc. Am., J., 46: 588-592.

DeConinck, F., 1980. Major mechanism in formation of spodic horizons. Geoderma, 24: 101-128.

Delcourt, H.R. and Delcourt, P.A., 1986. Late Quaternary vegetational change in the Central Atlantic States. In: J.N. McDonald and S.O. Bird (Editors), The Quaternary of Virginia — A Symposium Volume. Va. Div. Miner. Resour., Publ., 75: 23-35.

Delcourt, P.A. and Delcourt, H.R., 1983. Late-Quaternary vegetational dynamics and community stability reconsidered. Quat. Res., 19: 265-271.

Everett, K.R., 1983. Histosols. In: L.P. Wilding, N.E. Smeck and G.F. Hall (Editors), Pedogenesis and Soil Taxonomy. II. The Soil Orders. Elsevier, New York, N.Y,., pp. 1-53.

Fanning, D.S. and Fanning, M.C.B., 1989. Soil morphology, genesis, and classification. Wiley, New York, N.Y., 395 pp.

Flach, K.W., 1960. Sols Bruns Acides in the United States: Genesis, morphology and relationships to associated soils. Ph.D. Thesis, Cornell University, Ithaca, N.Y., 210 pp.

Franzmeier, D.P. and Whiteside, E.P., 1963. A chronosequence of podzols in northern Michigan. I. Ecology and descriptions of pedons, II. Physical and chemical properties. Mich. State Univ., Agric. Exp. Stn., Q. Bull., 46: 1-36.

Froelich, A.J. and Heironimus, T.L., 1977. Thickness of overburden map of Fairfax County, VA. U.S. Geol. Surv., Open File Rep., 77-797, one page map.

Goldthwait, R.P., 1976. Frost sorted patterned ground. Quat. Res., 6: 27-35.

Goodman, K.V., 1953. Brown forest, polygenetic, and congeliturbate profiles of Potter County, Pennsylvania. Soil Sci. Soc. Am., Proc., 17: 399-402.

Grossman, R.B., 1983. Entisols. In: L.P. Wilding, N.E. Smeck and G.F. Hall (Editors), Pedogenesis and Soil Taxonomy II. The Soil Orders. Elsevier, New York, N.Y., pp. 55-90.

Hall, G.F., Daniels, R.B. and Foss, J.E., 1982. Rate of soil formation and renewal in the USA. In: B.L. Schmidt (Editor), Determinants of Soil Loss Tolerance. Am. Soc. Agron., Spec. Publ., 45: 23-39.

Hoover, M.T., 1983. Soil development in colluvium in footslope positions in the Ridge and Valley Physiographic Province in Pennsylvania. Ph.D. Thesis. Pennsylvania State University, University Park, Pa., 271 pp.

Hoover, M.T. and Ciolkosz, E.J., 1988. Colluvial soil parent material relationships in the Ridge and Valley Physiographic Province of Pennsylvania. Soil Sci., 145: 163-172.

Jacobson, R., 1989. The role of catastrophic geomorphic events in the evolution of the central Appalachian landscape. In: T.W. Gardner and W.D. Sevon (Editors), Appalachian Geomorphology. Geomorphology, 2: 257-284 (this volume).

Jenny, H., 1980. The Soil Resource: Origin and Behavior. Springer, New York, N.Y., 377 pp.

Lee, F.Y., Yuan, T.L. and Carlisle, V.W., 1988. Nature of cementing material in ortstein horizons of selected Florida Spodosols: II. Soil properties and chemical form(s) of aluminium. Soil Sci. Soc. Am., J., 52: 1796-1801.

Lewkowicz, A.G., 1988. Slope processes. In: M.J. Clark (Editor), Advances in Periglacial Geomorphology. Wiley, New York, N.Y., pp. 325-368.

Lietzke, D.A. and McGuire, G.A., 1987. Characterization and classification of soils with spodic morphology in the Southern Appalachians. Soil Sci. Soc. Am., J., 51: 165-170.

Lindbo, D.L. and Veneman, P.L.M., 1989. Fragipans in the northeastern United States. In: N.E. Smeck and E.J. Ciolkosz (Editors), Fragipans: Their Occurrence, Classification, and Genesis. Soil Sci. Soc. Am., Spec. Publ., 24: 11-31.

Lyford, W.H., 1952. Characteristics of some Podzolic soils of the Northeastern United States. Soil Sci. Soc. Am., Proc., 16: 231-235.

Markewich, H.W., Pavich, M.J., Mausbach, M.J., Hall, R.L., Johnson, R.G. and Hearn, P.P., 1987. Age relationships between soils and geology in the Coastal Plain of Maryland and Virginia. U.S. Geol. Surv., Bull., 1589, Chap. A, 34 pp.

Merritt, R.S., 1958. Depth of leaching in relation to carbonate content of glacial till in central New York. M.S. Thesis, Cornell University, Ithaca, N.Y., 129 pp.

McKeague, J.A., DeConinck, F. and Franzmeier, 1983. Spodosols. In: L.P. Wilding, N.E. Smeck and G.F. Hall (Editors), Pedogenesis and Soil Taxonomy. II The Soil Orders. Elsevier, New York, N.Y., pp. 217-252.

Mickelson, D.M., Clayton, L., Fullerton, D.S. and Borns, H.W., 1983. The late Wisconsin glacial record in the United States. In: H.E. Wright, Jr. (Editor), Late Quaternary Environment of the United States, Vol. 1, The Late Pleistocene. University of Minnesota Press, Minneapolis, M pp. 3-37.

Mills, H.H., Brakenridge, G.R. Jacoboson, R.B., Newell, W.L., Pavich, M.J. and Pomeroy, J.S., 1987. Appalachian Mountains and plateaus. In: W.L. Graf (Editor), Geomorphic Systems of North America. Geol. Soc. Am., Centen. Spec. Vol., 2: 5-50.

Mokma, D.L. and Buurman, P., 1982. Podzols and podzolization in temperate regions. Int. Soil. Mus., Monogr. 1, 126 pp.

Osterkamp, W.R., Fenton, M.M., Gustavson, T.C., Hadley, R.F., Holliday, V.T., Morrison, R.R. and Toy, J.J., 1987. Great plains. In: W.L. Graf (Editor), Geomorphic Systems of North America. Geol. Soc. Am., Centen. Spec., 2: 163-210.

Pavich, M.J., 1986. Processes and rates of saprolite production and erosion on a foliated granitic rock of the Virginian Piedmont. In: S.M. Colman and D.P. Dethier (Editors), Rates of Chemical Weathering of Rocks and Minerals. Academic Press, New York, N.Y., pp. 552-590.

Pavich, M., 1989. Regolith residence time and the concept of surface age of the Piedmont "peneplain". In: T.W. Gardner and W.D. Sevon (Editors), Appalachian Geomorphology. Geomorphology, 2: 181-196 (this volume).

Poag, W. and Sevon, W.D., 1989. A record of Appalachian denudation in post-rift Mesozoic and Cenozoic sedimentary deposits of the U.S. middle Atlantic continental margin. In: T.W. Gardner and W.D. Sevon (Editors), Appalachian Geomorphology. Geomorphology, 2: 119-157 (this volume).

Ranney, R.W., Ciolkosz, E.J., Petersen, G.W., Matelski, R.P., Johnson, L.J. and Cunningham, R.L., 1974. The pH base-saturation relationships in B and C horizons of Pennsylvania soils. Soil Sci., 118: 247-253.

Rieger, S., 1983. The Genesis and Classification of Cold Soil. Academic Press, New York, N.Y., 229 pp.

Rostad, H.P.W., Smeck, N.E. and Wilding, L.P., 1976. Genesis of argillic horizons derived from coarse-textured calcareous gravels. Soil Sci. Soc. Am., J., 40: 739-744.

Rourke, R.V., Brasher, B.R., Yeck, R.D. and Miller, F.T., 1988. Characteristic morphology of U.S. Spodosols. Soil Sci. Soc. Am., J., 52: 445-449.

Schafer, J.P. and Hartshorn, J.H., 1965. The Quaternary of New England. In: H.E. Wright (Editor), The Quaternary of the United States. Princeton University Press, Princeton, N.J., pp. 113-128.

Sevon, W.D., 1985. Pennsylvania's polygenetic landscape. Harrisburg Area Geol. Soc., Guideb. 4th Ann. Field Trip, Harrisburg, Pa., 55 pp.

Smeck, N.E. and Ciolkosz, E.J. (Editors), 1989. Fragipans: Their occurrence, classification, and genesis. Soil Sci. Soc. Am., Spec. Publ., 24, 153 pp.

Smeck, N.E. and Wilding, L.P., 1980. Quantitative evaluation of pedon formation in calcareous glacial deposits in Ohio. Geoderma, 24: 1-16.

Smith, G.D., 1983. Historical development of soil taxonomy-background. In: L.P. Wilding, N.E. Smeck and G.F. Hall (Editors), Pedogenesis and Soil Taxonomy I. Concepts and Interactions. Elsevier, N.Y., pp. 28-49.

Smith, H., 1984. Soil temperature regimes of the northeastern United States. USDA, Soil Conserv. Serv., Chester, Pa., 1 p. map.

Smith, H., Grice, D.G., Miller, F.P., Rourke, R.V., Sauffer, E.H. and Simpson, T.W., 1984. General soil map of the northeastern United States. USDA, Soil Conserv. Serv., Chester, Pa, one page map.

Soil Survey Staff, 1968. Soil survey laboratory data and description for some soils of New England States, Soil Surv. Invest. Rep. No. 20. USDA, Soil Conservation Service, Washington, D.C., 295 pp.

Soil Survey Staff, 1975. Soil Taxonomy. USDA, Soil Conserv. Serv., Agric. Handb. No. 436, 754 pp.

Soil Survey Staff, 1987. Keys to Soil Taxonomy. Soil Manage. Support Serv., Tech. Monogr. No. 6, Cornell Univ., Ithaca, N.Y. 3rd ed., 280 pp.

Stanley, S.R. and Ciolkosz, E.J., 1981. Classification and genesis of Spodosols in the Central Appalachians. Soil Sci. Soc. Am., J., 45: 912-917.

Steele, F., Daniels, R.B., Gamble, E.E. and Nelson, L.A., 1969. Fragipan horizons and Be masses in the middle Coastal Plain of north central North Carolina. Soil Sci. Soc. Am., Proc., 33: 752-755.

Tedrow, J.C.F., 1977. Soils of the Polar Landscapes. Rutgers University Press, New Brunswick, N.J., 638 pp.

Waltman, S.W., 1988. The genesis of prairie soils in northwestern Pennsylvania. M.S. Thesis, Pennsylvania State University, University Park, Pa, 220 pp.

Waltman, W.J., 1981. Fragipan morphology in late Wisconsin and pre-Wisconsin age soils of Pennsylvania. M.S. Thesis, Pennsylvania State University, University Park, Pa., 189 pp.

Waltman, W.J., 1985. The stratigraphy and genesis of pre-Wisconsin soils in the Allegheny Plateau. Ph.D. Thesis, The Pennsylvania State University, University Park, Pa., 353 pp.

White, G.W., Totten, S.M. and Gross, D.L., 1969. Pleistocene stratigraphy of northwestern Pennsylvania. Pa. Geol. Surv. Gen. Geol. Rept. G. 55, 88 pp.

Whittecar, G.R., 1985. Stratigraphy and soil development in upland alluvium and colluvium in the northcentral Virginia Piedmont. Southeast Geol., 26: 117-129.

Wilding, L.P., Smeck, N.E. and Hall, G.F. (Editors), 1983a. Pedogenesis and Soil Taxonomy. I. Concepts and Interactions. Elsevier, New York, N.Y., 303 pp.

Wilding, L.P., Smeck, N.E. and Hall, G.F. Editors, 1983b. Pedogenesis and Soil Taxonomy. II. The Soil Orders. Elsevier, New York, N.Y., 410 pp.

Witty, J.E. and Arnold, R.W., 1987. Soil Taxonomy: An overview. Outlook Agric., 16: 8-13.

Geomorphology, 2 (1989) 303-318
Elsevier Science Publishers B.V., Amsterdam — Printed in The Netherlands

Erosion in the Juniata River Drainage Basin, Pennsylvania

W.D. SEVON

Pennsylvania Geological Survey, P.O. Box 2357, Harrisburg, PA 17105 (U.S.A.)

(Received December 10, 1988; accepted after revision May 5, 1989)

Abstract

Sevon, W.D., 1989. Erosion in the Juniata River drainage basin, Pennsylvania. In: T.W. Gardner and W.D. Sevon (Editors), Appalachian Geomorphology, Geomorphology, 2: 303-318.

Previously calculated erosion rates throughout the Appalachians range from 1.2 to 203 m Myr^{-1}. Calculation of erosion rates has been accomplished by: (1) evaluation of riverine solute and sediment load in either large or small drainage basins; (2) estimation from the volume of derived sediments; and (3) methods involving either ^{10}Be or fission-track dating. Values of specific conductance and suspended sediment collected at the Juniata River gauging station at Newport, Pennsylvania are used, with corrections, along with a bedload estimate to determine the total amount eroded from the 8687 km^2 drainage basin during the water years 1965-1986. The amount eroded is used to calculate a present erosion rate of 27 m Myr^{-1}.

Introduction

The purpose of this paper is to review the many erosion rates calculated for the Appalachian Mountains, to briefly analyze variations in the discharge, chemical-character, specific-conductance, and suspended-sediment data used in this report, and to calculate an erosion rate for the Juniata River drainage basin in Pennsylvania.

Previous work

Edward Waller Claypole (1885, pp. 39–40) was one of the first to calculate a sediment erosion rate for the Appalachians. Using a value of 8 grains of sediment per gallon of water (160 mg l^{-1}) in the Juniata River during ordinary weather, he calculated an erosion rate of 203 m Myr^{-1}.

About the same time, Ewing (1885) calculated a chemical erosion rate for limestone in Centre County, Pennsylvania. He determined through evaporation a value of 155.3 g m^{-3} of dissolved solids in stream water and used that value and a stream flow velocity to derive an erosion rate of 34.3 m Myr^{-1}.

Since then there have been many estimates of the rate of erosion in the Appalachian Mountains (Table 1). The approaches used to determine the various erosion rates fall into three main categories: (1) those which utilized stream sediment and/or dissolved load data for either large or small drainage basins; (2) those which used the volume of derived sediments; and (3) those which estimated erosion by other means, mainly fission-track dating or ^{10}Be.

The most commonly quoted erosion rates (e.g., Dole and Stabler, 1909; Hack, 1965) are those which were derived by evaluating one or all of riverine bedload, suspended load, and dissolved load for large drainage basins (Table

TABLE 1

Erosion rates calculated for the Eastern United States

Source	Bedrock	Data used: Bedload	Data used: Suspended load	Data used: Dissolved load	Years of record	Erosion rate (m Myr^{-1})	Area
A							
Claypole, 1885	Diverse		X		<1	203	PA
Dole and Stabler, 1909	Diverse		X	X	1	22	Northern Appalachians
Dole and Stabler, 1909	Diverse		X	X	1	36	Southern Appalachians
Gilluly, 1964	Diverse	X[a]	X	X	1	22	Appalachians
Judson and Ritter, 1964	Diverse	X	X	X[a]	>1	48	North Atlantic drainage
Hack, 1965	Diverse		X	X	1	40	VA, South Fork, Shenandoah River
Williams and Reed, 1972	Diverse		X		5	15	PA, NY Susquehanna River basin
Trimble, 1977	Diverse	X[a]	X[c]	X[c]	1	53	AL, GA, SC, NC, VA (stream sediment)
Hack, 1980	Diverse	Various published sources				40	Eastern North America
Fishel, 1984	Diverse	X[a]	X	X	1	30	PA, NY Susquehanna River basin
Sevon, 1985	Shale	X[a]	X	X		23–38	PA
Sevon (this report)	Diverse		X	X[b]	22	27	PA
B							
Bricker et al., 1968	Felsic schist		X	X	1	1.2	MD
Cleaves et al., 1970	Felsic schist		X	X	1	1.2	MD
Cleaves et al., 1974	Serpentinite		X	X	1	2.2	MD
Godfrey, 1975	Quartzite		X	X		2.5	MD
Godfrey, 1975	Metabasalt		X	X		6.5	MD
Goldfarb et al., 1989	>50% limestone		X	X	0.25–1.5	41–137	PA
Goldfarb et al., 1989	>60% shale		X	X	1.5–3	10–64	PA
Goldfarb et al., 1989	>95% sandstone		X	X	3	5	PA
C							
Ewing, 1885	Limestone			X	<1	34	PA
Berry, 1977	?			X		4–9	NC
Reesman and Godfrey, 1981	Siliceous rock			X	1	11	TN
Reesman and Godfrey, 1981	Limestone			X	1	40	TN
Ogden, 1982	Limestone			X		19–23	WV
Affifi and Bricker, 1983	Sandstone			X	1 1	2	VA
Affifi and Bricker, 1983	Shale			X	1	10	VA
Parizek and White, 1985	Limestone	Denudation curve (White, 1984)				30	PA
D							
Menard, 1961	Diverse	Volume of derived sediments				62	Appalachians (past rate)
Mathews, 1975	Diverse	Volume of derived sediments				5	Eastern North America, Cenozoic only
Gordon, 1979	Diverse	Volume of estuarine sediment				4	New England (calculated by Gordon)
E							
Whitney, et al., 1976	Granite	Age and depth of intrusion				41	GA
Trimble, 1977	Diverse	Erosion depth estimates				950	AL, GA, SC, NC, VA (upland erosion)
Lakatos and Miller, 1983	Diverse	Fission-track dating				32[d]	NY
Way et al., 1986	Diverse	Apatite fission track dating				10	PA
Pavich et al., 1985	Metapelite	Loss of ^{10}Be				4.5	VA
Brown et al., 1988	Diverse	Erosion index using ^{10}Be				3.8	Pre-colonial Eastern USA estimate

[a]Estimate.
[b]Corrected for input from atmosphere and by man.
[c]Data corrected from Dole and Stabler (1909) by Trimble.
[d]Rate calculated by Sevon.

TABLE 2

Relationships of bedrock, land use, and denudation rate for nine small drainage basins in Union County, Pennsylvania (data is from Goldfarb et al., 1989, tables 1, 2, and 5)

Drainage basin	Area (km^2)	Basin bedrock (%)			Basin land use (%)			Denudation rate (m Myr^{-1})		
		Limestone	Shale	Sandstone	Forest	Plowed	Other	Chemical	Physical	Total
1	4.68	89	10	1	<1	11	88	131	6	137
2	12.25	60	39	1	5	68	27	66	4	70
3	3.17	95	4	1	7	86	7	37	4	41
4	1.14	15	80	5	4	64	32	50	14	64
5	4.02	20	60	20	16	61	23	8	2	10
6	3.90	5	80	15	42	40	18	22	4	26
7	25.88	0	5	95	100	0	0	4	1	5
8	26.55	50	30	20	37	44	19	100	9	109
9	0.54	60	30	10	59	13	28	42	10	52

1A). These rates range from a low of 15 m Myr^{-1} (Williams and Reed, 1972) to a high of 203 m Myr^{-1} (Claypole, 1885) and have a mean of 34 m Myr^{-1} (excluding the disparate 203 m Myr^{-1} from the calculation). The variation in erosion rates reflects, at least in part, land-use differences as exemplified by the work of Judson and Ritter (1964, table 1, p. 3397). They calculated erosion rates of 20, 70, and 10 m Myr^{-1} for the Delaware, Schuylkill, and Juniata River drainage basins, respectively. The high value for the Schuylkill River drainage basin presumably reflects the presence of more strip-mining, more agricultural land, and less forest in that drainage basin that in the other two drainage basins.

Studies using suspended- and dissolved-load data from small drainage basins in Maryland (Table 1B) yield very low erosion rates which range from 1.2 to 6.5 m Myr^{-1}. These studies were done in forested areas and the erosion rates may be quite reasonable for the rock types and land use. In contrast, erosion rates of 5–137 m Myr^{-1} were similarly determined by Goldfarb et al. (1989) for nine small drainage basins with diverse bedrock and land use in Pennsylvania (Table 2). Two points in particular stand out in this latter work: (1) the 5 m Myr^{-1} erosion rate calculated for the only drainage basin totally in forest is of the same order of magnitude as the Maryland rates, and (2) of the extremes of 109 and 137 m Myr^{-1} are discounted, the mean erosion rate for the remaining seven drainage basins is 38 m Myr^{-1}, a rate almost the same as the mean (34 m Myr^{-1}) for large drainage basins with diverse rock types derived from Table 1A.

Erosion rates derived from evaluation of dissolved load only (Table 1C) yield a mean erosion rate of 32 m Myr^{-1} for limestone and 8 m Myr^{-1} for sandstones and shales. The rates are suprisingly comparable to those estimates which included both suspended sediment and dissolved load in some of their calculations (Table 1A, 1B).

Erosion rates determined by estimation of the volume of derived sediments (Table 1D) offer a useful alternative to the other methods because the volume method allows evaluation for a long time period (thousands to millions of years) and a large drainage area (multiple, large drainage basins). However, the method is very dependent on the precision of age and volume determinations which may change with the recovery of new data. Certainly the estimates of Menard (1961) and Mathews (1975) would be more accurate using new offshore data not available to them. When sufficient time and volume data are available, it is possible to evaluate variations in past erosion rates (Poag and Sevon, this volume).

Fission-track dating (Table 1E) allows generalizations about long-term erosion rates and is of particular interest because, when adequate regional data are available, regional variations in erosion rates can be approached (Roden and Miller, this volume). Loses of ^{10}Be appear to offer "...a datum with which present erosion rates can be quantitatively compared with past rates..." (Brown et al., 1988, p. 456). The ^{10}Be methodology was used to determine erosion rates in Virginia (Table 1E) which are comparable to those determined in other forested sites by other methods (Table 1B and Table 2).

The erosion depth estimates of Trimble (1977) indicate an exceedingly high erosion rate, 950 m Myr^{-1}, during the past 200 years, presumably attributable to the erosion-enhancing acts of man since European settlement. Trimble emphasizes that much of this material is stored within the drainage basin because stream-load data for the same area in the southern Piedmont indicate an erosion rate of only 55 m Myr^{-1} at the present time, thus not giving a true indication of the real rate of landscape lowering.

Attention is called to the work of Ritchie and McHenry (1989) who provide a thorough discussion and an extensive bibliography on the use of radioactive fallout Cesium-137 for measuring soil erosion. Measurable Censium-137 has been falling from the atmosphere only since 1952 and, as with the use of ^{10}Be, certain assumptions are made about its landscape cycle.

Three other works are of interest even though they do not specifically address erosion rates. Glenn (1911) describes vividly the extensive erosion resulting from deforestation in the southern Appalachian region. Meade (1969) describes the amount of estuarine sedimentation which has resulted from postcolonial land use and estimates the amount of atmospheric and man-induced chemical inputs into the solute system. Trimble (1974) cites pre colonial descriptions of the landscape when streams in flood ran crystal clear and then discusses the erosion induced by man.

Present work

Character of the drainage basin

The Juniata River drainage basin (JRDB) (Fig. 1) upstream from the gauging station at the Pennsylvania Route 34 bridge at Newport, Perry County, comprises 8687 km^2 of diverse topography and bedrock, most of which lies within the Appalachian Mountain Section of the Ridge and Valley Province. A small part of the western edge of the basin is in the Allegheny Mountain Section of the Appalachian Plateaus Province and comprises mainly the Allegheny Front and short tributary reaches which head west of the Front.

The drainage basin is underlain by sedimentary rocks composed mainly of sandstone, siltstone, shale, and carbonate. Cambrian through Silurian carbonate lithologies, limestone, dolomite, and calcareous shale, comprise 25% of the rocks in the drainage basin. The remainder of the rocks vary from relatively soft Ordovician and Devonian black shales to very hard siliceous sandstones and conglomerates of Silurian, Mississippian, and Pennsylvanian age. The rocks occur in a series of anticlines and synclines formed during the Alleghanian orogeny.

The topography is typical of the Appalachian Mountains and is characterized by long, linear ridges with siliceous sandstones at their crests and broad to narrow valleys underlain by shale and/or carbonate. Local relief from valley floor to ridge top is usually between 150 and 300 m. Relief on the valley floor is generally between 30 and 150 m.

Raystown Lake, dammed by Raystown Dam on the Raystown Branch Juniata River in the southern part of the area, has a drainage area of 2406 km^2 and normal storage capacity of 634 ha^3. The significance of the lake as a sediment trap will be discussed later.

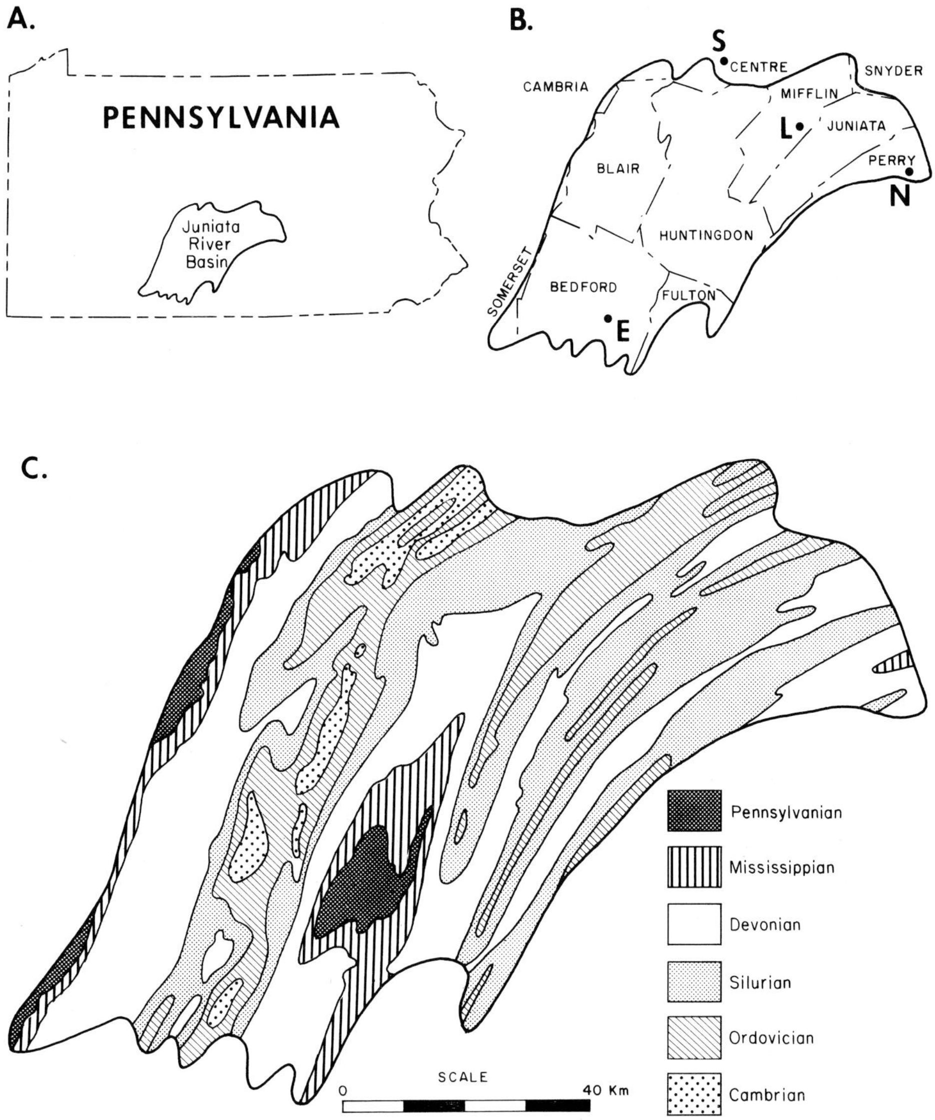

Fig. 1. Maps showing (A) the location of the Juniata River basin within Pennsylvania, (B) the counties within the drainage basin and the location of places mentioned elsewhere in the text: *E*, Everett; *L*, Lewistown; *N*, Newport; *S*, State College, and (C) the generalized geology of the drainage basin as delineated by the age of the rocks (geology from Geologic Map of Pennsylvania, Pennsylvania Geological Survey, 4th ser., Map 7).

Much of the JRDB is in forest cover and land-use estimates are: forest, 66%; cultivated, 14%; grass, 17%; urban, 3% (Williams and Reed, 1972) and forest, 68%; agriculture, 23%; urban, 5%; other, 4% (Taylor et al., 1982). The present trend is toward more urbanization.

Mean annual precipitation within the JRDB is 97.43 cm. Weather data for the area (Table 3) indicates some climatic variability for the area, but overall similarity. The water budget for the Juniata River at Newport is: precipitation, 92.2 cm; total runoff, 39.4 cm (14 cm surface runoff and 25.4 cm groundwater discharge), and evapotranspiration, 52.8 cm (Taylor et al., 1982). Data from other parts of the basin are similar.

TABLE 3

Summary of climatic data from four weather stations in or near the Juniata River basin (data from National Oceanic and Atmospheric Administration, 1985; see Fig. 1B for location of the weather stations)

	Daily temperature means (°C)								Precipitation monthly means (mm)				Snow monthly means (mm)			
	Maximum				Minimum				Eve	Lew	New	Sta	Eve	Lew	New	Sta
	Eve[1]	Lew[2]	New[3]	Sta[4]	Eve	Lew	New	Sta								
Jan	2.7	2.6	2.4	0.8	−7.3	−7.0	−7.4	−7.6	64	61	72	66	264	201	229	297
Feb	3.7	4.1	3.9	2.0	−6.8	−6.4	−7.1	−7.1	60	57	63	60	261	201	213	297
Mar	8.9	9.6	9.7	7.3	−2.3	−1.9	−2.3	−2.6	85	82	90	86	211	152	150	282
Apr	11.4	17.0	17.2	14.8	3.3	3.4	3.3	3.6	89	78	88	82	10	13	13	13
May	22.3	23.1	22.8	21.1	8.3	8.4	8.2	8.9	85	95	95	92				
Jun	26.9	27.5	27.6	25.6	12.8	13.3	13.6	13.8	88	103	100	95				
Jul	29.1	29.9	29.9	27.9	15.2	15.9	16.1	16.1	93	94	93	90				
Aug	28.4	28.9	28.9	26.9	14.6	15.4	15.4	15.3	82	86	92	86				
Sep	24.8	25.2	24.9	23.1	10.6	11.6	11.3	11.2	72	82	97	80				
Oct	18.3	18.8	18.3	16.6	4.0	5.1	4.8	5.2	77	75	83	72	3			5
Nov	11.1	11.4	11.2	9.3	−0.3	0.6	0.0	0.5	73	77	87	79	41	41	56	97
Dec	4.5	4.5	4.4	2.8	−4.8	−4.6	−4.9	−4.9	70	70	80	67	203	152	147	241
Year mean	16.4	16.9	16.8	14.8	3.9	4.5	4.3	4.4	939	962	1041	956	993	759	808	1262

[1]Everett; [2]Lewistown; [3]Newport; [4]State College.

Analysis of variations in discharge, chemical-character, specific-conductance, and suspended-sediment data

The data used for this study consist of records of discharge, specific conductance, and suspended-sediment concentration collected daily at the Newport, Pennsylvania Juniata River gauging station (Fig. 1B) during the water years 1965–1986 (U.S. Dep. Interior, 1965–1974, U.S. Geological Survey, 1975–1986). Data from periodic chemical analyses of Juniata River water are used also. These data are examined briefly to call attention to short-term variations which are masked by long-term averaging.

Discharge

Discharge is variable throughout the year with the highest daily average, 321 $m^3\ s^{-1}$, occurring in April and the lowest, 44 and 43 $m^3\ s^{-1}$, occurring in August and September, respectively (Table 4). Discharge, as shown in Table 4, has a regular progression between the extremes except for January which has lower discharge than December or February, presumably a reflection of cold temperatures inhibiting runoff. High-discharge events occur at any time during the year, but occur most often during February, March, and April. The highest-discharge event, 4673 $m^3\ s^{-1}$ during Tropical Storm Agnes, occurred in June, 1972. High-discharge events follow a pattern that has a rapid rise to a peak followed by a gradual decline which may not return to the pre-event level before another high-discharge event occurs (Fig. 2).

Chemical character

The Juniata River contains a dissolved load which is characterized by the major constituents indicated in Table 5A. Other minor constituents such as Al, Pb, Fe, and P are present in very small amounts and are not reported here. Comparison of Juniata River water with that of averages for other rivers (Table 5A) indicates

TABLE 4

Summary of Juniata River discharge at the Newport, Pennsylvania gauging station for the water years 1965–1986; values in the body of the table are the number of days that discharge was within the amount indicated on the left (data from U.S. Dep. of Interior, 1965-1974 and U.S. Geological Survey, 1975-1986)

Discharge		Oct	Nov	Dec	Jan	Feb	Mar	Apr	May	Jun	Jul	Aug	Sep	Total
(cfs)	($m^3 s^{-1}$)													
1- 250	1 - 7		1								1	1	5	8
251- 500	7.1- 14	62	38	3	19					2	42	41	44	251
501- 750	14.1- 21	141[a]	70	39	21	5				18	36	89	165[a]	584
751- 1000	21.1- 28	169[b]	84[a]	33	27	8				21	49	114[a]	156[b]	661
1001- 1500	28.1- 42	116	87	43	106[a]	64	18		14	90	189[a]	207[b]	148[c]	1082[a]
1501- 2000	42.1- 57	32[c]	45[b]	98[a]	113	60	27	18	55	93[a]	138[b]	95[b]	43	817
2001- 3000	57.1- 85	25	90	107	87[b]	108[c]	49	85	155[a]	158[b]	117[c]	79	57	1117[b]
3001- 4000	85.1- 113	37	74	71[b]	93	92[b]	85[a]	100[a]	111	117[b]	56	34	18	888
4001- 5000	114 - 142	17	41[c]	71	68[c]	54	71	110	96[b]	60	18	10	9	625[c]
5001- 7500	143 - 212	25	55	87[c]	80	62	143[b]	114[b]	126[c]	48	18	5	4	767
7501- 10,000	213 - 283	19	32	51	30	53[c]	93	84[c]	49	23	10	6	1	451
10,001- 15,000	284 - 425	23	25	46	23	51	86[c]	78	49	13	5	1	4	404
15,001- 20,000	426 - 566	10	8	20	9	21	46	35	13	7	1		3	173
20,001- 25,000	567 - 708	4	6	8	4	14	28	19	5	1	1		1	91
25,001- 50,000	709 -1416	2	4	5	2	27	35	15	9	5	1		2	107
50,001- 75,000	1417 -2124					1	1	2		1				5
75,001-100,000	2125 -2832					1				1				2
100,001-150,000	2833 -4248													
150,001-200,000	4249 -5564									2				2
Daily average ($m^3 s^{-1}$)		70	101	145	108	199	266	321	159	123	61	44	43	128

[a]25th percentile falls in this group.
[b]50th percentile falls in this group.
[c]75th percentile falls in this group.

a higher concentration in the Juniata River for all constituents except silica.

The original data show that the chemical character of the water changed after 1972 and the data is shown in Table 5B divided into two groups, a 1965–1972 group and a 1973–1986 group. This division shows a significant decrease in concentrations of all constituents in the 1973–1986 group. The possible significance of this is discussed in the next section.

Specific conductance

Specific conductance is a measure of the amount of dissolved solids in water. For any specific water, such as the Juniata River, it is possible to determine a general relationship between the specific conductance and the total dissolved solids through comparison of specific conductance and chemical analyses. For this study, a linear regression for 102 values of specific conductance (in microsiemens cm^{-1}) and total dissolved solids by residue (in mg^{-1}) yielded a correlation coefficient of 0.921. Slope and intercept values from the regression fitted the equation:

$$Y = 0.593 \cdot (\text{specific conductance}) + 7.37 \qquad (1)$$

to obtain the solute load (Y) in mg^{-1}. The value of solute load (Y) was then used in the equation:

$$X = Y \cdot \text{discharge} \cdot 0.0027 \qquad (2)$$

to obtain the value of dissolved solids (X) in t d^{-1} using original data values of discharge in $ft^3 s^{-1}$ (cfs). This equation was used to calcu-

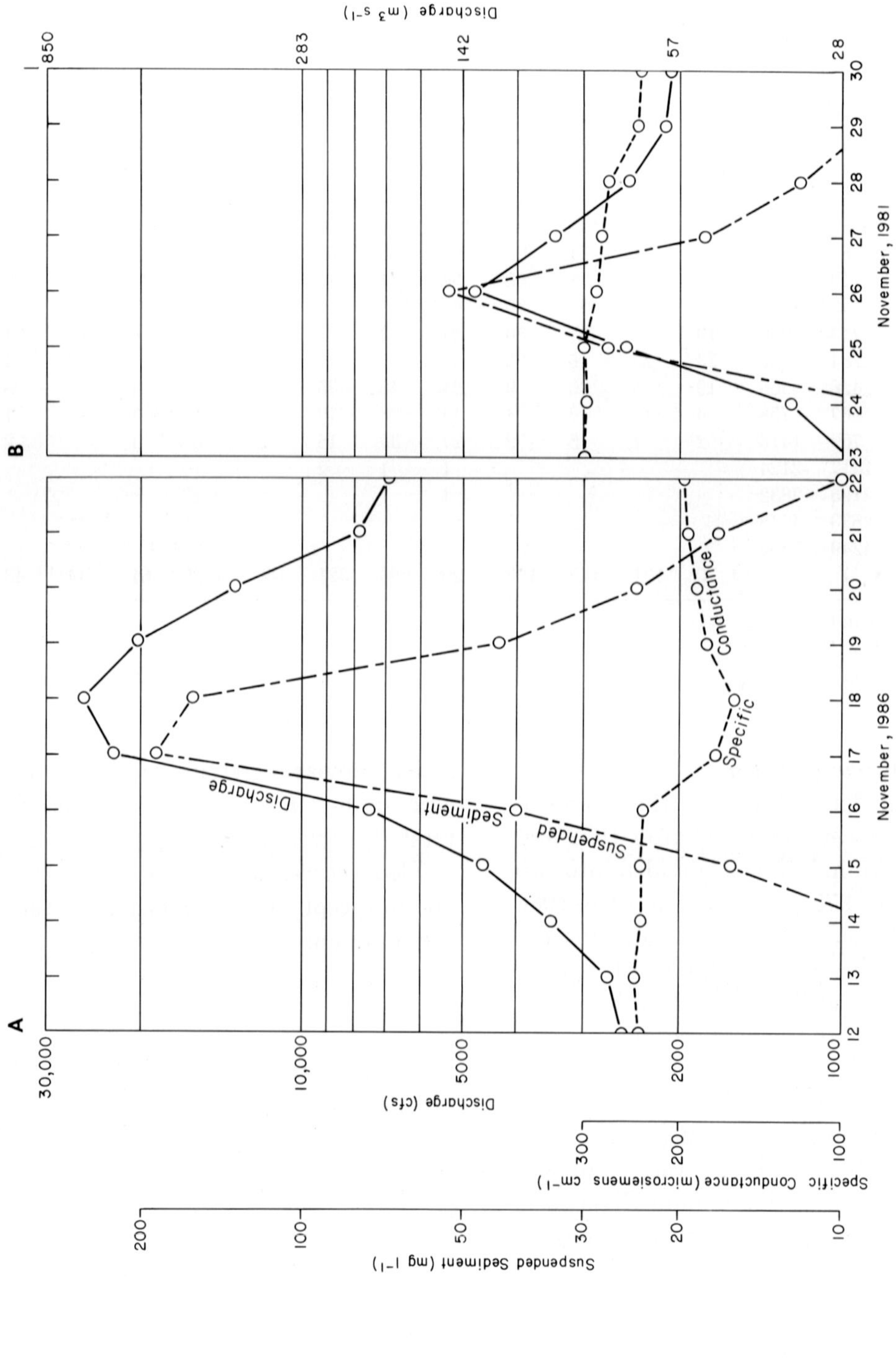

Fig. 2. Graphs showing the variation of values of discharge, specific conductance, and suspended-sediment concentration during two above normal discharge events recorded at the Newport gauging station. Graphs are plotted on a semilogarithmic scale.

TABLE 5

Concentration of major constituents in (A) solute load in riverwater of the Juniata River, North American rivers, and world rivers, (B) comparison of major constituents of Juniata River water separated into a 1965–1972 group and a 1973–1986 group, and (C) the wet precipitation contribution to the Juniata River drainage basin of major constituents found in the riverwater

Concentration (mg l^{-1})									
Ca	Mg	Na	K	Cl	SO_4	SiO_2	HCO_3	NO_3	
A									
27.9	7.8	7.7	1.9	12.5	30.3	3.1	85.3	3.2	Juniata River[1]
20.1	4.9	6.5	1.5	7.0	14.9	7.2	71.4		North America[2]
14.0	3.7	6.3	1.7	7.7	9.4	11.3	54.1		World rivers[3]
B									
1965-1972									
59	59	38	53	59	50	50			Number of analyses
28.9	8.1	8.6	2.1	14.5	38.8	3.6			Mean
8.1	2.8	3.5	0.6	8.1	13.2	2.4			Standard deviation (st. dev.)
1973-1986									
64	55	55	55	69	68	54			Number of analyses
26.8	7.5	7.1	1.7	10.8	24.1	2.7			Mean
5.4	1.8	2.9	0.3	3.7	5.3	1.9			St. dev.
−2.1	−0.6	−1.5	−0.4	−3.7	−14.7	−0.9			Difference: mean
−2.7	−1.0	−0.6	−0.3	−4.4	−7.9	−0.05			Difference: st. dev.
7	22	17	19	26	38	25			Percent difference: mean
33	36	17	50	54	60	21			Percent difference: st. dev.
C									
0.13	0.03	0.05	0.03	0.20	3.2			2.2	Wet precipitation[4]

[1]Means from 128 chemical analyses (1965–1986); data from U.S. Dep. of Interior, 1965–1974 and U.S. Geological Survey, 1975–1986.
[2]Walling and Webb (1986, p. 257, table 7.2).
[3]Walling and Webb (1986, p. 256, table 7.1, average by Sevon).
[4]Data from Lynch et al., 1988.

late the solute load which passed the Newport station during the 22-yr period.

Values for specific conductance were not available for every day during the 22-yr period of record at the Newport gauging station. Those times when data were not recorded were estimated by assigning values approximately equally divided between the enclosing values, for example, recorded October 26, 1966 μS value of 430 and October 30 value of 422 enclosed estimated values of: October 27, 428; October 28, 426; and October 29, 424. The longest period of nonrecord was May 29–July 31, 1977. Most periods of nonrecord were less than 20 days.

A summary of the 22-yr record of specific conductance (Table 6) shows a concentration of low values as well as a narrower range of values during the months of February through May. This probably results from a combination of factors: (1) frozen ground which inhibits groundwater recharge in February and March; (2) snow cover; (3) decreased solute contribution from fertilizer which is applied in spring; (4) possible decreased solute contribution from industrial discharge; (5) dilution of solute concentration by winter snowmelt and spring rainfall; and (6) other unknowns. Specific conductance is lowered temporarily by dilution during

TABLE 6

Summary of 22 years of daily specific conductance values for the stream gauging station at Newport, Pennsylvania; values in body of the table indicate the number of days during period of record that a specific conductance value fell within a defined range, (data from U.S. Dep. of Interior, 1965-1974 and U.S. Geological Survey, 1975-1986)

Month	Value range (μS cm^{-1})									
	101-150	151-200	201-250	251-300	301-350	351-400	401-450	451-500	501-550	551-600
Oct	2	66	85	143	197	70	49	59	9	2
Nov	3	105	207	136	97	15	31	48	18	
Dec	19	202	263	115	26	17	33	7		
Jan	9	230	271	120	40	6	3	3		
Feb	34	267	239	74	7					
Mar	57	415	116	33	11					
Apr	52	428	168	12						
May	60	364	241	17						
Jun	16	129	354	114	34	10	2	1		
Jul	2	45	245	222	74	60	34			
Aug	6	23	124	309	110	49	47	14		
Sep	6	18	72	241	169	60	53	36	5	
Totals	266	2292	2435	1536	765	287	252	168	32	2

high-discharge events (Fig. 2), but returns to pre-event levels shortly thereafter.

Of particular interest is the specific-conductance data separated as shown in Table 7 (1965–1973 group, and 1974–1986 group), a separation suggested by perusal of the data as a whole. The table shows an absence of high specific conductance values after 1973. A similar temporal separation was previously noted for the chemical constituents of the river water. An explanation is not apparent from the data nor are there any known quantified changes within the JRDB to explain the separation. The fact that storage in Raystown Lake began in November, 1972 probably does not have any relationship because reservoirs do not entrap appreciable amounts of dissolved materials (Fishel, 1984, p. 48). A possible explanation is enforced compliance to clean stream laws, mainly since 1972, for industrial waste discharge into the Juniata River (T. Kaktins, pers. commun., 1989).

Suspended sediment

Values of suspended-sediment concentration (in mg l^{-1}) were available for the whole 22-yr period and no estimates were necessary. The values are summarized in Table 8 in terms of the number of days a value occurred within a specific range. This table shows that 80% of the time suspended-sediment concentration was 25 mg l^{-1} or less and 55% of the time the concentration was 10 mg l^{-1} or less. These values are associated with low to moderate discharge. However, large quantities of suspended sediments are carried by high-discharge events when the suspended-sediment concentration increases dramatically for a short period of time (Fig. 2). The significance of such rainfall events is illustrated by Hurricane Agnes. During the interval June 21–29, 1972 the Juniata River carried 64% of the suspended-sediment load for the water year, while discharge accounted for only 31% of the year's total. Similarly, three relatively moderate rainfall events during the 1975 water year accounted for 58% of the year's suspended-sediment load during a period of 19 days, but only 22% of the year's discharge. Note also in Fig. 2 that the suspended-sediment concentration peaks early in the high discharge event and the drops rapidly to much lower levels. This relationship precludes direct correlation of discharge and suspended-sediment load.

TABLE 7

Summaries of two groupings of daily specific conductance values for the stream gauging station at Newport, Pennsylvania; values in body of table are numbers of days during period of record that a specific conductance value fell within a specific range (data from U.S. Dep. of Interior, 1965–1974 and U.S. Geological Survey, 1975–1986)

Month	Values range (μS cm^{-1})									
	101–150	151–200	201–251	251–300	301–350	351–400	401–450	451–500	501–550	551–600
Water years 1965–1973										
Oct	0	6	12	16	58	68	49	59	9	2
Nov	0	26	60	55	19	13	31	48	18	
Dec	9	69	86	42	16	17	33	7		
Jan	0	62	111	55	39	6	3	3		
Feb	17	90	98	44	5					
Mar	30	167	41	30	11					
Apr	29	139	90	12						
May	35	133	96	15						
Jun	13	34	99	83	32	9				
Jul	0	6	47	71	61	60	34			
Aug	1	2	16	81	70	49	47	13		
Sep	0	0	17	32	68	59	53	36	5	
Totals	134	734	773	536	379	281	250	166	32	2
Water years 1974–1986										
Oct	2	60	63	127	139	2				
Nov	3	79	147	81	78	2				
Dec	10	133	177	73	10					
Jan	9	168	160	65	1					
Feb	17	177	141	30	2					
Mar	27	248	125	3						
Apr	23	289	78							
May	25	231	145	2						
Jun	3	95	255	31	2	1	2	1		
Jul	2	39	198	151	13					
Aug	5	21	108	228	40			1		
Sep	6	18	55	209	101	1				
Totals	132	1558	1662	1000	386	6	2	1		

Available samples indicate that suspended sediment transported by the Juniata River is mainly clay and fine silt (Table 9). However, the data shown in Table 9 represents samples taken during times of high discharge and may not be a reasonable reflection of the material carried 80% of the time during lower velocity flow.

Perusal of the data as a whole indicated a real variation of suspended sediment during the months of June, July, August, September, and October when suspended-sediment concentrations less than 10 mg l^{-1} are common. Table 10 indicates that during the years 1975–1979 the suspended-sediment concentration was consistently higher than it was in earlier or later years. Table 11 presents some data which may be relevant. The numbers of building permits for houses show a trend similar to that of suspended-sediment concentration and a correlation between house building and the increased amount of suspended sediment in the Juniata River seems a distinct possibility. House construction temporarily disrupts the land surface and significant quantities of sediment will be contributed to runoff unless adequate erosion-

TABLE 8

Summary of 22 years of daily suspended sediment values for the stream gauging station at Newport, Pennsylvania; numbers in body of table are numbers of days during period of record which had suspended-sediment values within a specific range (data from U.S. Dep. of Interior, 1965–1974 and U.S. Geological Survey, 1975–1986)

Month	Value range (mg l^{-1})									
	1	2–10	11–25	26–50	51–75	76–100	101–250	251–500	501–750	751–1000
Oct	137	345	122	41	9	7	19	2		
Nov	110	347	117	46	17	8	14	1		
Dec	72	394	124	51	14	12	15			
Jan	64	430	121	35	11	9	11	1		
Feb	23	344	109	51	29	19	35	9	2	
Mar	7	229	181	102	29	19	35	10		
Apr	34	265	202	91	32	9	22	4	1	
May	8	151	284	161	45	12	20	1		
Jun	0	206	224	130	48	19	28	4		1
Jul	15	308	221	96	25	11	13	2	1	
Aug	29	372	199	48	17	3	14			
Sep	71	394	148	26	7	6	6	2		
Totals	570	3855	2042	878	283	134	232	36	4	1

TABLE 9

Particle size-analysis data for suspended sediment and bedload in the Juniata River at Newport, Pennsylvania between February 9, 1965 and March 15, 1986 (data from U.S. Dep. of Interior 1965–1974 and U.S. Geological Survey, 1975–1986)

Suspended sediment	Discharge ($ft^2 s^{-1}$)	Sediment concentration (mg l^{-1})	Percent finer than size indicated (mm)									
			0.002	0.004	0.008	0.016	0.031	0.062	0.125	0.250	0.5	1.0
n	51	51	6	51	51	51	51	51	48	32	7	
Mean	25,129	257	45	51.4	66.9	78.8	87.7	93.5	96.4	98.1	99	
St.dev.	16,968	165	7.95	11.24	9.5	7.3	4.8	3.4	2.5	1.1	0	

Bedload

Date	Percent finer than size indicated (mm)										
	0.25	0.5	1	2	4	8	16	32	64	128	256
9/16/65	0	1	1	2	3	16	27	49	76	92	97

control procedures are invoked. The suggested correlation may be fortuitous, but the variation in stream-sediment concentration indicates a sensitivity of the drainage system to changes in environmental conditions.

Bedload

Only one sample of bedload is available (Table 9) and no information is given about the method or location of sample collection. Sampling of bedload is difficult and one sample does not reflect even partially the material carried by the Juniata River as bedload. Bedload ma-

TABLE 10

Variations in suspended sediment load at the Newport gauging station for selected months during the period 1965–1986; values in body of table are number of days during which the concentration of suspended sediment (mg l^{-1}) was less than 10 (data from U.S. Dep. of Interior, 1965–1974 and U.S. Geological Survey, 1975–1986)

Year	Month					
	Jun	Jul	Aug	Sep	Oct	Nov
1965	21	31	28	29	31	28
1966	15	8	29	28	29	22
1967	21	10	9	26	20	18
1968	7	24	28	20	0	12
1969	11	13	16	30	26	8
1970	0	8	16	18	29	13
1971	21	22	22	14	19	55
1972	2	6	31	30	21	22
1973	30	3	0	3	31	20
1974	10	12	0	4	20	9
1975	0	10	29	23	11	30
1976	0	9	0	11	7	26
1977	2	0	14	1	1	27
1978	2	0	0	0	2	14
1979	4	0	4	7	11	30
1980	21	24	16	30	18	22
1981	0	12	21	18	31	26
1982	2	17	24	30	27	30
1983	8	22	24	30	30	30
1984	11	18	15	30	31	26
1985	1	8	27	30	31	28
1986	7	22	24	30	31	11

terial occurs in the modern floodplain and gives some information about the range of sizes of material normally transported. Kaktins (1986, pp. 152–160) indicates that Juniata River floodplain and terrace materials comprise cobble–boulder channel lag deposits, cobble–pebble–sand bar deposits, and sand–silty–clayey overbank deposits. The channel lag and bar deposits presumably represent the bed-transported material. The total thickness of these deposits is generally less than 3 m. Materials presumably formed in these environments occur on five terrace levels indicating the continued presence of comparable size bedload throughout the terrace-recorded history of Juniata River sediment transport. However, no answer about how much bedload is annually transported out of the JRDB is apparent from the available data.

Calculation of erosion rate

The data obtained from the Newport gauging station is used to determine a present erosion rate for the Juniata River drainage basin. However, the data cannot be used without some modification.

The values of specific conductance were used to calculate a total dissolved load of 10,500,000 metric tons passing the Newport gauging station during water years 1965–1986. This large quantity does not represent only material weathered from rock. Meade (1969, p. 1272) suggests that 25% of this volume was contributed by the atmosphere and that 10% was added by human activity (fertilizer, road salt, indus-

TABLE 11

Numbers of building permits issued in six counties in the Juniata River basin during the years 1971–1981 (data from U.S. Bureau of the Census, 1973–1982)

County	1971	1972	1973	1974	1975	1976	1977	1978	1979	1980	1981
Bedford	30	34	5	6	2	14	38	42	55	41	69
Blair	232	571	411	140	155	515	710	506	349	270	230
Fulton							49	19	25	9	12
Huntington	3	27	11	5	9	5	36	513	574	127	103
Juniata		1	2	1	4	4	6	101	54	52	43
Perry	3	15	24	39	137	139	169	219	172	100	96
Totals	268	891	453	191	307	677	1008	1400	1229	599	553

trial waste, etc.). Cleaves et al. (1970, p. 3027) noted that 38% of dissolved material leaving the Pond Branch watershed and 37% of the dissolved load in the Soldiers Delight watershed (Cleaves et al., 1974, p. 439), both in Maryland, came from precipitation.

Data on wet deposition in Pennsylvania (Lynch et al., 1988) were used to calculate the total amount of atmospheric input to the JRDB during the 22-yr period for the constituents indicated in Table 5B using a mean annual precipitation of 97.43 cm. The amount contributed to the JRDB was 1,150,000 metrict, only 11% of the total dissolved load (TDL) leaving the basin. Information on other solutes (e.g., F, Pb, Fe, Al) is inadequate to allow evaluation of their wet deposition quantity, but, because their quantities are generally reported as small amounts of micrograms, it is assumed that their total contribution was small.

Dry deposition also contributes to the quantity of dissolved load leaving the drainage basin. Lynch et al. (1988, pp. 93–96) discuss the problems of measuring dry deposition and its general character within Pennsylvania, but do not present data which allows quantification. They indicate that the spatial pattern of dry deposition is very similar to that of wet deposition and that the concentration of all ions is higher than in wet deposition.

No attempt has been made to quantify the contribution of human activity (e.g., chemical fertilizer, road salt, industrial waste), but it is assumed that these make a significant contribution to the total dissolved load in the Juniata River water.

For the purposes of this study, the derived values of 11% of TDL for wet deposition, an arbitrary value of 11% of TDL for dry deposition input, and an arbitrary value of 8% of TDL for human activity input, a total of 30% of TDL, is used as a reduction factor for TDL. The use of a 22% value for wet and dry deposition assumes that 100% of atmospheric input is exported from the JRDB by the Juniata River. The complex biogeochemical reactions involved with the input ions (Likens et al., 1977) almost certainly falsifies this assumption, but the lack of data required to correct the 22% value necessitates its use. The calculated 10,500,000 metric of TDL is reduced by 30% to a value of 7,350,000 metric for the period of record.

The total suspended sediment load for the 1965–1986 period was 5,100,000 metrict at the Newport gauging station. However, this values does not include the amount of suspended sediment trapped by the Raystown Dam whose reservoir began filling in November 1972. The capacity-inflow ratio method of Brune (1953) was utilized to determine that Raystown Lake has a 96% trap efficiency with regard to suspended sediment. During the 1973–1986 period, the annual discharge below the Raystown Dam varied between 22% and 29% of the total discharge at Newport. The assumption is made that the percent contribution of suspended sediment by the Raystown Branch to the total at Newport, if all the suspended sediment were delivered, would be the same as the percent contribution of discharge. Therefore, the yearly totals of suspended sediment at Newport for the years 1973–1986 were increased by an amount determined by the equation:

$$X=(Y\cdot 0.96)\cdot Z \qquad (3)$$

where X=sediment trapped by Raystown Lake, Y=percentage contribution of discharge made by the Raystown Branch (as measured immediately below the dam) to total discharge at Newport, and Z=measured suspended-sediment load at Newport. The total amount of suspended sediment passing Newport during the 1965–1986 period is calculated to be 5,800,000 metrict.

Because of the lack of information about the bedload, an estimate that bedload is 5% of total particle sediment load (Gregory and Walling, 1973, p. 203, table 4.4C) was used to obtain a total of 300,000 metric of bedload.

The above values sum to a total load of 13,450,000 metric leaving the Juniata River basin during the 1965–1986 period. This yields an average of 611,000 metrict yr^{-1} 8687 km^{-2} or 70 metrict yr^{-1} km^{-2}. Assuming that 1 m^3 of rock equals 2643 kg of eroded sediment, then the equation:

$$\frac{70 \text{ metric t km}^{-2} \text{ yr}^{-1}}{2643 \text{ kg m}^{-3} \text{ yr}^{-1}} = \text{erosion rate} \qquad (4)$$

yields a present erosion rate of 27 m Myr^{-1} for the JRDB.

Discussion

The calculated erosion rate of 27 m Myr^{-1} falls within the lower range of rates previously calculated for large drainage basins with diverse rock types (Table 1A), but is significantly higher than the 10 m Myr^{-1} rate calculated by Judson and Ritter (1964, table 1, p. 3397) for the same drainage basin. The reason for the latter difference is not apparent, but probably results from use of different values for calculation of the erosion rate.

The 27 m Myr^{-1} erosion rate is presumed to be representative of the JRDB and its land use for the 1965–1986 period. The current trend towards increasing urbanization could change the short-term erosion rate unless adequate erosion-control measures are used. The environment of the Juniata River drainage basin is very sensitive to disruptive changes which are quickly reflected in the suspended-sediment and dissolved load of the river.

References

Affifi, A.A. and Bricker, O.P., 1983. Weathering reactions, water chemistry and denudation rates in drainage basins of different bedrock types: I — sandstone and shale. In: B.W. Webb (Editor), Dissolved Loads of Rivers and Surface Water Quantity/Quality Relationships. Proc. Hamburg Symp., IAHS Publication, 141: 193–203.

Berry, J.L., 1977. Chemical weathering and geomorphological processes at Coweeta, North Carolina. Geol. Soc. Am., Abstr. Programs, 9: p. 120.

Bricker, O.P., Godfrey, A.E. and Cleaves, E.T., 1968. Mineral-water interaction during the chemical weathering of silicates. Am. Chem. Soc. Adv. Chem. Ser., 73: 128–142.

Brown, L., Pavich, M.J. Hickman, R.E., Klein, J. and Middleton, R., 1988. Erosion of the eastern United States observed with ^{10}Be. Earth Surf. Processes Landforms, 13: 441–457.

Brune, G.M., 1953. Trap efficiency of reservoirs. Am. Geophys. Union, Trans., 34: 407–418.

Claypole, E.W., 1885. A preliminary report on the palaeontology of Perry County. Pennsylvania Geol. Surv. 2nd Ser., Rep. Prog., F2, 437 pp.

Cleaves, E.T., Godfrey, A.E. and Bricker, O.P., 1970. Geochemical balance of a small watershed and its geomorphic implications. Geol. Soc. Am., Bull., 81: 3015–3032.

Cleaves, E.T., Fisher, D.W. and Bricker, O.P., 1974. Chemical weathering of serpentinite in the eastern Piedmont of Maryland. Geol. Soc. Am., Bull., 85: 437–444.

Dole, R.B. and Stabler, H., 1909. Denudation. U.S. Geol. Surv., Water Supply Pap., 234: 78–93.

Ewing, A.L., 1885. On the chemical erosion of the Nittany Valley limestone. In: I.C. White (Editor), The geology of Huntingdon County. Pennsylvania Geol. Surv., 2nd Ser., Rep. Prog., T3: 451–454.

Fishel, D.K., 1984. Water-quality and chemical loads of the Susquehanna River at Harrisburg, Pennsylvania, April 1980 to March 1981. U.S. Geol. Surv., Water Resour. Invest. Rep., 83-4164, 90 pp.

Gilluly, J., 1964. Atlantic sediments, erosion rates, and the evolution of the Continental Shelf; some implications. Geol. Soc. Am., Bull., 75: 483–492.

Glenn, L.C., 1911. Denudation and erosion in the southern Appalachian region and the Monongahela basin. U.S. Geol. Surv., Prof. Pap., 72, 237 pp.

Godfrey, A.E., 1975. Chemical and physical erosion in the South Mountain anticlinorium, Maryland. Maryland Geol. Surv., Inf. Circ., 19, 35 pp.

Goldfarb, R.J., Marchand, D.E., Dean, B.W., Mattocks, T. and Monahan, J., 1989. Land use effects on surface runoff, water chemistry, and erosion rates in some small central Pennsylvania watersheds. Pennsylvania Geol. Surv., 4th Ser., Open File Rep.

Gordon, R.B., 1979. Denudation rate of central New England determined from estuarine sedimentation. Am. J. Sci., 279: 632–642.

Gregory, K.J. and Walling, D.E., 1973. Drainage basin form and process — a geomorphological approach. Wiley, New York, N.Y., 456 pp.

Hack, J.T., 1965. Geomorphology of the Shenandoah Valley, Virginia and West Virginia and origin of the residual ore deposits. U.S. Geol. Surv., Prof. Pap., 484, 84 pp.

Hack, J.T., 1980. Rock control and tectonism — their importance in shaping the Appalachian highlands. U.S. Geol. Surv., Prof. Pap., 1126-B, 17 pp.

Judson, S. and Ritter, D.F., 1964. Rates of regional denudation in the United States. J. Geophys. Res., 69: 3395–3401.

Kaktins, T.L., 1986. Fluvial terraces of the Juniata River valley in central Pennsylvania. Unpubl. M.S. Thesis, The Pennsylvania State University, University Park, Pa., 283 pp.

Lakatos, S. and Miller, D.S., 1983. Fission-track anlaysis of apatite and zircon defines a burial depth of 4 to 7 km for lowermost Upper Devonian, Catskill Mountains, New York. Geology, 11: 103-109.

Likens, G.E., Bormann, F.H., Pierce, R.S., Eaton, J.S. and Johnson, N.M., 1977. Biogeochemistry of a forested ecosystem. Springer, New York, N.Y., 146 pp.

Lynch, J.A., Corbett, E.S. and Grimm, J.W., 1988. Atmospheric deposition: spatial and temporal variations in Pennsylvania — 1987. Environ. Resour. Res. Inst., Pennsylvania State University, University Park, Pa., ER8806, 103 pp.

Mathews, W.H., 1975. Cenozoic erosion and erosion surfaces of eastern North America. Am. J. Sci., 275: 818-824.

Meade, R.H., 1969. Errors in modern stream-load data to estimate natural rates of denudation. Geol. Soc. Am., Bull., 80: 1265-1274.

Menard, H.W., 1961. Some rates of regional erosion. J. Geol., 69: 154-161.

National Oceanic and Atmospheric Administration, 1985. Climatography of the United States No. 20: Pennsylvania. Natl. Clim. Data Cent., Ashville, North Carolina, 84 pp.

Ogden, A.E., 1982. Karst denudation rates for selected spring basins in West Virginia. Natl. Speleol. Soc., Q. J., NSS Bull., 44: 6-10.

Parizek, R.R. and White, W.B., 1985. Application of Quaternary and Tertiary geological factors to environmental problems in central Pennsylvania. In: D.P. Gold (Editor), Central Pennsylvania Geology Revisited. 50th Annu. Field Conf. Pa., Geol., Guidebook, 63-109 pp.

Pavich, M.J., Brown, L., Valette-Silver, J.N., Klein, J. and Middleton, R., 1985. ^{10}Be analysis of a Quaternary weathering profile in the Virginia Piedmont. Geology, 13: 39-41.

Poag, C.W. and Sevon, W.D., 1989. A record of Appalachian denudation in post-lift Mesozoic and Cenozoic sedimentary deposits of the U.S. Middle Atlantic continental margin. In: T.W. Gardner and W.D. Sevon (Editors), Appalachian Geomorphology. Geomorphology, 2: 119-157 (this volume).

Reesman, A.L. and Godfrey, A.E., 1981. Development of the Central Basin of Tennessee by chemical denudation. Z. Geomorphol., 25: 437-456.

Ritchie, J.C. and McHenry, J.R., 1989. Application of radioactive fallout cesium-137 for measuring soil erosion and sediment accumulation rates and patterns: a review with bibliography. U.S. Dep. Agric., Hydrol. Lab. Tech. Rep., HL-15, 67 pp.

Roden, M.K. and Miller, D.S., 1989. Apatite fission-track thermochronology of the Pennsylvania Appalachian basin. In: T.W. Gardner and W.D. Sevon (Editors), Appalachian Geomorphology. Geomorphology, 2: 39-51 (this volume).

Sevon, W.D., 1985. Pennsylvania's polygenetic landscape. 4th Annu. Field Trip, Harrisburg Area Geol. Soc., Guidebook, 55 pp.

Taylor, L.E., Werkheiser, W.H., DuPont, N.S. and Kriz, M.L., 1982. Groundwater resources of the Juniata River basin, Pennsylvania. Pennsylvania Geol. Surv., 4th Ser., Water Resour. Rep., 54, 131 pp.

Trimble, S.W., 1974. Man-induced soil erosion on the southern Piedmont 1700-1970. Soil Conserv. Soc. Am., 180 pp.

Trimble, S.W., 1977. The fallacy of stream equilibrium in contemporary denudation studies. Am. J. Sci., 277: 876-887.

U.S. Bureau of the Census, 1973-1982. Construction reports — housing units authorized by building permits and public contracts. U.S. Gov. Print. Off., Washington, D.C. (annual 1972-1981).

U.S. Department of Interior, 1965-1974. Water resources data for Pennsylvania, Part 1. Surface water records: Part 2. Water quality records. U.S. Geol. Surv. Water Resour. Div.

U.S. Geological Survey, 1975-1986. Water resources data for Pennsylvania water year 1975 (-1986), Volume 2. Susquehanna and Potomac River basins. U.S. Geol. Surv., Water Data Rep. PA-75 (-86)-2.

Walling, D.E. and Webb, B.W., 1983. The dissolved loads of rivers: a global overview. In: B.W. Webb (Editor), Dissolved Loads of Rivers and Surface water Quantity/Quality Relationships. Proc. Hamburg Symp., IAHS Publ., 141: 3-20.

Way, J.H., Smith, R.C., II and Roden, M., 1986. Detailed correlations across 175 miles of the Valley and Ridge of Pennsylvania using 7 ash beds in the Tioga zone. In: W.D. Sevon (Editor), Selected Geology of Bedford and Huntingdon Counties. 51st Annu. Field Conf. Pa. Geol., Guidebook, pp. 55-72.

White, W.B., 1984. Rate processes: chemical kinetics and karst landform development. In: R.G. La Fleur (Editor), Groundwater as a Geomorphic Agent. Allen and Unwin, London, pp. 227-248.

Whitney, J.A., Jones, L.M. and Walker, R.L., 1976. Age and origin of the Stone Mountain Granite, Lithonia district, Georgia. Geol. Soc. Am., Bull., 87: 1067-1077.

Williams, K.F. and Reed, L.A., 1972. Appraisal of stream sedimentation in the Susquehanna River basin. U.S. Geol. Surv., Water Supply Pap., 1532-F, 24 pp.